Collins

Student Book, **Foundation 1**

Delivering the Edexcel Specification

NEW GCSE MATHS
Edexcel Linear

Fully supports the 2010 GCSE Specification

Brian Speed • Keith Gordon • Kevin Evans • Trevor Senior

CONTENTS

INTRODUCTION

Welcome to Collins New GCSE Maths for Edexcel Linear Foundation Book 1.

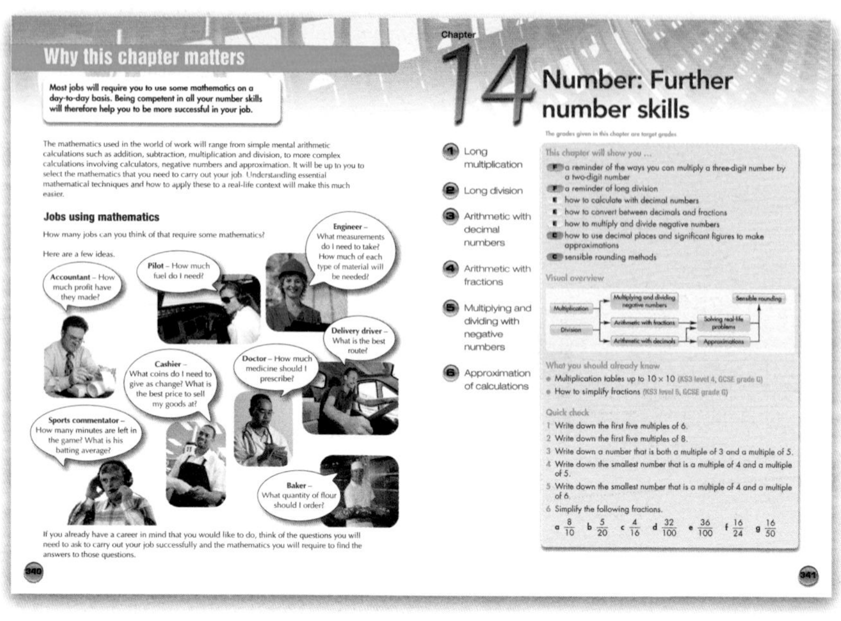

Why this chapter matters

Most jobs will require you to use some mathematics on a day-to-day basis. Being competent in all your number skills will therefore help you to be more successful in your job.

The mathematics used in the world of work will range from simple mental arithmetic calculations such as addition, subtraction, multiplication and division, to more complex calculations involving calculators, negative numbers and approximation. It will be up to you to select the mathematics that you need to carry out your job. Understanding essential mathematical techniques and how to apply these to a real-life context will make this much easier.

Jobs using mathematics

How many jobs can you think of that require some mathematics?

Here are a few ideas.

If you already have a career in mind that you would like to do, think of the questions you will need to ask to carry out your job successfully and the mathematics you will require to find the answers to those questions.

340

Chapter 14

Number: Further number skills

1. Long multiplication
2. Long division
3. Arithmetic with decimal numbers
4. Arithmetic with fractions
5. Multiplying and dividing with negative numbers
6. Approximation of calculations

The grades given in this chapter are target grades

This chapter will show you ...

- F a reminder of the ways you can multiply a three-digit number by a two-digit number
- F a reminder of long division
- E how to calculate with decimal numbers
- E how to convert between decimals and fractions
- E how to multiply and divide negative numbers
- C how to use decimal places and significant figures to make approximations
- C sensible rounding methods

Visual overview

What you should already know

- Multiplication tables up to 10 × 10 (KS3 level 4, GCSE grade G)
- How to simplify fractions (KS3 level 5, GCSE grade G)

Quick check

1 Write down the first five multiples of 6.
2 Write down the first five multiples of 8.
3 Write down a number that is both a multiple of 3 and a multiple of 5.
4 Write down the smallest number that is a multiple of 4 and a multiple of 5.
5 Write down the smallest number that is a multiple of 4 and a multiple of 6.
6 Simplify the following fractions.

a $\frac{8}{10}$ b $\frac{5}{20}$ c $\frac{4}{16}$ d $\frac{32}{100}$ e $\frac{36}{100}$ f $\frac{16}{24}$ g $\frac{16}{50}$

341

Why this chapter matters

Find out why each chapter is important through the history of maths, seeing how maths links to other subjects and cultures, and how maths is related to real life.

Chapter overviews

Look ahead to see what maths you will be doing and how you can build on what you already know.

Colour-coded grades

Know what target grade you are working at and track your progress with the colour-coded grade panels at the side of the page.

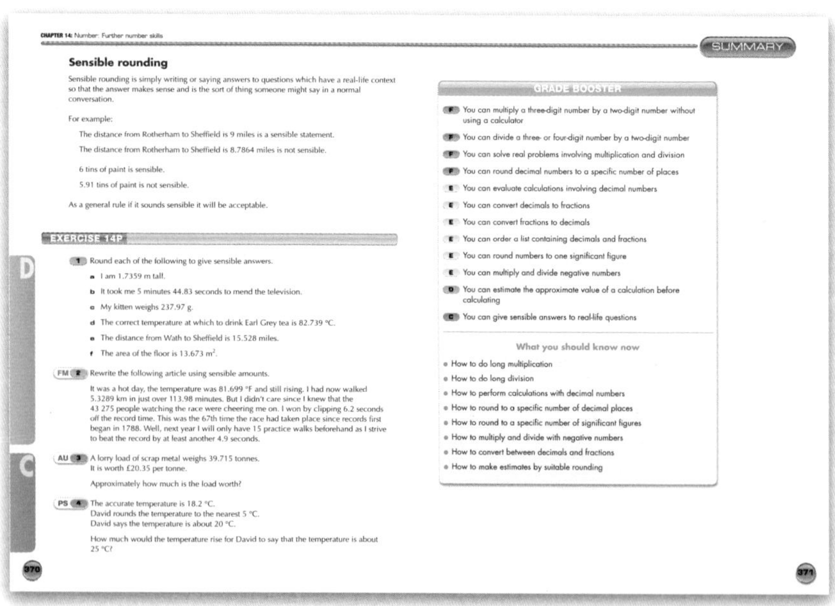

CHAPTER 14: Number: Further number skills

Sensible rounding

Sensible rounding is simply writing or saying answers to questions which have a real-life context so that the answer makes sense and is the sort of thing someone might say in a normal conversation.

For example:

The distance from Rotherham to Sheffield is 9 miles is a sensible statement.

The distance from Rotherham to Sheffield is 8.7864 miles is not sensible.

6 tins of paint is sensible.

5.91 tins of paint is not sensible.

As a general rule if it sounds sensible it will be acceptable.

EXERCISE 14P

D

1 Round each of the following to give sensible answers.

a I am 1.7359 m tall.

b It took me 5 minutes 44.83 seconds to mend the television.

c My kitten weighs 237.97 g.

d The correct temperature at which to drink Earl Grey tea is 82.739 °C.

e The distance from Wath to Sheffield is 15.528 miles.

f The area of the floor is 13.673 m².

FM 2 Rewrite the following article using sensible amounts.

It was a hot day, the temperature was 81.699 °F and still rising. I had now walked 5.3289 km in just over 113.98 minutes. But I didn't care since I knew that the 43 275 people watching the race were cheering me on. I won by clipping 6.2 seconds off the record time. This was the 67th time the race had taken place since records first began in 1788. Well, next year I will only have 15 practice walks beforehand as I strive to beat the record by at least another 4.9 seconds.

C

AU 3 A lorry load of scrap metal weighs 39.715 tonnes. It is worth £20.35 per tonne.

Approximately how much is the load worth?

PS 4 The accurate temperature is 18.2 °C. David rounds the temperature to the nearest 5 °C. David says the temperature is about 20 °C.

How much would the temperature rise for David to say that the temperature is about 25 °C?

370

SUMMARY

GRADE BOOSTER

- F You can multiply a three-digit number by a two-digit number without using a calculator
- F You can divide a three- or four-digit number by a two-digit number
- F You can solve real problems involving multiplication and division
- F You can round decimal numbers to a specific number of places
- E You can evaluate calculations involving decimal numbers
- E You can convert decimals to fractions
- E You can convert fractions to decimals
- E You can order a list containing decimals and fractions
- E You can round numbers to one significant figure
- E You can multiply and divide negative numbers
- D You can estimate the approximate value of a calculation before calculating
- C You can give sensible answers to real-life questions

What you should know now

- How to do long multiplication
- How to do long division
- How to perform calculations with decimal numbers
- How to round to a specific number of decimal places
- How to round to a specific number of significant figures
- How to multiply and divide with negative numbers
- How to convert between decimals and fractions
- How to make estimates by suitable rounding

371

Use of calculators

Questions when you must or could use your calculator are marked with an icon. Explanations involving calculators are based on the *Casio fx-83ES*.

Grade booster

Review what you have learnt and how to get to the next grade with the Grade booster at the end of each chapter.

Worked examples

Understand the topic before you start the exercise by reading the examples in blue boxes. These take you through questions step by step.

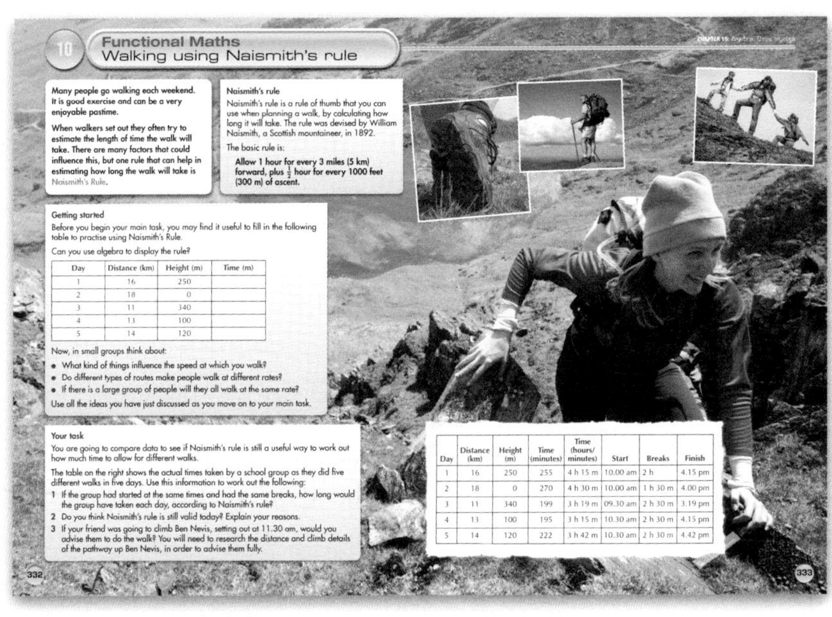

10 Functional Maths
Walking using Naismith's rule

Many people go walking each weekend. It is good exercise and can be a very enjoyable pastime.

When walkers set out they often try to estimate the length of time the walk will take. There are many factors that could influence this, but one rule that can help in estimating how long the walk will take is Naismith's Rule.

Naismith's rule

Naismith's rule is a rule of thumb that you can use when planning a walk, by calculating how long it will take. The rule was devised by William Naismith, a Scottish mountaineer, in 1892.

The basic rule is:

Allow 1 hour for every 3 miles (5 km) forward, plus $\frac{1}{2}$ hour for every 1000 feet (300 m) of ascent.

Getting started

Before you begin your main task, you may find it useful to fill in the following table to practise using Naismith's Rule.

Can you use algebra to display the rule?

Day	Distance (km)	Height (m)	Time (m)
1	16	250	
2	18	0	
3	11	340	
4	13	100	
5	14	120	

Now, in small groups think about:

- What kind of things influence the speed at which you walk?
- Do different types of routes make people walk at different rates?
- If there is a large group of people will they all walk at the same rate?

Use all the ideas you have just discussed as you move on to your main task.

Your task

You are going to compare data to see if Naismith's rule is still a useful way to work out how much time to allow for different walks.

The table on the right shows the actual times taken by a school group as they did five different walks in five days. Use this information to work out the following:

1 If the group had started at the same times and had the same breaks, how long would the group have taken each day, according to Naismith's rule?
2 Do you think Naismith's rule is still valid today? Explain your reasons.
3 If your friend was going to climb Ben Nevis, setting out at 11.30 am, would you advise them to do the walk? You will need to research the distance and climb details of the pathway up Ben Nevis, in order to advise them fully.

Day	Distance (km)	Height (m)	Time (minutes)	Time (hours/ minutes)	Start	Breaks	Finish
1	16	250	255	4 h 15 m	10.00 am	2 h	4.15 pm
2	18	0	270	4 h 30 m	10.00 am	1 h 30 m	4.00 pm
3	11	340	199	3 h 19 m	09.30 am	2 h 30 m	3.19 pm
4	13	100	195	3 h 15 m	10.30 am	2 h 30 m	4.15 pm
5	14	120	222	3 h 42 m	10.30 am	2 h 30 m	4.42 pm

332 333

Functional maths

Practise functional maths skills to see how people use maths in everyday life. Look out for practice questions marked FM.

There are also extra functional maths and problem-solving activities at the end of every chapter to build and apply your skills.

New Assessment Objectives

Practise new parts of the curriculum (Assessment Objectives AO2 and AO3) with questions that assess your understanding marked AU and questions that test if you can solve problems marked PS. You will also practise some questions that involve several steps and where you have to choose which method to use; these also test AO2. There are also plenty of straightforward questions (AO1) that test if you can do the maths.

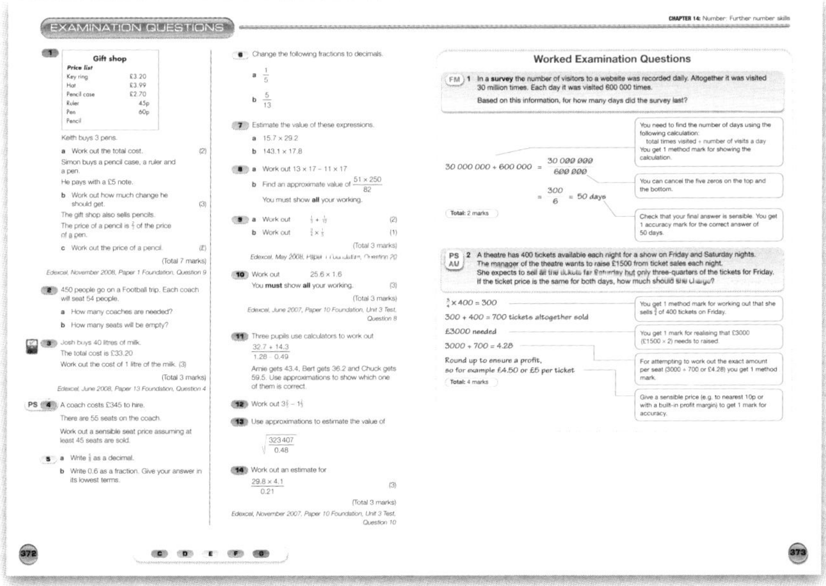

EXAMINATION QUESTIONS

CHAPTER 14: Number: Further number skills

1 Gift shop

Price list

Key ring	£3.20
Hat	£3.99
Pencil case	£2.70
Ruler	45p
Pen	60p
Pencil	

Keith buys 3 pens.

a Work out the total cost. (2)

Simon buys a pencil case, a ruler and a pen.

He pays with a £5 note.

b Work out how much change he should get. (3)

The gift shop also sells pencils.

The price of a pencil is $\frac{2}{3}$ of the price of a pen.

c Work out the price of a pencil. (2)

(Total 7 marks)

Edexcel, November 2008, Paper 1 Foundation, Question 9

2 450 people go on a Football trip. Each coach will seat 54 people.

a How many coaches are needed?

b How many seats will be empty?

3 Josh buys 40 litres of milk.

The total cost is £33.20

Work out the cost of 1 litre of the milk. (3)

(Total 3 marks)

Edexcel, June 2008, Paper 13 Foundation, Question 4

PS 4 A coach costs £345 to hire.

There are 55 seats on the coach.

Work out a sensible seat price assuming at least 45 seats are sold.

5 a Write [illegible] as a decimal.

b Write 0.6 as a fraction. Give your answer in its lowest terms.

6 Change the following fractions to decimals.

a $\frac{1}{5}$

b $\frac{5}{13}$

7 Estimate the value of these expressions.

a 15.7×29.2

b 143.1×17.8

8 a Work out $13 \times 17 - 11 \times 17$

b Find an approximate value of $\frac{51 \times 250}{82}$

You must show **all** your working.

9 a Work out [illegible] (2)

b Work out [illegible] (1)

(Total 3 marks)

Edexcel, May 2008, Paper [illegible] Foundation, Question 20

10 Work out 25.6×1.6

You **must** show **all** your working. (3)

(Total 3 marks)

Edexcel, June 2007, Paper 10 Foundation, Unit 3 Test, Question 8

11 Three pupils use calculators to work out

$\frac{32.7 + 14.3}{1.28 - 0.49}$

Arnie gets 43.4, Bert gets 36.2 and Chuck gets 59.5. Use approximations to show which one of them is correct.

12 Work out [illegible]

13 Use approximations to estimate the value of

$\sqrt{\frac{323\,407}{0.48}}$

14 Work out an estimate for

$\frac{29.8 \times 4.1}{0.21}$ (3)

(Total 3 marks)

Edexcel, November 2007, Paper 10 Foundation, Unit 3 Test, Question 10

Worked Examination Questions

FM 1 In a **survey** the number of visitors to a website was recorded daily. Altogether it was visited 30 million times. Each day it was visited 600 000 times.

Based on this information, for how many days did the survey last?

$30\,000\,000 \div 600\,000 = \frac{30\,000\,000}{600\,000}$

You need to find the number of days using the following calculation:
total times visited ÷ number of visits a day
You get 1 method mark for showing the calculation.

$= \frac{300}{6} = 50$ days

You can cancel the five zeros on the top and the bottom.

Check that your final answer is sensible. You get 1 accuracy mark for the correct answer of 50 days.

Total: 2 marks

PS AU 2 A theatre has 400 tickets available each night for a show on Friday and Saturday nights.
The manager of the theatre wants to raise £1500 from ticket sales each night.
She expects to sell all the tickets for Saturday but only three-quarters of the tickets for Friday.
If the ticket price is the same for both days, how much should she charge?

$\frac{3}{4} \times 400 = 300$

You get 1 method mark for working out that she sells $\frac{3}{4}$ of 400 tickets on Friday.

300 + 400 = 700 tickets altogether sold

£3000 needed

You get 1 mark for realising that £3000 (£1500 × 2) needs to raised.

3000 ÷ 700 = 4.28

For attempting to work out the exact amount per seat (3000 ÷ 700 or £4.28) you get 1 method mark.

Round up to ensure a profit, so for example £4.50 or £5 per ticket

Give a sensible price (e.g. to nearest 10p or with a built-in profit margin) to get 1 mark for accuracy.

Total: 4 marks

372 C D E F G 373

Exam practice

Prepare for your exams with past exam questions and detailed worked exam questions with examiner comments to help you score maximum marks.

Quality of Written Communication (QWC)

Practise using accurate mathematical vocabulary and writing logical answers to questions to ensure you get your QWC (Quality of Written Communication) marks in the exams. The Glossary and worked exam questions will help you with this.

Why this chapter matters

Thousands of years ago, many different civilisations developed different number systems. Most of these ways of counting were based on the number 10 (decimal numbers), simply because humans have 10 fingers and 10 toes.

The Egyptian number system (from about 3000BC) used different symbols to represent 1, 10, 100, 1000 and so on.

So the number 23 would be written: ∩∩ |||

Decimal number	Egyptian symbol	
1 =	|	staff
10 =	∩	heel bone
100 =	(coil)	piece of rope
1000 =	(flower)	flower
10 000 =	(finger)	pointing finger
100 000 =	(tadpole)	tadpole
1 000 000 =	(man)	man

The Chinese number system used sticks.

|	||	|||	||||	|||||	T	Π	⊤||	⊤|||
1	2	3	4	5	6	7	8	9

—	=	≡	≣	(5 bars)	⊥	(⊥ with bar)	(⊥ with 2 bars)	(⊥ with 3 bars)
10	20	30	40	50	60	70	80	90

So, the number 23 would be written: = |||

The Roman number system is still used today, often on clocks, and is based on the numbers five and ten.

I	V	X	L	C	D	M
1	5	10	50	100	500	1000

So, the number 23 would be written: XXIII

The system we use today is called the **Hindu-Arabic system** and uses the symbols 0, 1, 2, 3, 4, 5, 6, 7, 8 and 9. This has been widely used since about 900AD.

This system provides an almost universal 'language' of maths that has allowed us to make sense of the world around us and communicate ideas to others, even when their spoken language may differ from ours.

Here you can see the Roman numerals for 1–12 on a clock face.

Roman numerals also appear on Big Ben.

Chapter

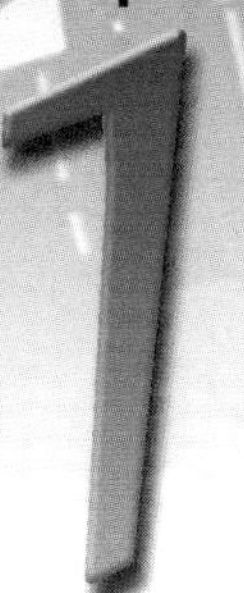

Number: Basic number

The grades given in this chapter are target grades.

This chapter will show you ...

to
G
F
how to use basic number skills without a calculator

Visual overview

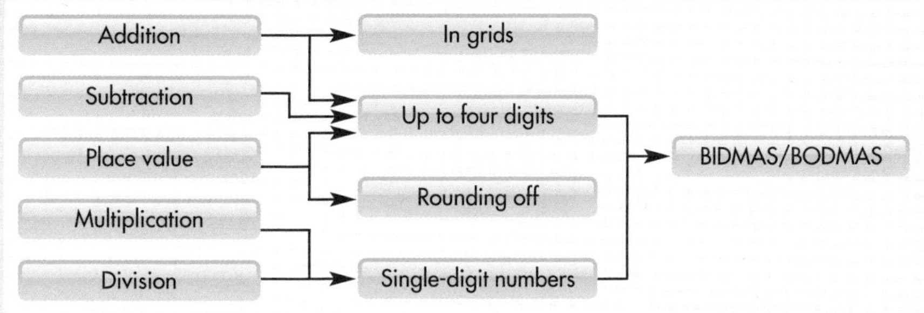

What you should already know

- Multiplication tables up to 10 × 10 **(KS3 level 4, GCSE grade G)**
- Addition and subtraction of numbers less than 20 **(KS3 level 2, GCSE grade G)**
- Simple multiplication and division **(KS3 level 3, GCSE grade G)**
- How to multiply numbers by 10 and 100 **(KS3 level 3, GCSE grade G)**

Quick check

How quickly can you complete these?

1 4×6	2 3×7	3 5×8	4 9×2
5 6×7	6 $13 + 14$	7 $15 + 15$	8 $18 - 12$
9 $19 - 7$	10 $11 - 6$	11 $50 \div 5$	12 $48 \div 6$
13 $35 \div 7$	14 $42 \div 6$	15 $36 \div 9$	16 8×10
17 9×100	18 3×10	19 14×100	20 17×10

1.1 Adding with grids

This section will show you how to:
- add and subtract single-digit numbers in a grid
- use row and column totals to find missing numbers in a grid

Key words
add
column
grid
row

ACTIVITY

Adding with grids

You need a set of cards marked 0 to 9.

0 1 2 3 4 5 6 7 8 9

Shuffle the cards and lay them out in a three by three **grid**.

You will have one card left over.

3	5	0
7	6	4
8	2	9

Copy your grid onto a piece of paper. Then **add** up the numbers in each **row** and each **column** and write down their totals. Finally, find the grand total and write it in the box at the bottom right.

3	5	0	8
7	6	4	17
8	2	9	19
18	13	13	44

Look out for things that help. For example:

- in the first column, 3 + 7 make 10 and 10 + 8 = 18
- in the last column, 9 + 4 = 9 + 1 + 3 = 10 + 3 = 13

Reshuffle the cards, lay them out again and copy the new grid. Copy the new grid again on a fresh sheet of paper, leaving out some of the numbers.

4	5	8
0	2	6
9	1	7

4	5	8	17
0	2	6	8
9	1	7	17
13	8	21	42

4	□	8	17
□	2	□	8
9	□	7	□
□	8	21	42

Pass this last grid to a friend to work out the missing numbers. You can make it quite hard because you are using only the numbers from 0 to 9. Remember: once a number has been used, it *cannot* be used again in that grid.

Example Find the numbers missing from this grid.

□	□	9	17
□	2	□	11
8	□	□	□
19	3	17	□

Clues The two numbers missing from the second column must add up to 1, so they must be 0 and 1. The two numbers missing from the first column add up to 11, so they could be 7 and 4 or 6 and 5. Now, 6 or 5 won't work with 0 or 1 to give 17 across the top row. That means it has to be:

7	1	9	17
4	2	□	11
8	0	□	□
19	3	17	□

giving

7	1	9	17
4	2	5	11
8	0	3	11
19	3	17	39

as the answer.

You can use your cards to try out your ideas.

EXERCISE 1A

G

1 Find the row and column totals for each of these grids.

a

1	3	7	□
9	2	8	□
6	5	4	□
□	□	□	□

b

0	6	7	□
8	1	4	□
9	5	3	□
□	□	□	□

c

0	8	7	□
1	6	2	□
9	3	4	□
□	□	□	□

d

2	4	6	□
3	5	7	□
8	9	1	□
□	□	□	□

e

5	9	3	□
6	1	8	□
2	7	4	□
□	□	□	□

f

0	8	3	□
7	2	4	□
1	6	5	□
□	□	□	□

g

9	4	8	□
7	0	5	□
1	6	3	□
□	□	□	□

h

0	8	6	□
7	1	4	□
5	9	2	□
□	□	□	□

i

1	8	7	□
6	2	5	□
0	9	3	□
□	□	□	□

FM **2** **a** Adam has two £10 notes, one £5 note and eight £1 coins in his pocket. Does he have enough money to buy a shirt costing £32?

b Belinda puts some plants into pots. There are six roses, eight sunflowers and nine lilies. How many plants altogether does she put into pots?

AU 3 Here is a list of numbers.

4 5 7 8 11 12

a From the list write down **two** numbers that add up to 18.

b From the list work out the largest total that can be made using **three** numbers.

c From the list work out the largest **even** total that can be made using **three** numbers.

PS 4 What number could be being described?

'It is bigger than three'
'It is less than 10'
'It is an even number'

5 Find the numbers missing from each of these grids. Remember: the numbers missing from each grid must be chosen from 0 to 9 without any repeats.

a

1	7	□	16
□	3	6	9
5	□	2	11
6	14	16	36

b

1	□	3	6
□	5	4	15
7	8	□	24
14	15	16	35

c

9	3	□	18
4	□	5	9
□	2	8	11
14	5	19	38

d

□	□	□	16
2	□	4	13
8	5	0	13
19	13	10	42

e

2	□	6	17
□	1	□	□
5	□	8	13
11	□	17	38

f

1	□	□	16
□	2	4	12
□	9	3	□
12	□	15	□

g

0	2	□	3
9	□	□	□
□	4	5	17
17	□	13	42

h

□	□	3	4
□	7	4	□
9	6	□	20
18	□	12	□

i

□	□	4	10
□	2	□	□
8	□	□	15
15	□	□	36

PS 6 Find **three different** numbers that add up to 36.

PS 7 Find **two different odd** numbers that add up to 24.

1.2 Multiplication tables check

This section will show you how to:

- recall and use your knowledge of multiplication tables

Key words
multiplication
multiplication tables
sign
times table

ACTIVITY

Special table facts

You need a sheet of squared paper.

Start by writing in the easy multiplication tables (or **times tables**). These are the 1 ×, 2 ×, 5 ×, 9 × and 10 × tables.

Now draw up a 10 by 10 table square before you go any further. (Time yourself doing this and see if you can get faster.)

Once you have a complete tables square, shade out all the **multiplications** that you already know. You should be left with something like the square on the right.

×	1	2	3	4	5	6	7	8	9	10
1										
2										
3			9	12		18	21	24		
4			12	16		24	28	32		
5										
6			18	24		36	42	48		
7			21	28		42	49	56		
8			24	32		48	56	64		
9										
10										

Now cross out **one** of each pair that have the same answer, such as 3 × 4 and 4 × 3. This leaves you with:

×	1	2	3	4	5	6	7	8	9	10
1										
2										
3			9							
4			12	16						
5										
6			18	24		36				
7			21	28		42	49			
8			24	32		48	56	64		
9										
10										

Now there are just 15 table facts. Do learn them.

The rest are easy tables, so you should know all of them. But keep practising!

EXERCISE 1B

G

1 Write down the answer to each of the following without looking at the multiplication square.

a 4×5 **b** 7×3 **c** 6×4 **d** 3×5 **e** 8×2

f 3×4 **g** 5×2 **h** 6×7 **i** 3×8 **j** 9×2

k 5×6 **l** 4×7 **m** 3×6 **n** 8×7 **o** 5×5

p 5×9 **q** 3×9 **r** 6×5 **s** 7×7 **t** 4×6

u 6×6 **v** 7×5 **w** 4×8 **x** 4×9 **y** 6×8

FM **z** Matt works for 6 hours each day and is paid £8 an hour.

He is saving for a bike that costs £75.

Will he have enough after two days?

Show how you worked out your answer.

2 Write down the answer to each of the following without looking at the multiplication square.

a $10 \div 2$ **b** $28 \div 7$ **c** $36 \div 6$ **d** $30 \div 5$ **e** $15 \div 3$

f $20 \div 5$ **g** $21 \div 3$ **h** $24 \div 4$ **i** $16 \div 8$ **j** $12 \div 4$

k $42 \div 6$ **l** $24 \div 3$ **m** $18 \div 2$ **n** $25 \div 5$ **o** $48 \div 6$

p $36 \div 4$ **q** $32 \div 8$ **r** $35 \div 5$ **s** $49 \div 7$ **t** $27 \div 3$

u $45 \div 9$ **v** $16 \div 4$ **w** $40 \div 8$ **x** $63 \div 9$ **y** $54 \div 9$

FM **z** Viki works for 7 hours and is paid £42. She wants to save £60 to buy two tickets to the ballet.

How many hours will she need to work in order to save enough?

3 Write down the answer to each of the following. Look carefully at the **signs**, because they are a mixture of ×, +, – and ÷ .

a $5 + 7$ **b** $20 - 5$ **c** 3×7 **d** $5 + 8$ **e** $24 \div 3$

f $15 - 8$ **g** $6 + 8$ **h** $27 \div 9$ **i** 6×5 **j** $36 \div 6$

k 7×5 **l** $15 \div 3$ **m** $24 - 8$ **n** $28 \div 4$ **o** $7 + 9$

p $9 + 6$ **q** $36 - 9$ **r** $30 \div 5$ **s** $8 + 7$ **t** 4×6

u 8×5 **v** $42 \div 7$ **w** $8 + 9$ **x** 9×8 **y** $54 - 8$

FM **z** Ahmed works for 3 hours and is paid £11 an hour. Ben works for 4 hours and is paid £8 an hour. Who is paid more?

G

AU 4 Here are four single-digit number cards.

[5] [8] [3] [6]

The cards are used for making calculations. Complete the following.

a [3] [5] + [...] [...] = 121

b [8] [3] − [...] [...] = 27

c [...] [...] [8] + [...] = 364

5 Write down the answer to each of the following.

a 3×10 **b** 5×10 **c** 8×10 **d** 10×10 **e** 12×10

f 18×10 **g** 24×10 **h** 4×100 **i** 7×100 **j** 9×100

k 10×100 **l** 14×100 **m** 24×100 **n** 72×100 **o** 100×100

p $20 \div 10$ **q** $70 \div 10$ **r** $90 \div 10$ **s** $170 \div 10$ **t** $300 \div 10$

u $300 \div 100$ **v** $800 \div 100$ **w** $1200 \div 100$ **x** $2900 \div 100$ **y** $5000 \div 100$

F

PS 6 **a** Two numbers when multiplied together give an answer of 24. One of the numbers is odd and greater than one. What is the other number?

b One whole number is divided by another whole number and gives an answer of 90. One of the numbers is 10. What is the other number?

PS 7 Consecutive numbers are numbers that are next to each other, for example 2 and 3.

Phil says that when two consecutive numbers are multiplied together the answer is always even.

Show, with two different examples, that this is true.

PS 8 Jenna says that when you multiply three consecutive numbers together you always get a number in the six times table, for example $2 \times 3 \times 4 = 24 = 4 \times 6$

Show, with two different examples, that this is true.

1.3 Order of operations and BIDMAS/BODMAS

This section will show you how to:

- work out the answers to a problem with a number of different mathematical operations

Key words
brackets
operation
order

Suppose you have to work out the answer to $4 + 5 \times 2$. You may say the answer is 18, but the correct answer is 14.

There is an **order** of **operations** which you *must* follow when working out calculations like this. The × is always done *before* the +.

In $4 + 5 \times 2$ this gives $4 + 10 = 14$.

Now suppose you have to work out the answer to $(3 + 2) \times (9 - 5)$. The correct answer is 20.

You have probably realised that the parts in the **brackets** have to be done *first*, giving $5 \times 4 = 20$.

So, how do you work out a problem such as $9 \div 3 + 4 \times 2$?

To answer questions like this, you *must* follow the BIDMAS (or BODMAS) rule. This tells you the order in which you *must* do the operations.

B	Brackets	**B**	Brackets
I	Indices (Powers)	**O**	pOwers
D	Division	**D**	Division
M	Multiplication	**M**	Multiplication
A	Addition	**A**	Addition
S	Subtraction	**S**	Subtraction

For example, to work out $9 \div 3 + 4 \times 2$:

First divide:	$9 \div 3 = 3$	giving	$3 + 4 \times 2$
Then multiply:	$4 \times 2 = 8$	giving	$3 + 8$
Then add:	$3 + 8 = 11$		

And to work out $60 - 5 \times 3^2 + (4 \times 2)$:

First, work out the brackets:	$(4 \times 2) = 8$	giving	$60 - 5 \times 3^2 + 8$
Then the index (power):	$3^2 = 9$	giving	$60 - 5 \times 9 + 8$
Then multiply:	$5 \times 9 = 45$	giving	$60 - 45 + 8$
Then add:	$60 + 8 = 68$	giving	$68 - 45$
Finally, subtract:	$68 - 45 = 23$		

ACTIVITY

Dice with BIDMAS/BODMAS

You need a sheet of squared paper and three dice.

Draw a five by five grid and write the numbers from 1 to 25 in the spaces.

The numbers can be in *any order*.

14	13	18	7	24
15	1	16	17	6
23	8	2	12	5
3	22	4	10	19
25	21	9	20	11

Now throw three dice. Record the score on each one.

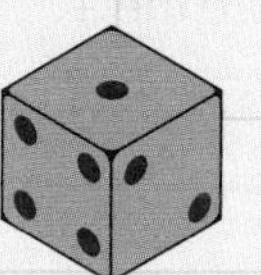
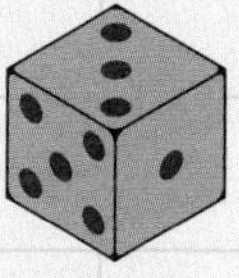

Use these numbers to make up a number problem.

You must use all three numbers, and you must not put them together to make a number (such as making 136 with the three dice shown above). For example, with 1, 3 and 6 you could make:

$1 + 3 + 6 = 10$ $\quad 3 \times 6 + 1 = 19$ $\quad (1 + 3) \times 6 = 24$

$6 \div 3 + 1 = 3$ $\quad 6 + 3 - 1 = 8$ $\quad 6 \div (3 \times 1) = 2$

and so on. The answer to the problem must be from 1 to 25. Remember to use BIDMAS/BODMAS.

You have to make only one problem with each set of numbers.

When you have made a problem, cross the answer off the grid and throw the dice again. Make up a problem with the next three numbers and cross that answer off the grid. Throw the dice again and so on.

The first person to make a line of five numbers across, down or diagonally is the winner.

You must write down each problem and its answer so that they can be checked.

Just put a line through each number on the grid, as you use it. Do not cross it out so that it cannot be read, otherwise your problem and its answer cannot be checked.

This might be a typical game.

14	13	18	7	24
15	1	16	17	6
23	8	2	12	5
3	22	4	10	19
25	21	9	20	11

First set (1, 3, 6)	$6 \times 3 \times 1 = 18$
Second set (2, 4, 4)	$4 \times 4 - 2 = 14$
Third set (3, 5, 1)	$(3 - 1) \times 5 = 10$
Fourth set (3, 3, 4)	$(3 + 3) \times 4 = 24$
Fifth set (1, 2, 6)	$6 \times 2 - 1 = 11$
Sixth set (5, 4, 6)	$(6 + 4) \div 5 = 2$
Seventh set (4, 4, 2)	$2 - (4 \div 4) = 1$

EXERCISE 1C

G

1 Work out each of these.

a $2 \times 3 + 5 =$
b $6 \div 3 + 4 =$
c $5 + 7 - 2 =$
d $4 \times 6 \div 2 =$
e $2 \times 8 - 5 =$
f $3 \times 4 + 1 =$
g $3 \times 4 - 1 =$
h $3 \times 4 \div 1 =$
i $12 \div 2 + 6 =$
j $12 \div 6 + 2 =$
k $3 + 5 \times 2 =$
l $12 - 3 \times 3 =$

2 Work out each of the following. Remember: first work out the bracket.

a $2 \times (3 + 5) =$
b $6 \div (2 + 1) =$
c $(5 + 7) - 2 =$
d $5 + (7 - 2) =$
e $3 \times (4 \div 2) =$
f $3 \times (4 + 2) =$
g $2 \times (8 - 5) =$
h $3 \times (4 + 1) =$
i $3 \times (4 - 1) =$
j $3 \times (4 \div 1) =$
k $12 \div (2 + 2) =$
l $(12 \div 2) + 2 =$

3 Copy each of these and put a loop round the part that you do first. Then work out the answer. The first one has been done for you.

a $(3 \times 3) - 2 = 7$
b $3 + 2 \times 4 =$
c $9 \div 3 - 2 =$
d $9 - 4 \div 2 =$
e $5 \times 2 + 3 =$
f $5 + 2 \times 3 =$
g $10 \div 5 - 2 =$
h $10 - 4 \div 2 =$
i $4 \times 6 - 7 =$
j $7 + 4 \times 6 =$
k $6 \div 3 + 7 =$
l $7 + 6 \div 2 =$

4 Work out each of these.

a $6 \times 6 + 2 =$
b $6 \times (6 + 2) =$
c $6 \div 6 + 2 =$
d $12 \div (4 + 2) =$
e $12 \div 4 + 2 =$
f $2 \times (3 + 4) =$
g $2 \times 3 + 4 =$
h $2 \times (4 - 3) =$
i $2 \times 4 - 3 =$
j $17 + 5 - 3 =$
k $17 - 5 + 3 =$
l $17 - 5 \times 3 =$
m $3 \times 5 + 5 =$
n $6 \times 2 + 7 =$
o $6 \times (2 + 7) =$
p $12 \div 3 + 3 =$
q $12 \div (3 + 3) =$
r $14 - 7 \times 1 =$
s $(14 - 7) \times 1 =$
t $2 + 6 \times 6 =$
u $(2 + 5) \times 6 =$
v $12 - 6 \div 3 =$
w $(12 - 6) \div 3 =$
x $15 - (5 \times 1) =$
y $(15 - 5) \times 1 =$
z $8 \times 9 \div 3 =$

5 Copy each of these and then put in brackets where necessary to make each answer true.

a $3 \times 4 + 1 = 15$
b $6 \div 2 + 1 = 4$
c $6 \div 2 + 1 = 2$
d $4 + 4 \div 4 = 5$
e $4 + 4 \div 4 = 2$
f $16 - 4 \div 3 = 4$
g $3 \times 4 + 1 = 13$
h $16 - 6 \div 3 = 14$
i $20 - 10 \div 2 = 5$
j $20 - 10 \div 2 = 15$
k $3 \times 5 + 5 = 30$
l $6 \times 4 + 2 = 36$
m $15 - 5 \times 2 = 20$
n $4 \times 7 - 2 = 20$
o $12 \div 3 + 3 = 2$
p $12 \div 3 + 3 = 7$
q $24 \div 8 - 2 = 1$
r $24 \div 8 - 2 = 4$

6 Three dice are thrown. They give scores of three, one and four.

A class makes the following questions with the numbers. Work them out.

a $3 + 4 + 1 =$ **b** $3 + 4 - 1 =$ **c** $4 + 3 - 1 =$

d $4 \times 3 + 1 =$ **e** $4 \times 3 - 1 =$ **f** $(4 - 1) \times 3 =$

g $4 \times 3 \times 1 =$ **h** $(3 - 1) \times 4 =$ **i** $(4 + 1) \times 3 =$

j $4 \times (3 + 1) =$ **k** $1 \times (4 - 3) =$ **l** $4 + 1 \times 3 =$

AU 7 Jack says that $5 + 6 \times 7$ is equal to 77.

Is he correct?

Explain your answer.

AU 8 This is Micha's homework.

Copy the questions where she has made mistakes and work out the correct answers.

a $2 + 3 \times 4 =$ 20 **b** $8 - 4 \div 4 =$ 7 **c** $6 + 3 \times 2 =$ 12

d $7 - 1 \times 5 =$ 30 **e** $2 \times 7 + 2 =$ 16 **f** $9 - 3 \times 3 =$ 18

9 Three different dice give scores of 2, 3, 5. Add ÷, ×, + or – signs to make each calculation work.

a 2 3 5 = 11 **b** 2 3 5 = 16 **c** 2 3 5 = 17

d 5 3 2 = 4 **e** 5 3 2 = 13 **f** 5 3 2 = 30

AU 10 Which is smaller

$4 + 5 \times 3$ or $(4 + 5) \times 3$?

Show your working.

PS 11 Here is a list of numbers, some signs and one pair of brackets.

2 5 6 18 – × = ( )

Use **all** of them to make a correct calculation.

PS 12 Here is a list of numbers, some signs and one pair of brackets.

3 4 5 8 – ÷ = ( )

Use **all** of them to make a correct calculation.

FM 13 Jeremy has a piece of pipe that is 10 m long.

He wants to use his calculator to work out how much pipe will be left when he cuts off three pieces, each of length 1.5 m.

Which calculations would give him the correct answer?

$10 - 3 \times 1.5$ $10 - 1.5 + 1.5 + 1.5$ $10 - 1.5 - 1.5 - 1.5$

1.4 Place value and ordering numbers

This section will show you how to:
- identify the value of any digit in a number

Key words
digit
order
place value

The ordinary counting system uses **place value**, which means that the value of a **digit** depends upon its place in the number.

In the number 5348

the 5 stands for five thousands or 5000

the 3 stands for three hundreds or 300

the 4 stands for four tens or 40

the 8 stands for eight units or 8

You write and say this number as:

five thousand, three hundred and forty-eight

In the number 4 073 520

the 4 stands for four millions or 4 000 000

the 73 stands for 73 thousands or 73 000

the 5 stands for five hundreds or 500

the 2 stands for two tens or 20

You write and say this number as:

four million, seventy-three thousand, five hundred and twenty

Note the use of narrow spaces between groups of three digits, starting from the right. All whole and mixed numbers with five or more digits are spaced in this way.

EXAMPLE 1

Put these numbers in **order**, putting the *smallest* first.

7031 3071 3701 7103 7130 1730

Look at the thousands column first and then each of the other columns in turn. The correct order is:

1730 3071 3701 7031 7103 7130

EXERCISE 1D

G

1 Write the value of each underlined digit.

a $3\underline{4}1$ **b** $47\underline{5}$ **c** $\underline{1}86$ **d** $2\underline{9}8$ **e** $\underline{8}3$

f $83\underline{9}$ **g** $23\underline{8}0$ **h** $1\underline{5}07$ **i** $653\underline{0}$ **j** $2\underline{5}\ 436$

k $29\ \underline{0}54$ **l** $18\ 25\underline{4}$ **m** $4\underline{3}08$ **n** $52\ 9\underline{9}4$ **o** $\underline{8}3\ 205$

2 Copy each of these sentences, writing the numbers in words.

a The last Olympic Games in Greece had only 43 events and 200 competitors.

b The last Olympic Games in Britain had 136 events and 4099 competitors.

c The last Olympic Games in the USA had 271 events and 10 744 competitors.

3 Write each of the following numbers in words.

a 5 600 000 **b** 4 075 200 **c** 3 007 950 **d** 2 000 782

4 Write each of the following numbers in numerals or digits.

a Eight million, two hundred thousand and fifty-eight

b Nine million, four hundred and six thousand, one hundred and seven

c One million, five hundred and two

d Two million, seventy-six thousand and forty

5 Write these numbers in order, putting the *smallest* first.

a 21, 48, 23, 9, 15, 56, 85, 54

b 310, 86, 219, 25, 501, 62, 400, 151

c 357, 740, 2053, 888, 4366, 97, 368

6 Write these numbers in order, putting the *largest* first.

a 52, 23, 95, 34, 73, 7, 25, 89

b 65, 2, 174, 401, 80, 700, 18, 117

c 762, 2034, 395, 6227, 89, 3928, 59, 480

7 Copy each sentence and fill in the missing word, *smaller* or *larger*.

a 7 is …… than 5

b 34 is …… than 29

c 89 is …… than 98

d 97 is …… than 79

e 308 is …… than 299

f 561 is …… than 605

g 870 is …… than 807

h 4275 is …… than 4527

i 782 is …… than 827

8 An estate agent advertises the following houses.

a Which house is the cheapest?

b Which house is the most expensive?

c George has just enough money to buy the cheapest house. How much more money would he need to buy the most expensive house?

PS 9 **a** Write as many three-digit numbers as you can, using the digits 3, 6 and 8. (Only use each digit once in each number.)

b Which of your numbers is the smallest?

c Which of your numbers is the largest?

PS 10 Using each of the digits 0, 4 and 8 only once in each number, write as many different three-digit numbers as you can. (Do not start any number with 0.) Write your numbers down in order, smallest first.

PS 11 Write down in order of size, smallest first, all the two-digit numbers that can be made using 3, 5 and 8. (Each digit can be repeated.)

AU 12 Using each of the digits 0, 4, 5 and 7, make a four-digit odd number greater than seven thousand.

13 Nick has these number cards

a Copy the blank calculation below and insert numbers into the boxes to make the largest possible total.

b Copy the blank calculation below and insert numbers into the boxes to make the smallest possible difference.

1.5 Rounding

This section will show you how to:
- round a number

Key words
approximation
rounded down
rounded up

You use rounded information all the time. Look at the examples on the right. All of these statements use rounded information. Each actual figure is either above or below the **approximation** shown here. But if the rounding is done correctly, you can find out what the maximum and the minimum figures really are. For example, if you know that the number of matches in the packet is rounded to the nearest 10, and it states that there are 30 matches in a packet:

- the smallest figure **rounded up** to 30 is 25, and
- the largest figure **rounded down** to 30 is 34 (because 35 would be rounded up to 40).

So, there could actually be from 25 to 34 matches in the packet.

What about the number of runners in the marathon? If you know that the number 23 000 is rounded to the nearest 1000:

- the smallest figure rounded up to 23 000 is 22 500, and
- the largest figure rounded down to 23 000 is 23 499 (because 23 500 would be rounded up to 24 000).

So, there could actually be from 22 500 to 23 499 people in the marathon.

EXERCISE 1E

G

1 Round each of these numbers to the nearest 10.

a 24	**b** 57	**c** 78	**d** 54	**e** 96
f 21	**g** 88	**h** 66	**i** 14	**j** 26
k 29	**l** 51	**m** 77	**n** 49	**o** 94
p 35	**q** 65	**r** 15	**s** 102	**t** 107

2 Round each of these numbers to the nearest 100.

a 240	**b** 570	**c** 780	**d** 504	**e** 967
f 112	**g** 645	**h** 358	**i** 998	**j** 1050
k 299	**l** 511	**m** 777	**n** 512	**o** 940
p 350	**q** 650	**r** 750	**s** 1020	**t** 1070

3 On the shelf of a sweetshop there are three jars like the ones below.

Jar 1

Jar 2

Jar 3

Look at each of the numbers below and write down which jar it could be describing. (For example, 76 sweets could be in jar 1.)

a 78 sweets	**b** 119 sweets	**c** 84 sweets	**d** 75 sweets
e 186 sweets	**f** 122 sweets	**g** 194 sweets	**h** 115 sweets
i 81 sweets	**j** 79 sweets	**k** 192 sweets	**l** 124 sweets

m Which of these numbers of sweets *could not* be in jar 1: 74, 84, 81, 76?

n Which of these numbers of sweets *could not* be in jar 2: 124, 126, 120, 115?

o Which of these numbers of sweets *could not* be in jar 3: 194, 184, 191, 189?

4 Round each of these numbers to the nearest 1000.

a 2400	**b** 5700	**c** 7806	**d** 5040	**e** 9670
f 1120	**g** 6450	**h** 3499	**i** 9098	**j** 1500
k 2990	**l** 5110	**m** 7777	**n** 5020	**o** 9400
p 3500	**q** 6500	**r** 7500	**s** 1020	**t** 1770

5 Round each of these numbers to the nearest 10.

a 234	**b** 567	**c** 718	**d** 524	**e** 906
f 231	**g** 878	**h** 626	**i** 114	**j** 296
k 279	**l** 541	**m** 767	**n** 501	**o** 942
p 375	**q** 625	**r** 345	**s** 1012	**t** 1074

AU **6**

Welcome to Elsecar	Welcome to Hoyland	Welcome to Jump
Population 800 (to the nearest 100)	Population 1200 (to the nearest 100)	Population 600 (to the nearest 100)

Which of these sentences could be true and which must be false?

a There are 789 people living in Elsecar.
b There are 1278 people living in Hoyland.
c There are 550 people living in Jump.
d There are 843 people living in Elsecar.
e There are 1205 people living in Hoyland.
f There are 650 people living in Jump.

FM **7** A sign maker is asked to create a sign similar to those shown in question 6 for Swinton, which has a population of 1385.

Make a diagram of the sign he should paint.

FM **8** These were the numbers of spectators in the crowds at nine Premier Division games on a weekend in May 2005.

Match	Crowd
Aston Villa v Man City	39 645
Blackburn v Fulham	18 991
Chelsea v Charlton	42 065
C. Palace v Southampton	26 066
Everton v Newcastle	40 438
Man.Utd v West Brom	67 827
Middlesbrough v Tottenham	34 766
Norwich v Birmingham	25 477
Portsmouth v Bolton	20 188

a Which match had the largest crowd?
b Which had the smallest crowd?
c Round all the numbers to the nearest 1000.
d Round all the numbers to the nearest 100.

9 Give these cooking times to the nearest 5 minutes.

a 34 min **b** 57 min **c** 14 min **d** 51 min **e** 8 min
f 13 min **g** 44 min **h** 32.5 min **i** 3 min **j** 50 s

PS **10** Matthew and Viki are playing a game with whole numbers.

a What is the smallest number Matthew could be thinking of?

I am thinking of a number. Rounded to the nearest 10 it is 380.

I am thinking of a different number. Rounded to the nearest 100 it is 400.

b Is Viki's number smaller than Matthew's? How many possible answers are there?

PS AU **11** The number of adults attending a comedy show is 80 to the nearest 10.
The number of children attending is 50 to the nearest 10.
Katie says that 130 adults and children attended the comedy show.
Give an example to show that she may **not** be correct.

1.6 Adding and subtracting numbers with up to four digits

This section will show you how to:

- add and subtract numbers with more than one digit

Key words
addition
column
digit
subtract

Addition

There are three things to remember when you are adding two whole numbers.

- The answer will always be larger than the bigger number.
- Always add the units **column** first.
- When the total of the **digits** in a column is more than nine, you have to carry a digit into the next column on the left, as shown in Example 2. It is important to write down the carried digit, otherwise you may forget to include it in the **addition**.

EXAMPLE 2

Add: **a** 167 + 25 **b** 2296 + 1173

a
```
  167
+  25
  ---
  192
   1
```

b
```
  2296
+ 1173
  ----
  3469
   1
```

Subtraction

These are four things to remember when you are subtracting one whole number from another.

- The bigger number must always be written down first.
- The answer will always be smaller than the bigger number.
- Always **subtract** the units column first.
- When you have to take a bigger digit from a smaller digit in a column, you must 'borrow' a 10 by taking one from the column to the left and putting it with the smaller digit, as shown in Example 3.

EXAMPLE 3

Subtract: **a** 874 − 215 **b** 300 − 163

a
```
  8⁶7̸¹4
− 2 1 5
  -----
  6 5 9
```

b
```
  ²3̸⁹0̸¹0
− 1 6 3
  -----
  1 3 7
```

EXERCISE 1F

G

1 Copy and work out each of these additions.

a $365 + 348$ **b** $95 + 56$ **c** $4872 + 1509$ **d** $317 + 416 + 235$ **e** $287 + 335$

f $483 + 832$ **g** $4676 + 3584$ **h** $438 + 147 + 233$ **i** $175 + 276$ **j** $562 + 93 + 197$

2 Copy and complete each of these additions.

a 128 + 518 **b** 563 + 85 + 178 **c** 3086 + 58 + 674

d 347 + 408 **e** 85 + 1852 + 659 **f** 759 + 43 + 89

g 257 + 93 **h** 605 + 26 + 2135 **i** 56 + 8407 + 395

j 89 + 752 **k** 6143 + 557 + 131 **l** 2593 + 45 + 4378

m 719 + 284 **n** 545 + 3838 + 67 **o** 5213 + 658 + 4073

3 Copy and complete each of these subtractions.

a $637 - 187$ **b** $908 - 345$ **c** $954 - 472$ **d** $572 - 158$ **e** $732 - 447$

f $673 - 187$ **g** $602 - 358$ **h** $638 - 354$ **i** $650 - 317$ **j** $580 - 364$

k $6254 - 3362$ **l** $8043 - 3626$ **m** $8432 - 4665$ **n** $8034 - 3947$ **o** $5375 - 3547$

4 Copy and complete each of these subtractions.

a 354 – 226 **b** 285 – 256 **c** 663 – 329

d 506 – 328 **e** 654 – 377 **f** 733 – 448

g 592 – 257 **h** 753 – 354 **i** 6705 – 2673

j 8021 – 3256 **k** 7002 – 3207 **l** 8700 – 3263

FM **5** The distance from Cardiff to London is 152 miles.
The distance from London to Edinburgh is 406 miles.

a How far is it to travel from London to Cardiff and then from Cardiff to Edinburgh?

b How much further is it to travel from London to Edinburgh than from Cardiff to London?

AU 6 Jon is checking the addition of two numbers.

His answer is 843.

One of the numbers is 591.

What should the other number be?

7 Copy each of these additions and fill in the missing digits.

a $\begin{array}{r} 5\,3 \\ +\,2\,\square \\ \hline \square\,9 \\ \hline \end{array}$ **b** $\begin{array}{r} \square\,7 \\ +\,3\,\square \\ \hline 8\,4 \end{array}$ **c** $\begin{array}{r} 4\,5 \\ +\,\square\,\square \\ \hline 9\,3 \end{array}$ **d** $\begin{array}{r} 4\,\square\,7 \\ +\,\square\,5\,\square \\ \hline 9\,3\,6 \end{array}$

e $\begin{array}{r} \square\,1\,8 \\ +\,2\,5\,\square \\ \hline 8\,\square\,7 \end{array}$ **f** $\begin{array}{r} 5\,4\,\square \\ +\,\square\,\square\,6 \\ \hline 8\,2\,2 \end{array}$ **g** $\begin{array}{r} 4\,6\,9 \\ +\,\square\,\square\,\square \\ \hline 7\,3\,5 \end{array}$ **h** $\begin{array}{r} \square\,\square\,\square \\ +\,3\,4\,8 \\ \hline 8\,0\,7 \end{array}$

i $\begin{array}{r} \square\,4\,\square \\ +\,3\,3\,7 \\ \hline 7\,\square\,5 \end{array}$ **j** $\begin{array}{r} 3\,5\,7\,8 \\ +\,\square\,\square\,\square\,\square \\ \hline 8\,0\,7\,6 \end{array}$

AU 8 Lisa is checking the subtraction: 614 – 258.

Explain how you know that her answer of 444 is incorrect without working out the whole calculation.

9 Copy each of these subtractions and fill in the missing digits.

a $\begin{array}{r} 7\,4 \\ -\,2\,\square \\ \hline \square\,1 \end{array}$ **b** $\begin{array}{r} \square\,7 \\ -\,3\,\square \\ \hline 5\,4 \end{array}$ **c** $\begin{array}{r} 8\,5 \\ -\,\square\,\square \\ \hline 2\,7 \end{array}$ **d** $\begin{array}{r} 6\,7\,\square \\ -\,\square\,\square\,3 \\ \hline 1\,3\,5 \end{array}$

e $\begin{array}{r} \square\,1\,4 \\ -\,2\,5\,\square \\ \hline 3\,\square\,7 \end{array}$ **f** $\begin{array}{r} 5\,4\,\square \\ -\,\square\,\square\,6 \\ \hline 3\,2\,5 \end{array}$ **g** $\begin{array}{r} 4\,6\,2 \\ -\,\square\,\square\,\square \\ \hline 1\,8\,5 \end{array}$ **h** $\begin{array}{r} \square\,\square\,\square \\ -\,2\,4\,7 \\ \hline 3\,0\,9 \end{array}$

i $\begin{array}{r} \square\,4\,\square \\ -\,5\,5\,8 \\ \hline 2\,\square\,5 \end{array}$ **j** $\begin{array}{r} 8\,0\,7\,6 \\ -\,\square\,\square\,\square\,\square \\ \hline 6\,1\,8\,7 \end{array}$

PS 10 A two-digit number is subtracted from a three-digit number.

The answer is 154.

Work out one pair of possible values for the numbers.

1.7 Multiplying and dividing by single-digit numbers

This section will show you how to:

- multiply and divide by a single-digit number

Key words
division
multiplication

Multiplication

There are two things to remember when you are multiplying two whole numbers.

- The bigger number must always be written down first.
- The answer will always be larger than the bigger number.

EXAMPLE 4

a Multiply 231 by 4.

$$\begin{array}{r} 213 \\ \times \quad 4 \\ \hline 852 \\ \hline {\scriptstyle 1} \end{array}$$

b Multiply 543 by 6.

$$\begin{array}{r} 543 \\ \times \quad 6 \\ \hline 3258 \\ \hline {\scriptstyle 2\,1} \end{array}$$

Note that in part **a** the first multiplication, 3×4, gives 12. So, you need to carry a digit into the next column on the left, as in the case of addition. Similar actions are carried out in part **b**.

Division

There are two things to remember when you are dividing one whole number by another whole number:

- The answer will always be smaller than the bigger number.
- Division starts at the *left-hand side*.

EXAMPLE 5

a Divide 417 by 3.

$417 \div 3$ is set out as:

$$\begin{array}{r} 1\;3\;9 \\ 3\,\overline{)\,4^{1}1^{2}7} \end{array}$$

b Divide 508 by 4.

$508 \div 4$ is set out as:

$$\begin{array}{r} 1\;2\;7 \\ 4\,\overline{)\,5^{1}0^{2}8} \end{array}$$

This is how the division was done in part **a**:

- First, divide 3 into 4 to get 1 and remainder 1. Note where to put the 1 and the remainder 1.
- Then, divide 3 into 11 to get 3 and remainder 2. Note where to put the 3 and the remainder 2.
- Finally, divide 3 into 27 to get 9 with no remainder, giving the answer 139.

A similar process takes place in part **b**.

G

EXERCISE 1G

1 Copy and work out each of the following multiplications.

a 14×4 **b** 13×5 **c** 17×3 **d** 19×2 **e** 18×6

f 23×5 **g** 34×6 **h** 42×7 **i** 53×4 **j** 85×5

k 50×3 **l** 200×4 **m** 320×3 **n** 340×4 **o** 253×6

2 Calculate each of the following multiplications by setting the work out in columns.

a 42 × 7 **b** 74 × 5 **c** 48 × 6

d 208 × 4 **e** 309 × 7 **f** 630 × 4

g 548 × 3 **h** 643 × 5 **i** 8 × 375

j 6 × 442 **k** 7 × 528 **l** 235 × 8

m 6043 × 9 **n** 5 × 4387 **o** 9 × 5432

3 Calculate each of the following divisions.

a 438 ÷ 2 **b** 634 ÷ 2 **c** 945 ÷ 3

d 636 ÷ 6 **e** 297 ÷ 3 **f** 847 ÷ 7

g 756 ÷ 3 **h** 846 ÷ 6 **i** 576 ÷ 4

j 344 ÷ 4 **k** 441 ÷ 7 **l** 5818 ÷ 2

m 3744 ÷ 9 **n** 2008 ÷ 8 **o** 7704 ÷ 6

AU **4** Dean, Sean and Andy are doing a charity cycle ride from Huddersfield to Southend. The distance from Huddersfield to Southend is 235 miles.

a How many miles do all three travel altogether?

b Dean is sponsored £6 per mile.
Sean is sponsored £5 per mile.
Andy is sponsored £4 per mile.

How much money do they raise altogether?

F

F

FM **5** The 235-mile charity cycle ride in question 4 is planned to take five days.

a What is the least number of miles each cyclist has to cover each day?

b The cyclists travel 60 miles on the first day and 50 miles on the second day.

Fill in a copy of the plan for the remaining days so they complete the ride on time. Each day the number of miles covered has to be fewer than for the previous day.

	Mileage
Day 3	
Day 4	
Day 5	

6 By doing a suitable multiplication, answer each of these questions.

a How many days are there in 17 weeks?

b How many hours are there in four days?

c Eggs are packed in boxes of six. How many eggs are there in 24 boxes?

d Joe bought five boxes of matches. Each box contained 42 matches. How many matches did Joe buy altogether?

e A box of Tulip Sweets holds 35 sweets. How many sweets are there in six boxes?

7 By doing a suitable division, answer each of these questions.

a How many weeks are there in 91 days?

b How long will it take me to save £111, if I save £3 a week?

c A rope, 215 m long, is cut into five equal pieces. How long is each piece?

d Granny has a bottle of 144 tablets. How many days will they last if she takes four each day?

e I share a box of 360 sweets among eight children. How many sweets will each child get?

PS **8** Here is part of the 38 times table.

×1	×2	×3	×4	×5
38	76	114	152	190

Show how you can use this table to work out

a 9×38 **b** 52×38 **c** 105×38.

Letter sets

Find the next letters in these sequences.

a **O, T, T, F, F, ...** **b** **T, F, S, E, T, ...**

Valued letters

In the three additions below, each letter stands for a single numeral. But a letter may not necessarily stand for the same numeral when it is used in more than one sum.

a
```
  ONE
+ ONE
  TWO
```

b
```
   TWO
+  TWO
  FOUR
```

c
```
  FOUR
+ FIVE
  NINE
```

Write down each addition in numbers.

Four fours

Write number sentences to give answers from 1 to 10, using only four fours and any number of the operations +, –, × and ÷ . For example:

1 = (4 + 4) ÷ (4 + 4) 2 = (4 × 4) ÷ (4 + 4)

Heinz 57

Pick any number in the grid on the right. Circle the number and cross out all the other numbers in the row and column containing the number you have chosen. Now circle another number that is not crossed out and cross out all the other numbers in the row and column containing this number. Repeat until you have five numbers circled. Add these numbers together. What do you get? Now do it again but start with a different number.

19	8	11	25	7
12	1	4	18	0
16	5	8	22	4
21	10	13	27	9
14	3	6	20	2

Magic squares

This is a magic square. Add the numbers in any row, column or diagonal. The answer is *always* 15.

8	1	6
3	5	7
4	9	2

Now try to complete this magic square using every number from 1 to 16.

1		14	
	6		9
8		11	
	3		16

Hints

Letter sets Think about numbers.

Valued letters **a** Try E = 2, N = 3 **b** Try O = 7, U = 3 **c** Try N = 5, O = 9

There are other answers to each sum.

GRADE BOOSTER

- **G** You can add columns and rows in grids
- **G** You know the multiplication tables up to 10 × 10
- **G** You can use BIDMAS/BODMAS to work out calculations in the correct order
- **G** You can identify the value of digits in different places
- **G** You can round numbers to the nearest 10 and 100
- **G** You can add and subtract numbers with up to four digits
- **G** You can multiply numbers by a single-digit number
- **F** You can solve problems involving multiplication and division by a single-digit number

What you should know now

- How to use BIDMAS/BODMAS
- How to put numbers in order
- How to round to the nearest 10, 100, 1000
- How to solve simple problems in context, using the four operations of addition, subtraction, multiplication and division

1 **a** Write the number 3187 to the nearest thousand. (1)

b Write the number **four thousand, six hundred and eighty one** in figures. (1)

c Write the number 5060 in words. (1)

(Total 3 marks)

Edexcel, May 2008, Paper 1 Foundation, Question 1

2 **a** Write the number **nine thousand, three hundred and seventy-four** in figures. (1)

b Write the number 62 500 in words. (1)

c Write down the value of the **8** in the number 3285. (1)

d Write the number 2174 to the nearest hundred. (1)

e Write the number 7362 to the nearest thousand. (1)

(Total 5 marks)

Edexcel, November 2008, Paper 1 Foundation, Question 2

3 Work out

i $3 \times 3 - 5$ (1)

ii $20 \div (12 - 2)$ (1)

iii $7 + 8 \div 4$ (1)

(Total 3 marks)

Edexcel, May 2008, Paper 1 Foundation, Question 10

4 **a** Write the number **seventeen thousand, two hundred and fifty-two** in figures. (1)

b Write the number 5367 correct to the nearest hundred. (1)

c Write down the value of the 4 in the number 274 863 (1)

(Total 3 marks)

Edexcel, June 2005, Paper 1 Foundation, Question 1

5 The number of people in a London Tube Station one morning was 29 765.

a Write the number 29 765 in words.

b In the number 29 765, write down the value of

i the figure 7

ii the figure 9.

c Write 29 765 to the nearest 100.

6 **a** **i** Write down the number **fifty-four thousand and seventy-three** in figures.

ii Write down **fifty-four thousand and seventy-three** to the nearest hundred.

b **i** Write down 21 809 in words.

ii Write down 21 809 to the nearest 1000.

7 Look at the numbers in the cloud.

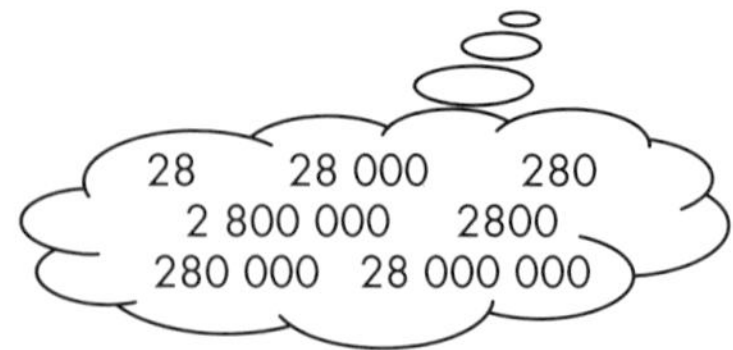

a Write down the number from the cloud which is

i twenty eight million

ii two thousand eight hundred.

b What number should go in the boxes to make the calculation correct?

i $28 \times \square = 2800$

ii $2 800 000 \div \square = 280 000$

8 Mount Everest is 8848 metres high.

Mount Snowdon is 1085 metres high.

How much higher is Mount Everest than Mount Snowdon?

Give your answer to the nearest 10 metres.

9 It costs £7 per person to visit a show.

a 215 people attend the show on Monday.
How much do they pay altogether?

b On Tuesday the takings were £1372.
How many fewer people attended on Tuesday than on Monday?

10 The 2009 population of Plaistow is given as 7800 to the nearest hundred.

a What is the lowest number that the population could be?

b What is the largest number that the population could be?

11 Murray and Harry both worked out $2 + 4 \times 7$.
Murray calculated this to be 42.
Harry worked this out to be 30.
Explain why they got different answers.

12 Work out the following. Be careful as they may involve addition, subtraction, multiplication and division. Decide what the calculation is and use a column method to work it out.

- **a** Trays of pansies contain 12 plants each. How many plants will I get in 8 trays?
- **b** There are 192 pupils in year 7. They are in 6 forms. How many pupils are in each form?
- **c** A school term consists of 42 days. If a normal school week is 5 days.
  - **i** How many full weeks will there be in the term?
  - **ii** How many odd dates will there be?
- **d** A machine produces 120 bolts every minute.
  - **i** How many bolts will be produced by the machine in 9 minutes?
  - **ii** The bolts are packed in bags of 8. How many bags will it take to pack 120 bolts?

13
- **a** There are 7 days in a week.
  - **i** How many days are there in 15 weeks?
  - **ii** How many weeks are there in 161 days?
- **b** Bulbs are sold in packs of 6.
  - **i** How many bulbs are there in 12 packs?
  - **ii** How many packs make up 186 bulbs?

14 A teacher asked her pupils to work out the following calculation without a calculator

$2 \times 3^2 + 6$

- **a** Alice got an answer of 42. Billy got an answer of 30. Chas got an answer of 24. Explain why Chas was correct.
- **b** Put brackets into these calculations to make them true.
  - **i** $2 \times 3^2 + 6 = 42$
  - **ii** $2 \times 3^2 + 6 = 30$

15 The following are two pupils' attempts at working out $3 + 5^2 - 2$

Adam $3 + 5^2 - 2 = 3 + 10 - 2 = 13 - 2 = 11$

Bekki $3 + 5^2 - 2 = 8^2 - 2 = 64 - 2 = 62$

- **a** Each pupil has made one mistake. Explain what this is for each of them.
- **b** Work out the correct answer to $3 + 5^2 - 2$.

Worked Examination Questions

1 Here are four number cards, showing the number 2745.

2	7	4	5

Using all four cards, write down:

a the largest possible number

b the smallest possible number

c the missing numbers from this problem. ☐ 7 × 2 = ☐ ☐

a 7542

Start with the largest number as the thousands digits, use the next largest as the hundreds digit and so on.
This is worth 1 mark.

1 mark

b 2457

Start with the smallest number as the thousands digits, use the next smallest as the hundreds digit and so on. Note the answer is the reverse of the answer to part **a**.
This is worth 1 mark.

1 mark

c 27 × 2 = 54

There are three numbers left, 2, 5, 4. The 2 must go into the first box and then you can work out that 2 × 27 is 54.
You get 1 mark for identifying the units digit of the answer as 4 and 2 marks if the whole answer is correct.

1 mark

Total: 4 marks

Worked Examination Questions

FM **2** I want to buy a burger, a portion of chips and a bottle of cola for my lunch. I have the following coins in my pocket: £1, 50p, 50p, 20p, 20p, 10p, 2p, 2p, 1p.

a Do I have enough money?

b What about if I replace the chips with beans?

Price List	
Burger	£1.20
Chips	90p
Beans	50p
Cola	60p

a Total cost for a burger, chips and a cola = £2.70.

Total of coins in my pocket = £2.55

No, as £2.70 > £2.55

3 marks

b Beans are 40p cheaper and £2.30 < £2.55.

So, I could afford this.

1 mark

Total: 4 marks

This is a question in which you could be assessed on your quality of written communication, so set out your answer clearly.

Show the total cost of the meal and the total of the coins in your pocket. This is worth 2 marks.

It is important to show a clear conclusion. Do not just say 'No'. Use the numbers to back up your answer. This is worth 1 mark.

Explain clearly why you could afford this meal. This is worth 1 mark.

1 Functional Maths
Activity holiday in Wales

You are planning to go on an activity holiday in Wales with a group of friends. You are arriving late on Sunday evening and will be staying all week until the end of Friday. You have chosen the area that you are staying in because there are lots of activities available and you want to make sure that everyone in the group can find something that they will enjoy on the holiday.

Your task

Work in groups of three or four.

Decide what the different interests in your group are and work together to make a timetable of activities that addresses all of these interests. You will need to make sure that everyone has the time to do all of the things that they would like to do.

You must then work out how much the holiday is going to cost for each of you.

Getting started

Use the following points to get you started:

- On your own, decide the activities you would like to do. You don't have to do everything.
- As a group, think of the questions you are going to ask as you do this task to find out the information that you will need.
- Think about how you might keep track of all the information that you need.
- Decide on how you will work out the cost for each member of your group.

Windsurfing Half-day £59 Full day £79	**Paragliding** Half-day £99 for 1 person £189 for 2 people	**Quadbikes** £21 per hour Race Event (2 hours) £100 for three or four people	**Horse riding** Half-day (Mondays, Wednesdays and Fridays) £32
Fishing Trip Half-day £18 per person	**Coast jumping** 2 hours £85 per group of up to 4 people	**Kayaking** Half-day £29 Full Day £49	**Raft racing** Half-day £60 per team (minimum 2 people)
Diving 3 hours £38 per person	**Spa** 2 hours £24 for 1 person £40 for 2 people	**Shopping Trip** Half-day £60 budget	**Steam train up mountain** 2 hours £5.60 per person (10% discount for groups of 4)

Why this chapter matters

The word 'fraction' comes from the Latin word *fractus*, meaning broken. Just think of when you fracture an arm or leg – you have cracked (or broken) it into parts.

We use fractions to break things – from measurements to shapes – into parts.

Chapter 1 (page 6) showed how there used to be several different number systems, developed by many different ancient civilisations. Each of these civilisations developed their own way of expressing fractions. Most of these fraction systems died out, but one – the Arabic system – directly led to the fractions that we use today.

The Babylonians – the first fractions

Fractions can be traced back to the Babylonians, in around 1800BC. This civilisation in Mesopotamia (modern day Iraq) was the first to develop a sensible way to represent a fraction.

Their fractions were based on the number 60 as was its whole number system. However, its number symbols could not accurately represent the fractions and the fractions had no symbol to show that they were fractions rather than standard numbers. This made their system of fractions very complicated.

Egyptian fractions

The Egyptians (around 1000BC) were known to use fractions but generally these were unit fractions (fractions with a numerator of 1), for example, $\frac{1}{2}, \frac{1}{3}, \frac{1}{4}$.

The Egyptians used this symbol to represent the "one":

Then, using the number symbols shown in Chapter 1, they made their fractions:

⌓ III $= \frac{1}{3}$

They also had special symbols for $\frac{1}{2}$, $\frac{2}{3}$, and $\frac{3}{4}$.

$= \frac{1}{2}$ | $= \frac{2}{3}$ | $= \frac{3}{4}$

Greek fractions

The Greeks also used unit fractions but the way they wrote them led to confusion. They had symbols for numbers, for example, the number 2 was β (beta) and to make the fraction $\frac{1}{2}$ they added a dash so $\frac{1}{2} = \beta'$.

Number	Symbol	Fraction	Symbol
2	β (beta)	$\frac{1}{2}$	β′
3	γ (gamma)	$\frac{1}{3}$	γ′
4	δ (delta)	$\frac{1}{4}$	δ′

Fractions in India

In India, around 500AD a system called *brahmi* was devised using symbols for the numbers. Fractions were written as a symbol above a symbol but without a line as used now.

Brahmi symbols

1	2	3	4	5	6	7	8	9
–	=	≡	+	h	4	7	ら	7

So $\frac{5}{9}$ was written as:

h
7

Arabic fractions

The Arabs, probably around 1200AD, built on the number system – including the fractions – developed by the Indians. It was in Arabia that the line was first introduced into fractions – sometimes drawn horizontally and sometimes slanting – leading to the clear fractions that we use today.

$\frac{3}{4}$ $\frac{1}{8}$ $\frac{1}{2}$ $\frac{2}{3}$ $\frac{1}{4}$ $\frac{5}{10}$

Chapter 2

Number: Fractions

The grades given in this chapter are target grades.

1 Recognise a fraction of a shape

2 Adding and subtracting simple fractions

3 Recognise equivalent fractions, using diagrams

4 Equivalent fractions and simplifying fractions by cancelling

5 Improper fractions and mixed numbers

6 Adding and subtracting fractions with the same denominator

7 Finding a fraction of a quantity

8 Multiplying and dividing fractions

9 One quantity as a fraction of another

This chapter will show you …

G to C how to add, subtract, multiply, divide and order simple fractions

G how to simplify fractions

F how to convert an improper fraction to a mixed number (and vice versa)

F how to calculate a fraction of a quantity

Visual overview

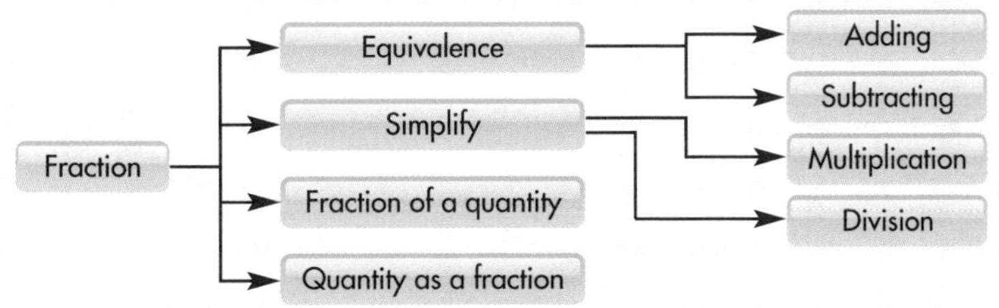

What you should already know

- Multiplication tables up to 10 × 10 **(KS3 level 4, GCSE grade G)**
- What a fraction is **(KS3 level 3, GCSE grade G)**

Quick check

How quickly can you calculate these?

1 2×4	2 5×3	3 5×2	4 6×3
5 2×7	6 4×5	7 3×8	8 4×6
9 9×2	10 3×7	11 half of 10	12 half of 12
13 half of 16	14 half of 8	15 half of 20	16 a third of 9
17 a third of 15	18 a quarter of 12	19 a fifth of 10	20 a fifth of 20

2.1 Recognise a fraction of a shape

This section will show you how to:

- recognise what fraction of a shape has been shaded
- shade a given simple fraction of a shape

Key words
fraction
shape

A **fraction** is a part of a whole. The top number is called the **numerator**. The bottom number is called the **denominator**. So, for example, $\frac{3}{4}$ means you divide a whole into four portions and take three of them.

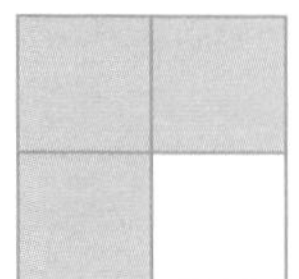

It really does help if you know the multiplication tables up to 10 × 10. They will be tested in the non-calculator paper, so you need to be confident about tables and numbers.

EXERCISE 2A

G

1 What **fraction** is shaded in each **shape** in these diagrams?

a 
b 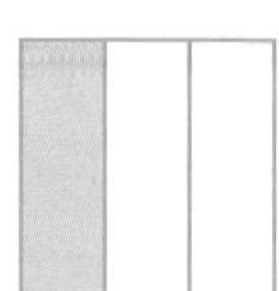
c 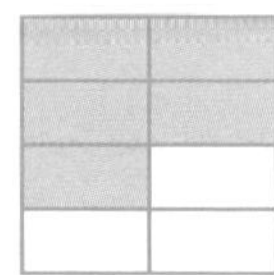
d 

e 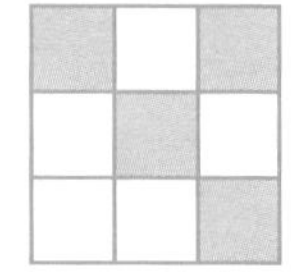
f 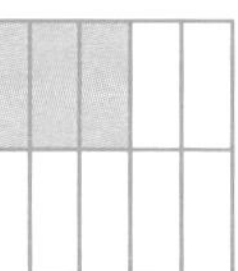
g 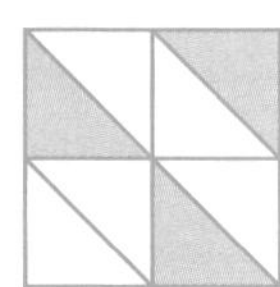
h 

i 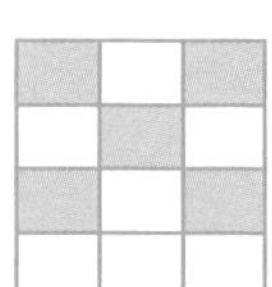
j 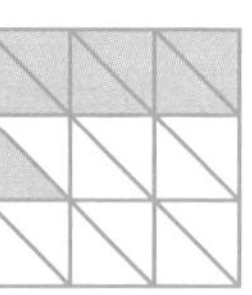
k 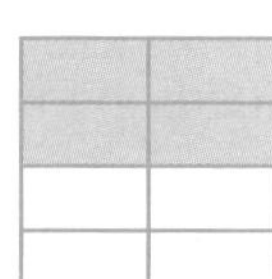
l 

m 
n 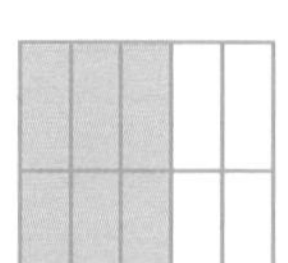
o 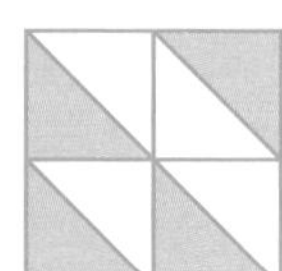
p 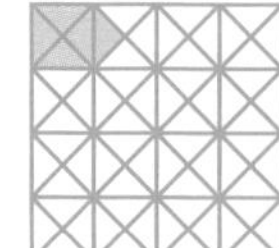

G

2 Draw diagrams as in question 1 to show these fractions.

a $\frac{3}{4}$ **b** $\frac{2}{3}$ **c** $\frac{1}{5}$ **d** $\frac{5}{8}$ **e** $\frac{1}{6}$ **f** $\frac{8}{9}$

g $\frac{1}{9}$ **h** $\frac{1}{10}$ **i** $\frac{4}{5}$ **j** $\frac{2}{7}$ **k** $\frac{3}{8}$ **l** $\frac{5}{6}$

3 A farmer decides to divide up a field into two parts so that one part is twice as big as the other part.

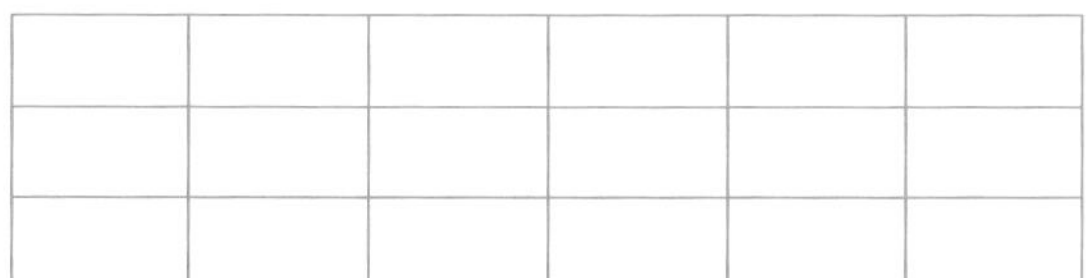

Make two copies of the grid and show two different ways that she can do this.

Shade the smaller area in each case.

F

PS 4 Look again at the diagrams in question 1.

For each question, decide which diagram, if any, has the greater proportion shaded.

a a and g **b** b and i **c** b and l **d** e and j

e k and o **f** e and m **g** g and k **h** e and i

AU 5 Three fractions are shown on the grids below.

A

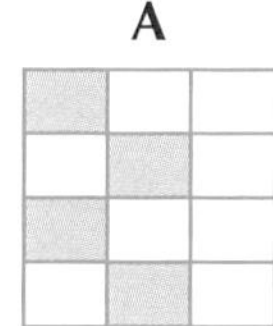

B

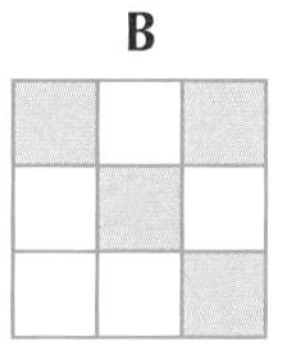

C

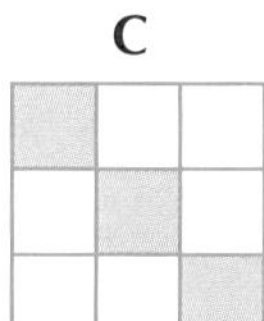

The fraction in grid A is the odd one out because it does not have a denominator of 9.

Give reasons as to why the fractions in grid B and grid C could be the odd one out.

FM 6 A field is divided up into six allotments, numbered 1 to 6.

1	2
3	4
5	6

Each allotment is the same shape and size.

The owners want to sell potatoes at market and decide that they need to plant half the field with potatoes to meet demand.

Potatoes are planted in half of allotment 4 and all of allotments 3 and 6. There are no potatoes in allotments 1 and 5.

The owner of allotment 2 agrees to plant some potatoes so that half the field will be potatoes.

What fraction of her allotment needs to be potatoes?

2.2 Adding and subtracting simple fractions

This section will show you how to:
- add and subtract two fractions with the same denominator

Key words
denominator
numerator

Fractions that have the same **denominator** (bottom number) can easily be added or subtracted.

EXAMPLE 1

Work out $\frac{3}{10} + \frac{4}{10}$

Add the **numerators** (top numbers). The bottom number stays the same.

$\frac{3}{10} + \frac{4}{10} = \frac{7}{10}$

Work out $\frac{7}{8} - \frac{2}{8}$

Subtract the **numerators** (top numbers). The bottom number stays the same.

$\frac{7}{8} - \frac{2}{8} = \frac{5}{8}$

EXERCISE 2B

G

1 Calculate each of the following.

a $\frac{1}{4} + \frac{2}{4}$ **b** $\frac{1}{8} + \frac{3}{8}$ **c** $\frac{2}{5} + \frac{1}{5}$ **d** $\frac{3}{10} + \frac{5}{10}$

e $\frac{1}{3} + \frac{1}{3}$ **f** $\frac{2}{7} + \frac{3}{7}$ **g** $\frac{2}{9} + \frac{5}{9}$ **h** $\frac{1}{6} + \frac{4}{6}$

i $\frac{3}{5} + \frac{1}{5}$ **j** $\frac{5}{8} + \frac{2}{8}$ **k** $\frac{2}{10} + \frac{3}{10}$ **l** $\frac{4}{7} + \frac{1}{7}$

m $\frac{3}{5} + \frac{1}{5}$ **n** $\frac{2}{6} + \frac{3}{6}$ **o** $\frac{4}{9} + \frac{1}{9}$ **p** $\frac{2}{11} + \frac{5}{11}$

2 Calculate each of the following.

a $\frac{3}{4} - \frac{1}{4}$ **b** $\frac{4}{5} - \frac{1}{5}$ **c** $\frac{7}{8} - \frac{4}{8}$ **d** $\frac{8}{10} - \frac{5}{10}$

e $\frac{2}{3} - \frac{1}{3}$ **f** $\frac{5}{6} - \frac{1}{6}$ **g** $\frac{5}{7} - \frac{2}{7}$ **h** $\frac{7}{9} - \frac{2}{9}$

i $\frac{3}{5} - \frac{2}{5}$ **j** $\frac{4}{7} - \frac{1}{7}$ **k** $\frac{8}{9} - \frac{5}{9}$ **l** $\frac{9}{10} - \frac{3}{10}$

m $\frac{4}{6} - \frac{1}{6}$ **n** $\frac{5}{8} - \frac{3}{8}$ **o** $\frac{7}{11} - \frac{5}{11}$ **p** $\frac{7}{10} - \frac{3}{10}$

G

AU 3 Copy and shade the diagrams to show the working for this question and then write down the answer.

$$\frac{1}{2} \quad + \quad \frac{1}{3} \quad = \quad \ldots\ldots$$

						+							=						

AU 4 **a** Draw a diagram to show $\frac{2}{4}$.

b Show on your diagram that $\frac{2}{4} = \frac{1}{2}$.

c Use the above information to write down the answers to these.

i $\frac{1}{4} + \frac{1}{2}$ **ii** $\frac{3}{4} - \frac{1}{2}$

AU 5 **a** Draw a diagram to show $\frac{5}{10}$.

b Show on your diagram that $\frac{5}{10} = \frac{1}{2}$.

c Use the above information to write down the answers to these.

i $\frac{1}{2} + \frac{1}{10}$ **ii** $\frac{1}{2} + \frac{3}{10}$ **iii** $\frac{1}{2} + \frac{2}{10}$

F

PS 6 On an egg tray there are 30 eggs.

Four-fifths of the eggs are removed.
Half of these eggs are now put back on the tray.

What fraction of the 30 eggs is now on the tray?

PS 7 An aeroplane can carry 240 passengers when full.

On a flight it is exactly half full. Two-thirds of the passengers are male. What fraction of the seats are occupied by females?

PS 8 In a class of 32 students, three-quarters walk to school.

Of the remainder, half come by car.

What fraction of the class comes by car?

FM 9 On a train there are two carriages, each with 180 seats.

The train operator estimates that 1 in every 20 people will buy a snack on the journey.

They carry 30 snacks.

a Will they have enough snacks for two journeys if the train is full?

b What fraction of the snacks will be used on the first journey if the train is full?

2.3 Recognise equivalent fractions, using diagrams

This section will show you how to:
- recognise equivalent fractions, using diagrams

Key words
equivalent
equivalent fractions

ACTIVITY

Making eighths

You need lots of squared paper and a pair of scissors.

Draw three rectangles, each 4 cm by 2 cm, on squared paper.

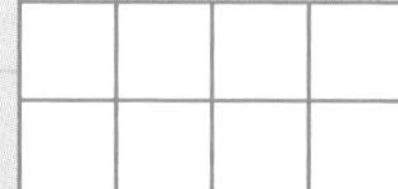

Each small square is called an *eighth* or $\frac{1}{8}$.

Cut one of the rectangles into halves, another into quarters and the last one into eighths.

$\frac{1}{2}$ $\frac{1}{2}$ $\frac{1}{4}$ $\frac{1}{4}$ $\frac{1}{4}$ $\frac{1}{4}$ $\frac{1}{8}$ $\frac{1}{8}$ $\frac{1}{8}$ $\frac{1}{8}$ $\frac{1}{8}$ $\frac{1}{8}$ $\frac{1}{8}$ $\frac{1}{8}$

You can see that the strip equal to one half takes up 4 squares, so:

$$\frac{1}{2} = \frac{4}{8}$$

These are called **equivalent fractions**.

1 Use the strips to write down the following fractions as eighths.

a $\frac{1}{4}$ **b** $\frac{3}{4}$

2 Use the strips to work out the following problems. Leave your answers as eighths.

a $\frac{1}{4} + \frac{3}{8}$ **b** $\frac{3}{4} + \frac{1}{8}$ **c** $\frac{3}{8} + \frac{1}{2}$

d $\frac{1}{4} + \frac{1}{2}$ **e** $\frac{1}{8} + \frac{1}{8}$ **f** $\frac{1}{8} + \frac{1}{4}$

g $\frac{3}{8} + \frac{3}{4}$ **h** $\frac{3}{4} + \frac{1}{2}$

Making twenty-fourths

You need lots of squared paper and a pair of scissors.

Draw four rectangles, each 6 cm by 4 cm, on squared paper.

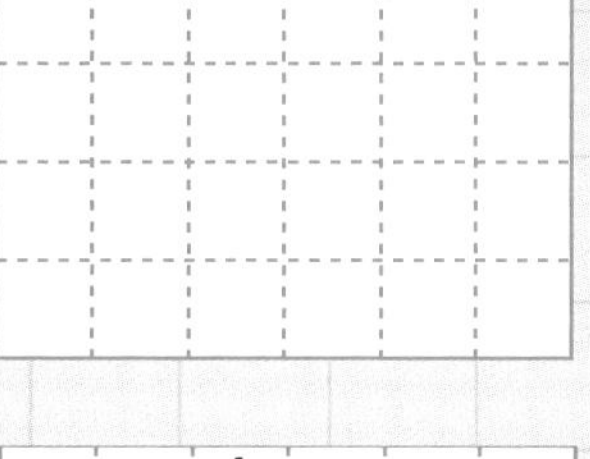

Each small square is called a *twenty-fourth* or $\frac{1}{24}$. Cut one of the rectangles into quarters, another into sixths, another into thirds and the remaining one into eighths.

You can see that the strip equal to a quarter takes up six squares, so:

$$\frac{1}{4} = \frac{6}{24}$$

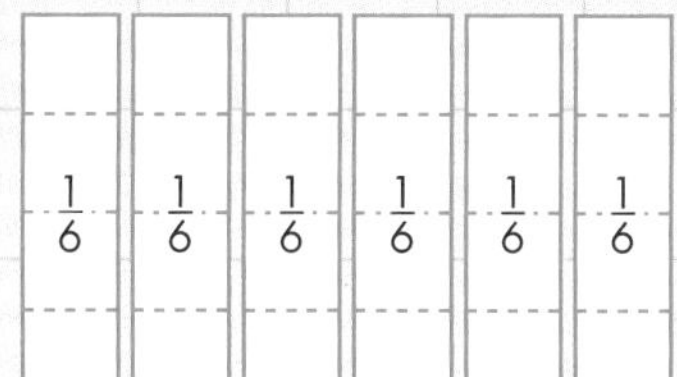

This is another example of equivalent fractions.

This idea is used to add fractions together. For example:

$$\frac{1}{4} + \frac{1}{6}$$

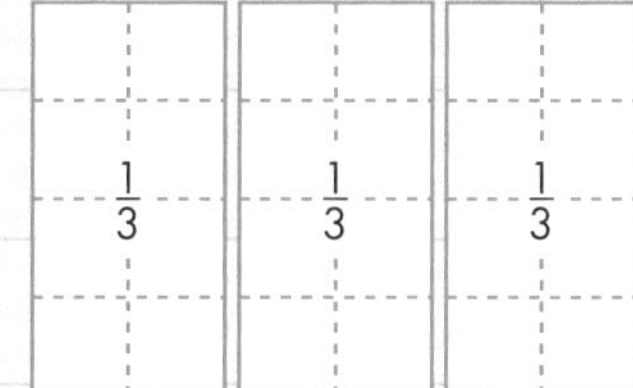

can be changed into:

$$\frac{6}{24} + \frac{4}{24} = \frac{10}{24}$$

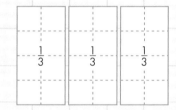

EXERCISE 2C

1 Use the strips from the activity to write down each of these fractions as twenty-fourths.

a $\frac{1}{6}$ **b** $\frac{1}{3}$ **c** $\frac{1}{8}$ **d** $\frac{2}{3}$ **e** $\frac{5}{6}$

f $\frac{3}{4}$ **g** $\frac{3}{8}$ **h** $\frac{5}{8}$ **i** $\frac{7}{8}$ **j** $\frac{1}{2}$

2 Use the strips from the activity to write down the answer to each of the following problems. Write each answer in twenty-fourths.

a $\frac{1}{3}+\frac{1}{8}$ **b** $\frac{1}{8}+\frac{1}{4}$ **c** $\frac{1}{6}+\frac{1}{8}$ **d** $\frac{2}{3}+\frac{1}{8}$ **e** $\frac{5}{8}+\frac{1}{3}$

f $\frac{1}{8}+\frac{5}{6}$ **g** $\frac{1}{2}+\frac{3}{8}$ **h** $\frac{1}{6}+\frac{3}{4}$ **i** $\frac{5}{8}+\frac{1}{6}$ **j** $\frac{1}{3}+\frac{5}{8}$

3 Draw three rectangles, each 5 cm by 4 cm. Cut one into quarters, another into fifths and the last into tenths.

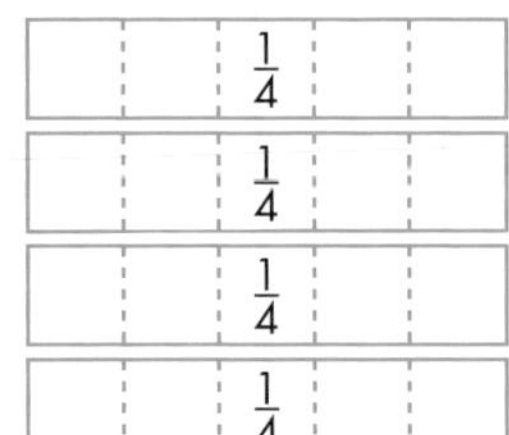

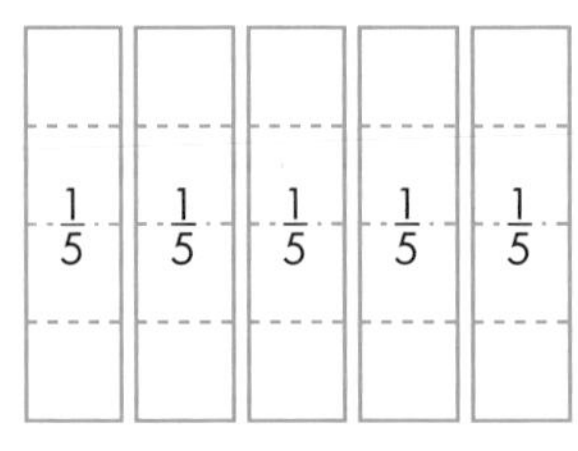

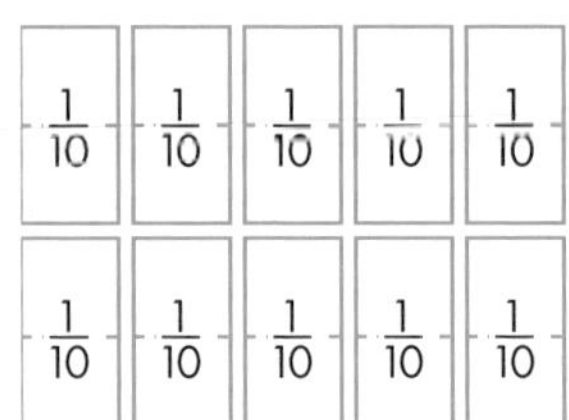

Use the strips to find the equivalent fraction, in twentieths, to each of the following.

a $\frac{1}{4}$ **b** $\frac{1}{5}$ **c** $\frac{3}{4}$ **d** $\frac{4}{5}$ **e** $\frac{1}{10}$

f $\frac{1}{2}$ **g** $\frac{3}{5}$ **h** $\frac{2}{5}$ **i** $\frac{7}{10}$ **j** $\frac{3}{10}$

4 Use the strips from question 3 to write down the answer to each of the following.

a $\frac{1}{4}+\frac{1}{5}$ **b** $\frac{3}{5}+\frac{1}{10}$ **c** $\frac{3}{10}+\frac{1}{4}$ **d** $\frac{3}{4}+\frac{1}{5}$ **e** $\frac{7}{10}+\frac{1}{4}$

PS 5 Use the strips from the activity to decide which fraction in each set of three, if any, is the odd one out.

a $\frac{6}{24}, \frac{2}{6}, \frac{2}{8}$ **b** $\frac{6}{8}, \frac{2}{3}, \frac{3}{4}$ **c** $\frac{1}{2}, \frac{4}{8}, \frac{3}{6}$ **d** $\frac{2}{3}, \frac{4}{6}, \frac{6}{8}$

G

AU 6 Here is a multiplication table.

×	2	3	4	5
2	4	6	8	10
5	10	15	20	25

The table can be used to pick out equivalent fractions.

For example, $\frac{2}{5} = \frac{6}{15}$.

a Use the table to write down three more fractions that are equivalent to $\frac{2}{5}$.

b Extend this table to write down a different fraction that is equivalent to $\frac{2}{5}$.

AU 7 **a** Complete this multiplication table.

×	2	3	4	5
3				
4				

b Use this table to write down four fractions that are equivalent to $\frac{3}{4}$.

F

8 You will need lots of squared paper and a pair of scissors.
Draw three rectangles on squared paper, each 6 cm by 2 cm.

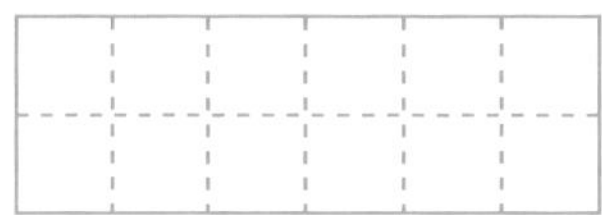

a Write down the fraction of a rectangle that each small square represents.

b Use your own choice of strips to write down the answer to each of the following.

i $\frac{1}{2} + \frac{1}{12}$ **ii** $\frac{5}{12} + \frac{1}{3}$ **iii** $\frac{1}{12} + \frac{5}{6}$ **iv** $\frac{2}{3} + \frac{1}{12}$ **v** $\frac{1}{3} + \frac{1}{4}$

2.4 Equivalent fractions and simplifying fractions by cancelling

This section will show you how to:

- create equivalent fractions
- simplify fractions by cancelling

Key words
cancel
denominator
lowest terms
numerator
simplest form
simplify

Equivalent fractions are two or more fractions that represent the same part of a whole.

EXAMPLE 2

Complete the following.

a $\frac{3}{4} \rightarrow \frac{\times 4}{\times 4} = \frac{\square}{16}$ **b** $\frac{2}{5} = \frac{\square}{15}$

a Multiplying the **numerator** by 4 gives 12. This means $\frac{12}{16}$ is an equivalent fraction to $\frac{3}{4}$.

b To change the **denominator** from 5 to 15, you multiply by 3. Do the same thing to the numerator, which gives $2 \times 3 = 6$. So, $\frac{2}{5} = \frac{6}{15}$.

The fraction $\frac{3}{4}$, in Example 2a, is in its **lowest terms** or **simplest form**.

This means that the only number that is a factor of both the numerator and denominator is 1.

EXAMPLE 3

Write these fractions in their simplest forms.

a $\frac{15}{35}$ **b** $\frac{24}{54}$

a Here is one reason why you need to know the multiplication tables. What is the biggest number that has both 15 and 35 in its multiplication table? You should know that this is the five times table. So, divide both the numerator and denominator by 5.

$$\frac{15}{35} = \frac{15 \div 5}{35 \div 5} = \frac{3}{7}$$

You can say that you have '**cancelled** by five'.

b The biggest number that has both 24 and 54 in its multiplication table is 6. So, divide both the numerator and denominator by 6.

$$\frac{24}{54} = \frac{24 \div 6}{54 \div 6} = \frac{4}{9}$$

Here, you have 'cancelled by six'.

You have **simplified** the fractions to write them in their lowest terms.

EXAMPLE 4

Put the following fractions in order, with the smallest first.

$\frac{5}{6}, \frac{2}{3}, \frac{3}{4}$

First write each fraction with the same denominator by using equivalent fractions.

$\frac{5}{6} = \frac{10}{\mathbf{12}}$

$\frac{2}{3} = \frac{4}{6} = \frac{6}{9} = \frac{8}{\mathbf{12}}$

$\frac{3}{4} = \frac{6}{8} = \frac{9}{\mathbf{12}}$

This shows that $\frac{5}{6} = \frac{10}{12}$, $\frac{2}{3} = \frac{8}{12}$ and $\frac{3}{4} = \frac{9}{12}$.

In order, the fractions are:

$\frac{2}{3}, \frac{3}{4}, \frac{5}{6}$

EXERCISE 2D

1 Copy and complete the following.

a $\frac{2}{5} \rightarrow \frac{\times 4}{\times 4} = \frac{\square}{20}$ **b** $\frac{1}{4} \rightarrow \frac{\times 3}{\times 3} = \frac{\square}{12}$ **c** $\frac{3}{8} \rightarrow \frac{\times 5}{\times 5} = \frac{\square}{40}$

d $\frac{4}{5} \rightarrow \frac{\times 3}{\times 3} = \frac{\square}{15}$ **e** $\frac{5}{6} \rightarrow \frac{\times 3}{\times 3} = \frac{\square}{18}$ **f** $\frac{3}{7} \rightarrow \frac{\times 4}{\times 4} = \frac{\square}{28}$

g $\frac{3}{10} \rightarrow \frac{\times \square}{\times 2} = \frac{\square}{20}$ **h** $\frac{1}{3} \rightarrow \frac{\times \square}{\times \square} = \frac{\square}{9}$ **i** $\frac{3}{5} \rightarrow \frac{\times \square}{\times \square} = \frac{\square}{20}$

j $\frac{2}{3} \rightarrow \frac{\times \square}{\times \square} = \frac{\square}{18}$ **k** $\frac{3}{4} \rightarrow \frac{\times \square}{\times \square} = \frac{\square}{12}$ **l** $\frac{5}{8} \rightarrow \frac{\times \square}{\times \square} = \frac{\square}{40}$

m $\frac{7}{10} \rightarrow \frac{\times \square}{\times \square} = \frac{\square}{20}$ **n** $\frac{1}{6} \rightarrow \frac{\times \square}{\times \square} = \frac{4}{\square}$ **o** $\frac{3}{8} \rightarrow \frac{\times \square}{\times \square} = \frac{15}{\square}$

G

2 Copy and complete the following.

a $\frac{1}{2} = \frac{2}{\square} = \frac{3}{\square} = \frac{\square}{8} = \frac{\square}{10} = \frac{6}{\square}$

b $\frac{1}{3} = \frac{2}{\square} = \frac{3}{\square} = \frac{\square}{12} = \frac{\square}{15} = \frac{6}{\square}$

c $\frac{3}{4} = \frac{6}{\square} = \frac{9}{\square} = \frac{\square}{16} = \frac{\square}{20} = \frac{18}{\square}$

d $\frac{2}{5} = \frac{4}{\square} = \frac{6}{\square} = \frac{\square}{20} = \frac{\square}{25} = \frac{12}{\square}$

e $\frac{3}{7} = \frac{6}{\square} = \frac{9}{\square} = \frac{\square}{28} = \frac{\square}{35} = \frac{18}{\square}$

3 Copy and complete the following.

a $\frac{10}{15} = \frac{10 \div 5}{15 \div 5} = \frac{\square}{\square}$

b $\frac{12}{15} = \frac{12 \div 3}{15 \div 3} = \frac{\square}{\square}$

c $\frac{20}{28} = \frac{20 \div 4}{28 \div 4} = \frac{\square}{\square}$

d $\frac{12}{18} = \frac{12 \div \square}{\square \div \square} = \frac{\square}{\square}$

e $\frac{15}{25} = \frac{15 \div 5}{\square \div \square} = \frac{\square}{\square}$

f $\frac{21}{30} = \frac{21 \div \square}{\square \div \square} = \frac{\square}{\square}$

FM 4 A shop manager is working out how much space to use for different items on a shelf.

The shelf has six equal sections.

He puts baked beans on three sections, tomatoes on two sections and spaghetti on one section.

a Baked beans are in boxes of 48 tins.
Each section of shelf holds 500 tins.
How many boxes will be needed to fill the baked bean sections?

b What fraction of the shelf has baked beans on it?
Give your answer in its simplest form.

c What fraction of the shelf does not have spaghetti on it?
Give your answer in its simplest form.

5 Cancel each of these fractions to its simplest form.

a $\frac{4}{6}$ **b** $\frac{5}{15}$ **c** $\frac{12}{18}$ **d** $\frac{6}{8}$ **e** $\frac{3}{9}$

f $\frac{5}{10}$ **g** $\frac{14}{16}$ **h** $\frac{28}{35}$ **i** $\frac{10}{20}$ **j** $\frac{4}{16}$

k $\frac{12}{15}$ **l** $\frac{15}{21}$ **m** $\frac{25}{35}$ **n** $\frac{14}{21}$ **o** $\frac{8}{20}$

p $\frac{10}{25}$ **q** $\frac{7}{21}$ **r** $\frac{42}{60}$ **s** $\frac{50}{200}$ **t** $\frac{18}{12}$

u $\frac{6}{9}$ **v** $\frac{18}{27}$ **w** $\frac{36}{48}$ **x** $\frac{21}{14}$ **y** $\frac{42}{12}$

6 Put the fractions in each set in order, with the smallest first.

a $\frac{1}{2}, \frac{5}{6}, \frac{2}{3}$

b $\frac{3}{4}, \frac{1}{2}, \frac{5}{8}$

c $\frac{7}{10}, \frac{2}{5}, \frac{1}{2}$

d $\frac{2}{3}, \frac{3}{4}, \frac{7}{12}$

e $\frac{1}{6}, \frac{1}{3}, \frac{1}{4}$

f $\frac{9}{10}, \frac{3}{4}, \frac{4}{5}$

g $\frac{4}{5}, \frac{7}{10}, \frac{5}{6}$

h $\frac{1}{3}, \frac{2}{5}, \frac{3}{10}$

AU 7 Here are four unit fractions.

$\frac{1}{2}$ $\frac{1}{3}$ $\frac{1}{4}$ $\frac{1}{5}$

a Which two of these fractions have a sum of $\frac{7}{12}$?

Show clearly how you work out your answer.

b Which fraction is the biggest?
Explain your answer.

PS 8 **a** Use the fact that $2 \times 2 \times 2 \times 2 \times 2 = 32$ to cancel the fraction $\frac{64}{320}$ to its lowest terms.

b Use the fact that $7 \times 11 \times 13 = 1001$ to cancel the fraction $\frac{13}{1001}$ to its simplest form.

PS 9 What fraction of each of these grids is shaded?

a

b

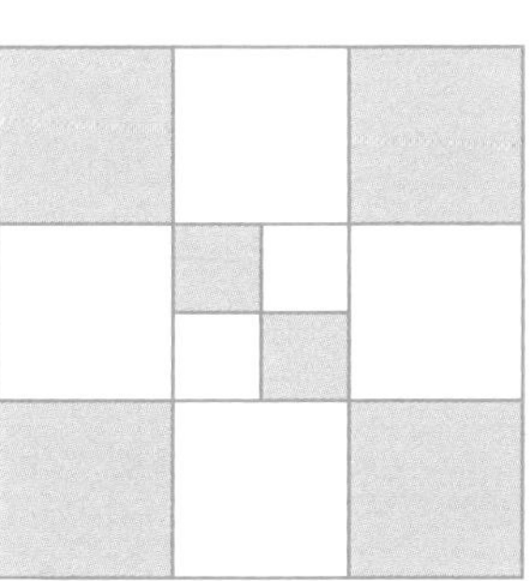

c

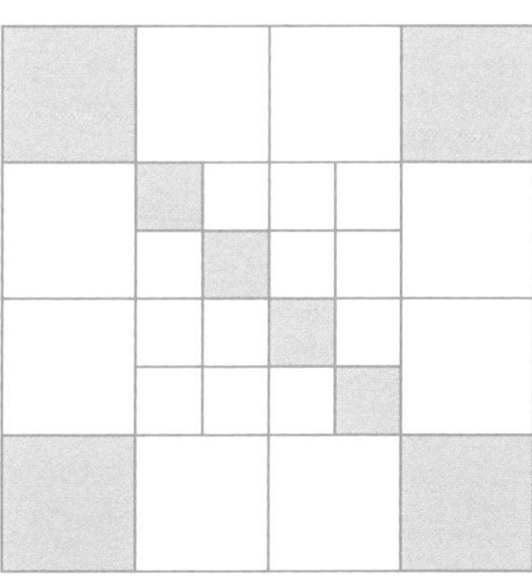

G F E

2.5 Improper fractions and mixed numbers

This section will show you how to:

- change improper fractions into mixed numbers
- change mixed numbers into improper fractions

Key words
improper fraction
mixed number
proper fraction
top-heavy

A fraction with the numerator (top number) smaller than the denominator (bottom number) is called a **proper fraction**. An example of a proper fraction is $\frac{4}{5}$.

An **improper fraction** has a bigger numerator (top number) than the denominator (bottom number). An example of an improper fraction is $\frac{9}{5}$. It is sometimes called a **top-heavy** fraction.

A **mixed number** is made up of a whole number and a proper fraction. An example of a mixed number is $1\frac{3}{4}$.

EXAMPLE 5

Convert $\frac{14}{5}$ into a mixed number.

$\frac{14}{5}$ means $14 \div 5$.

Dividing 14 by 5 gives 2 with a remainder of 4 (5 fits into 14 two times, with $\frac{4}{5}$ left over).

This means that there are 2 whole ones and $\frac{4}{5}$ left over.

So, $\frac{14}{5} = \frac{5}{5} + \frac{5}{5} + \frac{4}{5}$

$= 2\frac{4}{5}$

EXAMPLE 6

Convert $3\frac{1}{4}$ into an improper fraction.

Each whole one is four-quarters.

So, 3 whole ones = 12 quarters $\left(\frac{4}{4} + \frac{4}{4} + \frac{4}{4}\right)$

This means that $3\frac{1}{4} = \frac{12}{4} + \frac{1}{4} = \frac{13}{4}$

Using a calculator with improper fractions

Check that your calculator has a fraction key.

To key in a fraction press [fraction key].

Input the fraction so that it looks like this: $\frac{9}{5}$.

Now press the equals key **=** so that the fraction displays in the answer part of the screen.

Pressing shift and the key **S⟷D** will convert the fraction to a mixed number.

This is the mixed number $1\frac{4}{5}$.

Pressing the equals sign again will convert the mixed number back to an improper fraction.

- Can you see a way of converting an improper fraction to a mixed number without using a calculator?
- Test your idea. Then use your calculator to check it.

Using a calculator to convert mixed numbers to improper fractions

To input a mixed number, press the shift key first and then the fraction key .

Pressing the equals sign will convert the mixed number to an improper fraction.

- Now key in at least ten improper fractions and convert them to mixed numbers.
- Remember to press the equals sign to change the mixed numbers back to improper fractions.
- Now try inputting at least 10 mixed numbers and converting them to improper fractions.
- Look at your results. Can you see a way of converting a mixed number to an improper fraction without using a calculator?
- Test your idea. Then use your calculator to check it.

EXERCISE 2E

F

1 Change each of these improper fractions into a mixed number.

a $\frac{7}{3}$ **b** $\frac{8}{3}$ **c** $\frac{9}{4}$ **d** $\frac{10}{7}$ **e** $\frac{12}{5}$ **f** $\frac{7}{5}$

g $\frac{13}{5}$ **h** $\frac{15}{4}$ **i** $\frac{10}{3}$ **j** $\frac{15}{7}$ **k** $\frac{17}{6}$ **l** $\frac{18}{5}$

m $\frac{19}{4}$ **n** $\frac{22}{7}$ **o** $\frac{14}{11}$ **p** $\frac{12}{11}$ **q** $\frac{28}{5}$ **r** $\frac{19}{7}$

s $\frac{40}{7}$ **t** $\frac{42}{5}$ **u** $\frac{21}{10}$ **v** $\frac{5}{2}$ **w** $\frac{5}{3}$ **x** $\frac{25}{8}$

y $\frac{23}{10}$ **z** $\frac{23}{11}$

2 Change each of these mixed numbers into an improper fraction.

a $3\frac{1}{3}$ **b** $5\frac{5}{6}$ **c** $1\frac{4}{5}$ **d** $5\frac{2}{7}$ **e** $4\frac{1}{10}$ **f** $5\frac{2}{3}$

g $2\frac{1}{2}$ **h** $3\frac{1}{4}$ **i** $7\frac{1}{6}$ **j** $3\frac{5}{8}$ **k** $6\frac{1}{3}$ **l** $9\frac{8}{9}$

m $11\frac{4}{5}$ **n** $3\frac{1}{5}$ **o** $4\frac{3}{8}$ **p** $3\frac{1}{9}$ **q** $5\frac{1}{5}$ **r** $2\frac{3}{4}$

s $4\frac{2}{7}$ **t** $8\frac{1}{6}$ **u** $2\frac{8}{9}$ **v** $6\frac{1}{6}$ **w** $12\frac{1}{5}$ **x** $1\frac{5}{8}$

y $7\frac{1}{10}$ **z** $8\frac{1}{9}$

3 Check your answers to questions 1 and 2, using the fraction buttons on your calculator.

AU 4 Which of these improper fractions has the largest value?

$\frac{27}{4}$ $\frac{31}{5}$ $\frac{13}{2}$

Show your working to justify your answer.

PS 5 Find a mixed number that is greater than $\frac{85}{11}$ but smaller than $\frac{79}{10}$.

F

PS 6 Find a mixed number that is greater than $5\frac{9}{11}$ but smaller than $5\frac{7}{8}$.

PS 7 Here is a list of numbers.

4 9 16 25

Using one of these numbers for the numerator and one for the denominator, find an improper fraction with a value between two and three.

2.6 Adding and subtracting fractions with the same denominator

This section will show you how to:

- add and subtract two fractions with the same denominator, then simplify the result
- solve problems in words

Key words

improper fraction
lowest terms
mixed number
proper fraction
simplest form

When you add two fractions with the same denominator, you get one of the following:

- a **proper fraction** that cannot be simplified, for example:

$$\frac{1}{5} + \frac{2}{5} = \frac{3}{5}$$

- a proper fraction that can be simplified to its **lowest terms** or **simplest form**, for example:

$$\frac{1}{8} + \frac{3}{8} = \frac{4}{8} = \frac{1}{2}$$

- an **improper fraction** that cannot be simplified, so it is converted to a **mixed number**, for example:

$$\frac{6}{7} + \frac{2}{7} = \frac{8}{7} = 1\frac{1}{7}$$

- an improper fraction that can be simplified before it is converted to a mixed number, for example:

$$\frac{5}{8} + \frac{7}{8} = \frac{12}{8} = \frac{3}{2} = 1\frac{1}{2}$$

When you subtract two fractions with the same denominator, you get one of the following:

- a proper fraction that cannot be simplified, for example:

$\frac{3}{5} - \frac{1}{5} = \frac{2}{5}$

- a proper fraction that can be simplified, for example:

$\frac{1}{2} - \frac{1}{10} = \frac{5}{10} - \frac{1}{10} = \frac{4}{10} = \frac{2}{5}$

Note You must *always* simplify fractions by cancelling if possible.

EXAMPLE 7

In a box of chocolates, a quarter are truffles, half are orange creams and the rest are mints. What fraction are mints?

Truffles and orange creams together are $\frac{1}{4} + \frac{1}{2} = \frac{3}{4}$ of the box.

Take the whole box as 1. So, mints are $1 - \frac{3}{4} = \frac{1}{4}$ of the box.

EXERCISE 2F

F

1 Copy and complete each of these additions.

a $\frac{3}{7} + \frac{2}{7}$ **b** $\frac{5}{9} + \frac{2}{9}$ **c** $\frac{3}{5} + \frac{1}{5}$ **d** $\frac{3}{7} + \frac{3}{7}$

e $\frac{1}{11} + \frac{6}{11}$ **f** $\frac{4}{9} + \frac{3}{9}$ **g** $\frac{3}{13} + \frac{7}{13}$ **h** $\frac{2}{11} + \frac{8}{11}$

2 Copy and complete each of these subtractions.

a $\frac{4}{7} - \frac{1}{7}$ **b** $\frac{5}{9} - \frac{4}{9}$ **c** $\frac{7}{11} - \frac{3}{11}$ **d** $\frac{9}{13} - \frac{2}{13}$

e $\frac{6}{7} - \frac{2}{7}$ **f** $\frac{3}{5} - \frac{2}{5}$ **g** $\frac{8}{9} - \frac{4}{9}$ **h** $\frac{5}{11} - \frac{2}{11}$

3 Copy and complete each of these additions.

a $\frac{5}{8} + \frac{1}{8}$ **b** $\frac{3}{10} + \frac{1}{10}$ **c** $\frac{2}{9} + \frac{4}{9}$ **d** $\frac{1}{4} + \frac{1}{4}$

e $\frac{3}{10} + \frac{3}{10}$ **f** $\frac{5}{12} + \frac{1}{12}$ **g** $\frac{3}{16} + \frac{5}{16}$ **h** $\frac{7}{16} + \frac{3}{16}$

4 Copy and complete each of these subtractions.

a $\frac{7}{8} - \frac{3}{8}$ **b** $\frac{7}{10} - \frac{3}{10}$ **c** $\frac{5}{6} - \frac{1}{6}$ **d** $\frac{9}{10} - \frac{1}{10}$

e $\frac{7}{12} - \frac{1}{12}$ **f** $\frac{5}{8} - \frac{3}{8}$ **g** $\frac{9}{16} - \frac{7}{16}$ **h** $\frac{7}{18} - \frac{1}{18}$

5 Copy and complete each of these additions. Use equivalent fractions to make the denominators the same.

a $\frac{1}{2} + \frac{7}{10}$ **b** $\frac{1}{2} + \frac{5}{8}$ **c** $\frac{3}{4} + \frac{3}{8}$ **d** $\frac{3}{4} + \frac{7}{8}$

e $\frac{1}{2} + \frac{7}{8}$ **f** $\frac{1}{3} + \frac{5}{6}$ **g** $\frac{5}{6} + \frac{2}{3}$ **h** $\frac{3}{4} + \frac{1}{2}$

6 Copy and complete each of these additions.

a $\frac{3}{8} + \frac{7}{8}$ **b** $\frac{3}{4} + \frac{3}{4}$ **c** $\frac{2}{5} + \frac{3}{5}$ **d** $\frac{7}{10} + \frac{9}{10}$

e $\frac{5}{8} + \frac{5}{8}$ **f** $\frac{7}{16} + \frac{15}{16}$ **g** $\frac{5}{12} + \frac{11}{12}$ **h** $\frac{11}{16} + \frac{7}{16}$

7 Copy and complete each of these subtractions. Use equivalent fractions to make the denominators the same.

a $\frac{7}{8} - \frac{1}{4}$ **b** $\frac{7}{10} - \frac{1}{5}$ **c** $\frac{3}{4} - \frac{1}{2}$ **d** $\frac{5}{8} - \frac{1}{4}$

e $\frac{1}{2} - \frac{1}{4}$ **f** $\frac{7}{8} - \frac{1}{2}$ **g** $\frac{9}{10} - \frac{1}{2}$ **h** $\frac{11}{16} - \frac{3}{8}$

8 At a recent Premiership football match, $\frac{7}{8}$ of the crowd were home supporters. What fraction of the crowd were not home supporters?

PS 9 After Emma had taken a slice of cake, $\frac{3}{4}$ of the cake was left. Ayesha then had $\frac{1}{2}$ of what was left.

Who had more cake?

FM 10 Three friends share two pizzas. Each pizza is cut into six equal slices. What fraction of a pizza did each friend get?

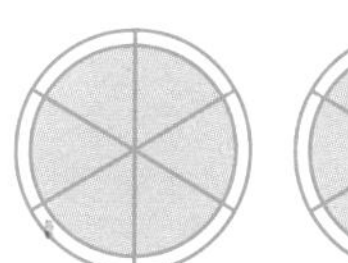

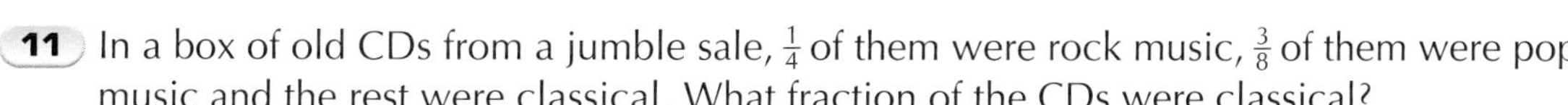

11 In a box of old CDs from a jumble sale, $\frac{1}{4}$ of them were rock music, $\frac{3}{8}$ of them were pop music and the rest were classical. What fraction of the CDs were classical?

PS 12 In a car park, $\frac{1}{5}$ of the cars were German makes. Half of the rest were Japanese makes. What fraction of the cars were Japanese makes?

E

PS 13 A fruit drink consists of $\frac{1}{2}$ orange juice, $\frac{1}{8}$ lemon juice and the rest is pineapple juice. What fraction of the drink is pineapple juice?

14 In a hockey team, $\frac{2}{11}$ of the team are French, $\frac{2}{11}$ are Italian, $\frac{3}{11}$ are Scottish and the rest are English. What fraction of the team is English?

15 In a packet of biscuits, $\frac{1}{6}$ are digestives, $\frac{2}{3}$ are Bourbons and the rest are Jammy Dodgers. What fraction are jammy dodgers?

FM 16 Jide pays $\frac{1}{4}$ of his wages in tax and $\frac{1}{8}$ of his wages in National Insurance. What fraction of his wages does he take home?

AU 17 Sally and Paul each buy a pizza.

Sally cuts hers into eight equal pieces and eats five.

Paul cuts his into six equal pieces and eats four.

Who eats the most pizza? Explain your answer.

2.7 Finding a fraction of a quantity

This section will show you how to:
- find a fraction of a given quantity

Key words
fraction
quantity

To do this, you simply multiply the **fraction** by the **quantity**, for example, $\frac{1}{2}$ of 30 is the same as $\frac{1}{2} \times 30$.

Remember: In mathematics 'of' is interpreted as ×.

For example, two lots of three is the same as 2×3.

EXAMPLE 8

Find $\frac{3}{4}$ of £196.

First, find $\frac{1}{4}$ by dividing by 4. Then find $\frac{3}{4}$ by multiplying your answer by 3:

$196 \div 4 = 49$ then $49 \times 3 = 147$

The answer is £147.

Of course, you can use your calculator to do this problem by either:

- pressing the sequence: 1 9 6 ÷ 4 × 3 =
- or using the [fraction] key: [fraction] 3 4 × 1 9 6 =

EXERCISE 2G

F

1 Calculate each of these.

a $\frac{3}{5}$ of 30 **b** $\frac{2}{7}$ of 35 **c** $\frac{3}{8}$ of 48 **d** $\frac{7}{10}$ of 40

e $\frac{5}{6}$ of 18 **f** $\frac{3}{4}$ of 24 **g** $\frac{4}{5}$ of 60 **h** $\frac{5}{8}$ of 72

2 Calculate each of these quantities.

a $\frac{3}{4}$ of £2400 **b** $\frac{2}{5}$ of 320 grams **c** $\frac{5}{8}$ of 256 kilograms

d $\frac{2}{3}$ of £174 **e** $\frac{5}{6}$ of 78 litres **f** $\frac{3}{4}$ of 120 minutes

g $\frac{4}{5}$ of 365 days **h** $\frac{7}{8}$ of 24 hours **i** $\frac{3}{4}$ of 1 day

j $\frac{5}{9}$ of 4266 miles

3 In each case, find out which is the larger number.

a $\frac{2}{5}$ of 60 or $\frac{5}{8}$ of 40 **b** $\frac{3}{4}$ of 280 or $\frac{7}{10}$ of 290

c $\frac{2}{3}$ of 78 or $\frac{4}{5}$ of 70 **d** $\frac{5}{6}$ of 72 or $\frac{11}{12}$ of 60

e $\frac{4}{9}$ of 126 or $\frac{3}{5}$ of 95 **f** $\frac{3}{4}$ of 340 or $\frac{2}{3}$ of 381

F

4 A director receives $\frac{2}{15}$ of his firm's profits. The firm made a profit of £45 600 in one year. How much did the director receive?

5 A woman left £84 000 in her will.

She left $\frac{3}{8}$ of the money to charity.

How much did she leave to charity?

AU 6 In the season 2008/2009, the attendance at Huddersfield Town versus Leeds United was 20 928. Of this crowd, $\frac{3}{8}$ were female. How many were male?

7 Two-thirds of a person's weight is water. Paul weighs 78 kg. How much of his body weight is water?

8 **a** Information from the first census in Singapore suggests that then $\frac{2}{25}$ of the population were Indian. The total population was 10 700. How many people were Indian?

b By 1990 the population of Singapore had grown to 3 002 800. Only $\frac{1}{16}$ of this population were Indian. How many Indians were living in Singapore in 1990?

9 Mark normally earns £500 a week. One week he is given a bonus of $\frac{1}{10}$ of his wage.

a Find $\frac{1}{10}$ of £500.

b How much does he earn altogether for this week?

10 The contents of a standard box of cereals weigh 720 g. A new larger box holds $\frac{1}{4}$ more than the standard box.

a Find $\frac{1}{4}$ of 720 g.

b How much do the contents of the new box of cereals weigh?

FM 11 The price of a new TV costing £360 is reduced by $\frac{1}{3}$ in a sale.

a Find $\frac{1}{3}$ of £360.

b How much does the TV cost in the sale?

E

FM 12 A car is advertised at Lion Autos at £9000 including extras but with a special offer of one-fifth off this price.

The same car is advertised at Tiger Motors for £6000 but the extras add one-quarter to this price.

Which garage is the cheaper?

D

FM 13 A jar of instant coffee normally contains 200 g and costs £2.

There are two special offers on a jar of coffee.

Offer A: $\frac{1}{4}$ extra for the same price.

Offer B: Same weight for $\frac{3}{4}$ of the original price.

Which offer is the best value?

2.8 Multiplying and dividing fractions

This section will show you how to:

- multiply a fraction by a fraction
- divide a fraction by a fraction

Key words
denominator
divide
multiply
numerator

What is $\frac{1}{2}$ of $\frac{1}{4}$?

The diagram shows the answer is $\frac{1}{8}$.

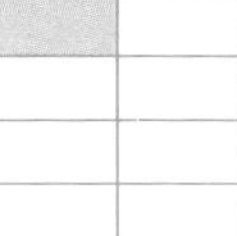

In mathematics, you always write $\frac{1}{2}$ of $\frac{1}{4}$ as $\frac{1}{2} \times \frac{1}{4}$

So you know that $\frac{1}{2} \times \frac{1}{4} = \frac{1}{8}$

To **multiply** fractions, you multiply the **numerators** together and you multiply the **denominators** together.

EXAMPLE 9

Work out $\frac{1}{4}$ of $\frac{2}{5}$.

$$\frac{1}{4} \times \frac{2}{5} = \frac{1 \times 2}{4 \times 5} = \frac{2}{20} = \frac{1}{10}$$

Dividing fractions

Look at the problem $3 \div \frac{3}{4}$. This is like asking, 'How many $\frac{3}{4}$s are there in 3?' Look at the diagram.

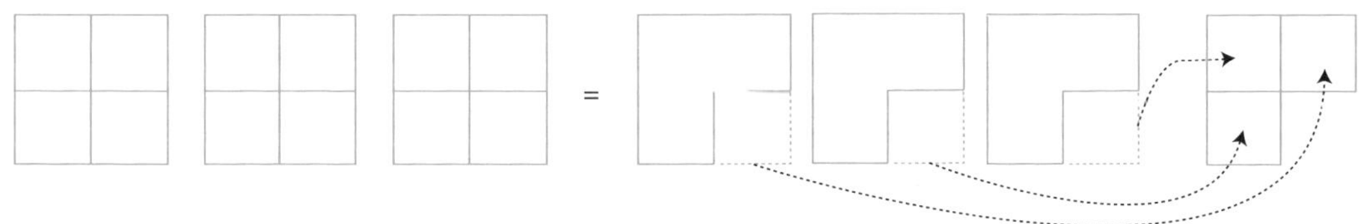

Each of the three whole shapes is divided into quarters. What is the total number of quarters divided by 3? Can you see that you could fit the four shapes on the right-hand side of the = sign into the three shapes on the left-hand side?

i.e. $3 \div \frac{3}{4} = 4$ or $3 \div \frac{3}{4} = 3 \times \frac{4}{3} = \frac{3 \times 4}{3} = \frac{12}{3} = 4$

So, to divide by a fraction, you turn the fraction upside down (finding its reciprocal), and then multiply.

E D C

EXERCISE 2H

1 Work out each of these multiplications.

a $\frac{1}{2} \times \frac{1}{3}$ **b** $\frac{1}{4} \times \frac{1}{5}$ **c** $\frac{1}{3} \times \frac{2}{3}$ **d** $\frac{1}{4} \times \frac{2}{3}$

e $\frac{1}{3} \times \frac{3}{4}$ **f** $\frac{2}{3} \times \frac{3}{5}$ **g** $\frac{3}{4} \times \frac{2}{3}$ **h** $\frac{5}{6} \times \frac{3}{5}$

i $\frac{2}{7} \times \frac{3}{4}$ **j** $\frac{5}{6} \times \frac{7}{8}$ **k** $\frac{4}{5} \times \frac{2}{3}$ **l** $\frac{3}{4} \times \frac{7}{8}$

FM 2 A farmer is converting a field into sheep pens. He divides the field in half and then divides each half into three equal parts.

a Each pen can hold 20 sheep.

The farmer already has 95 sheep.

How many more sheep should he buy in order to fill the pens?

b What fraction of the field does each sheep pen take up?

PS 3 There are 7.5 million people living in London. One-eighth of these are over 65. Of the over-65s, three-fifths are women

How many women over 65 live in London?

PS 4 Alec eats $\frac{1}{4}$ of a cake.

Ben eats $\frac{1}{3}$ of what is left.

Callum eats $\frac{1}{2}$ of what is left.

Dec gets the final piece.

Who got the most cake?

5 Evaluate the following, giving your answer as a mixed number where possible.

a $\frac{1}{4} \div \frac{1}{3}$ **b** $\frac{2}{5} \div \frac{2}{7}$ **c** $\frac{4}{5} \div \frac{3}{4}$ **d** $\frac{3}{7} \div \frac{2}{5}$

e $5 \div 1\frac{1}{4}$ **f** $6 \div 1\frac{1}{2}$ **g** $7\frac{1}{2} \div 1\frac{1}{2}$ **h** $3 \div 1\frac{3}{4}$

6 A grain merchant has only thirteen and a half tonnes in stock. He has several customers who are ordering three-quarters of a tonne. How many customers can he supply?

7 Evaluate the following, giving your answer as a mixed number where possible.

a $2\frac{2}{9} \times 2\frac{1}{10} \times \frac{16}{35}$ **b** $3\frac{1}{5} \times 2\frac{1}{2} \times 4\frac{3}{4}$

c $1\frac{1}{4} \times 1\frac{2}{7} \times 1\frac{1}{6}$ **d** $\frac{18}{25} \times \frac{15}{26} \div 2\frac{2}{5}$

2.9 One quantity as a fraction of another

This section will show you how to:
- express one quantity as a fraction of another

Key words
fraction
quantity

You may often be asked to give one amount or **quantity** as a **fraction** of another.

EXAMPLE 10

Write £5 as a fraction of £20.

As a fraction this is written $\frac{5}{20}$. This cancels to $\frac{1}{4}$.

So £5 is one-quarter of £20.

EXERCISE 2I

D

1 In each of the following, write the first quantity as a fraction of the second.

a 2 cm, 6 cm
b 4 kg, 20 kg
c £8, £20
d 5 hours, 24 hours
e 12 days, 30 days
f 50p, £3
g 4 days, 2 weeks
h 40 minutes, 2 hours

2 In a form of 30 pupils, 18 are boys. What fraction of the form are boys?

3 During March, it rained on 12 days. For what fraction of the month did it rain?

4 Reka wins £120 in a competition and puts £50 into her bank account. What fraction of her winnings does she keep to spend?

FM 5 Jon earns £90 and saves £30 of it.

Matt earns £100 and saves £35 of it.

Who is saving the greater proportion of their earnings?

AU 6 In two tests Isobel gets 13 out of 20 and 16 out of 25. Which is the better mark?

Explain your answer.

GRADE BOOSTER

- **G** You can state the fraction of a shape that is shaded
- **G** You can shade in a fraction of a shape
- **G** You can add and subtract simple fractions
- **G** You know how to identify equivalent fractions
- **G** You can simplify fractions by cancelling (when possible)
- **G** You can put simple fractions into order of size
- **F** You can change improper fractions to mixed numbers
- **F** You can find a fraction of a quantity
- **F** You can change mixed numbers to improper fractions
- **F** You can solve fraction problems expressed in words
- **E** You can add more difficult fractions
- **E** You can compare two fractions of quantities
- **E** You can multiply a fraction by a fraction
- **D** You can write a quantity as a fraction of another quantity
- **C** You can divide a fraction by a fraction

What you should know now

- How to recognise and draw fractions of shapes
- How to add, subtract, multiply, divide and cancel simple fractions without using a calculator
- How to work out equivalent fractions
- How to convert an improper fraction to a mixed number (and the other way)
- How to calculate a fraction of a quantity
- How to solve simple practical problems using fractions
- How to add, subtract, multiply and divide fractions with different denominators

1 **a** What fraction of the shape below is shaded?

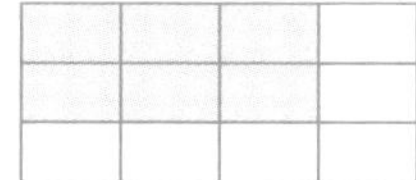

Give your answer as a fraction in its simplest form.

b Copy out and shade $\frac{3}{4}$ of the shape below.

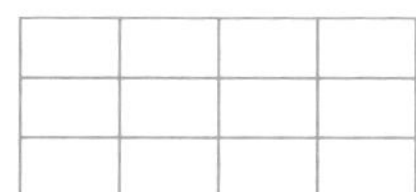

2 Put the following fractions into order, smallest first.

$\frac{3}{4}$ $\frac{1}{2}$ $\frac{2}{5}$ $\frac{1}{3}$

3 Work out

$\frac{7}{10} - \frac{2}{5}$

4 Alison travels by car to her meetings. Alison's company pays her 32p for each mile she travels.

One day Alison writes down the distance readings from her car.

Start of the day: 2430 miles

End of the day: 2658 miles

a Work out how much the company pays Alison for her day's travel. (2)

The next day Alison travelled a total of 145 miles.

She travelled $\frac{2}{5}$ of this distance in the morning.

b How many miles did she travel during the rest of the day? (2)

(Total 4 marks)

Edexcel, June 2005, Paper 2 Foundation, Question 9

5 Find $\frac{3}{5}$ of 45 kg.

6 **a** Here are some fractions.

$\frac{2}{4}$ $\frac{4}{8}$ $\frac{2}{5}$ $\frac{7}{14}$

Which one of these fractions is not equal to $\frac{1}{2}$? (1)

Give a reason for your answer. (1)

b Work out $\frac{3}{4}$ of 20 (2)

(Total 4 marks)

Edexcel, May 2008, Paper 1 Foundation, Question 11

7 Here are two fractions $\frac{4}{5}$ and $\frac{3}{4}$.

Explain which is the larger fraction.

You may use the grids to help with your explanation. (2)

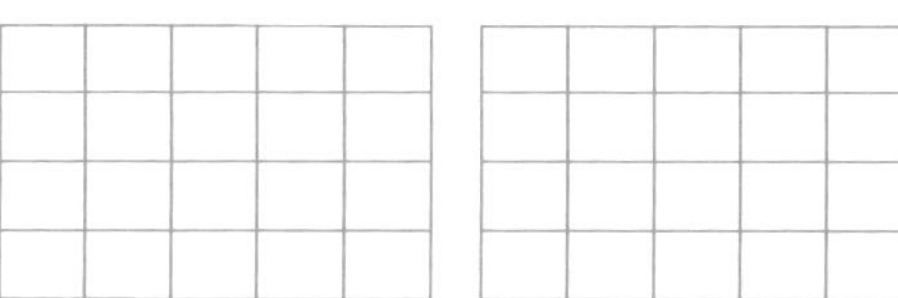

(Total 2 marks)

Edexcel, June 2007, Paper 10 Foundation, Question 5

8 A box contains 200 tissues.

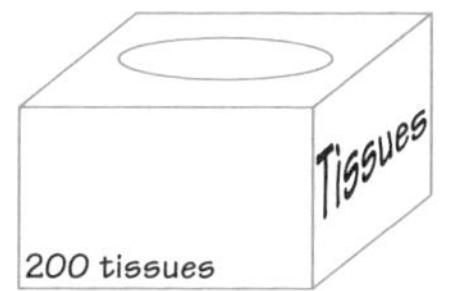

Toby takes $\frac{3}{5}$ of these tissues.

Work out how many tissues he takes. (2)

(Total 2 marks)

Edexcel, June 2008, Paper 13 Foundation, Question 5

9 A fruit punch was made using $\frac{1}{2}$ lemonade, $\frac{1}{5}$ orange juice with the rest lemon juice. What fraction of the drink is lemon juice?

10 Calculate the following, giving your answers as fractions in their simplest forms.

a $\frac{3}{8} + \frac{1}{8}$

b $\frac{9}{10} - \frac{1}{2}$

11 **a** Work out

i $\frac{3}{5}$ of 175 **ii** $\frac{3}{4} \times \frac{2}{3}$

b What fraction is 13 weeks out of a year of 52 weeks?

c What is $1 - \frac{2}{5}$?

12 Two-fifths of the price of a book goes to the bookshop. A book cost £12. How much goes to the bookshop?

13 When a cross is carved from a piece of wood, $\frac{4}{5}$ of the wood is cut away. The original block weighs 215 grams. What weight of wood is cut off?

14 **a** Work out $30 \times \frac{2}{3}$

b Work out the value of $\frac{14}{15} \times \frac{3}{4}$

Give your answer as a fraction in its simplest form.

15 Packets of Wheetix used to contain 550 grams of cereal. New packets contain one-fifth more. How much does a new packet contain?

16 Write 24 as a fraction of 36

Give your answer in its simplest form. (2)

(Total 2 marks)

Edexcel, May 2008, Paper 12 Foundation, Question 11(a)

17 The size of a packet of cereal is increased by one-fifth.

The new size is later reduced by one-fifth.

Is the new packet smaller, the same size or larger than the original packet?

Explain clearly how you worked out your answer.

Worked Examination Questions

1 Fred gets 3 pints of milk delivered on Saturday.

He uses $\frac{5}{8}$ of a pint a day.

On which day will he run out of milk?

$3 \div \frac{5}{8}$

$= 4.8$ or $4\frac{4}{5}$

The milk runs out on Wednesday.

Total: 3 marks

You get 1 method mark for identifying this as a division problem.

Use your calculator to work this out. This is worth 1 mark.

Count 5 days on from Saturday. The correct answer receives 1 mark.

PS **2** There are 900 students in a college.

$\frac{1}{10}$ are under 16.

$\frac{1}{4}$ are over 18.

What fraction of the students are neither under 16 nor over 18?

First method:

$\frac{1}{10}$ of $900 = \frac{1}{10} \times 900$

$\frac{1}{4}$ of $900 = \frac{1}{4} \times 900$

$= 90$ and 225

$900 - 90 - 225 = 585$

$\frac{585}{900}$ or $\frac{13}{20}$

1 method mark is available for identifying a method to work out the number under 16 or over 18.

You get 1 mark for obtaining both of these answers.

You get 1 mark for attempting to work out the number of people.

You get 1 accuracy mark for the correct answer.

Second method:

$\frac{1}{10} + \frac{1}{4}$

$\frac{2}{20} + \frac{5}{20}$

$\frac{7}{20}$

$\frac{13}{20}$

Total: 4 marks

1 method mark is available for identifying the valid method of adding the given fractions.

You get 1 method mark for getting a common denominator.

You get 1 mark for correctly getting the fraction under 16 and over 18.

You get 1 mark for the correct answer.

2 Functional Maths
In the trenches

During World War I, soldiers were often faced with very difficult circumstances. For long periods of time, they would live in trenches that were wet, dirty, and full of rats and disease. During periods either just before or during a big battle, food was also scarce.

Getting started

Think about the following questions.

- What are ounces and gills?
- With a partner, work out the metric equivalents of a soldier's rations.
- How much would each soldier receive each week?

Is there any other information that you need to be able to complete your task? Where might you be able to find this information?

Below is a list of what each soldier might be expected to receive each day.

20 oz bread
3 oz cheese
$\frac{5}{8}$ oz tea
4 oz jam
8 oz fresh vegetables
$\frac{1}{3}$ oz chocolate
$\frac{1}{2}$ gill rum

Current supplies

$\frac{3}{4}$ ton bread
300 lb cheese
100 lb tea
355 lb jam
$\frac{1}{2}$ ton fresh vegetables
30 lb chocolate
20 gallons rum

Your task

The logistics corps faced one of the biggest challenges of the war. They were in charge of making sure that the soldiers in the trenches got the food they were entitled to. Imagine you were working in the logistics corps.

Using the information below, decide if you have everything you need to supply the 200 troops on your part of the front line for the next week. If you have shortages, decide how you are going to ration what you have.

Then, write a letter to the field commander, ordering your supplies, so that you have everything you need for the following week.

JAM
BISCUITS
3 OUNCES
SALT
HER ROYAL HIGHNESS
THE PRINCESS MARY'S
CHRISTMAS FUND
1914
FRAY
BENTOS
CORNED BEEF
WD14
This is your combat
issue of Tea–
Use it wisely!
BISCUITS
3 OUNCES

Why this chapter matters

Life is full of pairs: up and down, hot and cold, left and right, light and dark, rough and smooth, to name a few. One pairing that is particularly relevant to maths is positive and negative.

You are already familiar with positive numbers, including where they appear in real life and how to carry out calculations with them. However, sometimes we need to use a set of numbers in addition to the positive counting numbers. This set of numbers is known as the negative numbers. Here are some examples of where you might encounter negative numbers.

A negative number on a bank statement will show by how much you are overdrawn (or, how much money you have spent above what you have in your bank account).

On the temperature scale of degrees Celsius zero is known as 'freezing point'. In many places – even in the UK! – temperatures fall below freezing point. At this stage we need negative numbers to represent the temperature.

Jet pilots experience g-forces when their aircraft accelerates or decelerates quickly. Negative g-forces can be felt when an object accelerates downwards very quickly and they are represented by negative numbers. These negative forces are responsible for the feeling of weightlessness that you have on rollercoasters!

$$5 - 9 = -4$$

When a bigger number is taken from a smaller one, the result is a negative number.

In lifts, negative numbers are used to represent floors below ground level. These are often called 'lower ground' floors.

Sea level can be given the value 'zero'. Mountains are described as being 'above sea level' and ocean floors as 'below sea level'. This means that when a submarine goes under the sea the depths that it reaches are given using negative numbers.

As you can see, negative numbers are just as important as positive numbers and you will encounter them in your everyday life.

Chapter

3 Number: Negative numbers

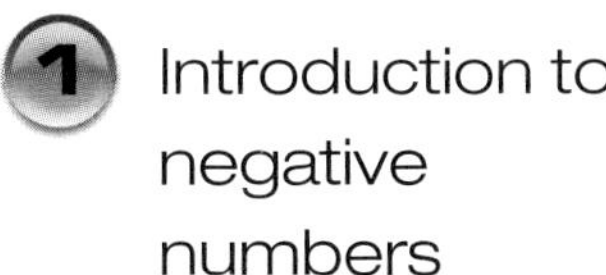

1. Introduction to negative numbers
2. Everyday use of negative numbers
3. The number line
4. Arithmetic with negative numbers

The grades given in this chapter are target grades.

This chapter will show you ...

- **G** how negative numbers are used in real life
- **G** what is meant by a negative number
- **G** how to use inequalities with negative numbers
- **F** how to do arithmetic with negative numbers

Visual overview

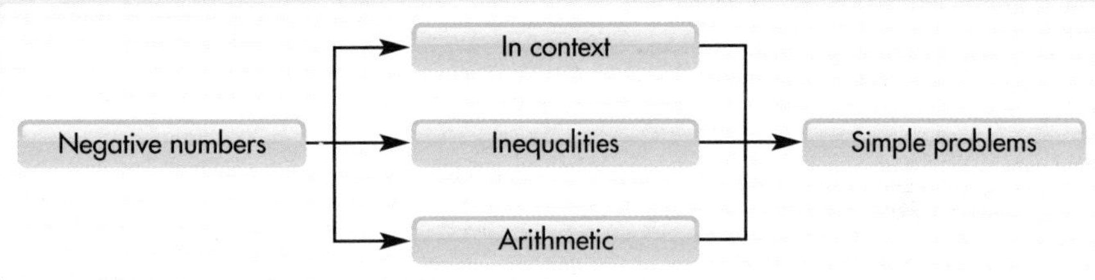

What you should already know

- What a negative number means (KS3 level 3, GCSE grade G)
- How to put numbers in order (KS3 level 3, GCSE grade G)

Quick check

Put the numbers in the following lists into order, smallest first.

1 8, 2, 5, 9, 1, 0, 4

2 14, 19, 11, 10, 17

3 51, 92, 24, 0, 32

4 87, 136, 12, 288, 56

5 5, 87, $\frac{1}{2}$, 100, 0, 50

3.1 Introduction to negative numbers

This section will show you how to:

- use negative numbers in real life

Key words
below
difference
negative number

Negative numbers are numbers **below** zero. You meet **negative numbers** often in winter when the temperature falls below freezing (0 °C).

The diagram below shows a thermometer with negative temperatures. The temperature is –3 °C. This means the temperature is three degrees below zero.

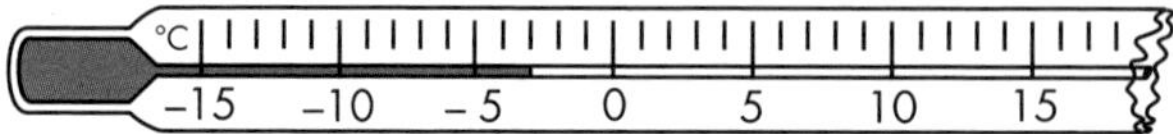

The number line below shows positive and negative numbers.

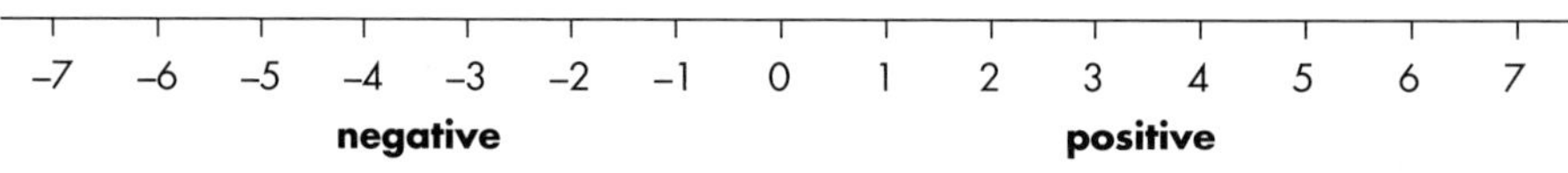

ACTIVITY

Seaport colliery

Top of winding gear 80 ft above ground
Ground 400 ft
South drift 310 ft
225 ft Closed gate
150 ft Collapsed tunnel
Zero level 0 ft
–50 ft North gate
A seam –325 ft
B seam –425 ft
–475 ft Dead Man's seam
C seam –550 ft
–600 ft D seam
Bottom gate –700 ft
Sea level

This is a section through Seaport Colliery.

The height above sea (zero) level for each tunnel is shown. Note that some of the heights are given as negative numbers.

The ground is 400 ft above sea level.

1 What is the **difference** in height between the following levels?

- **a** South drift and C seam
- **b** The ground and A seam
- **c** The closed gate and the collapsed tunnel
- **d** Zero level and Dead Man's seam
- **e** Ground level and the bottom gate
- **f** Collapsed tunnel and B seam
- **g** North gate and Dead Man's seam

h Zero level and the south drift

i Zero level and the bottom gate

j South drift and the bottom gate

2 How high above sea level is the top of the winding gear?

3 How high above the bottom gate is the top of the winding gear?

4 There are two pairs of tunnels that are 75 ft apart. Which two pairs are they?

5 How much cable does the engineman let out to get the cage from the south drift to D seam?

6 There are two pairs of tunnels that are 125 ft apart. Which two pairs are they?

7 Which two tunnels are 200 ft apart?

EXERCISE 3A

G

1 Write down the temperature for each thermometer.

a 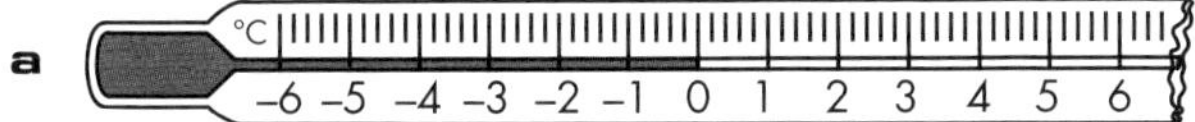

b 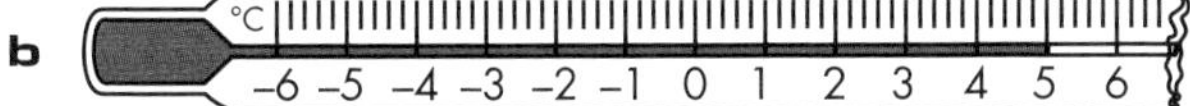

c 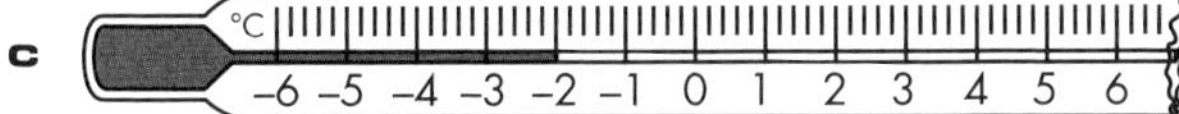

d 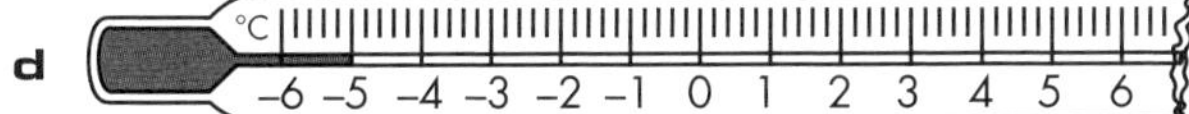

e 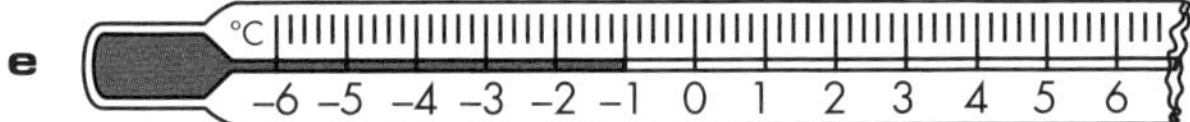

2 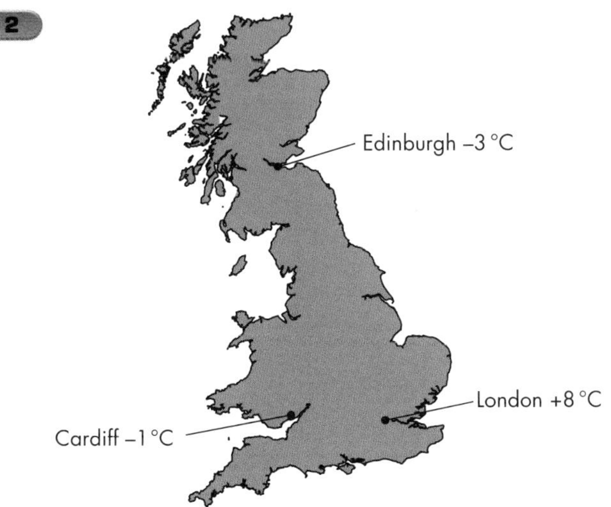

a How much colder is it in Edinburgh than in London?

b How much warmer is it in London than in Cardiff?

G

FM 3 The instructions on a bottle of de-icer say that it will stop water freezing down to –12 °C. The temperature is –4 °C.

How many more degrees does the temperature need to fall before the de-icer stops working?

F

FM 4 Chris Boddington, the famous explorer, is leading a joint expedition to the Andes mountains and cave system. The climbers establish camps at the various places shown and the cavers find the various caves shown.

To help him move supplies about, Chris needs a chart of the differences between the heights of the camps and caves. Help him by copying the chart below into your book and filling it in. (Some of the boxes have been done for you.)

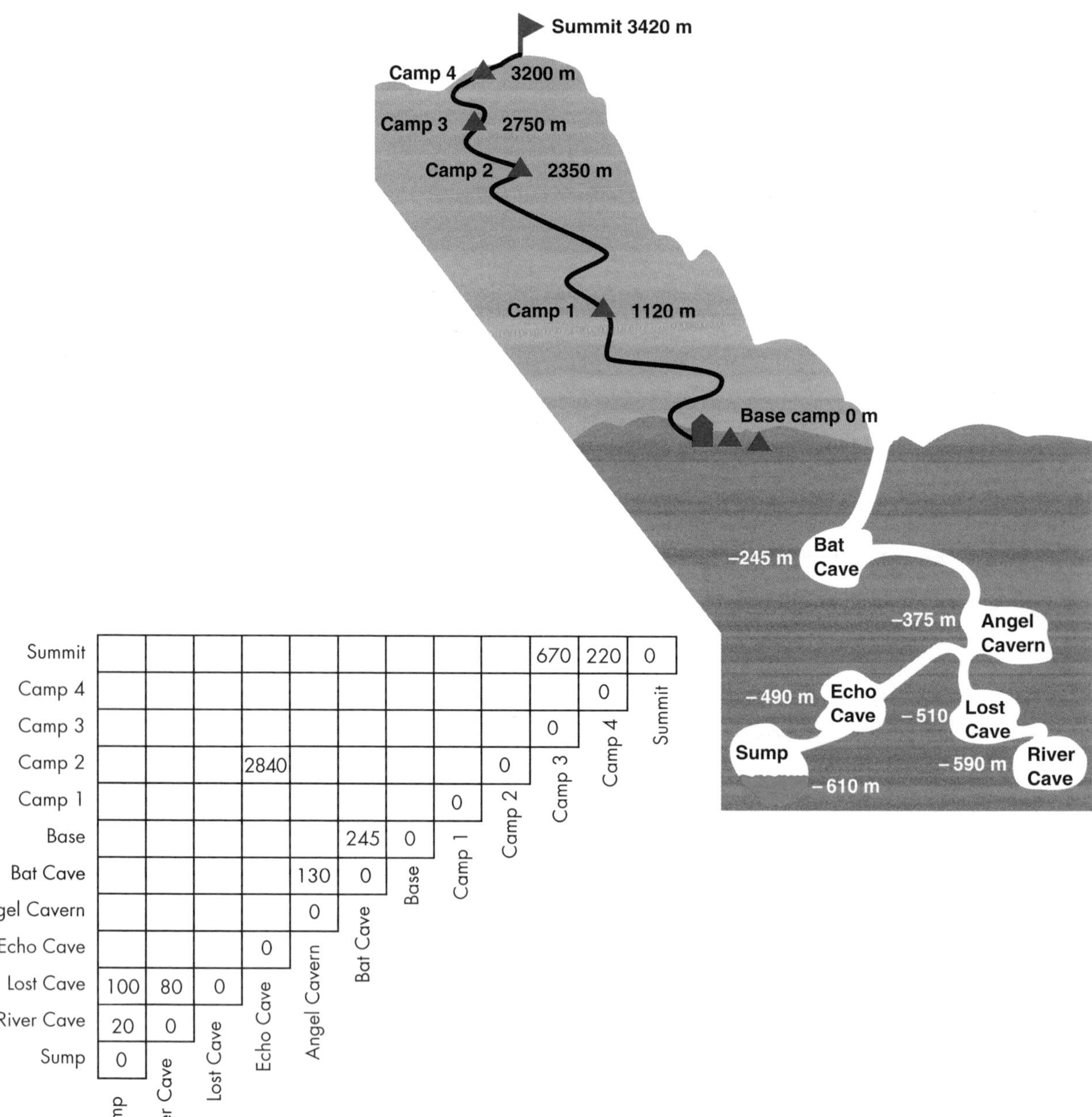

	Sump	River Cave	Lost Cave	Echo Cave	Angel Cavern	Bat Cave	Base	Camp 1	Camp 2	Camp 3	Camp 4	Summit
Summit										670	220	0
Camp 4											0	
Camp 3										0		
Camp 2				2840					0			
Camp 1								0				
Base						245	0					
Bat Cave					130	0						
Angel Cavern					0							
Echo Cave				0								
Lost Cave	100	80	0									
River Cave	20	0										
Sump	0											

3.2 Everyday use of negative numbers

This section will show you how to:

- use positive and negative numbers in everyday life

Key words

above/below
after/before
credit/debit
negative number/ positive number
profit/loss

There are many other situations where **negative numbers** are used. Here are three examples.

- When +15 m means 15 metres **above** sea level, then –15 m means 15 metres **below** sea level.
- When +2 h means 2 hours **after** midday, then –2 h means 2 hours **before** midday.
- When +£60 means a **profit** of £60, then –£60 means a **loss** of £60.

You also meet negative numbers on graphs, and you may already have plotted coordinates with negative numbers.

On bank statements and bills a negative number means you owe money. A **positive number** means they owe you money.

A **debit** is when you pay money out of your account.

A **credit** is when you pay money into your account.

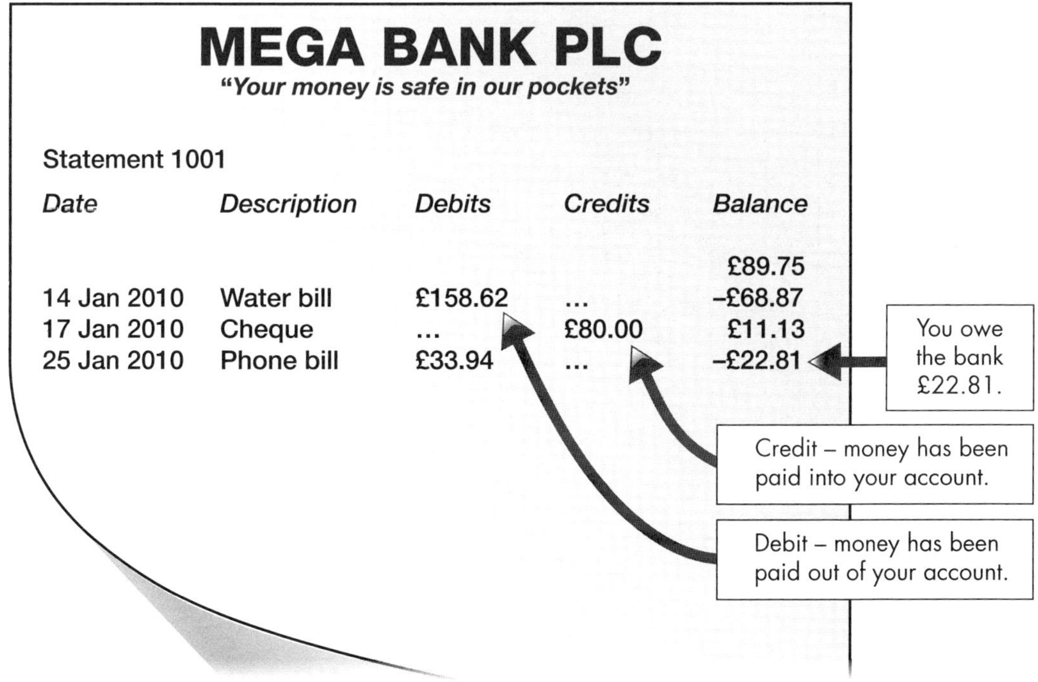

MEGA BANK PLC

"Your money is safe in our pockets"

Statement 1001

Date	*Description*	*Debits*	*Credits*	*Balance*
				£89.75
14 Jan 2010	Water bill	£158.62	...	–£68.87
17 Jan 2010	Cheque	...	£80.00	£11.13
25 Jan 2010	Phone bill	£33.94	...	–£22.81

You owe the bank £22.81.
Credit – money has been paid into your account.
Debit – money has been paid out of your account.

EXERCISE 3B

Copy and complete each of the following.

G

1. If +£5 means a profit of five pounds, then …… means a loss of five pounds.
2. If +£9 means a profit of nine pounds, then a loss of £9 is …… .
3. If –£4 means a loss of four pounds, then +£4 means a …… of four pounds.
4. If +200 m means 200 metres above sea level, then …… means 200 metres below sea level.
5. If +50 m means fifty metres above sea level, then fifty metres below sea level is written …… .
6. If –100 m means one hundred metres below sea level, then +100 m means one hundred metres …… sea level.
7. If +3 h means three hours after midday, then …… means three hours before midday.
8. If +5 h means 5 hours after midday, then …… means 5 hours before midday.
9. If –6 h means six hours before midday, then +6 h means six hours …… midday.
10. If +2 °C means two degrees above freezing point, then …… means two degrees below freezing point.
11. If +8 °C means eight degrees above freezing point, then …… means eight degrees below freezing point.
12. If –5 °C means five degrees below freezing point, then +5 °C means five degrees …… freezing point.
13. If +70 km means 70 kilometres north of the equator, then …… means 70 kilometres south of the equator.
14. If +200 km means 200 kilometres north of the equator, then 200 kilometres south of the equator is written …… .
15. If –50 km means fifty kilometres south of the equator, then +50 km means fifty kilometres …… of the equator.
16. If 10 minutes before midnight is represented by –10 minutes, then five minutes after midnight is represented by …… .
17. If a car moving forwards at 10 miles per hour is represented by +10 mph, then a car moving backwards at 5 miles per hour is represented by …… .

G

18 In an office building, the third floor above ground level is represented by +3. So, the second floor below ground level is represented by …… .

FM **19**

MEGA BANK PLC

"Your money is safe in our pockets"

Statement 2010

Date	*Description*	*Debits*	*Credits*	*Balance*
				£320.45
20 Sept 2009	Gas bill	£410.17	…	–£89.72
23 Sept 2009	Cheque	…	£140.00	£50.28
25 Sept 2009	Mobile phone bill	£63.48	…	–£13.20

a What does –£89.72 mean?

b What is a debit?

c What is a credit?

20 The temperature on three days in Moscow was –7 °C, –5 °C and –11 °C.

a Which temperature is the lowest?

b What is the difference in temperature between the coldest and the warmest days?

21 **a** Which is the smallest number in the cloud?

b Which is the largest number in the cloud?

c What is the difference between the smallest and largest numbers in the cloud?

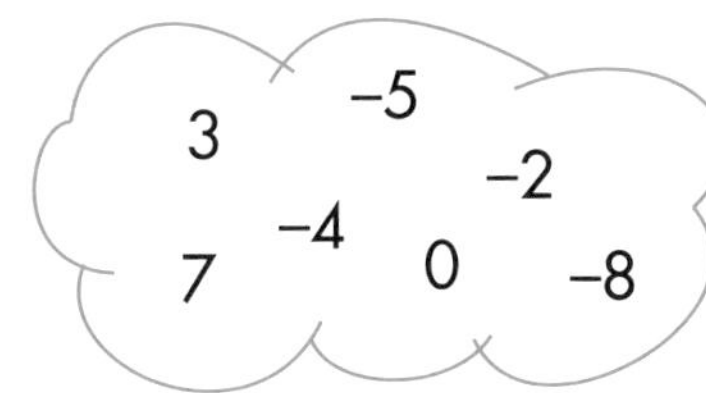

F

PS **22** Sydney the snail is at the bottom of a 10-foot well.

Each day he climbs 2 feet up the wall of the well.

Each night he slides 1 foot back down the wall of the well.

How many days does it take Sydney to reach the top of the well?

AU **23** A thermostat is set at 16 °C.

The temperature in a room at 1.00 am is –2 °C.

The temperature rises two degrees every 6 minutes.

At what time is the temperature on the thermostat reached?

3.3 The number line

This section will show you how to:

- use a number line to represent negative numbers
- use inequalities with negative numbers

Key words
greater than
inequality
less than
more than
negative
number line
positive

Look at the **number line**.

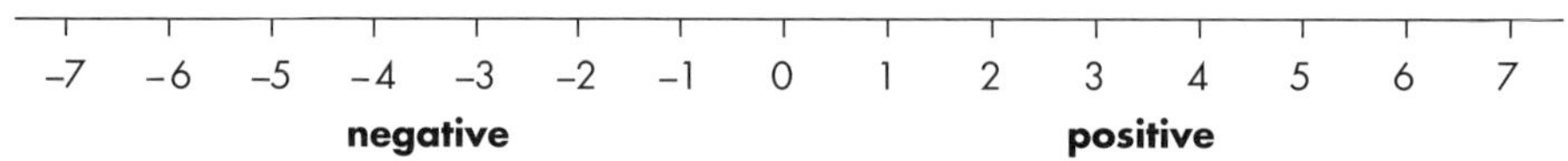

Notice that the **negative** numbers are to the left of 0 and the **positive** numbers are to the right of 0.

Numbers to the right of any number on the number line are always bigger than that number.

Numbers to the left of any number on the number line are always smaller than that number.

So, for example, you can see from a number line that:

2 is *smaller* than 5 because 2 is to the *left* of 5.

You can write this as $2 < 5$.

–3 is *smaller* than 2 because –3 is to the *left* of 2.

You can write this as $-3 < 2$.

7 is *bigger* than 3 because 7 is to the *right* of 3.

You can write this as $7 > 3$.

–1 is *bigger* than –4 because –1 is to the *right* of –4.

You can write this as $-1 > -4$.

Reminder The **inequality** signs:

$<$ means 'is **less than**'

$>$ means 'is **greater than**' or 'is **more than**'

EXERCISE 3C

1 Copy and complete each of the following by putting a suitable number in the box.

a □ is smaller than 3 **b** □ is smaller than 1 **c** □ is smaller than –3

d □ is smaller than –7 **e** –5 is smaller than □ **f** –1 is smaller than □

g 3 is smaller than □ **h** –2 is smaller than □ **i** □ is smaller than 0

j –4 is smaller than □ **k** □ is smaller than –8 **l** –7 is smaller than □

2 Copy and complete each of the following by putting a suitable number in the box.

a □ is bigger than –3 **b** □ is bigger than 1 **c** □ is bigger than –2

d □ is bigger than –1 **e** –1 is bigger than □ **f** –8 is bigger than □

g 1 is bigger than □ **h** –5 is bigger than □ **i** □ is bigger than –5

j 2 is bigger than □ **k** □ is bigger than –4 **l** –2 is bigger than □

3 Copy each of these and put the correct phrase in each space.

a –1 …… 3 **b** 3 …… 2 **c** –4 …… –1

d –5 …… –4 **e** 1 …… –6 **f** –3 …… 0

g –2 …… –1 **h** 2 …… –3 **i** 5 …… –6

j 3 …… 4 **k** –7 …… –5 **l** –2 …… –4

4 $-1 \quad -\frac{3}{4} \quad -\frac{1}{2} \quad -\frac{1}{4} \quad 0 \quad \frac{1}{4} \quad \frac{1}{2} \quad \frac{3}{4} \quad 1$

Copy each of these and put the correct phrase in each space.

a $\frac{1}{4}$ …… $\frac{3}{4}$ **b** $-\frac{1}{2}$ …… 0 **c** $-\frac{3}{4}$ …… $\frac{3}{4}$

d $\frac{1}{4}$ …… $-\frac{1}{2}$ **e** –1 …… $\frac{3}{4}$ **f** $\frac{1}{2}$ …… 1

5 In each case below, copy the statement and put the correct symbol, either < or >, in the box.

a 3 □ 5 **b** –2 □ –5 **c** –4 □ 3 **d** 5 □ 9

e –3 □ 2 **f** 4 □ –3 **g** –1 □ 0 **h** 6 □ –4

i 2 □ –3 **j** 0 □ –2 **k** –5 □ –4 **l** 1 □ 3

m –6 □ –7 **n** 2 □ –3 **o** –1 □ 1 **p** 4 □ 0

PS 6 Copy these number lines and fill in the missing numbers.

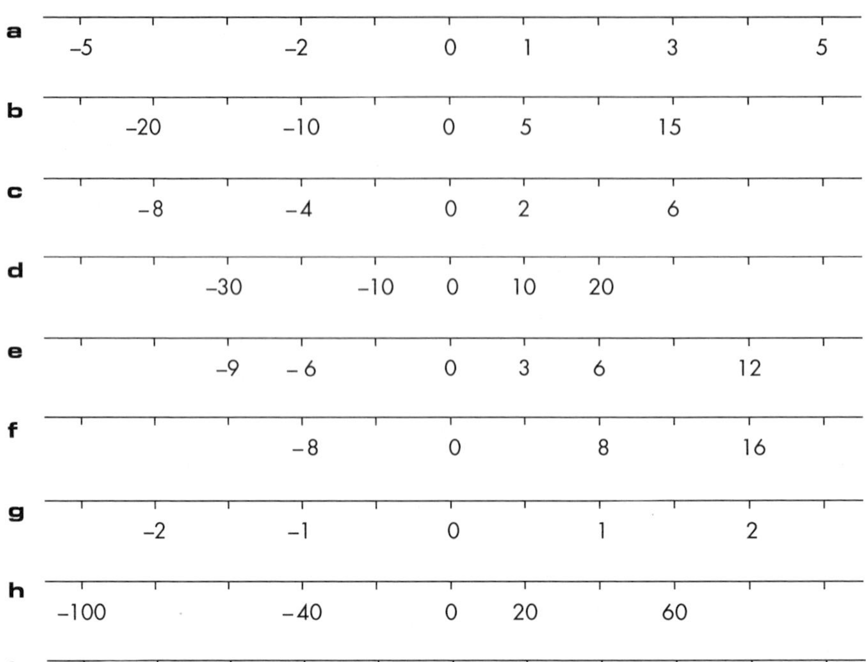

FM 7 Here are some temperatures.

2 °C   –2 °C   –4 °C   6 °C

Copy and complete the weather report, using these temperatures.

> The hottest place today is Barnsley with a temperature of ____, while in Eastbourne a ground frost has left the temperature just below zero at ____. In Bristol it is even colder at ____. Finally, in Tenby the temperature is just above freezing at ____.

AU 8 Here are some numbers.

$-4\frac{1}{2}$	$+3\frac{3}{4}$	$-\frac{1}{4}$

–6   –5   –4   –3   –2   –1   0   1   2   3   4   5   6

Copy the number line and mark the numbers on it.

3.4 Arithmetic with negative numbers

This section will show you how to:

- add and subtract positive and negative numbers to or from both positive and negative numbers

Key words
add
subtract

Adding and subtracting positive numbers

These two operations can be illustrated on a thermometer scale.

- **Adding** a positive number moves the marker *up* the thermometer scale.
  For example,

  $-2 + 6 = 4$

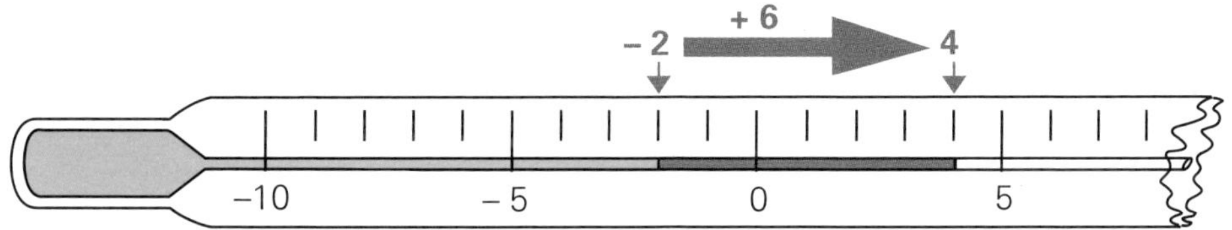

- **Subtracting** a positive number moves the marker *down* the thermometer scale.
  For example,

  $3 - 5 = -2$

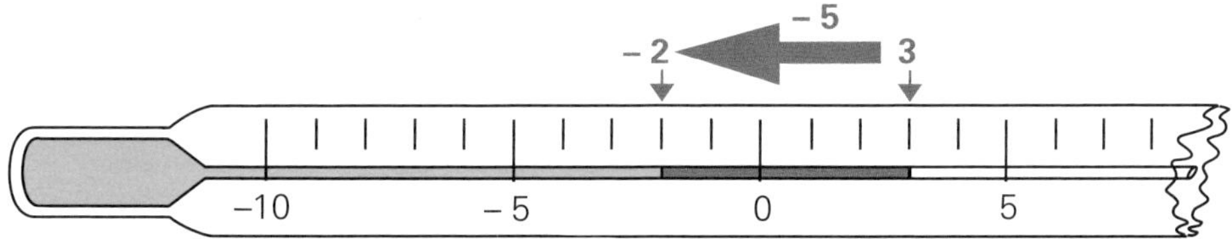

EXAMPLE 1

The temperature at midnight was 2 °C but then it fell by five degrees. What was the new temperature?

Falling five degrees means the calculation is $2 - 5$, which is equal to -3. So, the new temperature is -3 °C.

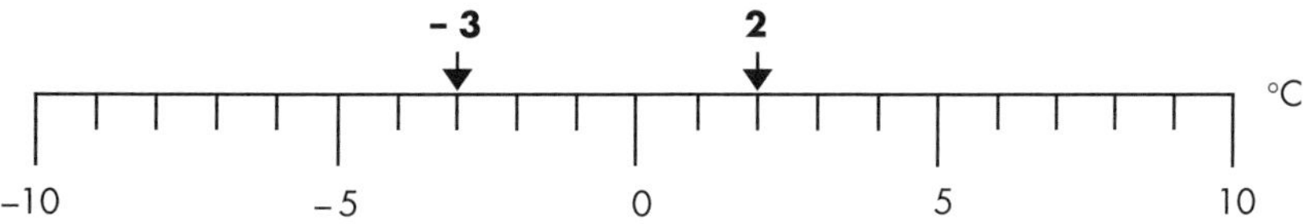

EXAMPLE 2

The temperature is −4 °C. It then falls by five degrees. What is the new temperature?

Falling five degrees means the calculation is −4 − 5, which is equal to −9. So, the new temperature is −9 °C.

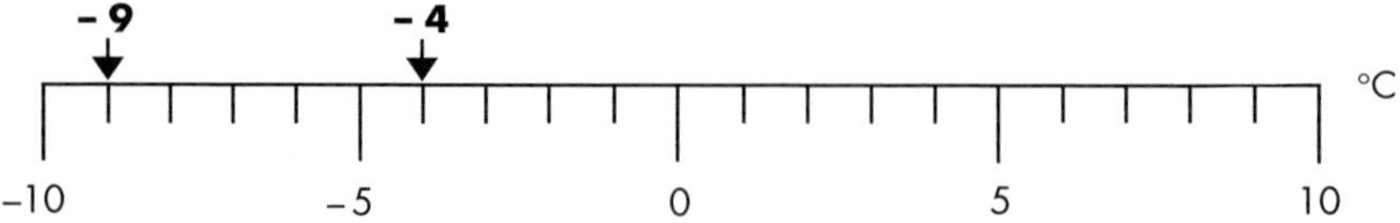

EXERCISE 3D

F

1 Use a thermometer scale to find the answer to each of the following.

a 2° – 4° = **b** 4° – 7° = **c** 3° – 5° = **d** 1° – 4° =
e 6° – 8° = **f** 5° – 8° = **g** –2 + 5 = **h** –1 + 4 =
i –4 + 3 = **j** –6 + 5 = **k** –3 + 5 = **l** –5 + 2 =
m –1 – 3 = **n** –2 – 4 = **o** –5 – 1 = **p** 3 – 4 =
q 2 – 7 = **r** 1 – 5 = **s** –3 + 7 = **t** 5 – 6 =
u –2 – 3 = **v** 2 – 6 = **w** –8 + 3 = **x** 4 – 9 =

2 Answer each of the following *without* the help of a thermometer scale.

a 5 – 9 = **b** 3 – 7 = **c** –2 – 8 = **d** –5 + 7 =
e –1 + 9 = **f** 4 – 9 = **g** –10 + 12 = **h** –15 + 20 =
i 23 – 30 = **j** 30 – 42 = **k** –12 + 25 = **l** –30 + 55 =
m –10 – 22 = **n** –13 – 17 = **o** 45 – 50 = **p** 17 – 25 =
q 18 – 30 = **r** –25 + 35 = **s** –23 – 13 = **t** 31 – 45 =
u –24 + 65 = **v** –19 + 31 = **w** 25 – 65 = **x** 199 – 300 =

3 Work out each of the following.

a 8 + 3 – 5 = **b** –2 + 3 – 6 = **c** –1 + 3 + 4 =
d –2 – 3 + 4 = **e** –1 + 1 – 2 = **f** –4 + 5 – 7 =
g –3 + 4 – 7 = **h** 1 + 3 – 6 = **i** 8 – 7 + 2 =
j –5 – 7 + 12 = **k** –4 + 5 – 8 = **l** –4 + 6 – 8 =
m 103 – 102 + 7 = **n** –1 + 4 – 2 = **o** –6 + 9 – 12 =
p –3 – 3 – 3 = **q** –3 + 4 – 6 = **r** –102 + 45 – 23 =
s 8 – 10 – 5 = **t** 9 – 12 + 2 = **u** 99 – 100 – 46 =

F

4 Use a calculator to check your answers to questions 1 to 3.

AU 5 At 5 am the temperature in London was –4 °C.

At 11 am the temperature was 3 °C.

a By how many degrees did the temperature rise?

b The temperature in Brighton was two degrees lower than in London at 5 am
What was the temperature is Brighton at 5 am?

PS 6 Here are five numbers.

4 7 8 2 5

a Use two of the numbers to make a calculation with an answer of –6.

b Use three of the numbers to make a calculation with an answer of –1.

c Use four of the numbers to make a calculation with an answer of –18.

d Use all five of the numbers to make a calculation with an answer of –12.

FM 7 A submarine is 1600 feet below sea level.

A radar system can detect submarines down to 900 feet below sea level.

To safely avoid detection, the submarine captain keeps the submarine 200 feet below the level of detection.

How many feet can the submarine climb to be safe from detection?

Adding and subtracting negative numbers

To *subtract a negative number* …

… treat the – – as a +

For example: 4 – (–2) = 4 + 2 = 6

To *add a negative number* …

… treat the + – as a –

For example: 3 + (–5) = 3 – 5 = –2

Using your calculator

Calculations involving negative numbers can be done by using the **(–)** key.

EXAMPLE 3

Work out –3 + 7.

Press **(–)** **3** **+** **7** **=**

The answer should be 4.

EXAMPLE 4

Work out −6 − (−2).

Press (−) 6 − (−) 2 =

The answer should be −4.

EXERCISE 3E

F

1 Answer each of the following. Check your answers on a calculator.

a 2 − (−4) = **b** 4 − (−3) = **c** 3 − (−5) = **d** 5 − (−1) =

e 6 − (−2) = **f** 8 − (−2) = **g** −1 − (−3) = **h** −4 − (−1) =

i −2 − (−3) = **j** −5 − (−7) = **k** −3 − (−2) = **l** −8 − (−1) =

m 4 + (−2) = **n** 2 + (−5) = **o** 3 + (−2) = **p** 1 + (−6) =

q 5 + (−2) = **r** 4 + (−8) = **s** −2 + (−1) = **t** −6 + (−2) =

u −7 + (−3) = **v** −2 + (−7) = **w** −1 + (−3) = **x** −7 + (−2) =

2 Write down the answer to each of the following, then check your answers on a calculator.

a −3 − 5 = **b** −2 − 8 = **c** −5 − 6 = **d** 6 − 9 =

e 5 − 3 = **f** 3 − 8 = **g** −4 + 5 = **h** −3 + 7 =

i −2 + 9 = **j** −6 + −2 = **k** −1 + −4 = **l** −8 + −3 =

m 5 − −6 = **n** 3 − −3 = **o** 6 − −2 = **p** 3 − −5 =

q −5 − −3 = **r** −2 − −1 = **s** −4 − 5 = **t** 2 − 7 =

u −3 + 8 = **v** −4 + − 5 = **w** 1 − −7 = **x** −5 − −5 =

3 The temperature at midnight was 4 °C. Find the temperature if it *fell* by:

a 1 degree **b** 4 degrees **c** 7 degrees

d 9 degrees **e** 15 degrees

4 What is the *difference* between the following temperatures?

a 4 °C and −6 °C

b −2 °C and −9 °C

c −3 °C and 6 °C

5 Rewrite the following list, putting the numbers in order of size, lowest first.

1 −5 3 −6 −9 8 −1 2

F

6 Write down the answers to each of the following, then check your answers on a calculator.

a 2 – 5 = **b** 7 – 11 = **c** 4 – 6 =
d 8 – 15 = **e** 9 – 23 = **f** –2 – 4 =
g –5 – 7 = **h** –1 – 9 = **i** –4 + 8 =
j –9 + 5 = **k** 9 – –5 = **l** 8 – –3 =
m –8 – –4 = **n** –3 – –2 = **o** –7 + –3 =
p –9 + 4 = **q** –6 + 3 = **r** –1 + 6 =
s –9 – –5 = **t** 9 – 17 =

7 Find what you have to *add to* 5 to get:

a 7 **b** 2 **c** 0
d –2 **e** –5 **f** –15

8 Find what you have to *subtract from* 4 to get:

a 2 **b** 0 **c** 5
d 9 **e** 15 **f** –4

9 Find what you have to *add to* –5 to get:

a 8 **b** –3 **c** 0
d –1 **e** 6 **f** –7

10 Find what you have to *subtract from* –3 to get:

a 7 **b** 2 **c** –1
d –7 **e** –10 **f** 1

AU 11 Write down *ten* different addition sums that give the answer 1.

AU 12 Write down *ten* different subtraction calculations that give the answer 1. There must be *one negative number* in each calculation.

13 Use a calculator to work out each of these.

a –7 + – 3 – –5 = **b** 6 + 7 – 7 = **c** –3 + –4 – –7 =
d –1 – 3 – –6 = **e** 8 – –7 + –2 = **f** –5 – 7 – –12 =
g –4 + 5 – 7 = **h** –4 + –6 – –8 = **i** 103 – –102 – –7 =
j –1 + 4 – –2 = **k** 6 – –9 – 12 = **l** –3 – –3 – –3 =
m –45 + –56 – –34 = **n** –3 + 4 – –6 = **o** 102 + –45 – 32 =

F

14 Give the outputs of each of these function machines.

a −4, −3, −2, −1, 0 → [+ 3] → ?, ?, ?, ?, ?

b −4, −3, −2, −1, 0 → [− 2] → ?, ?, ?, ?, ?

c −4, −3, −2, −1, 0 → [+ 1] → ?, ?, ?, ?, ?

d −4, −3, −2, −1, 0 → [− 4] → ?, ?, ?, ?, ?

e −4, −3, −2, −1, 0 → [− 5] → ?, ?, ?, ?, ?

f −4, −3, −2, −1, 0 → [+ 7] → ?, ?, ?, ?, ?

g −10, −9, −8, −7, −6 → [− 2] → ?, ?, ?, ?, ?

h −10, −9, −8, −7, −6 → [− 6] → ?, ?, ?, ?, ?

i −5, −4, −3, −2, −1, 0 → [+ 3] → ?, ?, ?, ?, ?, ? → [− 2] → ?, ?, ?, ?, ?, ?

j −5, −4, −3, −2, −1, 0 → [− 7] → ?, ?, ?, ?, ?, ? → [− 2] → ?, ?, ?, ?, ?, ?

k −5, −4, −3, −2, −1, 0 → [+ 3] → ?, ?, ?, ?, ?, ? → [+ 2] → ?, ?, ?, ?, ?, ?

l −3, −2, −1, 0, 1, 2, 3 → [− 5] → ?, ?, ?, ?, ?, ? → [+ 3] → ?, ?, ?, ?, ?, ?

m −3, −2, −1, 0, 1, 2, 3 → [− 7] → ?, ?, ?, ?, ?, ? → [+ 9] → ?, ?, ?, ?, ?, ?

n −3, −2, −1, 0, 1, 2, 3 → [+ 6] → ?, ?, ?, ?, ?, ? → [− 8] → ?, ?, ?, ?, ?, ?

15 What numbers are missing from the boxes to make the number sentences true?

a $2 + -6 = \square$

b $4 + \square = 7$

c $-4 + \square = 0$

d $5 + \square = -1$

e $3 + 4 = \square$

f $\square - -5 = 7$

g $\square - 5 = 2$

h $6 + \square = 0$

i $\square - -5 = -2$

j $2 + -2 = \square$

k $\square - 2 = -2$

l $-2 + -4 = \square$

m $2 + 3 + \square = -2$

n $-2 + -3 + -4 = \square$

o $\square - 5 = -1$

p $\square - 8 = -8$

q $-4 + 2 + \square = 3$

r $-5 + 5 = \square$

s $7 - -3 = \square$

t $\square - -5 = 0$

u $3 - \square = 0$

v $-3 - \square = 0$

w $-6 + -3 = \square$

x $\square - 3 - -2 = -1$

y $\square - 1 = -4$

z $7 - \square = 10$

AU **16** You have the following cards.

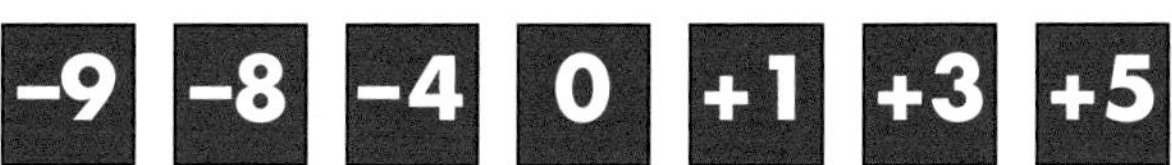

a Which card should you choose to make the answer to the following sum as large as possible? What is the answer?

+6 + ▇ = ...

b Which card should you choose to make the answer to part **a** as small as possible? What is the answer?

c Which card should you choose to make the answer to the following subtraction as large as possible? What is the answer?

+6 – ▇ = ...

d Which card should you choose to make the answer to part **c** as small as possible? What is the answer?

AU **17** You have the following cards.

-9 **-7** **-5** **-4** **0** **+1** **+2** **+4** **+7**

a Which cards should you choose to make the answer to the following calculation as large as possible? What is the answer?

+5 + ▇ – ▇ = ...

b Which cards should you choose to make the answer to part **a** as small as possible? What is the answer?

c Which cards should you choose to make the answer to the following number sentence zero? Give all possible answers.

▇ + ▇ = 0

FM AU **18** The thermometer in a car is inaccurate by up to two degrees.

An ice alert warning comes on at 3 °C, according to the thermometer temperature.

If the actual temperature is 2 °C, will the alert come on?

Explain how you decide.

PS **19** Two numbers have a sum of 5.

One of the numbers is negative.

The other number is even.

What are the two numbers if the even number is as large as possible?

F

E

Negative magic squares

Make your own magic square with negative numbers. You need nine small square cards and two pens or pencils of different colours.

This is perhaps the best known magic square.

8	3	4
1	5	9
6	7	2

But magic squares can be made from many different sets of numbers, as shown by this second square.

8	13	6
7	9	11
12	5	10

This square is now used to show you how to make a magic square with negative numbers. But the method works with any magic square. So, if you can find one of your own, use it!

- Arrange the nine cards in a square and write on them the numbers of the magic square. Picture **a** below.
- Rearrange the cards in order, lowest number first, to form another square. Picture **b** below.
- Keeping the cards in order, turn them over so that the blank side of each card is face up. Picture **c** below.

a

8	13	6
7	9	11
12	5	10

b

5	6	7
8	9	10
11	12	13

c

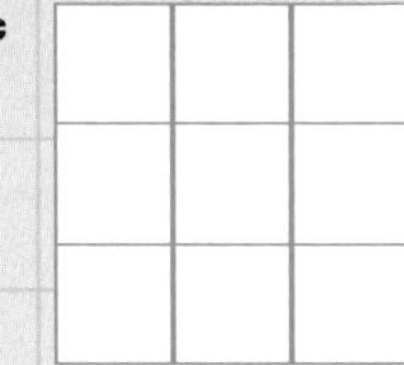

- Now use a different coloured pen to avoid confusion.
- Choose any number (say four) for the top left-hand corner of the square. Picture **d** below.
- Choose another number (say three) and subtract it from each number in the first row to get the next number. Picture **e** below.
- Now choose a third number (say two) and subtract it from each number in the top row to make the second row, and then again from each number in the second row. Picture **f** below.

d

4		

e

4	1	–2

f

4	1	–2
2	–1	–4
0	–3	–6

- Turn the cards over. Picture **g**.
- Rearrange the cards into the original magic square. Picture **h**.
- Turn them over again. Picture **i**.

g

5	6	7
8	9	10
11	12	13

h

8	13	6
7	9	11
12	5	10

i

2	–6	1
–2	–1	0
–3	4	–4

You should have a **magic square of negative numbers**.

Try it on any square. It works even with squares bigger than 3 × 3. Try it on this 4 × 4 square.

2	13	9	14
16	7	11	4
15	8	12	3
5	10	6	17

EXERCISE 3F

PS Copy and complete each of these magic squares. In each case, write down the 'magic number'.

1

–1		
–5	–4	–3
		–7

2

	–4	3
		–2
	4	–1

3

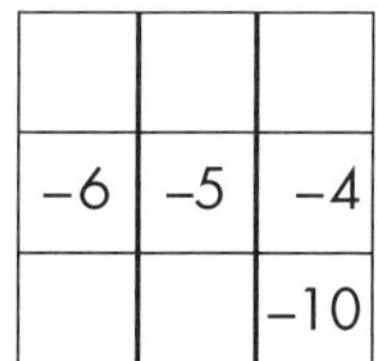

–6	–5	–4
		–10

4

2		
–4		
–7	6	–8

5

		–9
–3	–6	–9

6

		–1
	–7	
–13		–12

7

–4		
–8		–6
–9		

8

2	1	–3
	0	

9

–2		
	–5	
–7		–8

10

–8			–14
–8	–9		
		–4	–5
1	–10	–12	–5

11

–7		2	–16
	–8		
–11	–3	0	–2
		–13	–1

GRADE BOOSTER

- **G** You can use negative numbers in context
- **G** You can use < for less than and > for greater than
- **F** You can add positive and negative numbers to positive and negative numbers
- **F** You can subtract positive and negative numbers from positive and negative numbers
- **E** You can solve problems involving simple negative numbers

What you should know now

- How to order positive and negative numbers
- How to add and subtract positive and negative numbers
- How to use negative numbers in practical situations
- How to use a calculator when working with negative numbers

1 The temperature in a school yard was measured at 9am each morning for one week.

Day	Monday	Tuesday	Wednesday	Thursday	Friday
9am temperature	–1	–3	–2	1	2

a Which day was the coldest at 9am?

b Which day was the warmest at 9am?

2 The table shows the temperatures at midnight in 6 cities during one night in 2006.

City	Temperature
Berlin	5°C
London	10°C
Moscow	–3°C
New York	2°C
Oslo	–8°C
Paris	7°C

a Write down the city which had the lowest temperature. (1)

b Work out the difference in temperature between London and Moscow. (2)

(Total 3 marks)

Edexcel, June 2008, Paper 10 Foundation, Unit 3 Test, Question 2

3

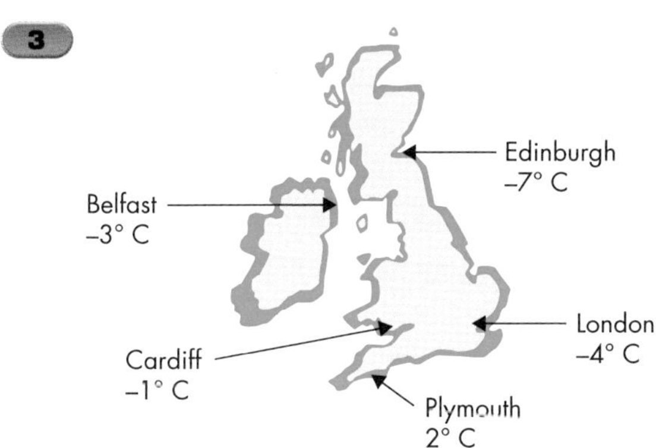

The map shows the temperature in 5 cities, at midnight, one night last year.

a Write down the name of the city with the highest temperature. (1)

b Write down the name of the city with the lowest temperature. (1)

(Total 2 marks)

Edexcel, June 2007, Paper 10 Foundation, Unit 3 Test, Question 3

4 This is a thermometer.

Write down the temperature shown. (1)

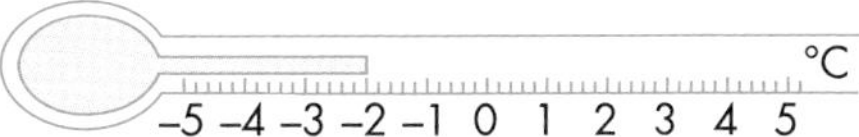

(Total 1 mark)

Edexcel, November 2007, Paper 10 Foundation, Unit 3 Test, Question 1(b)

5 Write out and complete the following to make the statement correct.

a 3 – ☐ = –5 **b** ☐ + 4 = –5

c 1 – ☐ = 8

6 The table shows the temperature on the surface of each of five planets.

Planet	Temperature
Venus	480 °C
Mars	–60 °C
Jupiter	–150 °C
Saturn	–180 °C
Uranus	–210 °C

a Work out the difference in temperature between Mars and Jupiter. (1)

b Work out the difference in temperature between Venus and Mars. (1)

c Which planet has a temperature 30 °C higher than the temperature on Saturn? (1)

The temperature on Pluto is 20 °C lower than the temperature on Uranus.

d Work out the temperature on Pluto. (1)

(Total 4 marks)

Edexcel, June 2005, Paper 2 Foundation, Question 8

7 The temperatures of the first few days of January were recorded as

1 °C, –1 °C, 0 °C and –2 °C

a Write down the four temperatures, in order, with the lowest first.

b What is the difference between the coldest and the warmest of these four days?

8 You have the following cards.

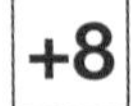

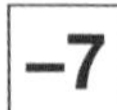

a i What card should you choose to make the answer to the following sum as large as possible?

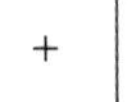
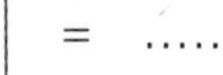

= ……

ii What is the answer to the sum in **i**?

iii What card would you have chosen to make the sum as small as possible?

b i What card should you choose to make the answer to the following subtraction as large as possible?

ii What is the answer to the subtraction sum in **i**?

iii What card would you have chosen to make the subtraction as small as possible?

9 a Work out –2 + 5 (1)

b Work out –3 – 5 (1)

(Total 2 marks)

Edexcel, May 2008, Paper 12 Foundation, Question 3

10 The table shows the temperatures in four cities at noon one day.

Oslo	–13°C
New York	–5°C
Cape Town	9°C
London	2°C

a Write down the **highest** temperature. (1)

b Work out the difference in temperature between Oslo and New York. (1)

At 8 pm the temperature in London was 3°C lower than the temperature at noon.

c Work out the temperature in London at 8 pm. (1)

(Total 3 marks)

Edexcel, June 2008, Paper 13 Foundation, Question 2

11 Nitrogen gas makes up most of the air we breathe.

Nitrogen freezes under –210 °C and is a gas above –196 °C. In between it is liquid.

Write down a possible temperature where nitrogen is

i a liquid **ii** a gas **iii** frozen.

12 The most common rocket fuel is liquid hydrogen and liquid oxygen. These two gases are kept in storage, as a liquid, at the following temperatures:

Liquid hydrogen –253 °C

Liquid oxygen –183 °C

a i Which of the two gases is kept at the coldest temperature?

ii What is the difference between the two storage temperatures?

b Scientists are experimenting with liquid methane as its liquid storage temperature is only –162 °C. How much warmer is the stored liquid methane than the

i stored liquid oxygen?

ii stored liquid hydrogen?

Worked Examination Questions

AU **1** The number $2\frac{1}{2}$ is halfway between 1 and 4.

What number is halfway between:

a −8 and −1 **b** $\frac{1}{4}$ and $1\frac{1}{4}$?

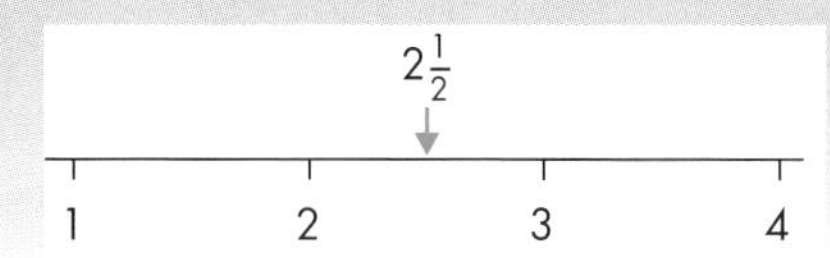

a The number halfway between −8 and −1 is $-4\frac{1}{2}$.

1 mark

The hint in the question is to sketch the numbers on a number line.

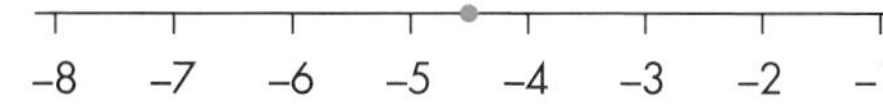

Just by counting from each end you can find the middle value. You get 1 mark for the correct answer.

b The number halfway between $\frac{1}{4}$ and $1\frac{1}{4}$ is $\frac{3}{4}$.

1 mark

Sketch the number line and mark on the quarters.

$\frac{1}{4}$ $\frac{1}{2}$ $\frac{3}{4}$ 1 $1\frac{1}{4}$

Count from each end to identify the middle value to get 1 mark.

Total: 2 marks

2 The table shows the temperatures in four cities.

a Write down the lowest temperature.

b Work out the difference in temperature between Birmingham and London.

c Three hours later the temperature in Manchester had risen by two degrees. Work out the new temperature in Manchester.

Birmingham	5 °C
London	−1 °C
Manchester	−3 °C
Exeter	1 °C

a Lowest temperature is −3 °C in Manchester

1 mark

From lowest to highest the temperatures are −3 °C, −1 °C, 1 °C, 5 °C.
You get 1 mark for the correct answer.

b 6 degrees

1 mark

Find the difference between temperatures in Birmingham (5°C) and London (−1 °C).
You get 1 mark for independent workings.

c −1 °C

1 mark

You get 1 mark for independent workings for the correct calculation −3 °C + 2 °C.

Total: 3 marks

FM **3** Maisie has £103.48 in her bank account.

On 11th December 2009 she takes £20 out of a cash machine and uses her debit card to pay for petrol that costs £25.60.

	Transaction	Paid In	Paid Out	Balance
11/12/09	Carried forward			£103.48
11/12/09	Cash ATM		£20	
11/12/09	Sovereign garage		£25.60	
11/12/09	Hair n' Nails	£152.70		
11/12/09	Closing Balance			

On the same day her weekly wages of £152.70 are paid in.

Fill in the statement to work out how much she has at the end of the day.

£83.48, £77.48, £230.58

Subtract any values in the 'Paid out' column and add any values in the 'Paid in' column. You get 1 mark for each correct answer.

Total: 3 marks

Problem Solving
The negative numbers game

In some sports you are required to keep a score card as you play and each player is responsible for filling in their card. In some sports these cards are handed in at the end of the game for checking.

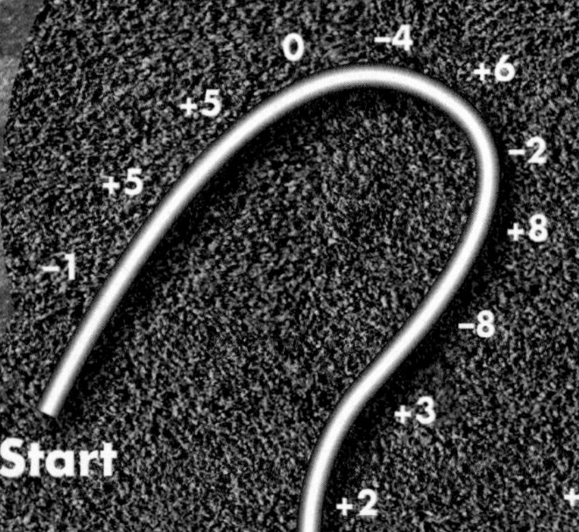

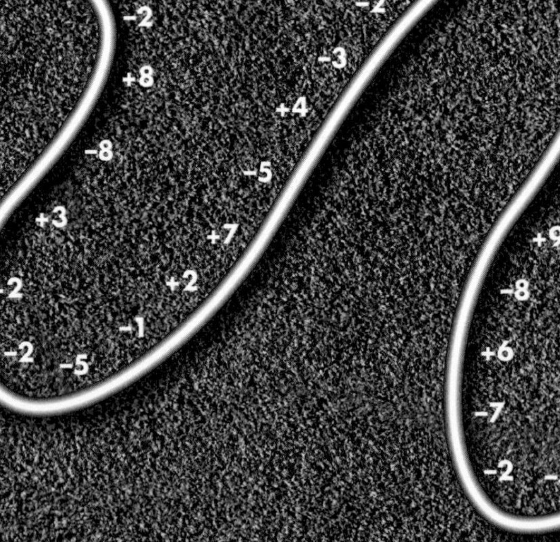

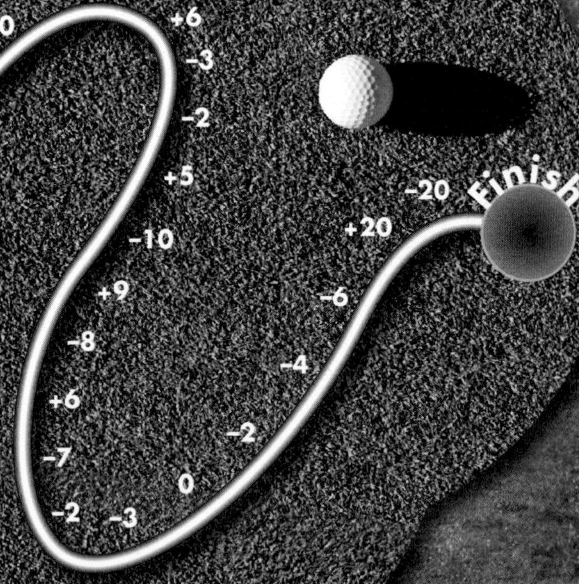

Your task

You are going to fill in a score card of your own but instead of positive numbers, your card will involve negative numbers.

How to play

- In groups of three or four, discuss the rules given below and check that you all understand how to play the game.
- Together, decide how many errors you can make before you are disqualified.
- Once you are sure of the rules, each player takes a score card and throws the dice. The person who rolls the highest number goes first.
- After all the players have reached the finish line, compare final scores.
- You should consider:
  - who has won
  - what the most popular scores in the game were
  - if there were some numbers that were better to roll than others.

Resources required

Coloured counters
A dice
Score cards

The rules

- Each player starts with 10 points.
- Each player must keep a record of their total but they must not reveal it to their opponents until the end of the game.
- In turn, players throw the dice and move their counter forward that number of spaces.
- Each time they land on a number they have to add it to their score.
- At the finish the player with the highest score wins.
- Other players have to check the winner's score card. Any errors should be corrected but too many errors can lead to a player being disqualified!

Score card

Number landed on	Running total
Start	+10

Why this chapter matters

A pattern is an arrangement of repeated uniform parts. You see patterns every day in clothes, art and home furnishings. Patterns can also occur in numbers.

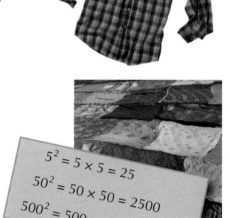

There are many mathematical problems that can be solved using patterns in numbers. Some numbers have fascinating features.

Here is a pattern.

$3 + 5 = 8$ (5 miles $\approx$ 8 km)

$5 + 8 = 13$ (8 miles $\approx$ 13 km)

$8 + 13 = 21$ (13 miles $\approx$ 21 km)

Approximately how many kilometres are there in 21 miles?

Note: $\approx$ means 'approximately equal to'.

In the boxes are some more patterns. Can you work out the next line of each pattern?

Now look at these numbers and see why they are special.

$4096 = (4 + 0^9)^6$

$81 = (8 + 1)^2$

$10 \times 10 = 100$

$10 \times 10 \times 10 = 1000$

$10 \times 10 \times 10 \times 10 = 10\,000$

Some patterns have special names.
Can you pair up these patterns and the names?

4, 8, 12, 16, …	Prime numbers
1, 4, 9, 16, …	Multiples (of 4)
2, 3, 5, 7, …	Cube numbers
1, 8, 27, 64, …	Square numbers

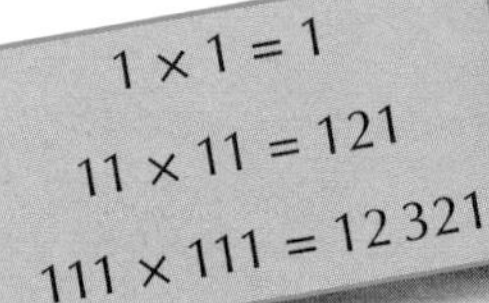

You will look at these in more detail in this chapter.

Below are four sets of numbers. Think about which number links together all the other numbers in each set. (The mathematics that you cover in 4.2 'Factors of whole numbers' will help you to work this out!)

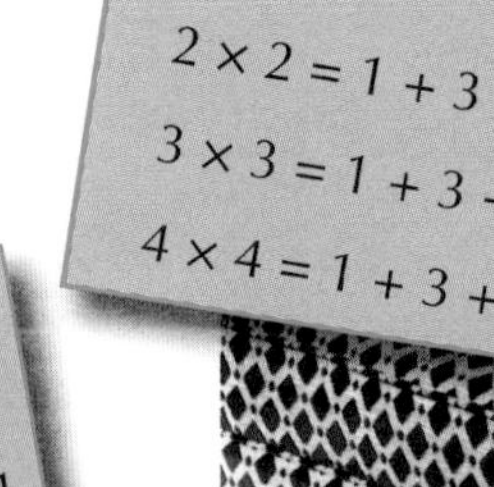

10, 5, 2, 1

18, 9, 6, 3, 2, 1

25, 5, 1

32, 16, 8, 4, 2, 1

$1089 \times 9 = 9801$

$10\,989 \times 9 = 98\,901$

$109\,989 \times 9 = 989\,901$

Chapter 4 Number: Number properties 1

The grades given in this chapter are target grades.

1. Multiples of whole numbers
2. Factors of whole numbers
3. Prime numbers
4. Square numbers
5. Square roots
6. Powers

This chapter will show you ...

- **G** the meaning of multiples
- **G** the meaning of factors
- **F** how to work out squares and square roots
- **E** the meaning of prime numbers
- **E** how to work out powers

Visual overview

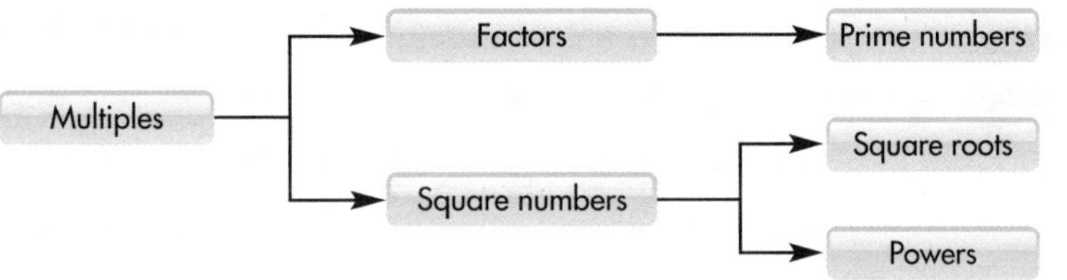

What you should already know

- Multiplication tables up to 10 × 10 **(KS3 level 4, GCSE grade G)**

Quick check

Write down the answers to the following.

1	**a** 2 × 3	**b** 4 × 3	**c** 5 × 3
	d 6 × 3	**e** 7 × 3	**f** 8 × 3
2	**a** 2 × 4	**b** 4 × 4	**c** 5 × 4
	d 6 × 4	**e** 7 × 4	**f** 8 × 4
3	**a** 2 × 5	**b** 9 × 5	**c** 5 × 5
	d 6 × 5	**e** 7 × 5	**f** 8 × 5
4	**a** 2 × 6	**b** 9 × 6	**c** 8 × 8
	d 6 × 6	**e** 7 × 9	**f** 8 × 6
5	**a** 2 × 7	**b** 9 × 7	**c** 8 × 9
	d 6 × 7	**e** 7 × 7	**f** 8 × 7

4.1 Multiples of whole numbers

This section will show you how to:
- find multiples of whole numbers
- recognise multiples of numbers

Key words
multiple
multiplication table
times table

When you multiply any whole number by another whole number, the answer is called a **multiple** of either of those numbers.

For example, $5 \times 7 = 35$, which means that 35 is a multiple of 5 and it is also a multiple of 7. Here are some other multiples of 5 and 7:

multiples of 5 are 5 10 15 20 25 30 35 …

multiples of 7 are 7 14 21 28 35 42 …

Multiples are the answers that appear in **multiplication tables**. These are also called **times tables**.

×	2	3	4	5	6
2	4	6	8	10	12
3	6	9	12	15	18
4	8	12	16	20	24
5	10	15	20	25	30
6	12	18	24	30	36

For example, 12 is a multiple of 2, 3, 4 and 6 as well as a multiple of 1 and 12.

Recognising multiples

You can recognise the multiples of 2, 3, 4, 5, 6 and 9 in the following ways.

- Multiples of 2 always end in an even number or 0. For example:

  12 34 96 1938 370

- Multiples of 3 are always made up of digits that add up to a multiple of 3. For example:

  15 because $1 + 5 = 6$ which is $2 \times \mathbf{3}$

  72 because $7 + 2 = 9$ which is $3 \times \mathbf{3}$

  201 because $2 + 0 + 1 = 3$ which is $1 \times \mathbf{3}$

- Multiples of 4 always give an even number when divided by 2. For example:

  64 because $64 \div 2 = 32$ which is even

  212 because $212 \div 2 = 106$ which is even

  500 because $500 \div 2 = 250$ which is even

Another check is to look at the last two digits of the number. If they give a multiple of 4, the whole numbers is a multiple of 4.

- Multiples of 5 always end in 5 or 0. For example:

  35 60 155 300

- Multiples of 6 are multiples of 2 and multiples of 3. For example:

  42 because it is even and $4 + 2 = 6$ which is $2 \times \mathbf{3}$

  54 because it is even and $5 + 4 = 9$ which is $3 \times \mathbf{3}$

- Multiples of 9 are always made up of digits that add up to a multiple of 9. For example,

  63 because $6 + 3 = 9$ which is $1 \times \mathbf{9}$

  738 because $7 + 3 + 8 = 18$ which is $2 \times \mathbf{9}$

You can find out whether numbers are multiples of 7 and 8 by using your calculator. For example, to find out whether 341 is a multiple of 7, you have to see whether 341 gives a whole number when it is divided by 7. You therefore key

The answer is 48.714 286, which is a decimal number, not a whole number. So, 341 is *not* a multiple of 7.

ACTIVITY

Grid locked

You need eight copies of this 10 × 10 grid.

1	2	3	4	5	6	7	8	9	10
11	12	13	14	15	16	17	18	19	20
21	22	23	24	25	26	27	28	29	30
31	32	33	34	35	36	37	38	39	40
41	42	43	44	45	46	47	48	49	50
51	52	53	54	55	56	57	58	59	60
61	62	63	64	65	66	67	68	69	70
71	72	73	74	75	76	77	78	79	80
81	82	83	84	85	86	87	88	89	90
91	92	93	94	95	96	97	98	99	100

Take one of the grids and shade in all the multiples of 2. You should find that they make a neat pattern.

Do the same thing for the multiples of 3, 4, … up to 9, using a fresh 10 × 10 grid for each number.

Next, draw a grid which is 9 squares wide and write the numbers from 1 to 100 in the grid, like this:

1	2	3	4	5	6	7	8	9
10	11	12	13	14	15	16	17	18
		21						27

Make seven more copies of this grid. Then shade in the multiples of 2, 3, … up to 9, using a fresh grid for each number.

Write out the numbers from 1 to 100 on grids of different widths and shade in the multiples of 2, 3, … up to 9, as before.

Describe the patterns that you get.

EXERCISE 4A

G

1 Write out the first five multiples of:

a 3 **b** 7 **c** 9 **d** 11 **e** 16

Remember: the first multiple is the number itself.

2 From the numbers below, write down those that are:

a multiples of 2 **b** multiples of 3

c multiples of 5 **d** multiples of 9.

111	254	255	108	73
68	162	711	615	98
37	812	102	75	270

HINTS AND TIPS

Remember the rules on pages 98–99.

3 Use your calculator to see which of the numbers below are:

a multiples of 4 **b** multiples of 7 **c** multiples of 6.

72	135	102	161	197
132	78	91	216	514
312	168	75	144	294

HINTS AND TIPS

There is no point testing odd numbers for multiples of even numbers such as 4 and 6.

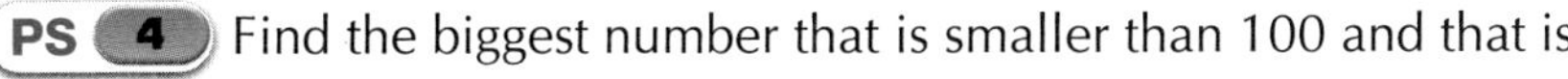

PS 4 Find the biggest number that is smaller than 100 and that is:

a a multiple of 2
b a multiple of 3
c a multiple of 4
d a multiple of 5
e a multiple of 7
f a multiple of 6.

PS 5 Find the smallest number that is bigger than 1000 and that is:

a a multiple of 6
b a multiple of 8
c a multiple of 9.

FM AU 6 Vishal is packing eggs into boxes of six. He has 50 eggs.
Will all the boxes be full?
Give a reason for your answer.

FM 7 A party of 20 people are getting into taxis. Each taxi holds the same number of passengers. If all the taxis fill up, how many people could be in each taxi? Give two possible answers.

8 Here is a list of numbers.

6 8 12 15 18 28

a From the list, write down a multiple of 9.
b From the list, write down a multiple of 7.
c From the list, write down a multiple of both 3 and 5.

PS 9 Find the lowest even number that is a multiple of 11 and a multiple of 3.

PS 10 How many numbers between 1 and 100 inclusive are multiples of both 6 and 9?
List the numbers.

PS 11 The number 24 is in the following times tables: 1, 2, 3, 4, 6, 8, 12 and 24.

That is a total of eight times tables.

a In which times tables is the number 10?
b In which times tables is the number 30?
c In which times table is the number 100?
d Which number less than 100 is in the most times tables?

4.2 Factors of whole numbers

This section will show you how to:
- identify the factors of a number

Key words
factor
factor pair

A **factor** of a whole number is any whole number that divides into it exactly. So:

the factors of 20 are 1 2 4 5 10 20

the factors of 12 are 1 2 3 4 6 12

This is where it is important to know your multiplication tables.

Factor facts

Remember these facts.

- 1 is always a factor and so is the number itself.
- When you have found one factor, there is always another factor that goes with it – unless the factor is multiplied by itself to give the number. For example, look at the number 20:

  $1 \times 20 = 20$ so 1 and 20 are both factors of 20

  $2 \times 10 = 20$ so 2 and 10 are both factors of 20

  $4 \times 5 = 20$ so 4 and 5 are both factors of 20.

  These are called **factor pairs**.

You may need to use your calculator to find the factors of large numbers.

EXAMPLE 1

Find the factors of 32.

Look for the factor pairs of 32. These are:

$1 \times 32 = 32$ $2 \times 16 = 32$ $4 \times 8 = 32$

So, the factors of 32 are 1, 2, 4, 8, 16, 32.

EXAMPLE 2

Find the factors of 36.

Look for the factor pairs of 36. These are:

$1 \times 36 = 36$ $2 \times 18 = 36$ $3 \times 12 = 36$ $4 \times 9 = 36$ $6 \times 6 = 36$

6 is a repeated factor which is counted only once.

So, the factors of 36 are 1, 2, 3, 4, 6, 9, 12, 18, 36.

EXERCISE 4B

1 What are the factors of each of these numbers?

a 10 **b** 28 **c** 18 **d** 17 **e** 25

f 40 **g** 30 **h** 45 **i** 24 **j** 16

2 How many different ways can 24 chocolate bars be packed into boxes so that there are exactly the same number of bars in each box?

3 Use your calculator to find the factors of each of these numbers.

a 120 **b** 150 **c** 144 **d** 180

e 169 **f** 108 **g** 196 **h** 153

i 198 **j** 199

HINTS AND TIPS

Remember that once you find one factor this will give you another, unless it is a repeated factor such as 5 × 5.

4 What is the biggest factor that is less than 100 for each of these numbers?

a 110 **b** 201 **c** 145 **d** 117

e 130 **f** 240 **g** 160 **h** 210

i 162 **j** 250

5 Find the largest common factor for each of the following pairs of numbers. (Do not include 1.)

a 2 and 4 **b** 6 and 10 **c** 9 and 12

d 15 and 25 **e** 9 and 15 **f** 12 and 21

g 14 and 21 **h** 25 and 30 **i** 30 and 50

j 55 and 77

HINTS AND TIPS

Look for the largest number that has both numbers in its multiplication table.

FM 6 A designer is making a box for 12 Christmas decorations.

The decorations are to be packed in rows.

How many decorations must be put in each row to make the top of the box a flat, square shape?

AU 7 Here are five numbers.

18 21 27 32 36

Use factors to explain why 32 could be the odd one out.

PS 8 Find the highest odd number that is a factor of 40 and a factor of 60.

G

F

4.3 Prime numbers

This section will show you how to:
- identify prime numbers

Key word
prime number

What are the factors of 2, 3, 5, 7, 11 and 13?

Notice that each of these numbers has only two factors: itself and 1. They are all examples of **prime numbers**.

So, a prime number is a whole number that has only two factors: itself and 1.

Note: 1 is *not* a prime number, since it has only one factor – itself.

The prime numbers up to 50 are:

2, 3, 5, 7, 11, 13, 17, 19, 23, 29, 31, 37, 41, 43, 47

You will need to know these for your GCSE examination.

ACTIVITY

Prime search

You need a 10 × 10 grid.

Cross out 1.

Leave 2 and cross out the rest of the multiples of 2.

Leave 3 and cross out the rest of the multiples of 3. Some of them will already have been crossed out.

Leave 5 and cross out the rest of the multiples of 5. Some of them will already have been crossed out.

Leave 7 and cross out the rest of the multiples of 7. All but three of them will already have been crossed out.

The numbers left are prime numbers.

1	2	3	4	5	6	7	8	9	10
11	12	13	14	15	16	17	18	19	20
21	22	23	24	25	26	27	28	29	30
31	32	33	34	35	36	37	38	39	40
41	42	43	44	45	46	47	48	49	50
51	52	53	54	55	56	57	58	59	60
61	62	63	64	65	66	67	68	69	70
71	72	73	74	75	76	77	78	79	80
81	82	83	84	85	86	87	88	89	90
91	92	93	94	95	96	97	98	99	100

The activity is known as the Sieve of Eratosthenes. (Eratosthenes, a Greek scholar, lived from about 275BC to 194BC.)

EXERCISE 4C

1. Write down the prime numbers between 20 and 30.

2. Write down the only prime number between 90 and 100.

AU 3. Using the rules for recognising multiples, decide which of these numbers are **not** prime numbers.

   462 108 848 365 711

PS 4. When three different prime numbers are multiplied together the answer is 105.

   What are the three prime numbers?

FM 5. A shopkeeper has 31 identical soap bars.

   He is trying to arrange the bars on a shelf in rows, each with the same number of bars.

   Is it possible?

   Explain your answer.

4.4 Square numbers

This section will show you how to:
- identify square numbers
- use a calculator to find the square of a number

Key words
square
square numbers

What is the next number in this sequence?

1, 4, 9, 16, 25, …

Write each number as:

$1 \times 1, 2 \times 2, 3 \times 3, 4 \times 4, 5 \times 5, \ldots$

These factors can be represented by **square** patterns of dots:

1 × 1   2 × 2   3 × 3   4 × 4   5 × 5

From these patterns, you can see that the next pair of factors must be 6 × 6 = 36, therefore 36 is the next number in the sequence.

Because they form square patterns, the numbers 1, 4, 9, 16, 25, 36, ... are called **square numbers**.

When you multiply any number by itself, the answer is called the *square of the number* or the *number squared*. This is because the answer is a square number. For example:

the square of 5 (or 5 squared) is 5 × 5 = 25

the square of 6 (or 6 squared) is 6 × 6 = 36

There is a short way to write the square of any number. For example:

5 squared (5 × 5) can be written as 5^2

13 squared (13 × 13) can be written as 13^2

So, the sequence of square numbers, 1, 4, 9, 16, 25, 36, ..., can be written as:

$1^2,\ 2^2,\ 3^2,\ 4^2,\ 5^2,\ 6^2,\ \ldots$

You are expected to know the square numbers up to 15 × 15 (= 225) for the GCSE examination.

EXERCISE 4D

G

1 The square number pattern starts:

1 4 9 16 25 ...

Copy and continue the pattern above until you have written down the first 20 square numbers. You may use your calculator for this.

PS **2** Work out the answer to each of these number sentences.

1 + 3 =

1 + 3 + 5 =

1 + 3 + 5 + 7 =

Look carefully at the pattern of the three number sentences. Then write down the next three number sentences in the pattern and work them out.

AU 3 Draw one counter. 

Now add more counters to your picture to make the next square number.

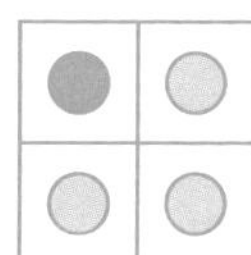

a How many extra counters did you add?

Now add more counters to your picture to make the next square number.

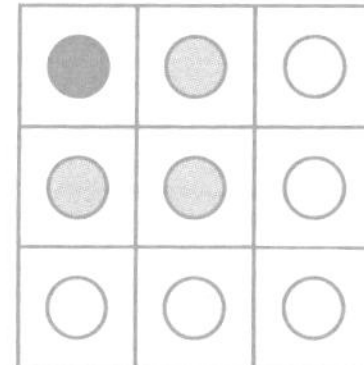

b How many extra counters did you add?

c Without drawing, how many more counters will you need to make the next square number?

d Describe the pattern of counters you are adding.

4 Write down the answer to each of the following. You will need to use your calculator. Look for the **x^2** key.

a 23^2 **b** 57^2 **c** 77^2 **d** 123^2 **e** 152^2

f 3.2^2 **g** 9.5^2 **h** 23.8^2 **i** $(-4)^2$ **j** $(-12)^2$

5 **a** Write down the value of 13^2.

b Write down the value of 14^2.

c Estimate the value of 13.2^2.

6 Find the next three numbers in each of these number patterns. (They are all based on square numbers.) You may use your calculator.

	1	**4**	**9**	**16**	**25**	**36**	**49**	**64**	**81**
a	2	5	10	17	26	37	…	…	…
b	2	8	18	32	50	72	…	…	…
c	3	6	11	18	27	38	…	…	…
d	0	3	8	15	24	35	…	…	…
e	101	104	109	116	125	136	…	…	…

HINTS AND TIPS

Look for the connection with the square numbers on the top line.

E

7 **a** Work out each of the following. You may use your calculator.

$3^2 + 4^2$	and	5^2
$5^2 + 12^2$	and	13^2
$7^2 + 24^2$	and	25^2
$9^2 + 40^2$	and	41^2
$11^2 + 60^2$	and	61^2

b Describe what you notice about your answers to part **a**.

8 Jasper's bill for his internet is £12 each month for 12 months.

How much does he pay for the whole year?

FM 9 A builder is using flagstones to lay a patio. He thinks that he will need 15 rows, each with 15 flagstones in them. However, he did not allow for the gaps between the flagstones and finds that he only needs 14 rows of each with 14 flagstones in them.

How many flagstones does he have left?

F

PS 10 4 and 81 are square numbers with a sum of 85.

Find two different square numbers with a sum of 85.

The following exercise will give you some practice on multiples, factors, square numbers and prime numbers.

EXERCISE 4E

G

1 Write out the first five multiples of each number.

a 6 **b** 13 **c** 8 **d** 20 **e** 18

Remember: the first multiple is the number itself.

2 Write down the square numbers up to 100.

3 What are the factors of these numbers?

a 12 **b** 20 **c** 9 **d** 32 **e** 24

f 38 **g** 13 **h** 42 **i** 45 **j** 36

F

4 Write out the first three numbers that are multiples of both of the numbers shown.

a 3 and 4 **b** 4 and 5 **c** 3 and 5 **d** 6 and 9 **e** 5 and 7

E

AU 5 In question 3, every number had an even number of factors except parts **c** and **j**. What sort of numbers are 9 and 36?

AU 6 In question 3, part **g**, there were only two factors. Why?

7 Write down the prime numbers up to 20.

AU 8 Copy these number sentences and write out the *next four* sentences in the pattern.

$1 = 1$

$1 + 3 = 4$

$1 + 3 + 5 = 9$

$1 + 3 + 5 + 7 = 16$

AU 9 Here are four numbers.

14 16 35 49

Copy and complete the table by putting each of the numbers in the correct box.

	Square number	Factor of 70
Even number		
Multiple of 7		

AU 10 Arrange these four number cards to make a square number.

1 4 6 7

D

PS 11 If hot-dog sausages are sold in packs of 10 and hot-dog buns are sold in packs of eight, how many of each must you buy to have complete hot dogs with no extra sausages or buns?

PS 12 Rover barks every 8 seconds and Spot barks every 12 seconds. If both dogs bark together, how many seconds will it be before they both bark together again?

PS 13 A bell chimes every 6 seconds. Another bell chimes every 5 seconds. If they both chime together, how many seconds will it be before they both chime together again?

PS 14 Fred runs round a running track in 4 minutes. Debbie runs round in 3 minutes.

If they both start together on the line at the end of the finishing straight, when will they both be on the same line together again?

How many laps will Debbie have run? How many laps will Fred have run?

D

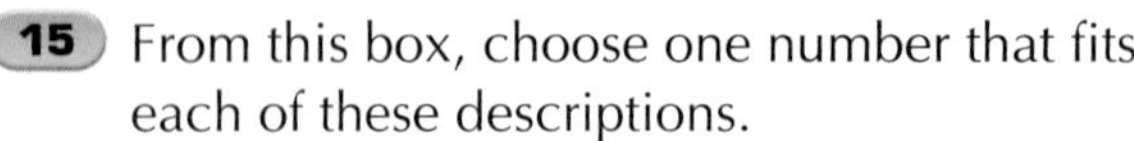

15 From this box, choose one number that fits each of these descriptions.

12 13 21 8 15 17 9 18 10 6 14 16

- **a** a multiple of 3 and a multiple of 4
- **b** a square number and an odd number
- **c** a factor of 24 and a factor of 18
- **d** a prime number and a factor of 39
- **e** an odd factor of 30 and a multiple of 3
- **f** a number with 4 factors and a multiple of 2 and 7
- **g** a number with 5 factors exactly
- **h** a multiple of 5 and a factor of 20
- **i** an even number and a factor of 36 and a multiple of 9
- **j** a prime number that is one more than a square number
- **k** written in order, the four factors of this number make a number pattern in which each number is twice the one before
- **l** an odd number that is a multiple of 7

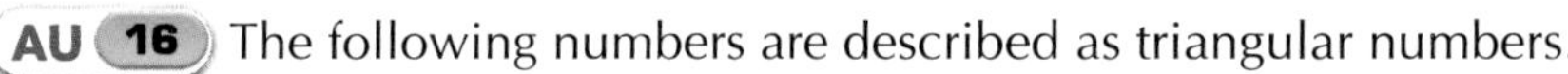

AU 16 The following numbers are described as triangular numbers.

1, 3, 6, 10, 15

- **a** Investigate why they are called triangular numbers.
- **b** Write down the next five triangular numbers.

4.5 Square roots

This section will show you how to:
- find a square root of a square number
- use a calculator to find the square roots of any number

Key word
square root

The **square root** of a given number is a number that, when multiplied by itself, produces the given number.

For example, the square root of 9 is 3, since $3 \times 3 = 9$.

Numbers also have a negative square root, since -3×-3 also equals 9.

A square root is represented by the symbol $\sqrt{\ }$. For example, $\sqrt{16} = 4$.

EXERCISE 4F

F

1 Write down the positive square root of each of these numbers.

a 4 **b** 25 **c** 49 **d** 1 **e** 81
f 100 **g** 64 **h** 9 **i** 36 **j** 16
k 121 **l** 144 **m** 400 **n** 900 **o** 169

2 Write down both possible values of each of these square roots.

a $\sqrt{25}$ **b** $\sqrt{36}$ **c** $\sqrt{100}$ **d** $\sqrt{49}$ **e** $\sqrt{64}$
f $\sqrt{16}$ **g** $\sqrt{9}$ **h** $\sqrt{81}$ **i** $\sqrt{1}$ **j** $\sqrt{144}$

3 Write down the value of each of these. You need only give positive square roots. You will need to use your calculator for some of them. Look for the $\sqrt{\square}$ key.

a 9^2 **b** $\sqrt{1600}$ **c** 10^2 **d** $\sqrt{196}$ **e** 6^2
f $\sqrt{225}$ **g** 7^2 **h** $\sqrt{144}$ **i** 5^2 **j** $\sqrt{441}$
k 11^2 **l** $\sqrt{256}$ **m** 8^2 **n** $\sqrt{289}$ **o** 21^2

4 Write down the positive value of each of the following. You will need to use your calculator.

a $\sqrt{576}$ **b** $\sqrt{961}$ **c** $\sqrt{2025}$ **d** $\sqrt{1600}$ **e** $\sqrt{4489}$
f $\sqrt{10\,201}$ **g** $\sqrt{12.96}$ **h** $\sqrt{42.25}$ **i** $\sqrt{193.21}$ **j** $\sqrt{492.84}$

AU 5 Put these in order, starting with the smallest value.

3^2 $\sqrt{90}$ $\sqrt{50}$ 4^2

AU 6 Between which two consecutive whole numbers does the square root of 20 lie?

E

PS 7 Use these number cards to make this calculation correct.

3

8

FM 8 A square wall in a kitchen is being tiled.
Altogether it needs 225 square tiles.
How many tiles are there in each row?

4.6 Powers

This section will show you how to:
- use powers

Key words
cube
indices
power
square

Powers are a convenient way of writing repetitive multiplications. (Powers are also called **indices** – singular, index.)

The power that you will use most often is 2, which has the special name **square**. The only other power with a special name is 3, which is called **cube**.

You are expected to know the cubes of numbers, $1^3 = 1$, $2^3 = 8$, $3^3 = 27$, $4^3 = 64$, $5^3 = 125$ and $10^3 = 1000$, for the GCSE examination.

EXAMPLE 3

a What is the value of:

i 7 squared **ii** 5 cubed?

b Write each of these numbers out in full.

i 4^6 **ii** 6^4 **iii** 7^3 **iv** 12^2

c Write the following multiplications as powers.

i $3 \times 3 \times 3 \times 3 \times 3 \times 3 \times 3 \times 3$

ii $13 \times 13 \times 13 \times 13 \times 13$

iii $7 \times 7 \times 7 \times 7$

iv $5 \times 5 \times 5 \times 5 \times 5 \times 5 \times 5$

a The value of 7 squared is $7^2 = 7 \times 7 = 49$

The value of 5 cubed is $5^3 = 5 \times 5 \times 5 = 125$

b **i** $4^6 = 4 \times 4 \times 4 \times 4 \times 4 \times 4$

ii $6^4 = 6 \times 6 \times 6 \times 6$

iii $7^3 = 7 \times 7 \times 7$

iv $12^2 = 12 \times 12$

c **i** $3 \times 3 \times 3 \times 3 \times 3 \times 3 \times 3 \times 3 = 3^8$

ii $13 \times 13 \times 13 \times 13 \times 13 = 13^5$

iii $7 \times 7 \times 7 \times 7 = 7^4$

iv $5 \times 5 \times 5 \times 5 \times 5 \times 5 \times 5 = 5^7$

Working out powers on your calculator

How would you work out the value of 5^7 on a calculator?

You could key it in as $5 \times 5 \times 5 \times 5 \times 5 \times 5 \times 5 =$. But as you key it in, you may miss a number or press a wrong key. If your calculator has one, you could use the power key, $x^{\blacksquare}$.

Insert the number 5, then press $x^{\blacksquare}$ and insert the number 7. The answer displayed is 78125, so

$5^7 = 78\,125$

Make sure you know where to find the power key on your calculator. It may be an INV or SHIFT function.

Try using your calculator to work out 3^4, 7^8, 23^4 and 72^3.

Check that you get 81, 5 764 801, 279 841 and 373 248.

EXERCISE 4G

1 Use your calculator to work out the value of each of the following.

a 3^3 **b** 5^3 **c** 6^3 **d** 12^3 **e** 2^4

f 4^4 **g** 5^4 **h** 2^5 **i** 3^7 **j** 2^{10}

2 Work out the values of the following powers of 10.

a 10^2 **b** 10^3 **c** 10^4 **d** 10^5 **e** 10^6

f Describe what you notice about your answers.

g Now write down the value of each of these.

i 10^8 **ii** 10^{10} **iii** 10^{15}

HINTS AND TIPS

When working out a power, make sure you multiply the number by itself and not by the power. A very common error is to write, for example, $2^3 = 6$ instead of $2^3 = 2 \times 2 \times 2 = 8$.

3 Rewrite each of these, using index notation. You do not have to work them out yet.

a $2 \times 2 \times 2 \times 2$ **b** $3 \times 3 \times 3 \times 3 \times 3$

c 7×7 **d** $5 \times 5 \times 5$

e $10 \times 10 \times 10 \times 10 \times 10 \times 10 \times 10$ **f** $6 \times 6 \times 6 \times 6$

g $4 \times 4 \times 4 \times 4$ **h** $1 \times 1 \times 1 \times 1 \times 1 \times 1 \times 1$

i $0.5 \times 0.5 \times 0.5 \times 0.5$ **j** $100 \times 100 \times 100$

4 Write the following out in full. You do not have to work them out yet.

a 3^4 **b** 9^3 **c** 6^2 **d** 10^5 **e** 2^{10}

f 8^6 **g** 0.1^3 **h** 2.5^2 **i** 0.7^3 **j** 1000^2

E

E

5 Using the power key on your calculator (or another method), work out the values of the power terms in question 3.

6 Using the power key on your calculator (or another method), work out the values of the power terms in question 4.

PS 7 Write the answer to question 3, part **j** as a power of 10.

PS 8 Write the answer to question 4, part **j** as a power of 10.

HINTS AND TIPS

Use the answer you found for question 2f to help you.

9 Copy this pattern of powers of 2 and continue it for another five terms.

2^2 2^3 2^4 … … … … …

4 8 16 … … … … …

10 Copy the pattern of powers of 10 and fill in the previous five and the next five terms.

… … … … … 10^2 10^3 … … … … …

… … … … … 100 1000 … … … … …

FM 11 A storage container is a cube.
The length of the container is 5 m.
To work out the volume of a cube, use the formula

volume = (length of edge)^3

Work out the total storage space in the container.

D

AU 12 Write each number as a power of a different number.
The first one has been done for you.

a $32 = 2^5$

b 100

c 8

d 25

AU 13 Marie says that as $3^4 = 81$, 6^2 must also equal 81.

Explain why Marie is wrong.

C

AU 14 Find the value of the letters that makes the following true.

a $3^4 = x^2$

b $2^6 = y^2$

c $2^6 = z^3$

d $8^2 = w^3$

e $100^3 = s^2$

f 1 million $= r^2$

GRADE BOOSTER

G You can recognise multiples of the first ten whole numbers

G You can find factors of numbers less than 100

G You can recognise the square numbers up to 100

F You can write down any square number up to $15 \times 15 = 225$

F You can find the square root of any number, using a calculator

E You can write down the cubes of 1, 2, 3, 4, 5 and 10

E You can calculate simple powers of whole numbers

E You can recognise two-digit prime numbers

What you should know now

- What multiples are
- How to find the factors of any whole number
- What a prime number is
- What square numbers are
- What square roots are
- How to find powers of numbers

1 **a** Write down the largest multiple of 3 smaller than 100.

b Write down the smallest multiple of 6 larger than 100.

2 Look at the numbers in this cloud.

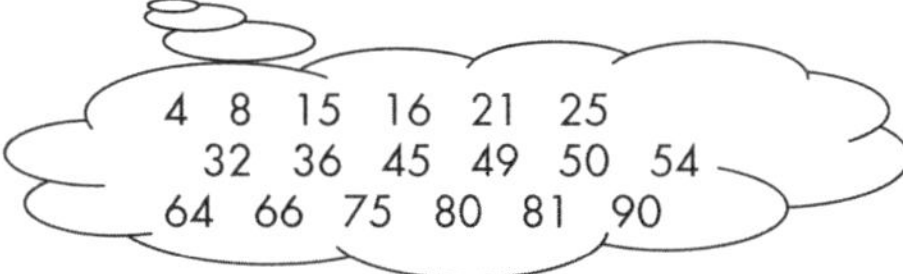

4 8 15 16 21 25
32 36 45 49 50 54
64 66 75 80 81 90

Write down all the square numbers that are inside the cloud.

3 **a** Write down the largest factor of 360, smaller than 100.

b Write down the smallest factor of 315 larger than 100.

4 **a** Copy and complete the missing numbers in the pattern below.

	Last digit
$4^1 = 4 = 4$	4
$4^2 = 4 \times 4 = 16$	6
$4^3 = 4 \times 4 \times 4 = 64$	
$4^4 = 4 \times 4 \times 4 \times 4 = 256$	
$4^5 =$ ☐ $=$ ☐	

b What will the last digit of 4^{17} be?

5 **a** Write down the first five multiples of 6.

b Write down the factors of 12.

c Write down a square number between 20 and 30.

d Write down two prime numbers between 20 and 30.

6 Here are six number cards.

a Which of the numbers are multiples of 4?

b Which of the numbers are factors of 10?

c Which of the numbers are prime numbers?

d Which numbers are square numbers?

e Which number is a cube number?

7 **a** Find the value of

3.7^2

The table shows some numbers.

51	52	53	54	55	56	57	58	59

Two of the numbers are prime numbers.

b Which two numbers are these?

8 Write down the value of

a 2^3

b $\sqrt{64}$

9 Find the value of

a 8^2 (1)

b $\sqrt{100}$ (1)

c the cube of 5 (1)

(Total 3 marks)

Edexcel, June 2008, Paper 10 Foundation, Unit 3 Test, Question 5

10 John set up two computer virus checkers on his computer on January 1st.

Checker A would check every 8 days.

Checker B would check every 10 days.

After how many days will both checkers be checking on the same day again?

11 Mary set up her Christmas Tree with two sets of twinkling lights.

Set A would twinkle every 3 seconds.

Set B would twinkle every 4 seconds.

How many times in a minute will both sets be twinkling at the same time?

12 Small pies are sold in packs of 4

Bread sticks are sold in packs of 10

What is the least number of each pack that needs to be bought to have the same number of pies and bread sticks?

Worked Examination Questions

1 A PE teacher is organising 20 students into equal-sized teams for a competition.
How many can she have in each team?

2, 4, 5 or 10

Total: 2 marks

By identifying at least three factors of 20 you will get 1 method mark. Although 1 and 20 are factors they cannot form part of the final answer as this is not addressing the team requirement. You will get 1 mark for accuracy for showing this in your final answer.

PS **2** Give a reason why each of the following could be the odd one out.

123 144 169

123 because it is not a square number

144 because it is not odd

169 because it is not a multiple of 3

Total: 3 marks

Answer the whole question by giving a reason for each number being the odd one out and not just identifying one of the numbers as the odd one out. Each answer is for any valid reason, so there could be other solutions. For each reason you will earn 1 mark.

Functional Maths
Creating your own allotment

Today, many people living in towns and cities do not have their own gardens. Allotments give these people the opportunity to enjoy gardening regardless of not having their own garden. Allotments are also increasingly popular as a way of producing home-grown cheap fruit and vegetables. You have just started renting an allotment so that can grow your own fruit and vegetables. The allotment is divided into rows for planting. The whole plot is 3 m wide and each row is 5 m long.

Getting started

Before you begin your main task, you may find it useful to fill in a copy of the following table using the fact box opposite to help you.

Vegetable	Distance between plants (cm)	Number of vegetables per row
Potatoes	30 cm	
Carrots	10 cm	
Broad beans	10 cm	
Onions	10 cm	
Cucumber	30 cm	
Lettuce	20 cm	

Your task

Design the plant layout for the allotment using the information that you will gather from the tables. Consider as many different arrangements as possible.

You must explain your choices, stating the assumptions that you have made.

Facts

This table shows how much distance is needed between plants and between rows for the key vegetables in your allotment.

Vegetable	Distance between plants (cm)	Distance between rows (cm)
Potatoes	30 cm	50 cm
Carrots	10 cm	20 cm
Broad beans	10 cm	30 cm
Onions	10 cm	20 cm
Cucumber	30 cm	60 cm
Lettuce	20 cm	30 cm

Here is an example for you to discuss.

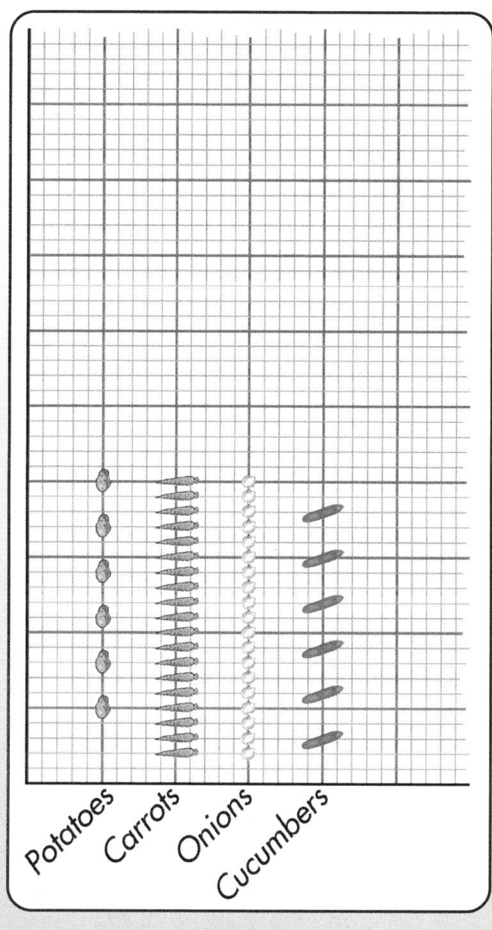

Handy hints

You may find these rules helpful when planning your allotment:

- use one type of vegetable for each complete row
- do not have more than two rows of the same vegetable
- if different plants are next to each other, make sure that the largest necessary distance between rows is left between them
- make sure that vegetables are not planted too near to the edge of the allotment
- use graph paper to represent the allotment
- use a code to represent each plant.

Why this chapter matters

For centuries statistical graphs such as bar charts, line graphs and pictograms have been used in many different areas, from science to politics. They provide a way to represent, analyse and interpret information.

Developing statistical analysis

The development of statistical graphs was spurred on by:

- the need in the 17th and 18th centuries to base policies on demographic and economic data and for this information to be shared with a large number of people
- greater understanding of measures and numbers in the 19th century
- increasing interest in analysing and understanding social conditions in the 19th and 20th centuries.

Since the later part of the 20th century, statistical graphs have become an important way of analysing information, and computer-generated statistical graphs are seen every day on TV, in newspapers and in magazines.

Who invented statistical diagrams?

William Playfair was a Scottish engineer, who is thought to be the founder of representing statistics in a graphical way.

He invented three types of diagrams: in 1786 the line graph and bar chart, then in 1801 the pie chart (which you will come across in later chapters).

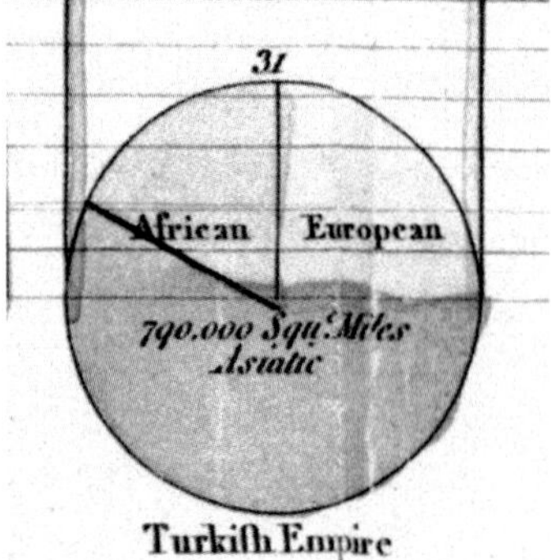

William Playfair pioneered charts and graphs such as this pie chart.

Florence Nightingale was born in 1820. She was very good at mathematics from an early age, becoming a pioneer in presenting information visually. She developed a form of the pie chart now known as the 'polar area diagram' or the 'Nightingale rose diagram', which was like a modern circular bar chart. It illustrated monthly patient deaths in military field hospitals. She called these diagrams 'coxcombs' and used them a great deal to present reports on the conditions of medical care in the Crimean War to Parliament and to civil servants who may not have fully understood traditional statistical reports.

In 1859, Florence Nightingale was elected the first female member of the Royal Statistical Society.

This chapter introduces you to some of the most common forms of statistical representation. They fall into two groups: graphical diagrams, such as pictograms and bar charts; and quantitive diagrams, such as frequency tables and stem-and-leaf diagrams.

Florence Nightingale was a nurse and hospital reformer. She used charts and graphs in her work.

Chapter 5

Statistics: Statistical representation

The grades given in this chapter are target grades.

This chapter will show you …

- **F** how to collect and organise data, and how to represent data on various types of diagram
- **F** to **D** how to draw diagrams for data, including line graphs for time series and frequency diagrams
- **G** to **D** how to draw conclusions from statistical diagrams
- **E** to **D** how to draw diagrams for discrete data, including stem-and-leaf diagrams

Visual overview

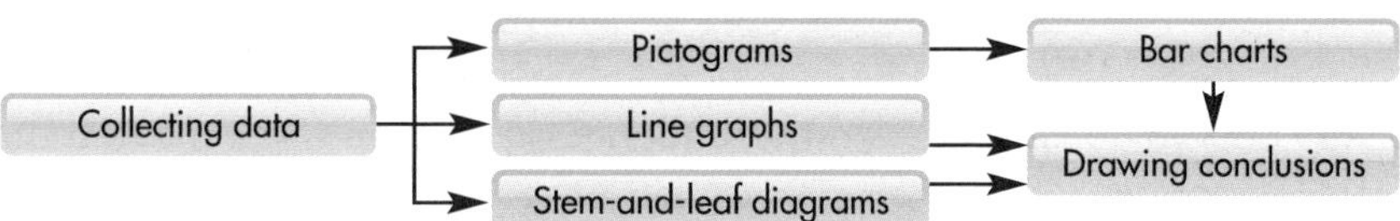

What you should already know

- How to use a tally for recording data **(KS3 level 3, GCSE grade G)**
- How to read information from charts and tables **(KS3 level 3, GCSE grade G)**

Quick check

Zoe works in a dress shop. She recorded the sizes of all the dresses sold during a week. The table shows the results.

Day	Size of dresses sold									
Monday	12	8	10	8	14	8	12	8	8	
Tuesday	10	10	8	12	14	16	8	12	14	16
Wednesday	16	8	12	10						
Thursday	12	8	8	10	12	14	16	12	8	
Friday	10	10	8	10	12	14	14	12	10	8
Saturday	10	8	8	12	10	12	8	10		

a Use a tallying method to make a table showing how many dresses of each size were sold in the week.

b Which dress size had the most sales?

5.1 Frequency diagrams

This section will show you how to:

- collect and represent discrete and grouped data using tally charts and frequency tables

Key words
class
class interval
data collection sheet
experiment
frequency
frequency table
grouped data
grouped frequency table
observation
sample
tally chart

Statistics is concerned with the collection and organisation of data, the representation of data on diagrams and the interpretation of data.

When you are collecting data for simple surveys, it is usual to use a **data collection sheet**, also called a **tally chart**. For example, data collection sheets are used to gather information on how people travel to work, how students spend their free time and the amount of time people spend watching TV.

It is easy to record the data by using tally marks, as shown in Example 1. Counting up the tally marks in each row of the chart gives the **frequency** of each category. By listing the frequencies in a column on the right-hand side of the chart, you can make a **frequency table** (see Example 1). Frequency tables are an important part of making statistical calculations, as you will see in Chapter 15.

EXAMPLE 1

Sandra wanted to find out about the ways in which students travelled to school. She carried out a survey. Her frequency table looked like this:

Method of travel	Tally	Frequency
Walk	~~IIII~~ ~~IIII~~ ~~IIII~~ ~~IIII~~ ~~IIII~~ III	28
Car	~~IIII~~ ~~IIII~~ II	12
Bus	~~IIII~~ ~~IIII~~ ~~IIII~~ ~~IIII~~ III	23
Bicycle	~~IIII~~	5
Taxi	II	2

By adding together all the frequencies, you can see that 70 students took part in the survey. The frequencies also show you that more students travelled to school on foot than by any other method of transport.

Three methods are used to collect data.

- **Taking a sample:** For example, to find out which 'soaps' students watch, you would need to take a sample from the whole school population by asking at random an equal number of boys and girls from each year group. In this case, a good sample size would be 50.
- **Observation:** For example, to find how many vehicles a day use a certain road, you would need to count and record the number of vehicles passing a point at different times of the day.
- **Experiment:** For example, to find out how often a six occurs when you throw a dice, you would need to throw the dice 50 times or more and record each score.

EXAMPLE 2

Andrew wanted to find out the most likely outcome when two coins are tossed. He carried out an experiment by tossing two coins 50 times. His frequency table looked like this.

Number of heads	Tally	Frequency
0	卌 卌 II	12
1	卌 卌 卌 卌 卌 II	27
2	卌 卌 I	11

From Andrew's table, you can see that a single head appeared the highest number of times.

Grouped data

Many surveys produce a lot of data that covers a wide range of values. In these cases, it is sensible to put the data into groups before attempting to compile a frequency table. These groups of data are called **classes** or **class intervals**.

Once the data has been grouped into classes, a **grouped frequency table** can be completed. The method is shown in Example 3.

EXAMPLE 3

These marks are for 36 students in a Year 10 mathematics examination.

31	49	52	79	40	29	66	71	73	19	51	47
81	67	40	52	20	84	65	73	60	54	60	59
25	89	21	91	84	77	18	37	55	41	72	38

a Construct a frequency table, using classes of 1–20, 21–40 and so on.

b What was the most frequent interval of marks?

a Draw the grid of the table shown below and put in the headings.

Next, list the classes, in order, in the column headed 'Marks'.

Using tally marks, indicate each student's score against the class to which it belongs. For example, 81, 84, 89 and 91 belong to the class 81–100, giving five tally marks, as shown below.

Finally, count the tally marks for each class and enter the result in the column headed 'Frequency'. The table is now complete.

Marks	Tally	Frequency
1–20	III	3
21–40	~~IIII~~ III	8
41–60	~~IIII~~ ~~IIII~~ I	11
61–80	~~IIII~~ IIII	9
81–100	~~IIII~~	5

b From the grouped frequency table, you can see that the highest number of students obtained a mark in the 41–60 interval.

EXERCISE 5A

G

1 Philip kept a record of the number of goals scored by Burnley Rangers in the last 20 matches. These are his results:

0 1 1 0 2 0 1 3 2 1

0 1 0 3 2 1 0 2 1 1

a Draw a frequency table for his data.

b Which was the most frequent score?

c How many goals were scored in total for the 20 matches?

F

FM **2** Monica was doing a geography project on the weather. As part of her work, she kept a record of the daily midday temperatures in June.

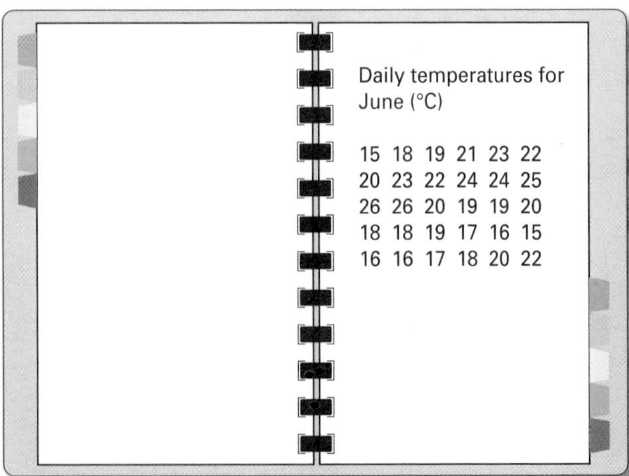

Daily temperatures for June (°C)
15 18 19 21 23 22
20 23 22 24 24 25
26 26 20 19 19 20
18 18 19 17 16 15
16 16 17 18 20 22

F

a Copy and complete the grouped frequency table for her data.

Temperature (°C)	Tally	Frequency
14–16		
17–19		
20–22		
23–25		
26–28		

b In which interval do the most temperatures lie?

c Describe what the weather was probably like throughout the month.

E

3 For the following surveys, decide whether the data should be collected by:

i sampling **ii** observation **iii** experiment.

a The number of people using a new superstore.

b How people will vote in a forthcoming election.

c The number of times a person scores double top in a game of darts.

d Where people go for their summer holidays.

e The frequency of a bus service on a particular route.

f The number of times a drawing pin lands point up when dropped.

Look back to page 119 where each method of collecting data is discussed.

4 In a game, Mitesh used a six-sided dice. He decided to keep a record of his scores to see whether the dice was fair. His scores were:

2 4 2 6 1 5 4 3 3 2 3 6 2 1 3

5 4 3 4 2 1 6 5 1 6 4 1 2 3 4

a Draw a frequency table for his data.

b How many throws did Mitesh have during the game?

c Do you think the dice was a fair one? Explain why.

5 The data shows the heights, in centimetres, of a sample of 32 Year 10 students.

172 158 160 175 180 167 159 180

167 166 178 184 179 156 165 166

184 175 170 165 164 172 154 186

167 172 170 181 157 165 152 164

a Draw a grouped frequency table for the data, using class intervals 151–155, 156–160, …

b In which interval do the most heights lie?

c Does this agree with a survey of the students in your class?

PS 6 Kathy used a stopwatch to time how long it took her rabbit to find food left in its hutch.

The following is her record in seconds.

7	30	14	27	8	31	8	28	10	41	51	37	15	21	37	16	38
23	20	9	11	55	9	33	8	35	45	35	25	25	49	23	43	55
45	8	13	9	39	12	57	16	37	26	32	19	48	29	37		

Find the best way to put this data into a frequency chart to illustrate the length of time it took the rabbit to find the food.

AU 7 David was doing a survey to find the ages of people at a football competition.

He said that he would make a frequency table with the regions 15–20, 20–25, 25–30.

Explain what difficulty David could have with these class divisions.

8 Conduct some surveys of your own choice and draw frequency tables for your data.

Double dice

This is an activity for two or more players. Each player needs two six-sided dice.

Each player throws their two dice together 100 times. For each throw, add together the two scores to get a total score.

What is the lowest total score anyone can get? What is the highest total score?

Everyone keeps a record of their 100 throws in a frequency table.

Compare your frequency table with someone else's and comment on what you notice. For example: Which scores appear the most often? What about 'doubles'?

How might this information be useful in games that use two dice?

Repeat the activity in one or more of the following ways.

- For each throw, multiply the score on one dice by the score on the other.
- Use two four-sided dice (tetrahedral dice), adding or multiplying the scores.
- Use two different-sided dice, adding or multiplying the scores.
- Use three or more dice, adding and/or multiplying the scores.

5.2 Statistical diagrams

This section will show you how to:
- show collected data as pictograms

Key words
key
pictograms
symbol

Data collected from a survey can be presented in pictorial or diagrammatic form to help people to understand it more quickly. You see plenty of examples of this in newspapers and magazines and on TV, where every type of visual aid is used to communicate statistical information.

Pictograms

A **pictogram** is a frequency table in which frequency is represented by a repeated **symbol**. The symbol itself usually represents a number of items, as Example 5 shows. However, sometimes it is more sensible to let a symbol represent just a single unit, as in Example 4. The **key** tells you how many items are represented by a symbol.

EXAMPLE 4

The pictogram shows the number of phone calls made by Mandy from her mobile phone during a week.

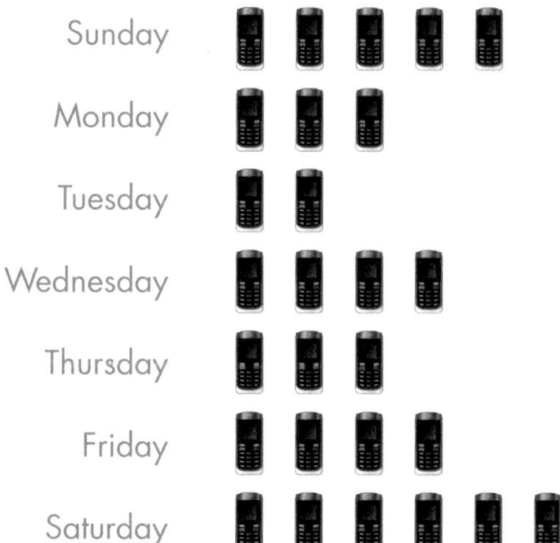

Key represents 1 call

How many calls did Mandy make in the week?

From the pictogram, you can see that Mandy made a total of 27 calls.

Although pictograms can have great visual impact (particularly as used in advertising) and are easy to understand, they have a serious drawback. Apart from a half, fractions of a symbol cannot usually be drawn accurately and so frequencies are often represented only approximately by symbols.

Examples 5 and 6 highlight this difficulty.

EXAMPLE 5

The pictogram shows the number of Year 10 students who were late for school during a week.

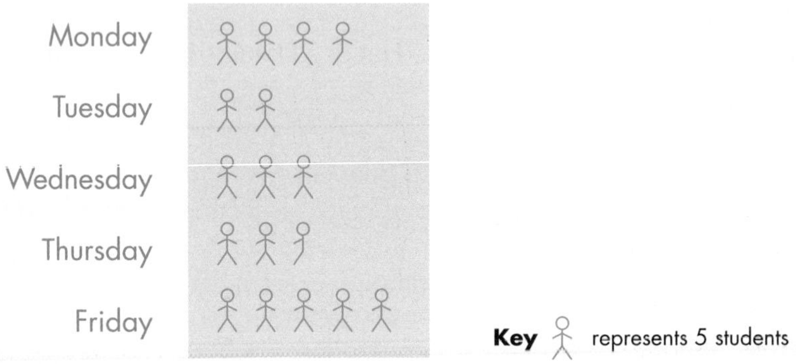

Monday
Tuesday
Wednesday
Thursday
Friday

Key represents 5 students

How many students were late on:

a Monday **b** Thursday?

Precisely how many students were late on Monday and Thursday respectively?

If you assume that each 'limb' of the symbol represents one student and its 'body' also represents one student, then the answers are:

a 19 students were late on Monday. **b** 13 on Thursday.

EXAMPLE 6

This pictogram is used to show how many trains ran late in the course of one weekend.

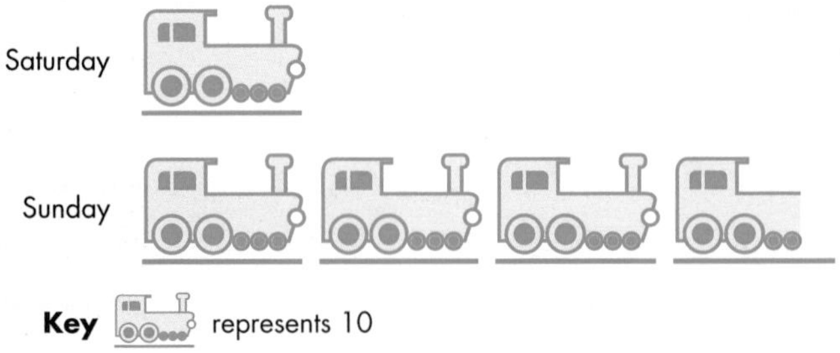

Saturday

Sunday

Key represents 10

Give a reason why the pictogram is difficult to read.

The last train of those representing the number of trains late on Sunday is drawn as a fraction of 10. However, we cannot easily make out what this fraction is specifically.

EXERCISE 5B

1 The frequency table shows the numbers of cars parked in a supermarket's car park at various times of the day. Draw a pictogram to illustrate the data. Use a key of 1 symbol = 5 cars.

Time	9 am	11 am	1 pm	3 pm	5 pm
Frequency	40	50	70	65	45

2 Mr Weeks, a milkman, kept a record of how many pints of milk he delivered to 10 flats on a particular morning. Draw a pictogram for the data. Use a key of 1 symbol = 1 pint.

Flat 1	Flat 2	Flat 3	Flat 4	Flat 5	Flat 6	Flat 7	Flat 8	Flat 9	Flat 10
2	3	1	2	4	3	2	1	5	1

3 The pictogram, taken from a Suntours brochure, shows the average daily hours of sunshine for five months in Tenerife.

a Write down the average daily hours of sunshine for each month.

b Which month had the most sunshine?

c Give a reason why pictograms are useful in holiday brochures.

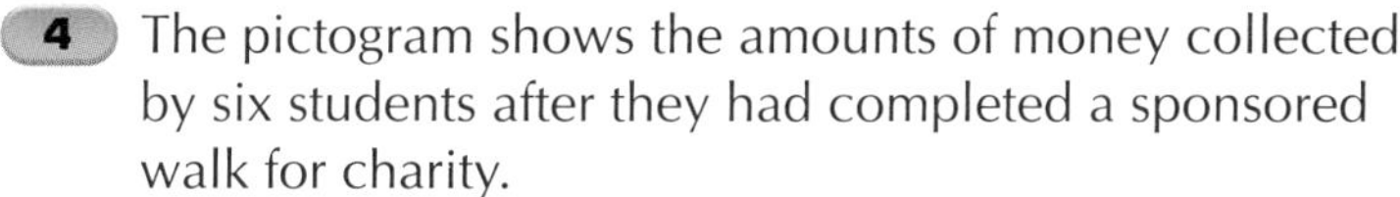

4 The pictogram shows the amounts of money collected by six students after they had completed a sponsored walk for charity.

Anthony	£ £ £ £ £
Ben	£ £ £ £ £ £
Emma	£ £ £ £ £
Leanne	£ £ £ £
Reena	£ £ £ £ £ £
Simon	£ £ £ £ £ £ £

Key £ represents £5

a Who raised the most money?

b How much money was raised altogether by the six students?

c Robert also took part in the walk and raised £32. Why would it be difficult to include him on the pictogram?

FM 5 A newspaper showed the following pictogram about one of its team member's family and the number of emails they each received during one Sunday.

Key represents 4 emails

		Frequency
Dad		
Mum		
Teenage son		
Teenage daughter		23
Young son		9

F

a How many emails did:

i Dad receive

ii Mum receive

iii the teenage son receive?

b Copy and complete the pictogram.

c How many emails were received altogether?

PS 6 A survey was taken on the types of books read by students of a school.

	Frequency
Thriller	51
Romance	119
Science fiction	187
Historical	136

This information is to be put into a pictogram.

Design a pictogram to show this information with as few symbols as possible.

AU 7 A pictogram is to be made from this frequency table which shows the ways students travel to school.

Car	342
Bus	336
Walk	524

Explain why a key of four students to a symbol is not a good idea.

8 Draw pictograms of your own to show the following data.

a The number of hours for which you watched TV every evening last week.

b The magazines that students in your class read.

c The favourite colours of students in your class.

AU 9 A children's charity produces an advert to show how much money they have raised over the last two years.

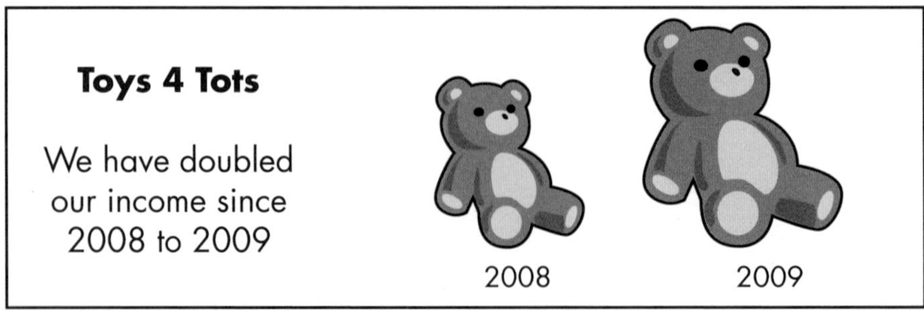

Toys 4 Tots
We have doubled our income since 2008 to 2009
2008
2009

Why is the advert misleading?

5.3 Bar charts

This section will show you how to:
- draw bar charts to represent statistical data

Key words
axis
bar chart
class interval
dual bar chart

A **bar chart** consists of a series of bars or blocks of the *same* width, drawn either vertically or horizontally from an **axis**.

The heights or lengths of the bars always represent *frequencies*.

Sometimes, the bars are separated by narrow gaps of equal width, which makes the chart easier to read.

EXAMPLE 7

The grouped frequency table below shows the marks of 24 students in a test. Draw a bar chart for the data.

Marks	1–10	11–20	21–30	31–40	41–50
Frequency	2	3	5	8	6

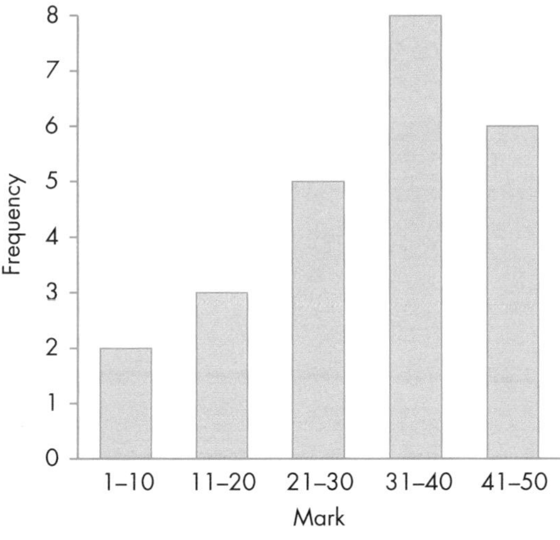

Note:

- Both axes are labelled.
- The **class intervals** are written under the middle of each bar.
- The bars are separated by equal spaces.

By using a **dual bar chart**, it is easy to compare two sets of related data, as Example 8 shows.

EXAMPLE 8

This dual bar chart shows the average daily maximum temperatures for England and Turkey over a five-month period.

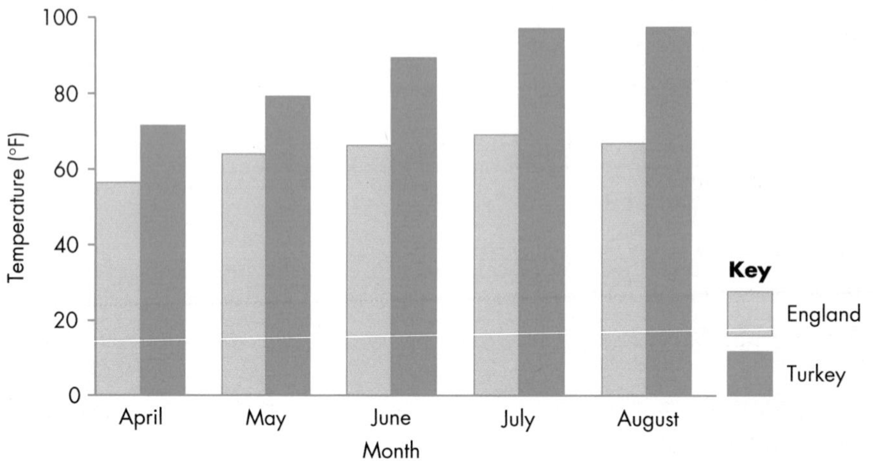

In which month was the difference between temperatures in England and Turkey the greatest?

The largest difference can be seen in August.

Note: You must always include a key to identify the two different sets of data.

EXERCISE 5C

F

1 For her survey on fitness, Maureen asked a sample of people, as they left a sports centre, which activity they had taken part in. She then drew a bar chart to show her data.

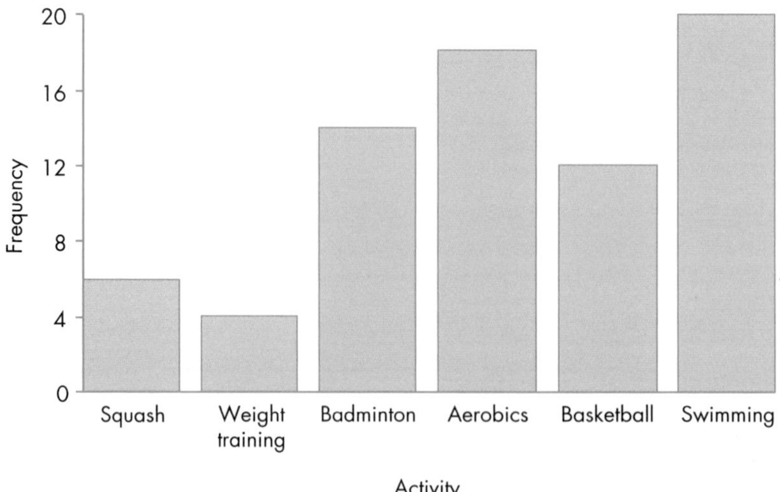

F

a Which was the most popular activity?

b How many people took part in Maureen's survey?

c Give a probable reason why fewer people took part in weight training than in any other activity.

d Is a sports centre a good place in which to do a survey on fitness? Explain why.

2 The frequency table below shows the levels achieved by 100 Year 10 students in their mock GCSE examinations.

Grade	F	E	D	C	B	A
Frequency	12	22	24	25	15	2

a Draw a suitable bar chart to illustrate the data.

b What fraction of the students achieved a grade C or grade B?

c Give one advantage of drawing a bar chart rather than a pictogram for this data.

3 This table shows the number of points Richard and Derek were each awarded in eight rounds of a general knowledge quiz.

Round	**1**	**2**	**3**	**4**	**5**	**6**	**7**	**8**
Richard	7	8	7	6	8	6	9	4
Derek	6	7	6	9	6	8	5	6

a Draw a dual bar chart to illustrate the data.

b Comment on how well each of them did in the quiz.

4 Kay did a survey on the time it took students in her form to get to school on a particular morning. She wrote down their times to the nearest minute.

15 23 36 45 8 20 34 15 27 49

10 60 5 48 30 18 21 2 12 56

49 33 17 44 50 35 46 24 11 34

a Draw a grouped frequency table for Kay's data, using class intervals 1–10, 11–20, …

b Draw a bar chart to illustrate the data.

c What conclusions can Kay draw from the bar chart?

5 This table shows the number of accidents at a dangerous crossroads over a six-year period.

Year	2000	2001	2002	2003	2004	2005
No. of accidents	6	8	7	9	6	4

a Draw a pictogram for the data.

b Draw a bar chart for the data.

AU **c** Which diagram would you use if you were going to write to your local council to suggest that traffic lights should be installed at the crossroads? Explain why.

F

AU **6** The diagram below shows the minimum and maximum temperatures, in degrees Celsius, for one day in August in five cities.

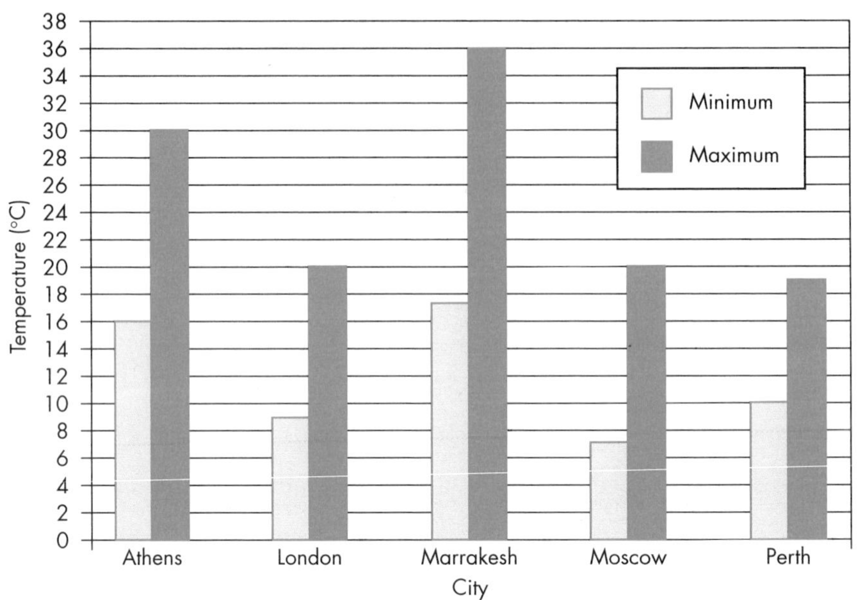

Chris says that the minimum temperature is always about half the maximum temperature for most cities.

Is Chris correct?

Give reasons to justify your answer.

AU **7** The bar chart shows the average daily temperatures, in degrees Celsius, in England and Scotland.

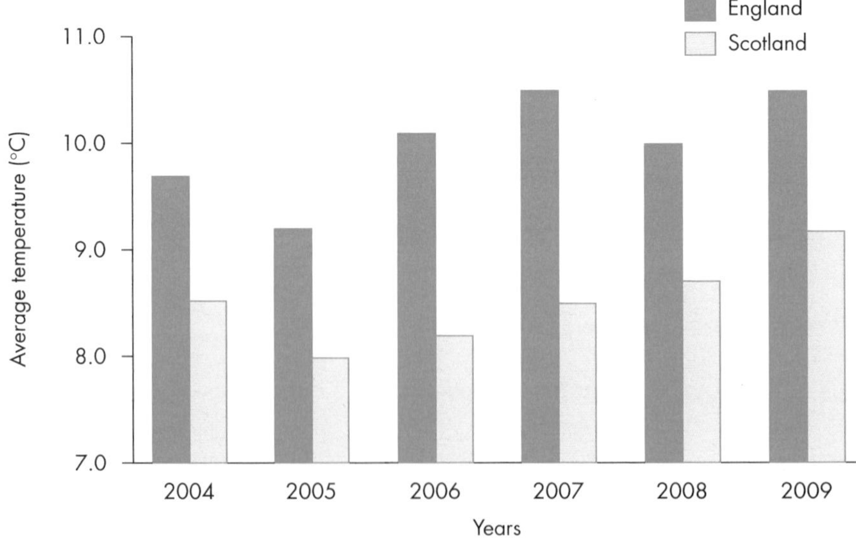

Derek says that the graph shows that the temperature in England is always more than double the temperature in Scotland.

Explain why Derek is wrong.

F

FM **8** Conduct a survey to find the colours of cars that pass your school or your home.

a Draw pictograms and bar charts to illustrate your data.

b Compare your results with someone else's in your class and comment on any conclusions you can draw concerning the colours of cars in your area.

FM **9** Choose a broadsheet newspaper, such as *The Times* or the *Guardian* and a tabloid newspaper, such as the *Sun* or the *Mirror*. Take a fairly long article from both papers, preferably on the same topic. Count the number of words in the first 50 sentences of each article.

a For each article, draw a grouped frequency table for the number of words in each of the first 50 sentences.

b Draw a dual bar chart for your data.

c Do your results support the hypothesis that

'Sentences in broadsheet newspapers are longer than the sentences in tabloid newspapers'?

E

FM AU **10** The first bar chart shows the stock of crisps in the school tuck shop at the start of break.

The second bar chart shows how many of each flavour were left at the end of break.

40 bags of prawn cocktail crisps were sold.

Each bag of crisps was 30p.

How much money did the tuck shop make selling crisps?

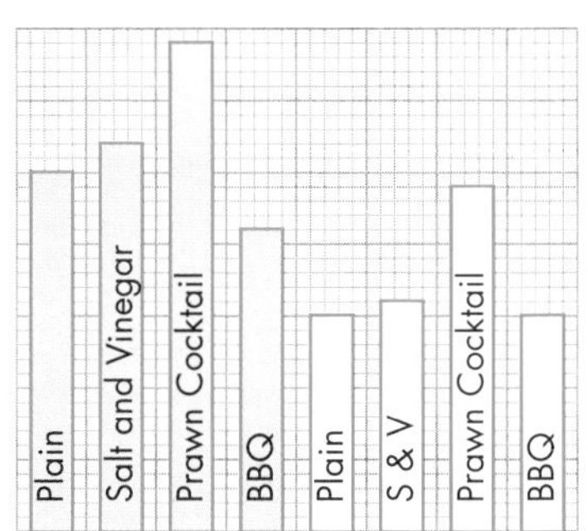

5.4 Line graphs

This section will show you how to:
- draw a line graph to show trends in data

Key words
line graphs
trends

Line graphs are usually used in statistics to show how data changes over a period of time. One such use is to indicate **trends**, for example, whether the Earth's temperature is increasing as the concentration of carbon dioxide builds up in the atmosphere, or whether a firm's profit margin is falling year-on-year.

Line graphs are best drawn on graph paper.

EXAMPLE 9

This line graph shows the outside temperature at a weather station, taken at hourly intervals. Estimate the temperature at 3.30 pm.

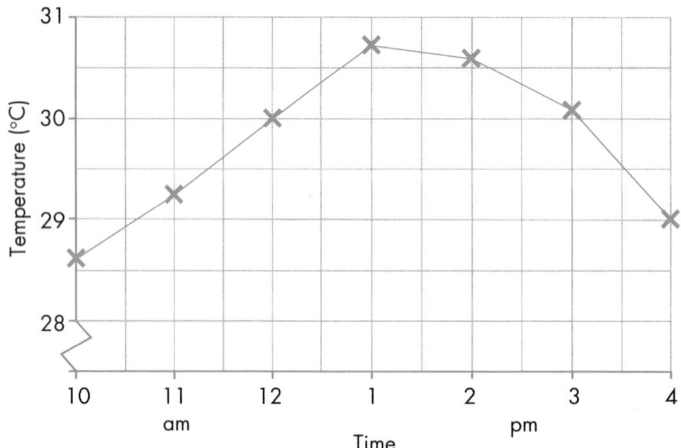

At 3.30 the temperature is approximately 29.5 °C.

Note: The temperature axis starts at 28 °C rather than 0 °C. This allows the use of a scale which makes it easy to plot the points and then to read the graph. The points are joined with lines so that the intermediate temperatures can be estimated for other times of the day.

EXAMPLE 10

This line graph shows the profit made each year by a company over a six-year period. Between which years did the company have the greatest increase in profits?

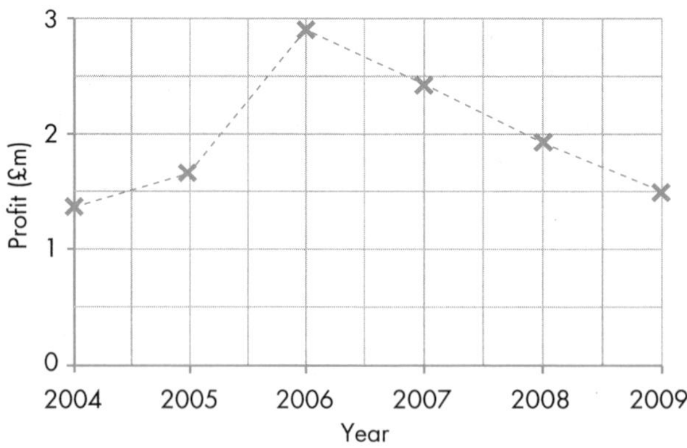

The greatest increase in profits was between 2005 and 2006.

For this graph, the values between the plotted points have no meaning because the profit of the company would have been calculated at the end of every year. In cases like this, the lines are often dashed. Although the trend appears to be that profits have fallen after 2006, it would not be sensible to predict what would happen after 2009.

EXERCISE 5D

1 This line graph shows the value of Spevadon shares on seven consecutive trading days.

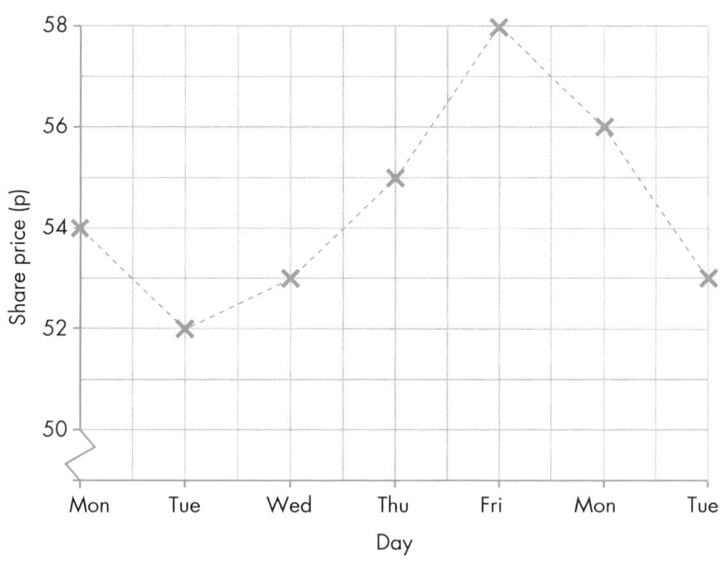

a On which day did the share price have its lowest value and what was that value?

b By how much did the share price rise from Wednesday to Thursday?

c Which day had the greatest rise in the share price from the previous day?

FM **d** Mr Hardy sold 500 shares on Friday. How much profit did he make if he originally bought the shares at 40p each?

2 The table shows the population of a town, rounded to the nearest thousand, after each census.

Year	1941	1951	1961	1971	1981	1991	2001
Population (1000s)	12	14	15	18	21	25	23

a Draw a line graph for the data.

b From your graph estimate the population in 1966.

c Between which two consecutive censuses did the population increase the most?

AU **d** Can you predict the population for 2011? Give a reason for your answer.

D

3 The number of ants in an ants' nest are counted at the end of each week.

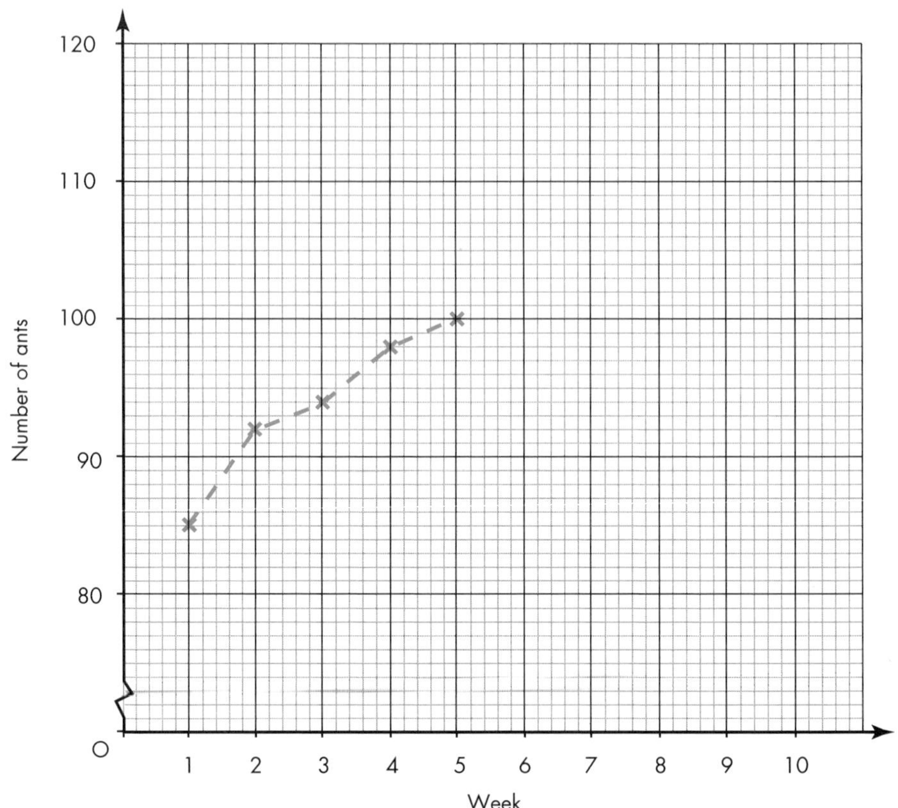

The graph shows the number of ants. At the end of week 5 the number is 100.

a At the end of week 6 the number is 104. At the end of week 10 the number is 120.

  i Copy the graph and plot these points on the graph.

  ii Complete the graph with straight lines.

b Use your graph to estimate the number of ants at the end of week 8.

FM **4** The table shows the estimated number of tourists worldwide.

Year	1970	1975	1980	1985	1990	1995	2000	2005
No. of tourists (millions)	100	150	220	280	290	320	340	380

a Draw a line graph for the data.

b Use your graph to estimate the number of tourists in 2010.

c Between which two consecutive years did tourism increase the most?

d Explain the trend in tourism. What reasons can you give to explain this trend?

D

5 The table shows the maximum and minimum daily temperatures for London over a week.

Day	Sunday	Monday	Tuesday	Wednesday	Thursday	Friday	Saturday
Maximum (°C)	12	14	16	15	16	14	10
Minimum (°C)	4	5	7	8	7	4	3

a Draw line graphs on the *same* axes to show the maximum and minimum temperatures.

b Find the smallest and greatest differences between the maximum and minimum temperatures.

PS 6 A puppy is weighed at the end of each week as shown in the table.

Week	1	2	3	4	5
Weight (g)	850	920	940	980	1000

Estimate how much the puppy would weigh after eight weeks.

AU 7 When plotting a graph to show the summer midday temperatures in Spain, Abbass decided to start his graph at the temperature 20 °C.

Explain why he might have done that.

ACTIVITY

Diagrams from the press

This is an activity for a group of two or more people. You will need a large selection of recent newspapers and magazines.

In a group, look through the newspapers and magazines.

Cut out any statistical diagrams and stick them on large sheets of coloured paper.

Underneath each diagram, explain what the diagram shows and how useful the diagram is in showing that information.

If any of the diagrams appears to be misleading, explain why.

You now have a lot of up-to-date statistics to display in your classroom.

5.5 Stem-and-leaf diagrams

This section will show you how to:

- draw and read information from an ordered stem-and-leaf diagram

Key words
discrete data
ordered data
raw data
unordered data

Raw data

If you were recording the ages of the first 20 people who line up at a bus stop in the morning, the **raw data** might look like this.

23, 13, 34, 44, 26, 12, 41, 31, 20, 18, 19, 31, 48, 32, 45, 14, 12, 27, 31, 19

This data is **unordered** and is difficult to read and analyse. When the data is **ordered**, it will look like this.

12, 12, 13, 14, 18, 19, 19, 20, 23, 26, 27, 31, 31, 31, 32, 34, 41, 44, 45, 48

This is easier to read and analyse.

Another method for displaying **discrete data** is a stem-and-leaf diagram. The tens digits will be the 'stem' and the units digits will be the 'leaves'.

Key 1 | 2 represents 12

1	2	2	3	4	8	9	9
2	0	3	6	7			
3	1	1	1	2	4		
4	1	4	5	8			

This is called an ordered stem-and-leaf diagram and gives a better idea of how the data is distributed.

A stem-and-leaf diagram should always have a key.

EXAMPLE 11

Put the following data into an ordered stem-and-leaf diagram.

45, 62, 58, 58, 61, 49, 61, 47, 52, 58, 48, 56, 65, 46, 54

a What is the largest value?

b What is the most common value?

c What is the difference between the largest and smallest values?

First decide on the stem and the leaf.

In this case, the tens digit will be the stem and the units digit will be the leaf.

Key 4 | 5 represents 45

4	5	6	7	8	9	
5	2	4	6	8	8	8
6	1	1	2	5		

a The largest value is 65.

b The most common value is 58 which occurs three times.

c The difference between the largest and the smallest is 65 – 45 = 20.

EXERCISE 5E

1 The following stem-and-leaf diagram shows the times taken for 15 students to complete a mathematical puzzle.

Key 1 | 7 represents 17 seconds

1	7	8	8	9		
2	2	2	2	5	6	9
3	3	4	5	5	8	

a What is the shortest time to complete the puzzle?

b What is the most common time to complete the puzzle?

c What is the difference between the longest time and the shortest time to complete the puzzle?

E

E

FM 2 This stem-and-leaf diagram shows the marks for the boys and girls in form 10E in a maths test.

Key Boys: 2 | 4 means 42 marks

Girls: 3 | 5 means 35 marks

HINTS AND TIPS

Read the boys' marks from right to left.

Boys					Girls					
6	4	2	3	3	5	7	9			
9	9	6	2	4	2	2	3	8	8	8
7	6	6	6	5	1	1	5			

a What was the highest mark for the boys?

b What was the highest mark for the girls?

c What was the most common mark for the boys?

d What was the most common mark for the girls?

e What overall conclusions can you draw from this data?

D

3 The heights of 15 sunflowers were measured.

43 cm, 39 cm, 41 cm, 29 cm, 36 cm,

34 cm, 43 cm, 48 cm, 38 cm, 35 cm,

41 cm, 38 cm, 43 cm, 28 cm, 48 cm

a Show the results in an ordered stem-and-leaf diagram, using this key:

Key 4 | 3 represents 43 cm

b What was the largest height measured?

c What was the most common height measured?

d What is the difference between the largest and smallest heights measured?

4 A student records the number of text messages she receives each day for two weeks.

12, 18, 21, 9, 17, 23, 8, 2, 20, 13, 17, 22, 9, 9

a Show the results in an ordered stem-and-leaf diagram, using this key:

Key 1 | 2 represents 12 messages

b What was the largest number of text messages received in a day?

c What is the most common number of text messages received in a day?

AU 5 The number of matches in a set of boxes were each counted with the following results.

50, 52, 51, 53, 52, 51, 51, 53, 54, 55, 54, 52, 52, 51, 50, 53

Explain why a stem-and-leaf diagram is not a good way to represent this information.

GRADE BOOSTER

F You can draw and read information from bar charts, dual bar charts and pictograms

F You can work out the total frequency from a frequency table and compare data in bar charts

E You can read information from a stem-and-leaf diagram

D You can draw an ordered stem-and-leaf diagram

What you should know now

- How to draw frequency tables for grouped and ungrouped data
- How to draw and interpret pictograms, bar charts and line graphs
- How to read information from statistical diagrams, including stem-and-leaf diagrams

1 Mary threw a dice 24 times.

Here are the 24 scores.

3	5	3	4	1	2	4	5
6	2	3	4	3	1	4	3
2	3	5	5	3	4	2	1

Complete the frequency table. (3)

Score	Tally	Frequency
1		
2		
3		
4		
5		
6		

(Total 3 marks)

Edexcel, March 2007, Paper 8 Foundation, Unit 2 Test, Question 1a

2 The bar chart shows the number of TVs sold by a shop six days last week.

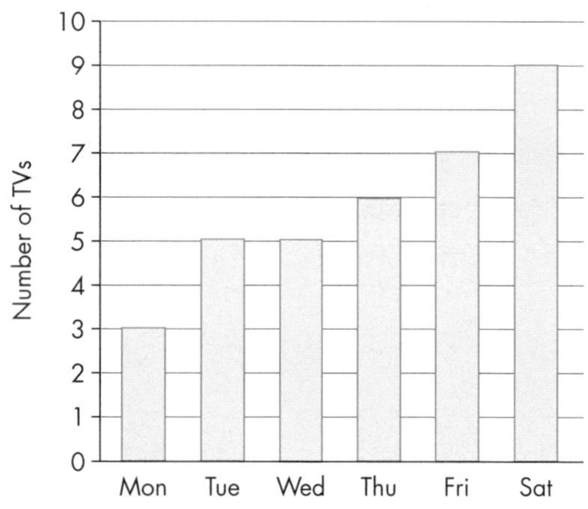

a How many TVs were sold on Friday? (1)

b On which day was the **least** number of TVs sold? (1)

c On which two days were the same number of TVs sold? (1)

(Total 3 marks)

Edexcel, March 2009, Paper 5 Foundation, Unit 1 Test, Question 1

3 The pictogram shows the number of plates sold by a shop on Monday, Tuesday, Wednesday and Thursday of one week.

Monday	◯ ◯
Tuesday	◯ ◖
Wednesday	◯ ◯ ◯
Thursday	◯
Friday	
Saturday	

Key: ◯ represents 10 plates

a Work out the number of plates sold on Monday. (1)

b Work out the number of plates sold on Tuesday. (1)

The shop sold 40 plates on Friday.

The shop sold 25 plates on Saturday.

c Use this information to complete the pictogram. (2)

(Total 4 marks)

Edexcel, November 2008, Paper 2 Foundation, Question 1

4 Steve asked his friends to tell him their favourite colour.

Here are his results.

Favourite colour	Tally	Frequency
Red	ⅠⅠⅠⅠ̸ Ⅰ	6
Blue	ⅠⅠⅠⅠ̸ ⅠⅠⅠ	8
Green	ⅠⅠⅠⅠ̸	5
Yellow	ⅠⅠⅠ	3

a Complete the bar chart to show his results. (2)

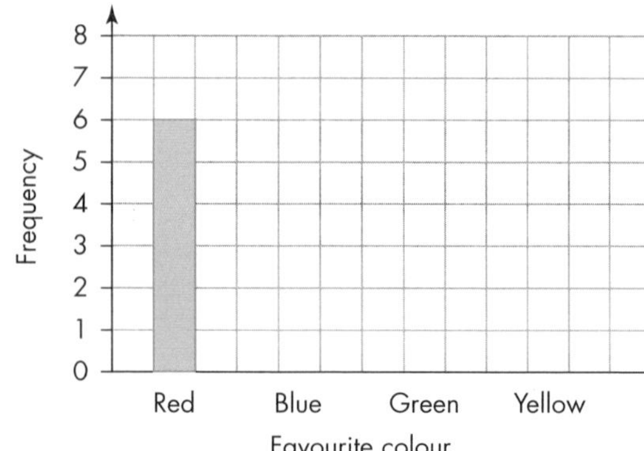

D

E

F

G

b Which colour did most of his friends say? (1)

(Total 3 marks)

Edexcel, May 2008, Paper 1 Foundation, Question 8

5 Here is a pictogram.

It shows the number of books read by Asad, by Betty, and by Chris.

Asad	
Betty	
Chris	
Diana	
Erikas	

Key: represents 4 books

a Write down the number of books read by

i Asad **ii** Chris (2)

Diana read 12 books.
Erikas read 9 books.

b Show this information on the pictogram. (2)

(Total 4 marks)

Edexcel, March 2009, Paper 5 Foundation, Unit 1 Test, Question 1

6 The bar chart shows information about the amount of time, in minutes, that Andrew and Karen spent watching television on four days last week.

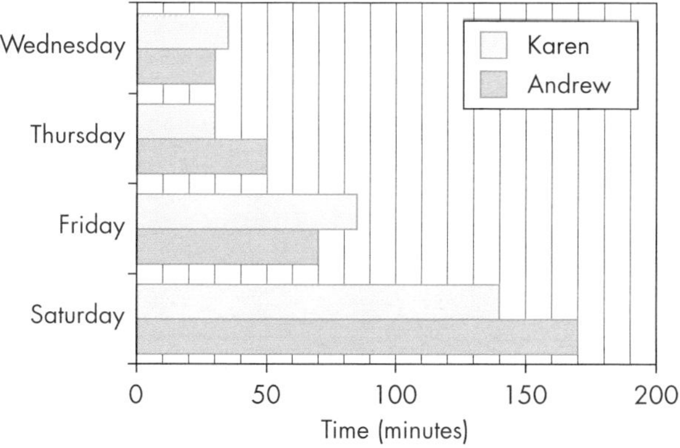

Karen spent more time watching television than Andrew on two of these four days.

a Write down these two days. (2)

b Work out the total amount of time Andrew spent watching television on these four days. (2)

(Total 4 marks)

Edexcel, June 2008, Paper 5 Foundation, Unit 1 Test, Question 2

7 The bar chart shows the number of buses going to different towns from Speedville each day.

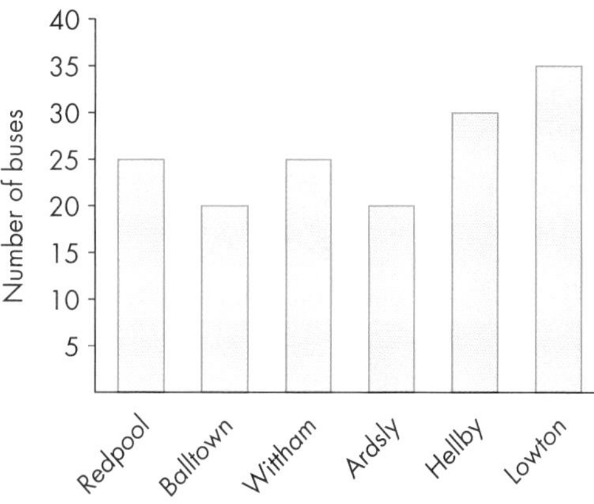

The bar chart was redrawn as a pictogram for the newspaper, as shown below.

Town		Frequency
Redpool		25
Balltown		20
Wittham		25
Ardsley		20
Hellby		30
Lowton		35

Explain what is wrong with the pictogram.

8 The stem-and-leaf diagram shows the test scores of 13 students.

Key 2 | 5 is a score of 25

1	9
2	2 2 3 5 7
3	3 4 4 8 8 9
4	0

How many students scored less than 20?

9 Here are the amounts, in £, spent by some shoppers at a supermarket.

37	56	23	40
38	56	31	48
25	49	32	46

Draw an ordered stem-and-leaf diagram for these amounts.

You must include a key. (3)

2	
3	
4	
5	

Key:

(Total 3 marks)

Edexcel, March 2008, Paper 5 Foundation, Unit 1 Test, Question 4

10 Here are the ages, in years, of some members of a swimming club.

9	12	18	10	9	7	21	30	23	16
19	32	17	28	15	8	10	15	21	10

Draw an ordered stem-and-leaf diagram for these ages.

You must include a key. (3)

0	
1	
2	
3	

Key:

(Total 3 marks)

Edexcel, March 2007, Paper 8 Foundation, Unit 2 Test, Question 3

11 Tom and Barbara grew tomatoes. They compared their tomatoes by selecting 100 of each one weekend. The table shows the mean weight of Tom's tomatoes.

Weight, w (grams)	Tom's Tomatoes
$50 \leqslant w < 100$	21
$100 \leqslant w < 150$	28
$150 \leqslant w < 200$	26
$200 \leqslant w < 250$	14
$250 \leqslant w < 300$	9
$300 \leqslant w < 350$	2

a Which class interval contains the median weight for Tom's Tomatoes?

b The graph for Barbara's Tomatoes is drawn on the following grid. Copy it on to graph paper. On the same grid draw the graph for Tom's Tomatoes.

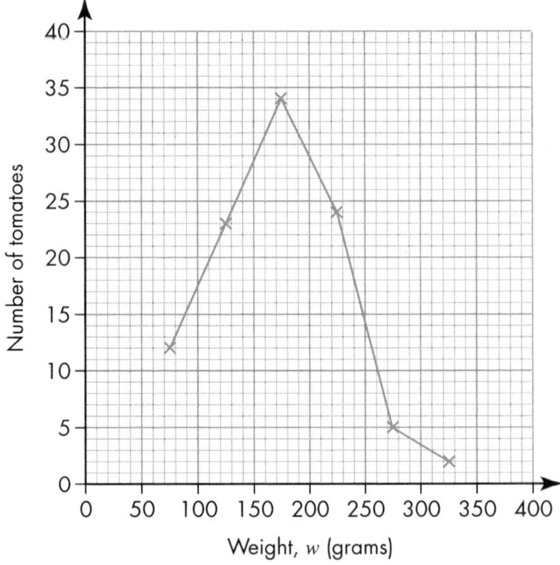

c Use the graphs to write down one comparison between Tom and Barbara's Tomatoes.

Worked Examination Questions

1 The temperature is recorded in 20 towns on one day.

8 14 21 15 2 10 11 17 7 24

23 18 5 11 4 20 18 23 4 19

Draw a **stem-and-leaf diagram** to represent these data.

Key 1 | 4 represents 14

0	2 4 4 5 7 9
1	0 1 1 4 5 7 8 8 9
2	0 1 3 3 4

Draw a basic stem-and-leaf diagram when the stem is the tens digits and the leaves are the units digits. Create a key, using any value for 1 method mark.

Now complete the stem-and-leaf diagram, keeping the data in order. You get 2 marks for accuracy and 1 mark if you make only one error.

Total: 3 marks

PS **2** Andrew was born on 27 March and had his weight recorded regularly as shown in the table.

Day	1	5	9	13	17
Weight (g)	4100	3800	4000	4500	4900

Estimate how much Andrew would weigh after 3 weeks.

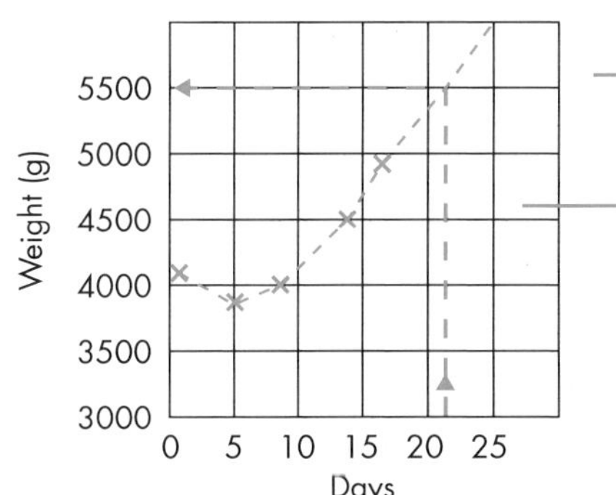

Use the table to draw the given points and plot a line graph.

Plot the points on a graph so that a prediction can be made by extending the graph. You get 1 mark for accuracy for this.

You get 1 method mark for showing 21 days (3 weeks) being read vertically to the extended graph.

You get 1 method mark for showing a horizontal reading from the graph to the weight.

Estimated weight is 5500 g

You get 1 mark for accuracy for an answer close to 5500 g.

Total: 4 marks

AU **3** When drawing a bar chart to show the summer midday temperatures in various capital cities of Europe, Joy decided to start her graph at the temperature 10 °C.

Explain why she might have done that.

The lowest temperature to be shown would be about 10 degrees and she wanted to emphasise the differences between the cities.

State a lowest temperature of 10 °C or just higher. This is worth 1 mark.

Suggest that she wanted to emphasise the differences. This is worth 1 mark.

Total: 2 marks

5 Functional Maths
Reporting the weather

The weather often appears in the news headlines and a weather report is given regularly on the television. In this activity you are required to look at the data supplied in a weather report and interpret its meaning.

Your task

The type of information given on the map and in the table is found in most newspapers in the UK each day. The map gives a forecast of the weather for the day in several large towns, while the table summarises the weather on the previous day.

It is your task to use appropriate statistical diagrams and measures to summarise the data given in the map and table.

Then, you must write a report to describe fully the weather in the UK on Friday 21st April 2010. You should use your statistical analysis to support your descriptions.

Getting started

Look at the data provided in the table.

- Which cities had the most sun, which had the most rain and which were the warmest on this particular day?
- Is there any other information you can add?

Friday 21st April 2010

	Sun (hrs)	Rain (mm)	Max (°C)	Min (°C)	Daytime weather
Aberdeen	5	3	10	5	rain
Barnstaple	5	0	12	8	sunny
Belfast	5	0	10	4	cloudy
Birmingham	5	0	11	5	sunny
Bournemouth	5	0	12	6	sunny
Bradford	5	1	10	5	rain
Brighton	5	0	11	5	mixed
Bristol	5	2	11	6	rain
Cardiff	5	2	11	6	rain
Carlisle	5	3	10	5	rain
Chester	5	3	10	4	rain
Eastbourne	5	0	9	5	windy
Edinburgh	5	0	9	5	cloudy
Falmouth	5	0	12	8	sunny
Glasgow	5	0	9	4	cloudy
Grimsby	5	1	10	4	mixed
Holyhead	5	0	8	3	windy
Ipswich	5	1	11	6	mixed
Isle of Man	5	0	9	3	windy
Isle of Wight	5	0	13	6	sunny

	Sun (hrs)	Rain (mm)	Max (°C)	Min (°C)	Daytime weather
Leeds	5	0	11	6	cloudy
Lincoln	5	3	10	4	rain
Liverpool	5	1	12	6	mixed
London	5	2	13	6	rain
Manchester	5	0	13	7	sunny
Middlesbrough	5	1	11	6	rain
Newcastle	5	0	11	6	cloudy
Newquay	5	0	12	8	sunny
Nottingham	5	0	12	6	cloudy
Oxford	5	2	12	5	rain
Plymouth	6	0	12	9	sunny
Rhyl	5	1	9	4	mixed
Scunthorpe	5	1	10	4	mixed
Sheffield	5	0	12	7	sunny
Shrewsbury	5	0	9	4	windy
Southampton	5	0	12	6	sunny
Swindon	5	1	11	5	mixed
Weymouth	5	0	12	6	sunny
Windermere	5	2	10	4	mixed
York	5	0	11	5	cloudy

Why this chapter matters

Civil and mechanical engineers, decorators, plumbers, industrial and structural engineers, town planners, landscape architects, surveyors, interior designers, builders and computer designers all need to work with great precision. Some projects involve measures in units that seem too big to work with; some require accuracy to within infinitesimally small margins. So how do they manage to work with these difficult measures?

The answer is that they all work with drawings drawn to scale. This allows them to represent the lengths they cannot easily measure with standard equipment. Most products and structures in everyday life were probably designed originally by someone who used a scale drawing.

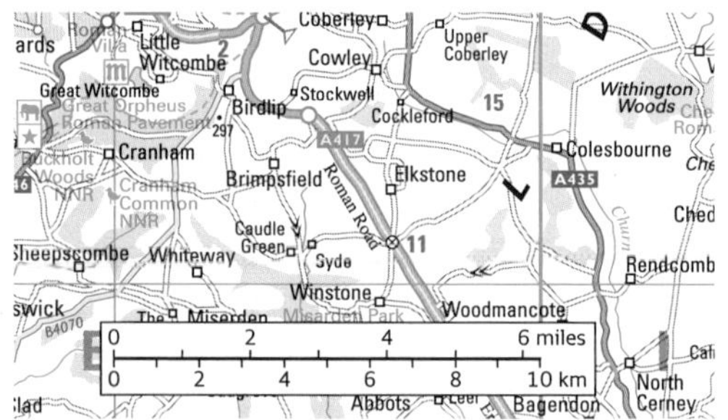

In a scale drawing, one length is used to represent another. For example, a map cannot be drawn to the same size as the area it represents. The measurements are scaled down, to make a map of a size that can be conveniently used by motorists, tourists and walkers.

Satellite navigation can now bring maps to life when you are driving.

Architects use scale drawings to show views of a planned house from different directions.

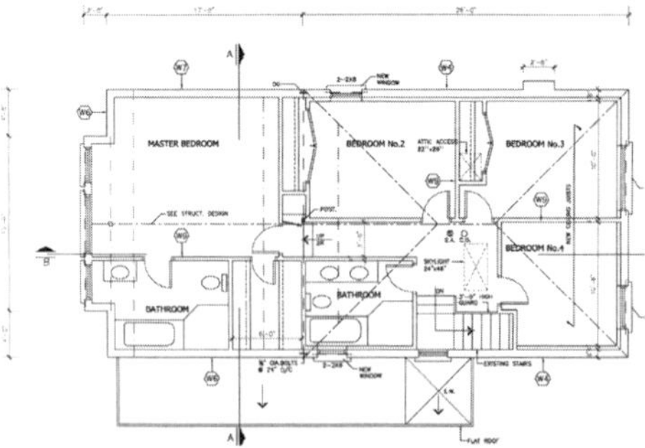

Computers can be programmed to design nets to produce boxes and cartons that use the minimum possible amount of material.

In computer design, people who design microchips need to scale up their drawings, as the dimensions they work with are so small. Use the Internet to research more about how scale drawings are used.

Chapter 6

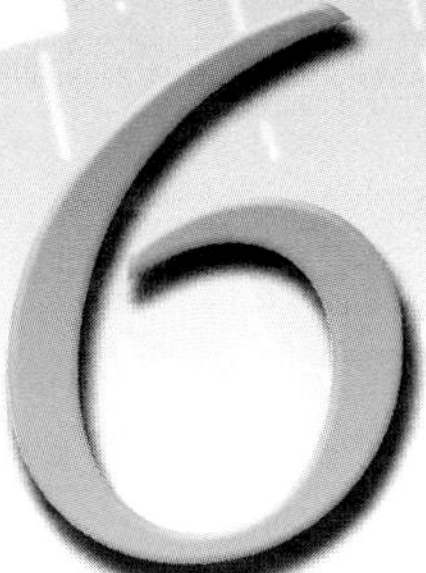

Geometry and measures: Scale and drawing

The grades given in this chapter are target grades.

1 Reading scales

2 Sensible estimates

3 Scale drawings

4 Nets

5 Using an isometric grid

This chapter will show you …

- **G** how to read scales
- **F** how to make estimations
- **E** how to read map scales
- **E** how to draw nets of 3D shapes
- **D** how to draw plans and elevations

Visual overview

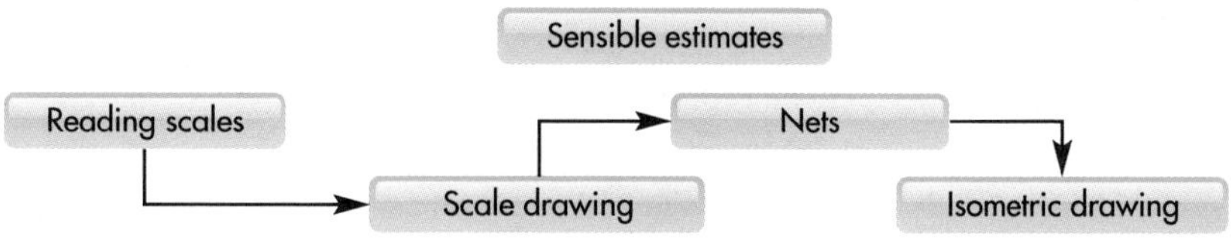

What you should already know

- The names of common 3D shapes **(KS3 level 3, GCSE grade G)**
- How to measure lengths of lines **(KS3 level 4, GCSE grade G)**
- The metric units for length **(KS3 level 4, GCSE grade G)**

Quick check

Name these 3D shapes.

1

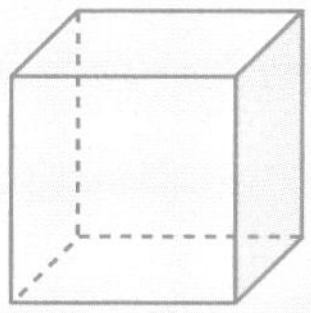

2

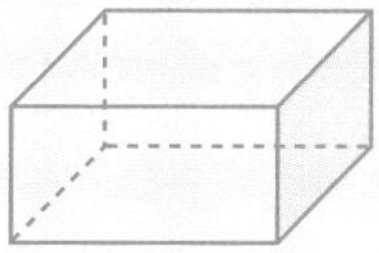

3

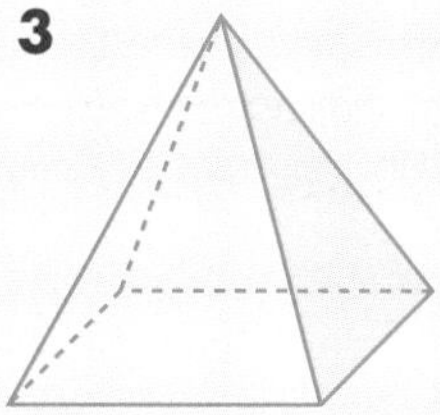

4

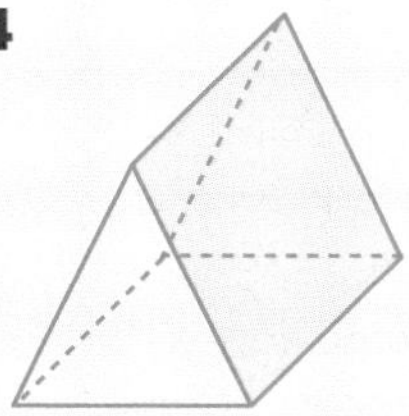

5

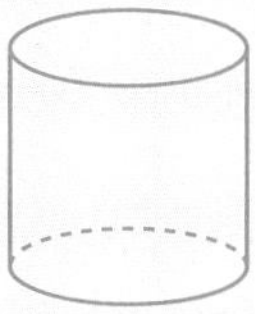

6

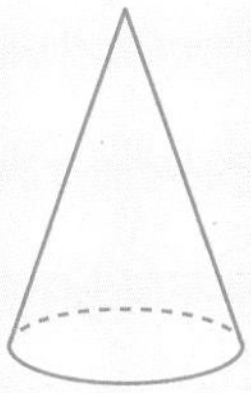

6.1 Reading scales

This section will show you how to:

- read and interpret scales

Key words
divisions
scales
units

You will come across **scales** in a lot of different places.

For example, there are scales on thermometers, car speedometers and weighing scales. It is important that you can read scales accurately.

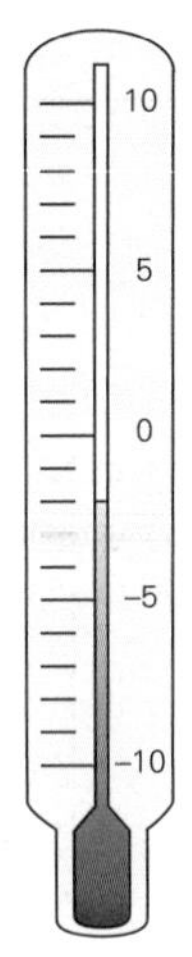

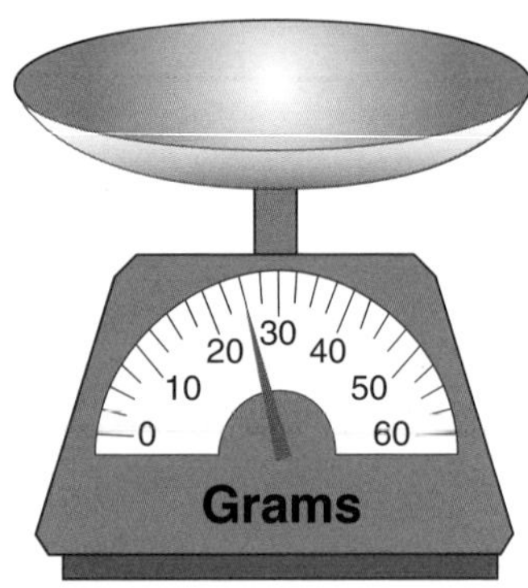

There are two things to do when reading a scale. First, make sure that you know what each **division** on the scale represents. Second, make sure you read the scale in the right direction, for example some scales read from right to left.

Also, make sure you note the **units**, if given, and include them in your answer.

EXAMPLE 1

Read the values from the following scales.

a 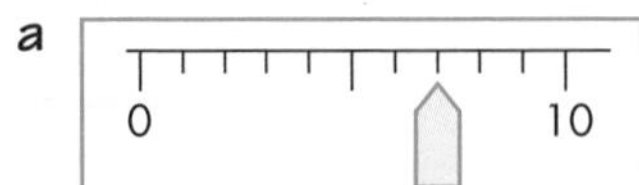

b 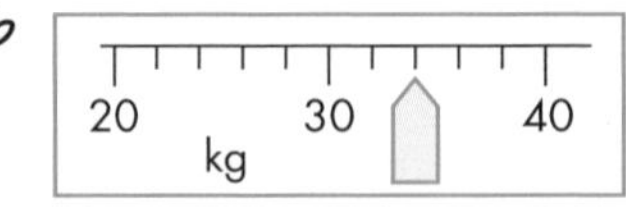

c 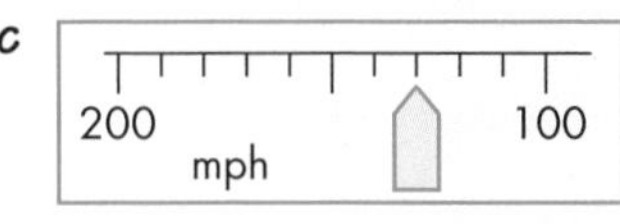

a The scale shows 7. This is a very straightforward scale. It reads from left to right and each division is worth 1 unit.

b The scale shows 34 kg. The scale reads from left to right and each division is worth 2 units.

c The scale shows 130 mph. The scale reads from right to left and each division is worth 10 units. You should know that mph stands for miles per hour. This is a unit of speed found on most British car speedometers.

EXERCISE 6A

1 Read the values from the following scales.
Remember to state the units if they are shown.

HINTS AND TIPS
Remember to check what each division is worth.

a **i**

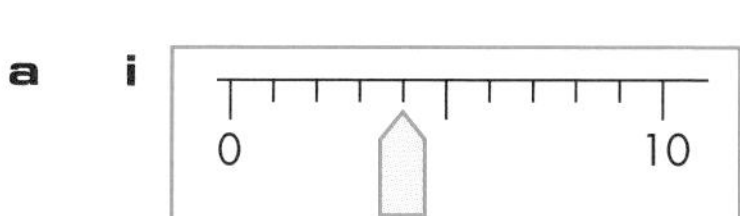

ii

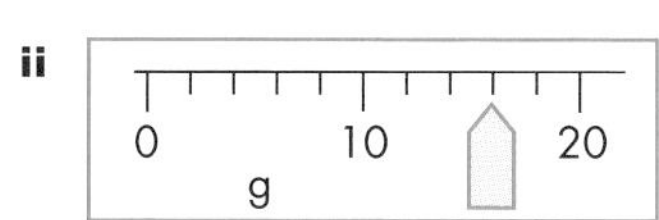

iii

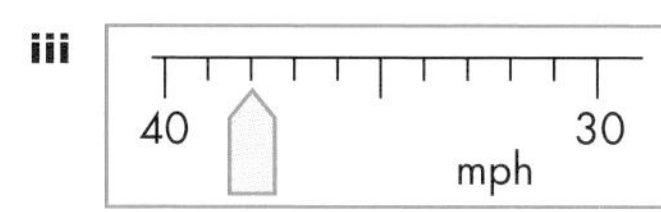

b **i**

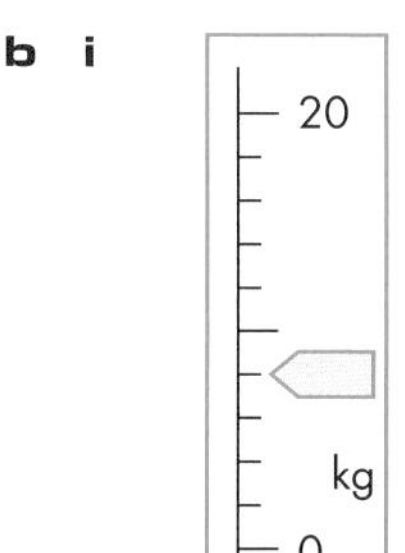

ii

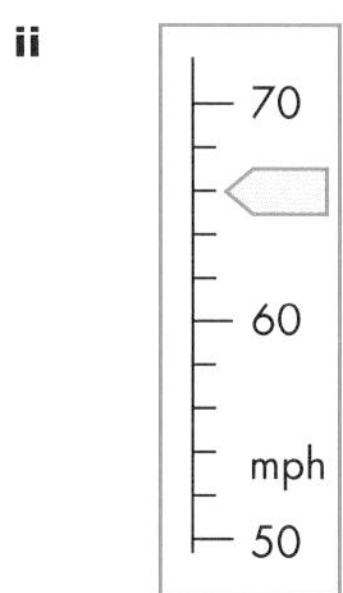

iii

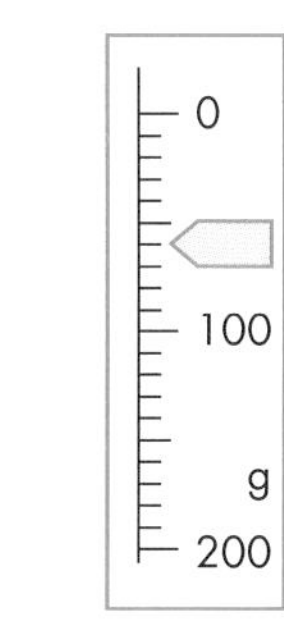

c **i**

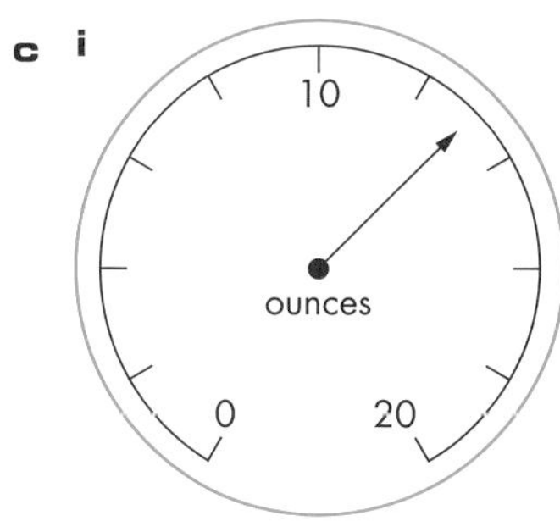

ii

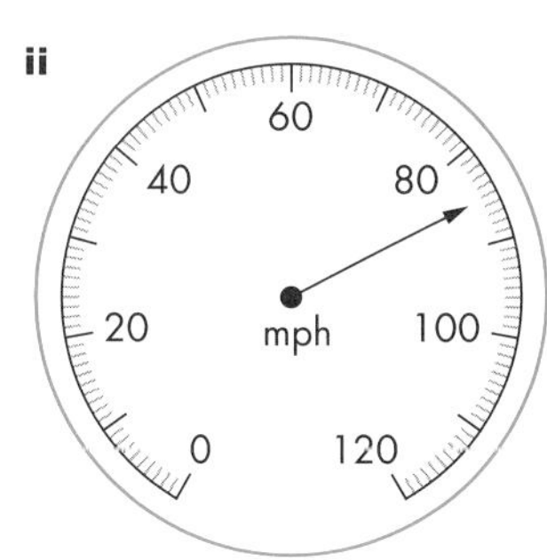

iii

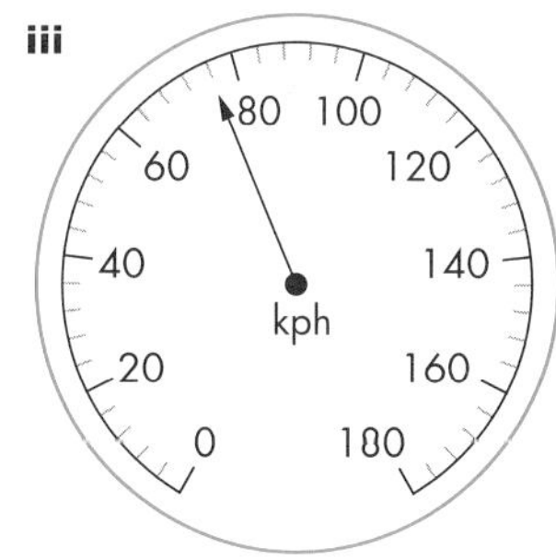

d **i**

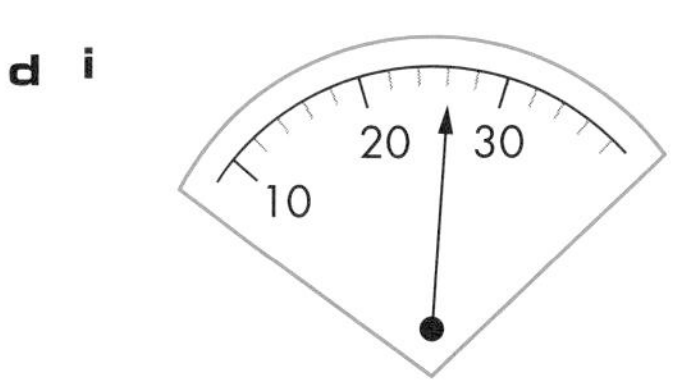

ii

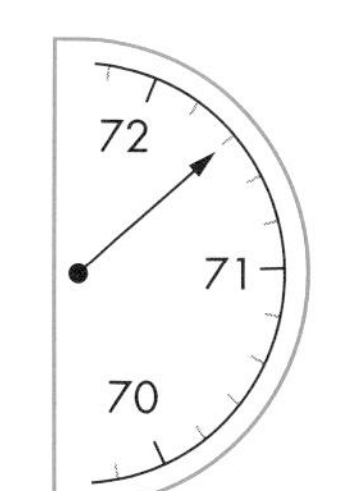

iii

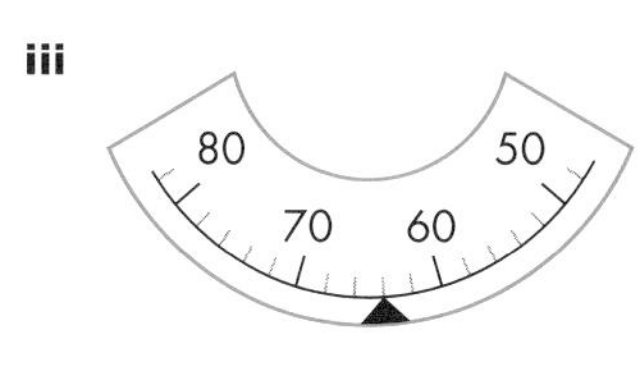

G

2 Copy (or trace) the following dials and mark on the values shown.

a

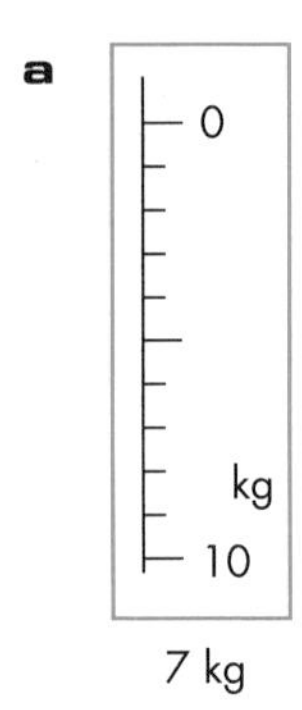

7 kg

b

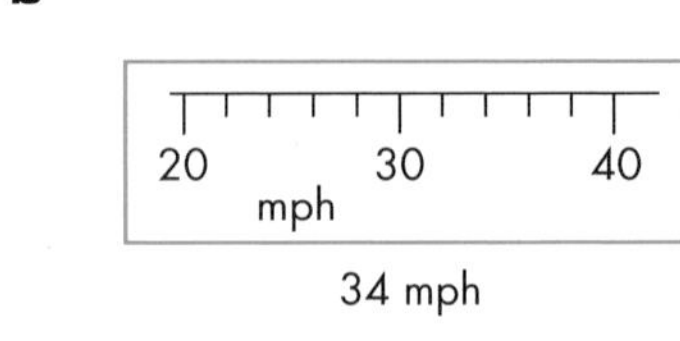

34 mph

c

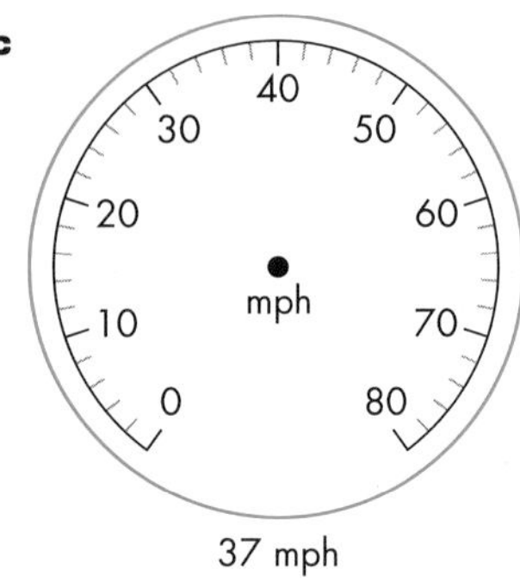

37 mph

d

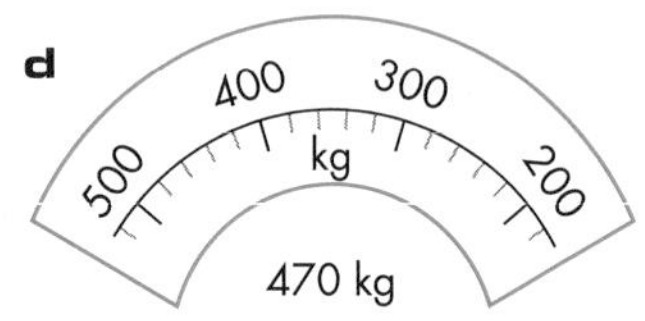

470 kg

e

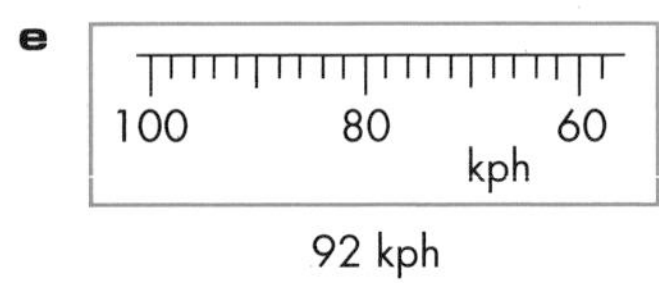

92 kph

f

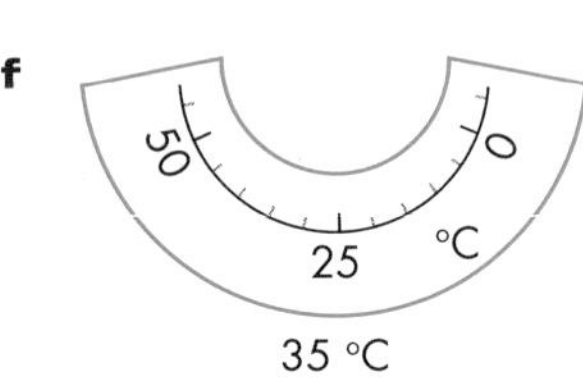

35 °C

3 Read the temperatures shown by each of these thermometers.

a

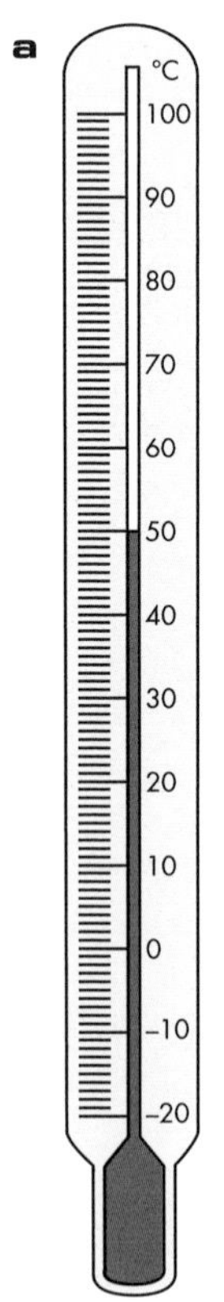

b

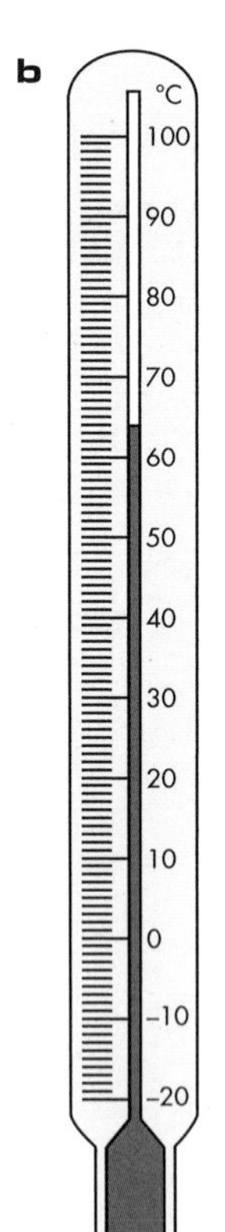

c

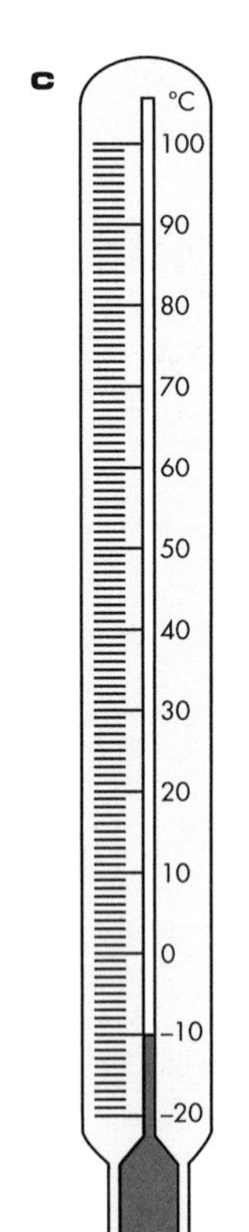

d

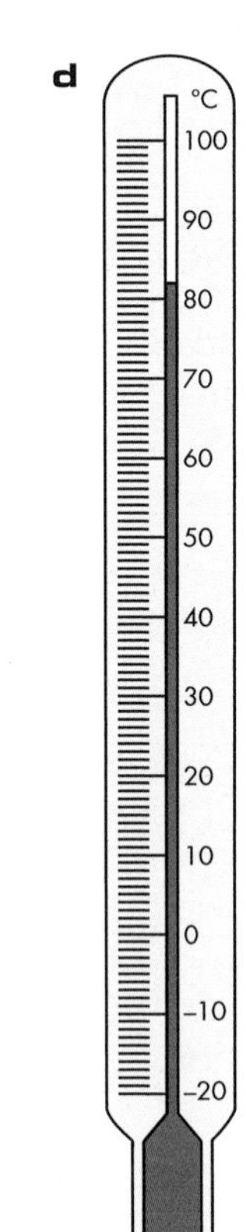

e

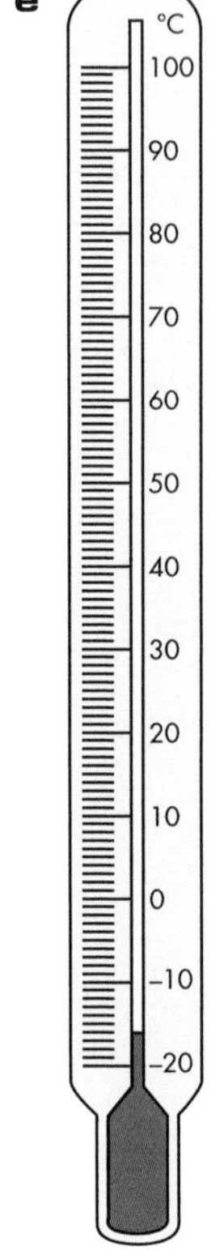

G

4 Read the values shown on these scales.

a

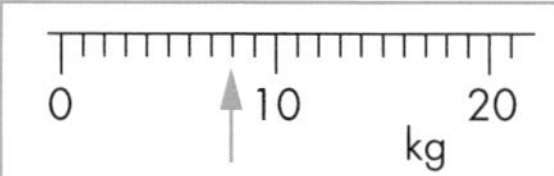

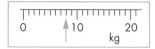

b

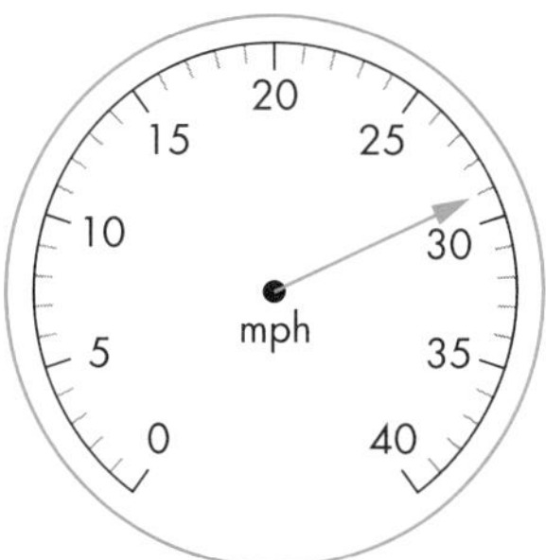

c

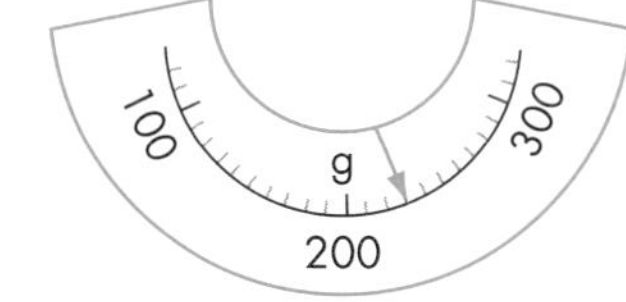

d

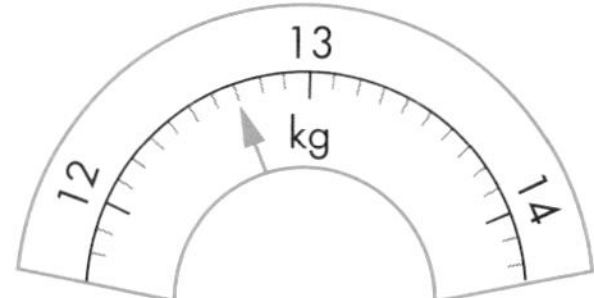

5 Susie is using kitchen scales to weigh out flour.

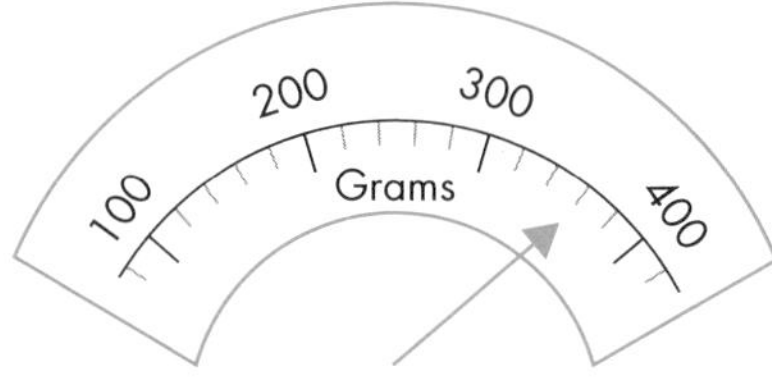

a What is the weight of the flour shown on the scales?

b These scales can weigh items up to 400 g. Susie needs to weigh 700 g of currants using these scales. Explain how she could do this.

F

6 A pineapple was weighed.

A pineapple and an orange were weighed together.

a How much does the pineapple weigh?
Give your answer in kilograms.

b How much does the orange weigh?
Give your answer in grams.

F

PS 7 **a** What speed is shown on the scale?

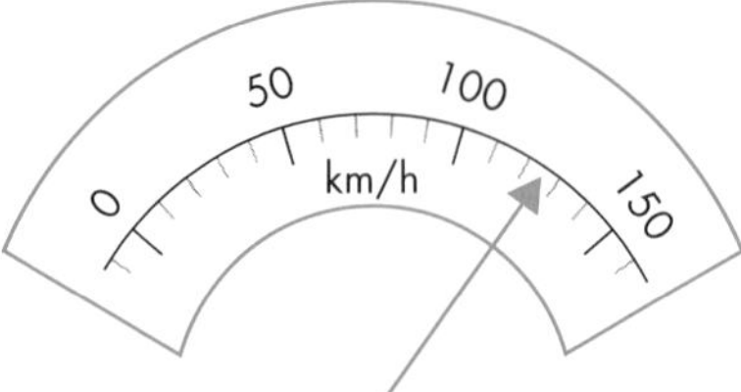

b Copy this scale and show the same speed as in part **a**.

AU 8 Dean says that the arrow on this scale is pointing to 7.49 m.

Dean is not correct. Explain the mistake that he has made.

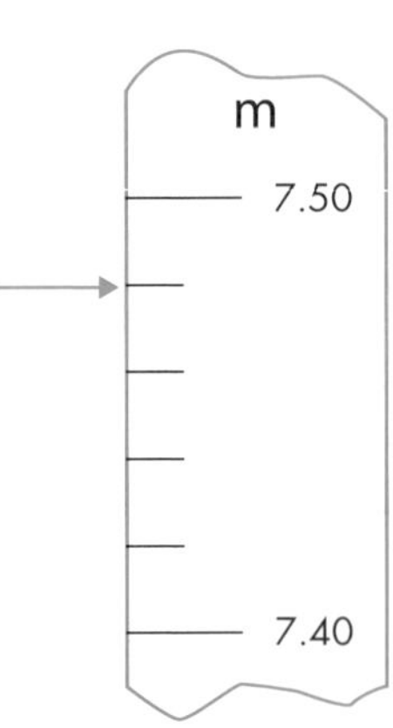

6.2 Sensible estimates

This section will show you how to:

- make sensible estimates using standard measures

Key word

estimate

The average height of a man is 1.78 m. For a sensible **estimate**, we would usually say that the height of a man is about 1.8 m. From this information, you can estimate the lengths or heights of other objects.

EXAMPLE 2

Look at the picture.

Estimate, in metres, the height of the lamppost and the length of the bus.

Assume the man is about 1.8 m tall. The lamppost is about three times as high as he is. **Note:** One way to check this is to use tracing paper to mark off the height of the man and then measure the other lengths against this. This makes the lamppost about 5.4 m high, or close to 5 m high. As it is an estimate, there is no need for an exact value.

The bus is about four times as long as the man so the bus is about 7.2 m long, or close to 7 m long.

EXAMPLE 3

Look at the picture.

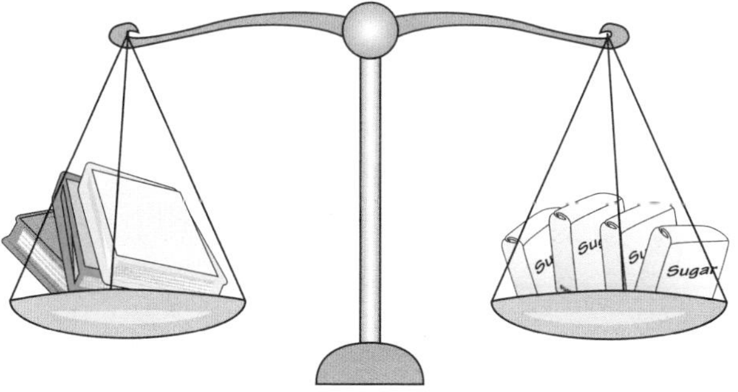

It shows three maths textbooks balanced by four bags of sugar. Estimate the mass of one textbook.

You should know that a bag of sugar weighs 1 kilogram, so the three maths books weigh 4000 grams. This means that each one weighs about 1333 grams or about 1.3 kg.

EXERCISE 6B 

F

1 The car in the picture is 4 metres long. Use this to estimate the length of the bicycle, bus and the train.

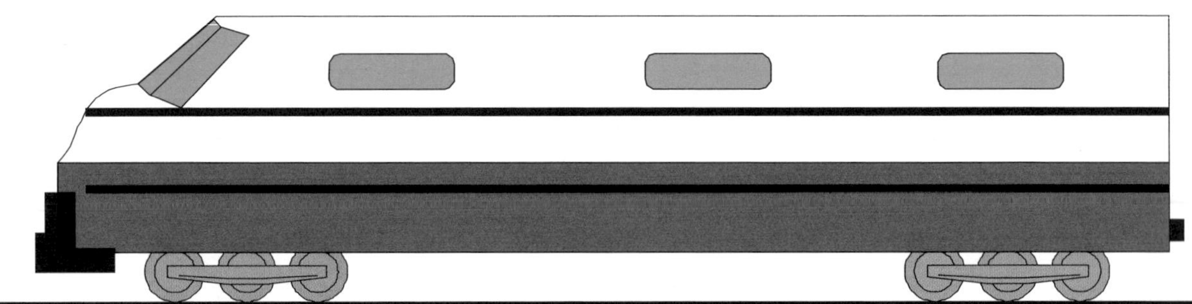

2 Estimate the greatest height and length of the whale.

HINTS AND TIPS

Remember, a man is about 1.8 m tall.

3 Estimate the weight of one apple.

HINTS AND TIPS

Remember, a bag of sugar weighs 1 kg.

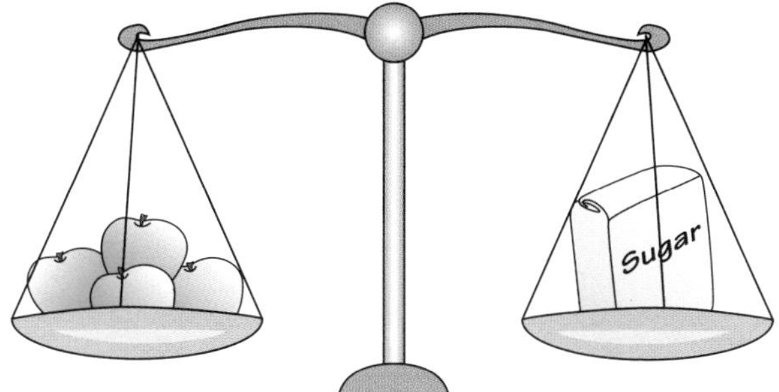

4

Estimate the following.

a the height of the traffic lights

b the width of the road

c the height of the flagpole

5

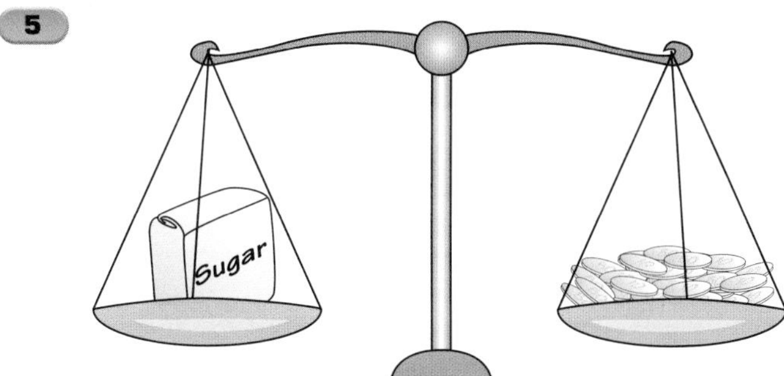

A charity collection balances pound coins against a bag of sugar. It take £105 to balance the bag of sugar. Estimate the weight of one pound coin.

F

6 This is an illustration of Joel standing next to a statue.

Joel's height is 1.5 m. Explain how he could use this information to estimate the height of the statue.

AU 7 Estimate the height of Tyrannosaurus Rex.

HINTS AND TIPS

Remember, a man is about 1.8 m tall.

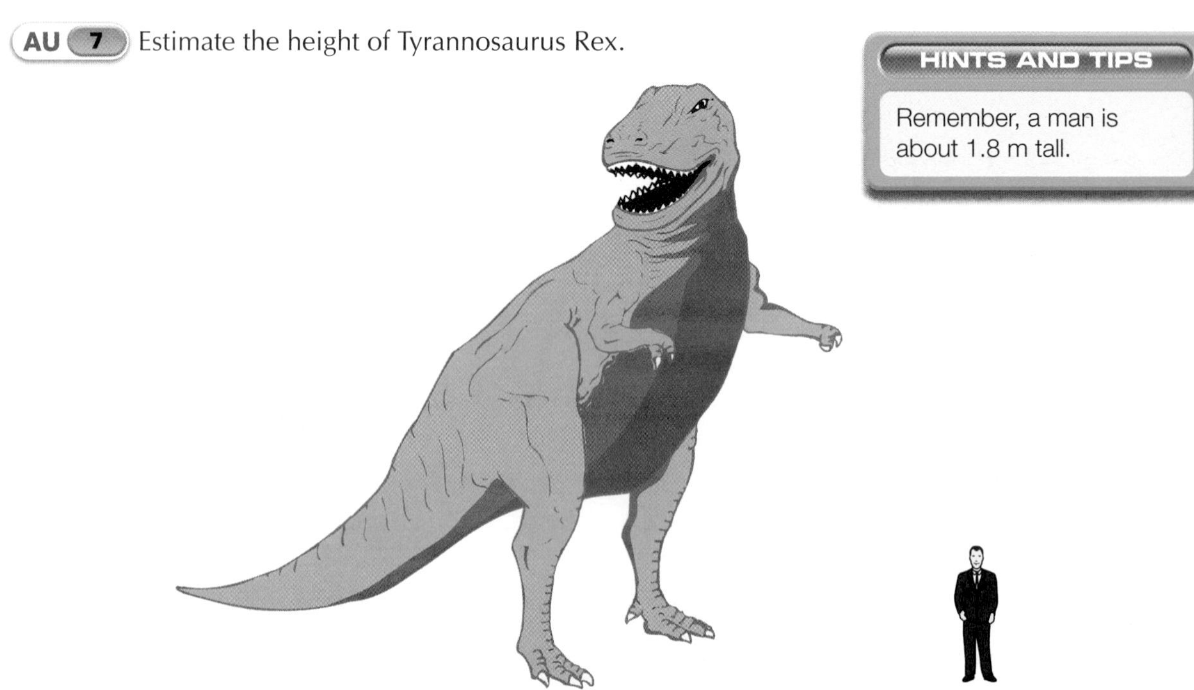

6.3 Scale drawings

This section will show you how to:

- read scales and draw scale drawings

Key words

measurement
ratio
scale drawing
scale factor

A **scale drawing** is an accurate representation of a real object.

Scale drawings are usually smaller in size than the original objects. However, in certain cases, they have to be enlargements, typical examples of which are drawings of miniature electronic circuits and very small watch movements.

In a scale drawing:

- all the measurements must be in proportion to the corresponding measurements on the original object
- all the angles must be equal to the corresponding angles on the original object.

To obtain the measurements for a scale drawing, all the actual measurements are multiplied by a common **scale factor**, usually referred to as a scale. (See the section on enlargements in Book 2, Chapter 10.)

Scales are often given as **ratios**, for example, 1 cm : 1 m.

When the units in a ratio are the *same*, they are normally not given. For example, a scale of 1 cm : 1000 cm is written as 1 : 1000.

Note When making a scale drawing, take care to express *all* **measurements** in the *same* unit.

EXAMPLE 4

The diagram shows the front of a kennel.
It is drawn to a scale of 1 : 30. Find:

a the actual width of the front

b the actual height of the doorway.

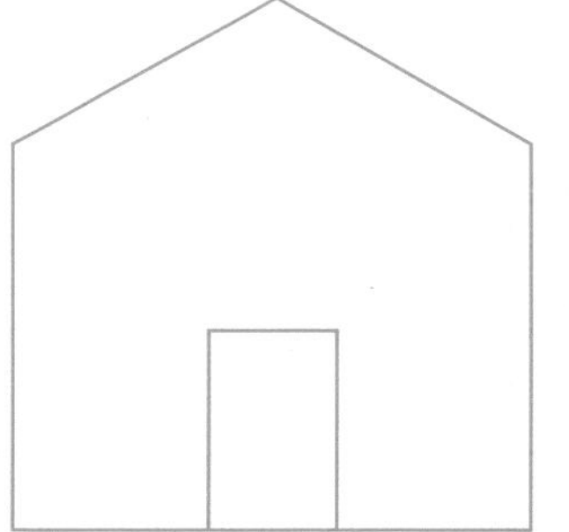

The scale of 1 : 30 means that a measurement of 1 cm on the diagram represents a measurement of 30 cm on the actual kennel.

a So, the actual width of the front is
$4 \text{ cm} \times 30 = 120 \text{ cm}$

b The actual height of the doorway is
$1.5 \text{ cm} \times 30 = 45 \text{ cm}$

Map scales are usually expressed as ratios, such as 1 : 50 000 or 1 : 200 000.

The first ratio means that 1 cm on the map represents 50 000 cm or 500 m on the land.
The second ratio means that 1 cm represents 200 000 cm or 2 km.

EXAMPLE 5

Find the actual distances between the following towns.

a Bran and Kelv   **b** Bran and Daid   **c** Daid and Malm

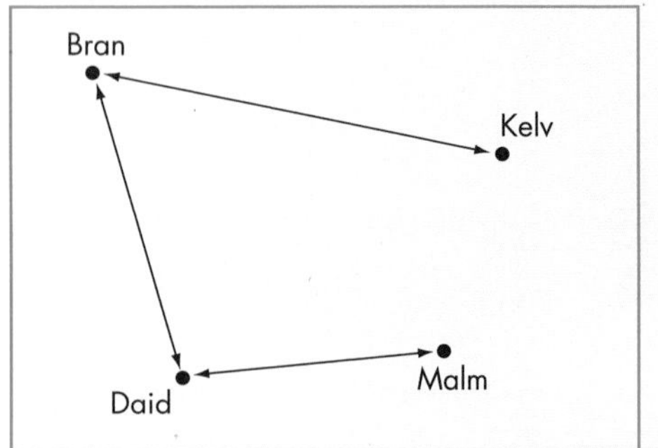

This map is drawn to a scale of 1 : 2 000 000.

2 000 000 cm = 20 000 m = 20 km so the scale means that a distance of 1 cm on the map represents a distance of 20 km on the land.

So, the actual distances are:

a Bran and Kelv: 4×20 km = 80 km

b Bran and Daid: 3×20 km = 60 km

c Daid and Malm: 2.5×20 km = 50 km

EXERCISE 6C

F

1 Look at this plan of a garden.

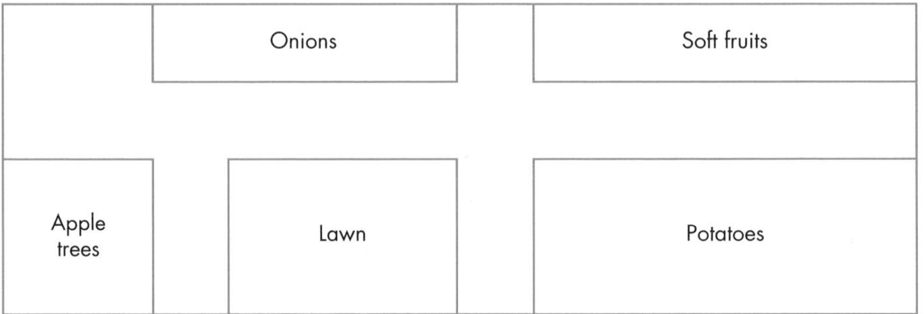

Scale: 1 cm represents 10 m

a State the actual dimensions of each plot of the garden.

b Calculate the actual area of each plot.

2 Below is a plan for a mouse mat.

HINTS AND TIPS

Remember to check the scale.

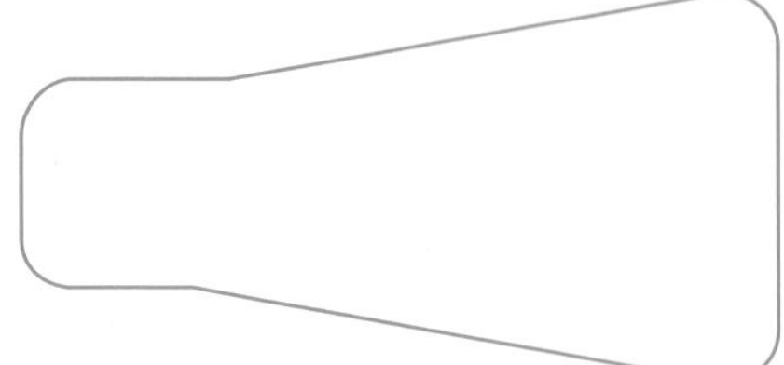

Scale: 1 cm represents 6 cm

a How long is the actual mouse mat?

b How wide is the narrowest part of the mouse mat?

F

3 Below is a scale plan of the top of Derek's desk, where the scale is 1 : 10.

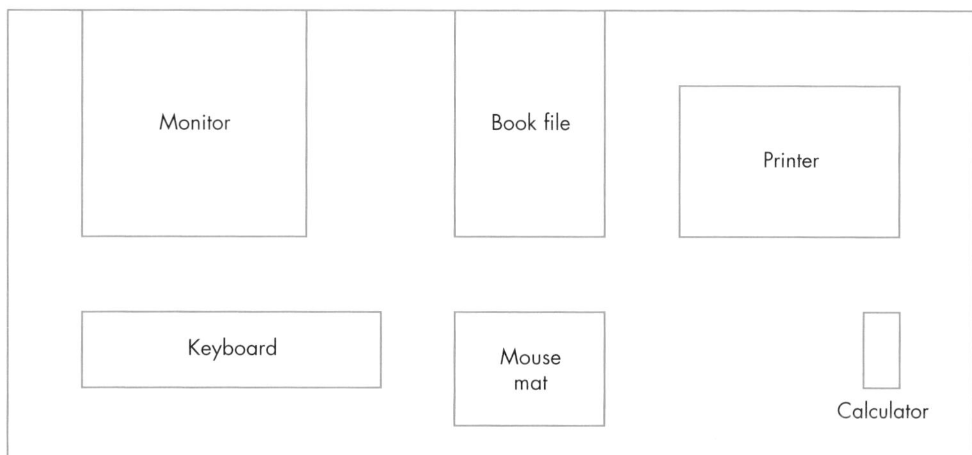

What are the actual dimensions of each of these objects?

a monitor **b** keyboard **c** mouse mat

d book file **e** printer **f** calculator

FM **4** The diagram shows a sketch of a garden.

a Make an accurate scale drawing of the garden.

Use a scale of 1 cm to represent 2 m.

b Marie wants to plant flowers along the side marked x on the diagram. The flowers need to be planted 0.5 m apart. Use your scale drawing to work out how many plants she needs.

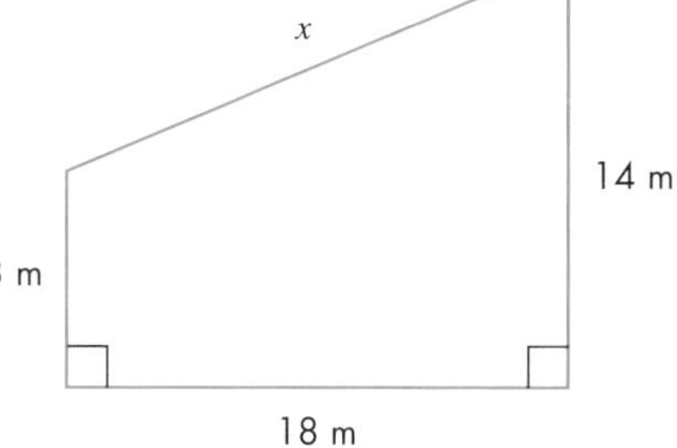

E

5 Look at the map below, drawn to a scale of 1 : 200 000.

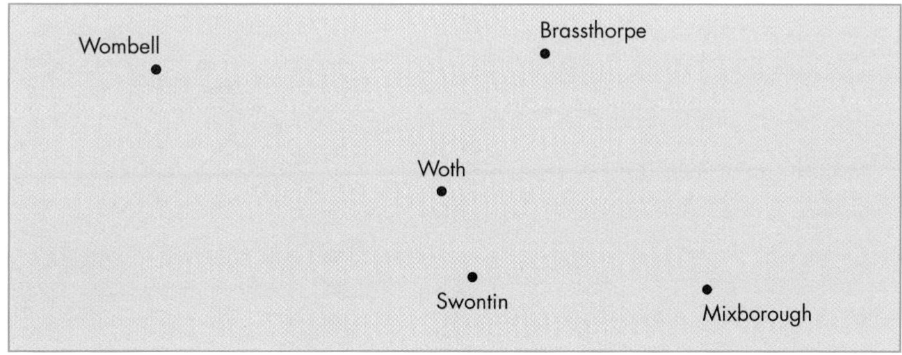

State the following actual distances to the nearest tenth of a kilometre.

a Wombell to Woth **b** Woth to Brassthorpe

c Brassthorpe to Swontin **d** Swontin to Mixborough

e Mixborough to Woth **f** Woth to Swontin

E

6 This map is drawn to a scale of 1 : 4 000 000.

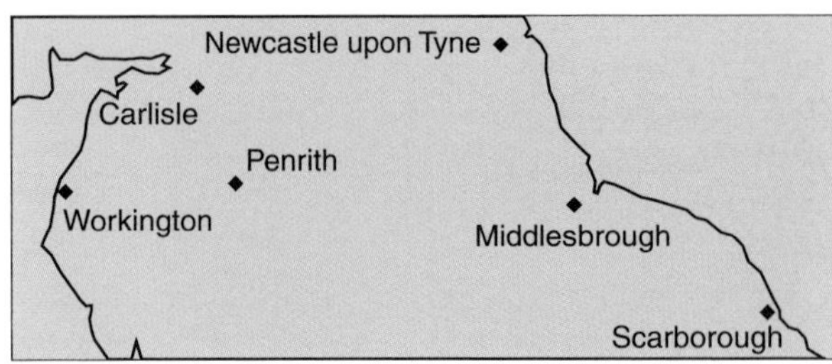

Give the approximate direct distances for each of the following.

a Penrith to:

i Workington **ii** Scarborough

iii Newcastle upon Tyne **iv** Carlisle

b Middlesbrough to:

i Scarborough **ii** Workington

iii Carlisle **iv** Penrith

7 This map is drawn to a scale of 1 : 2 000 000.

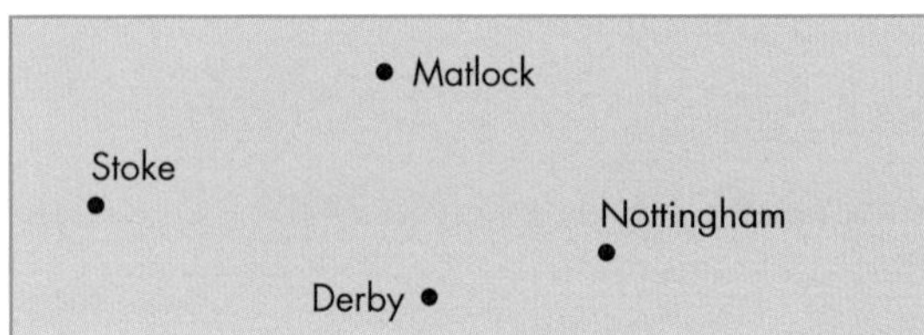

State the direct distance, to the nearest 5 kilometres, from Matlock to the following.

a Stoke **b** Derby **c** Nottingham

AU 8 Here is a scale drawing of the Great Beijing Wheel in China.

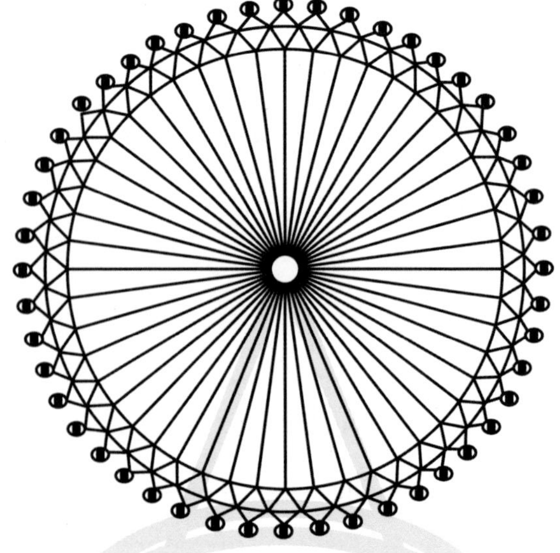

The height of the wheel is 210 m.

Which of the following is the correct scale?

a 1 : 30

b 1 : 700

c 1 : 3000

d 1 : 30 000

E

PS 9 Maps for walkers often give a scale of 1 inch represents 1 mile. Use these measurements to write this scale as a ratio.

12 inches = 1 foot

3 feet = 1 yard

1760 yards = 1 mile

ACTIVITY

Little and large!

Ask your teacher for a map of Britain.

- Use the scale that is given in miles.
- Use the scale on the map to find the direct distance between:
  - Sheffield and London
  - Birmingham and Oxford
  - Glasgow and Bristol.
- Give your answers to the nearest 10 miles.

6.4 Nets

This section will show you how to:
- draw and recognise shapes from their nets

Key words
net
3D shape

Many of the **3D shapes** that you come across can be made from **nets**.

A net is a flat shape that can be folded into a 3D shape.

EXAMPLE 6

Sketch the net for each of these shapes.

a cube **b** square-based pyramid

a This is a sketch of a net for a cube.

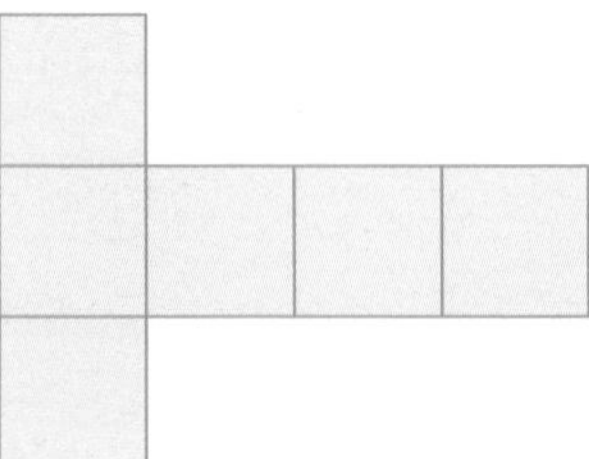

b This is a sketch of a net for a square-based pyramid.

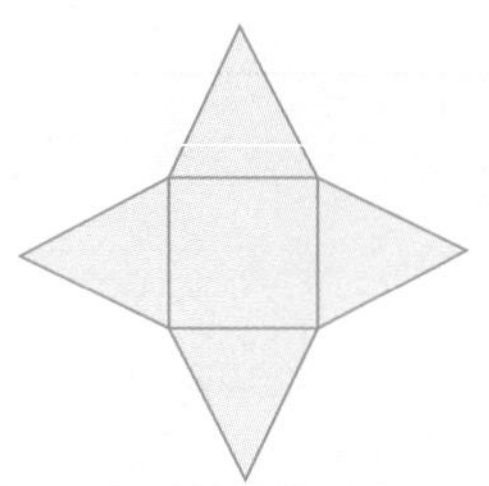

EXERCISE 6D

1 Draw, on squared paper, an accurate net for each of these cuboids.

a

2 cm
3 cm
4 cm

b

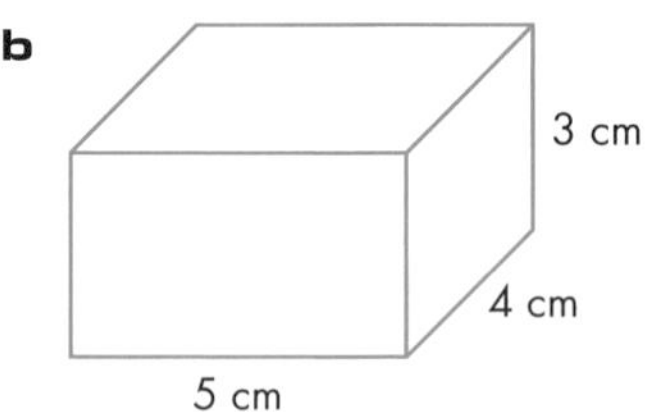

c

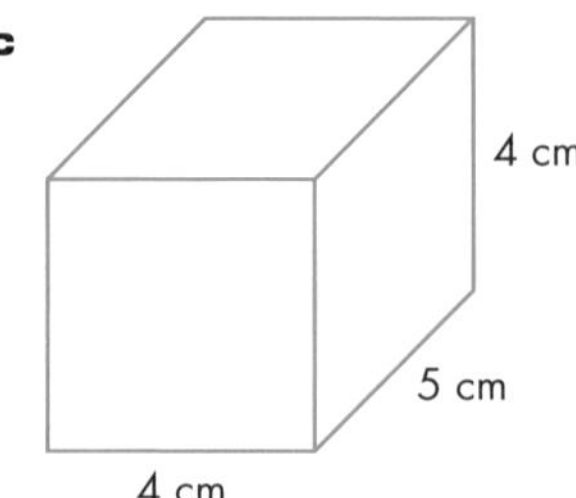

FM 2 Jenny is making an open box from card.

This is a sketch of the box.

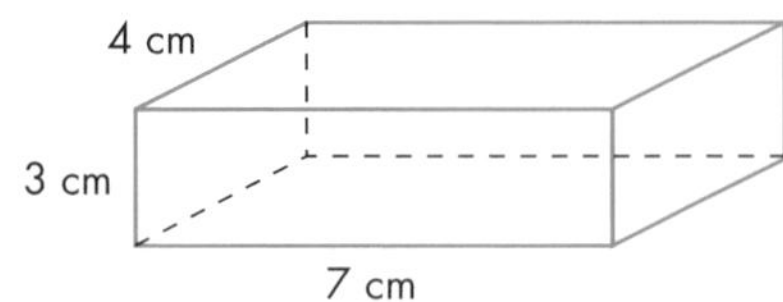

Jenny has a piece of card that measures 15 cm by 21 cm. Can she make the box using this card?

3 The shape on the right is a triangular prism. Its ends are isosceles triangles and its other faces are rectangles. Draw an accurate net for this prism on squared paper.

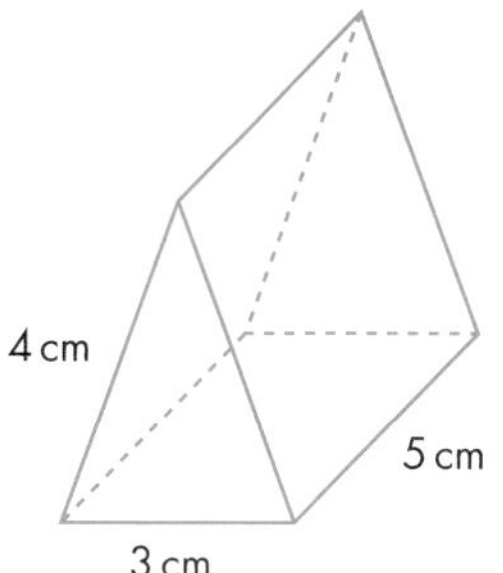

4 Sketch the nets of these shapes.

a

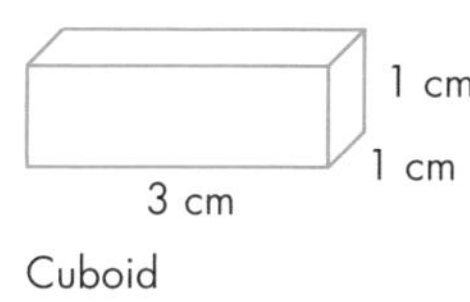

Cuboid

b

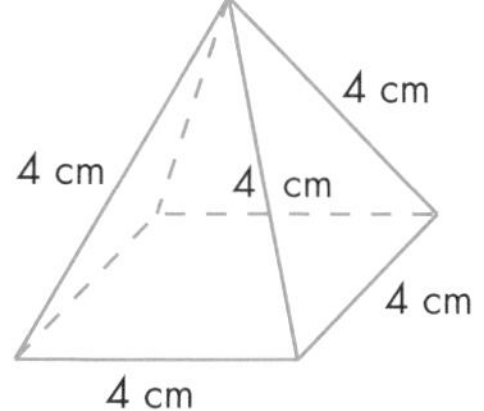

Square-based pyramid

c

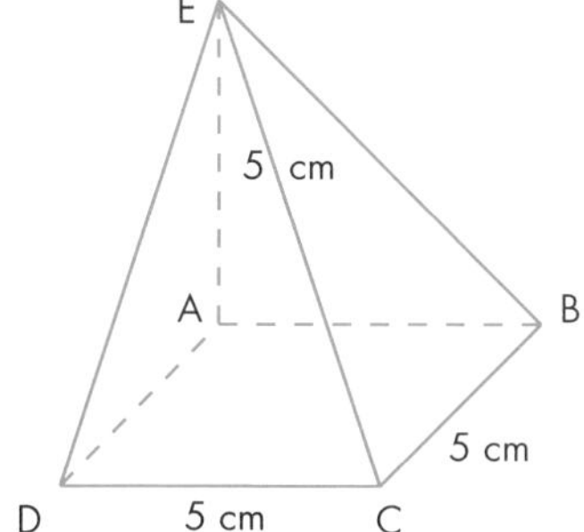

Square-based pyramid, with point E directly above point A

d

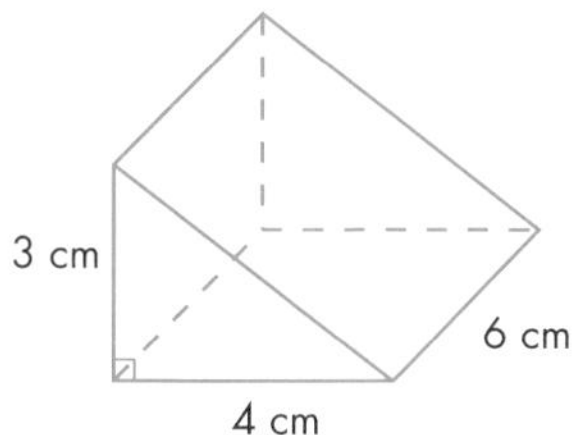

Right-angled triangular prism

PS 5 Here is a net for a cube.

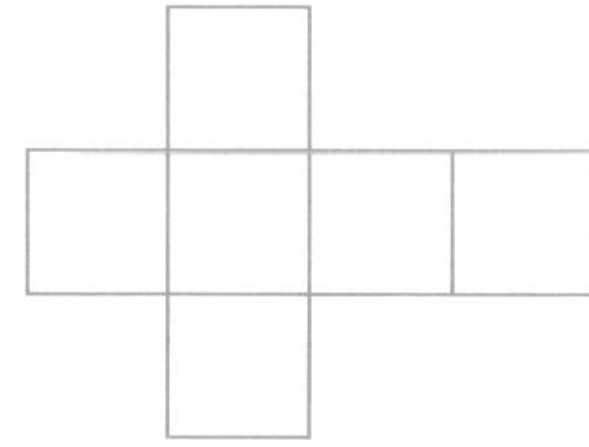

How many different nets can you draw for a cube?

> **HINTS AND TIPS**
> There are 11 altogether. How many can you find?

AU 6 Which of the following are nets for a square-based pyramid?

a

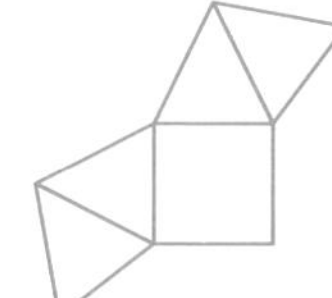

b

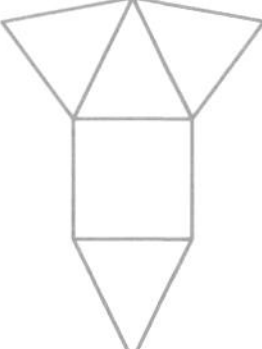

c

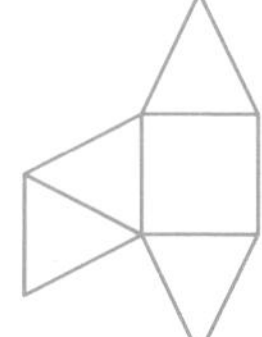

6.5 Using an isometric grid

This section will show you how to:
- read from and draw on isometric grids
- interpret diagrams to find plans and elevations

Key words
elevation
front elevation
isometric grid
plan
side elevation

Isometric grids

The problem with drawing a 3D shape is that you have to draw it on a flat (2D) surface so that it looks like the original 3D shape. The drawing is given the appearance of depth by slanting the view.

One easy way to draw a 3D shape is to use an **isometric grid** (a grid of equilateral triangles).

EXAMPLE 7

Below are two drawings of the same cuboid, one on squared paper, the other on isometric paper. The cuboid measures $5 \times 4 \times 2$ units.

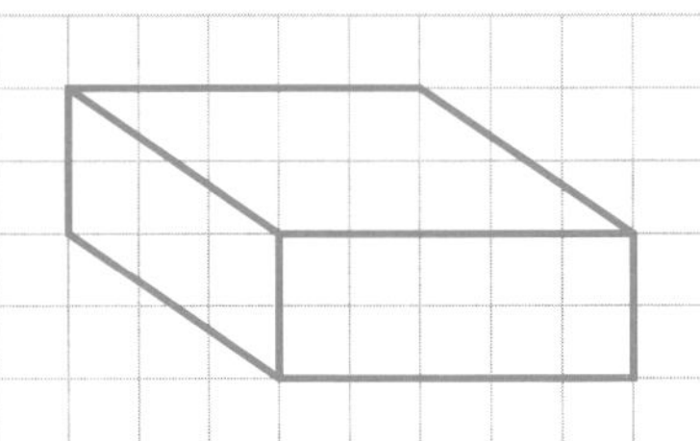

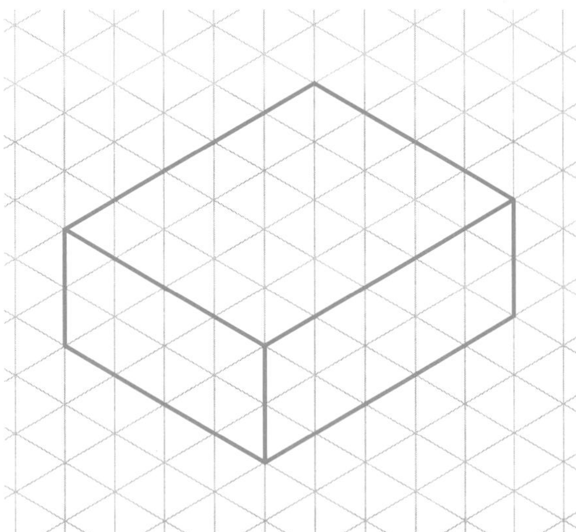

Note: The dimensions of the cuboid can be taken straight from the isometric drawing, whereas they cannot be taken from the drawing on squared paper.

You can use a triangular dot grid instead of an isometric grid but you *must* make sure that it is the correct way round – as shown here.

Plans and elevations

A **plan** is the view of a 3D shape when it is seen from above.

An **elevation** is the view of a 3D shape when it is seen from the front or from another side.

EXAMPLE 8

The 3D shape below is drawn on an triangular dot grid.

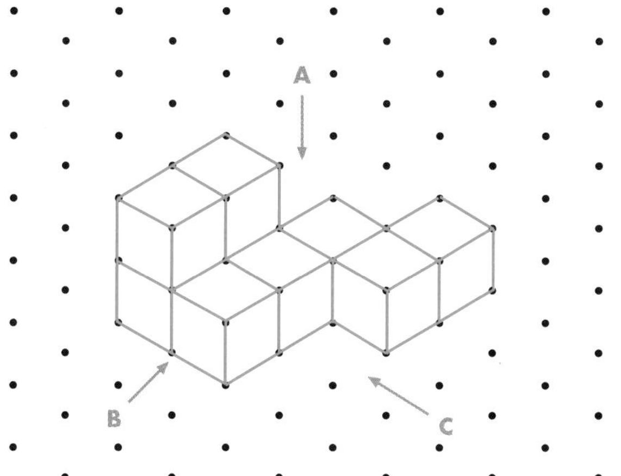

Its plan, front elevation and side elevation can be drawn on squared paper.

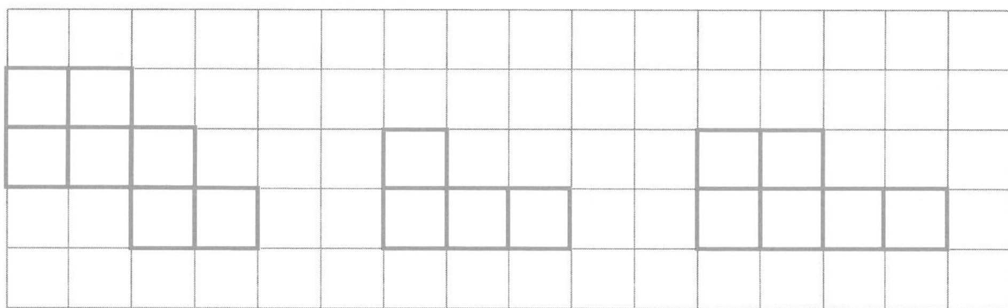

Plan from **A** | **Front elevation** from **B** | **Side elevation** from **C**

EXERCISE 6E

E

1 Draw each of these cuboids on an isometric grid.

a

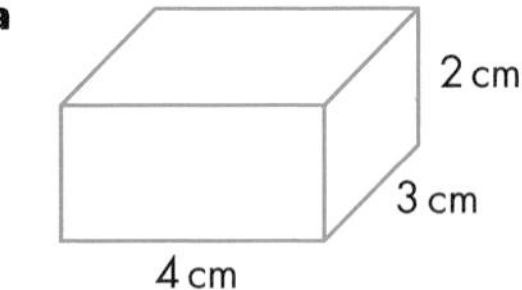

b

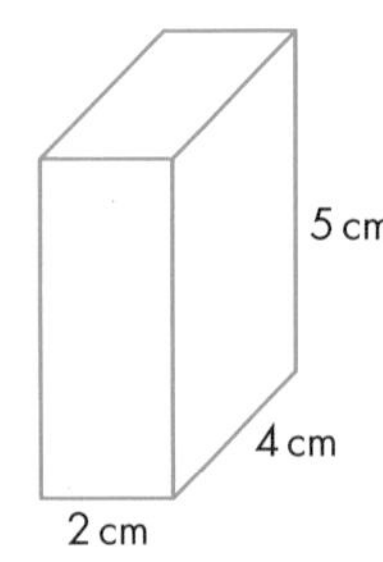

c

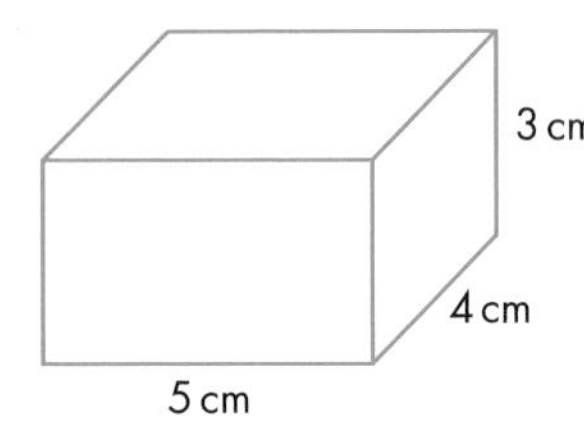

D

AU 2 The diagram shows an L-shaped prism.

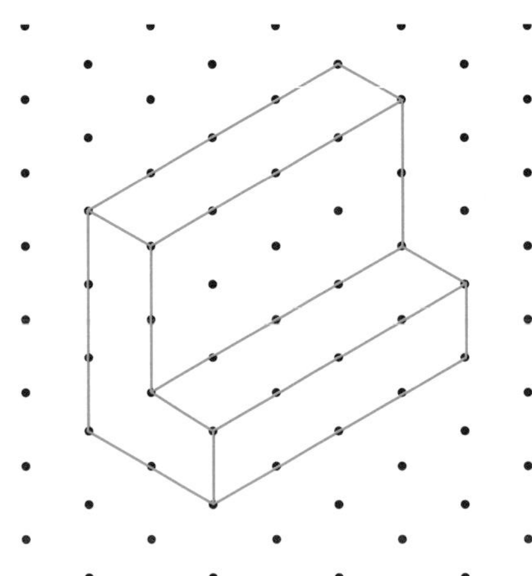

On squared paper, draw a front elevation and a side elevation.

3 For each of the following 3D shapes, draw the following on squared paper.

i the plan **ii** the front elevation **iii** the side elevation

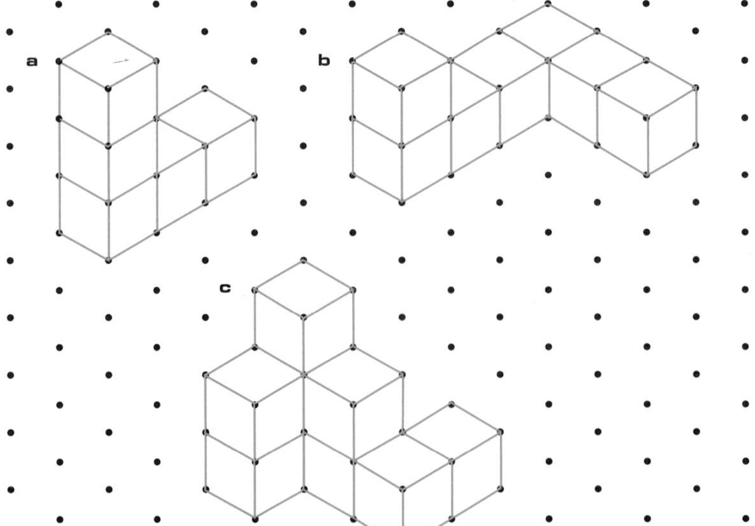

1 **a** Write down the number marked by the arrow. (1)

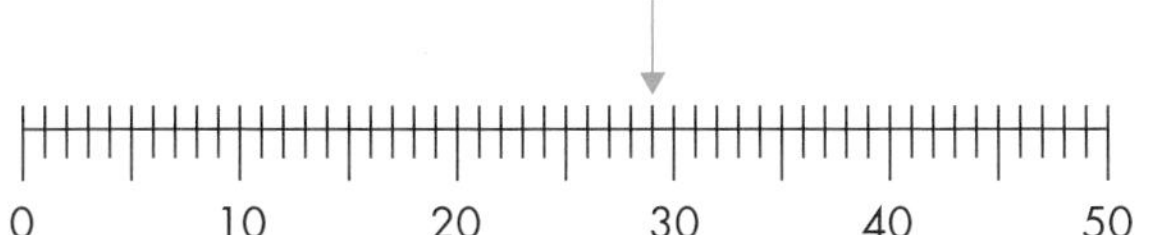

b Find the number 120 on the number line. Mark it with an arrow. (↓) (1)

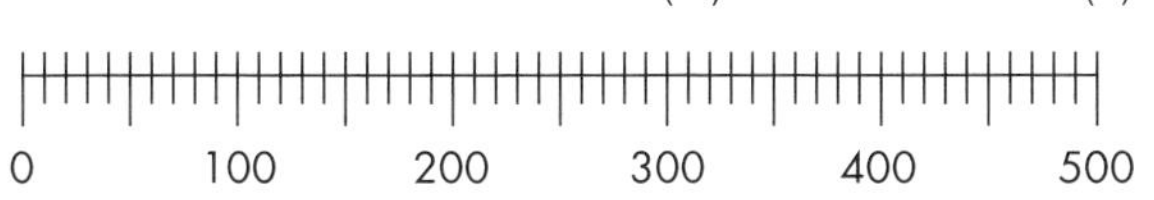

(Total 2 marks)

Edexcel, November 2008, Paper 10 Foundation, Unit 3 Test, Question 2

2 This is part of a ruler.

a Write down the length marked with an arrow. (1)

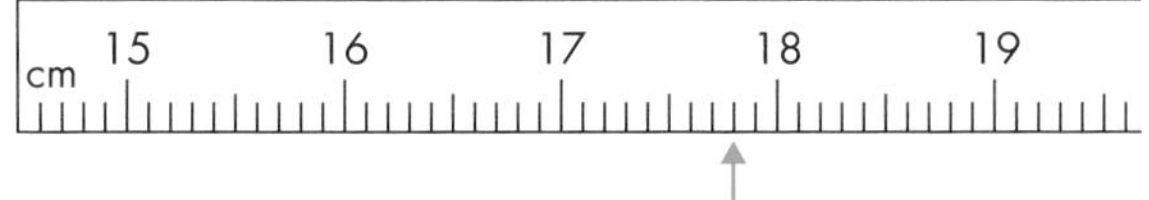

This is a thermometer.

b Write down the temperature shown. (1)

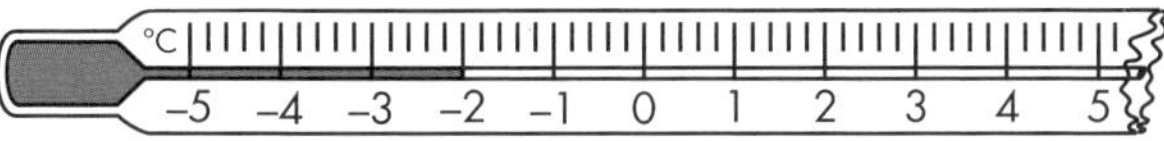

This is a parcel on some scales.

c Write down the weight of the parcel. (1)

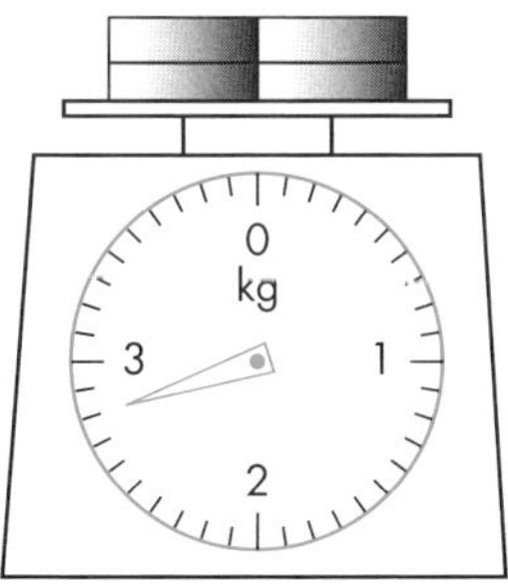

(Total 3 marks)

Edexcel, November 2007, Paper 10 Foundation, Unit 3 Test, Question 1

3 The diagram shows a 1 litre measuring flask.
700 ml of milk are needed for a recipe.
Copy the scale. Draw an arrow to show where 700 ml is on the scale.

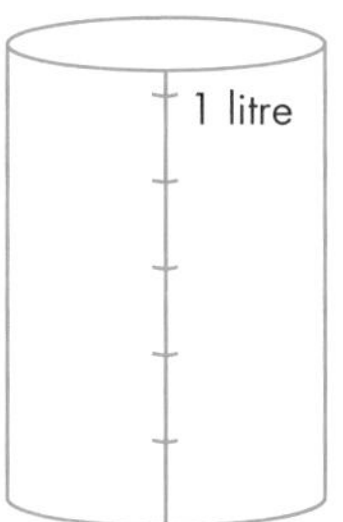

4

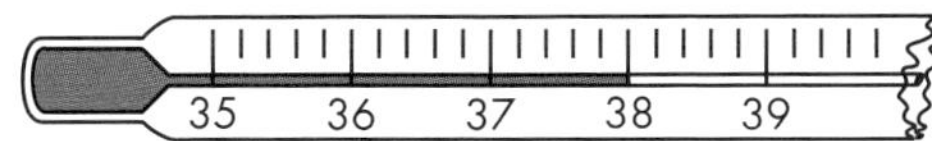

a The thermometer shows Peter's temperature in degrees Celsius. What is his temperature?

b

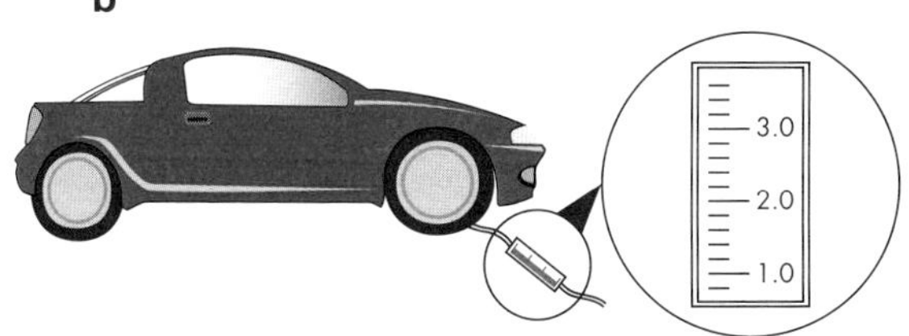

The tyre pressure for Peter's car is 2.7 units.

Copy the scale. Use an arrow to show 2.7 on your scale.

5 Six nets are shown below.
List the nets that would not make a cube.

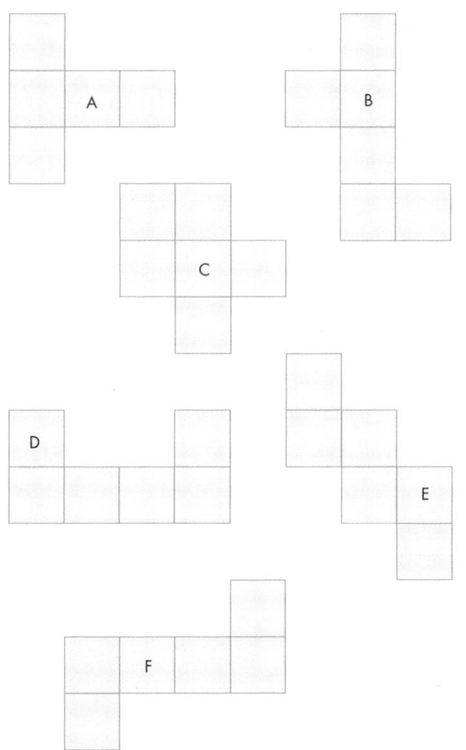

6 **a** This is the petrol gauge on Patricia's car.

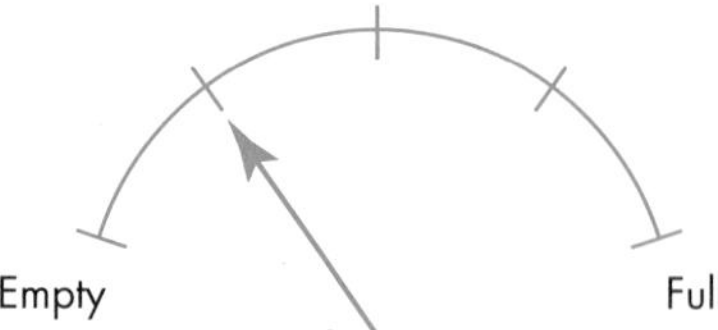

When full, the tank contains 32 litres.
Estimate the amount of petrol in the tank.

b Petrol costs 97p per litre.
Calculate the cost of 30 litres.

7

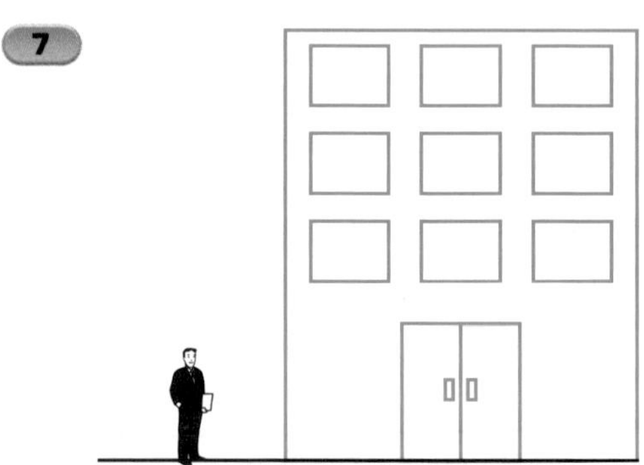

The diagram shows a building and a man.

The man is of normal height.

The man and the building are drawn to the same scale.

a Write down an estimate for the height of the man. (1)

b Write down an estimate for the height of the building. (2)

(Total 3 marks)

Edexcel, November 2008, Paper 2 Foundation, Question 10

8 A model of the Eiffel Tower is made to a scale of 2 millimetres to 1 metre.

The width of the base of the real Eiffel Tower is 125 metres.

Eiffel Tower

a Work out the width of the base of the model.

Give your answer in millimetres. (2)

The height of the model is 648 millimetres.

b Work out the height of the real Eiffel Tower.

Give your answer in metres. (2)

(Total 4 marks)

Edexcel, June 2008, Paper 2 Foundation, Question 17(b), (c)

9 On the grid, enlarge the shape with a scale factor of 2. (2)

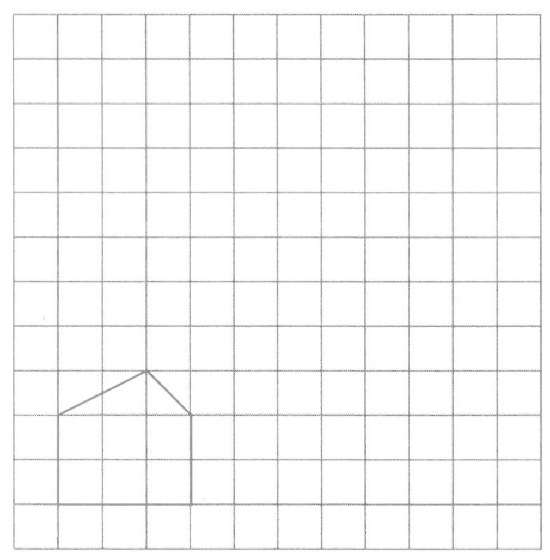

(Total 2 marks)

Edexcel, November 2008, Paper 2 Foundation, Question 2

10 The diagram shows some kitchen scales.

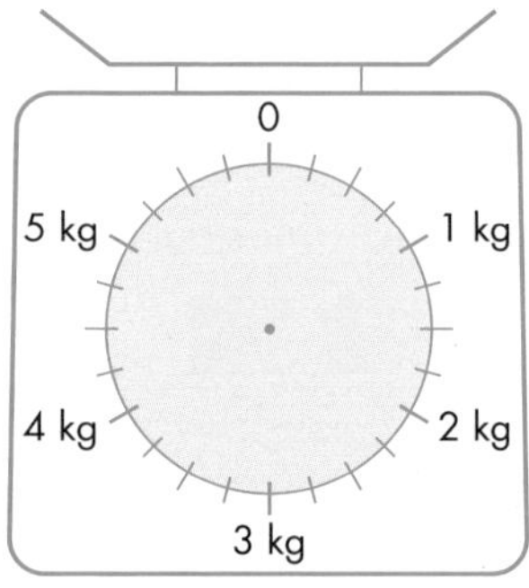

a Mrs Hall weighs a chicken on the scales. The chicken weights $3\frac{1}{2}$ kilograms.

Copy the scale and draw an arrow on your diagram to show $3\frac{1}{2}$ kilograms.

b Mrs Kitchen weighs a pumpkin on the scales. The pumpkin weighs $2\frac{3}{4}$ kg.

Draw an arrow on your diagram to show $2\frac{3}{4}$ kilograms.

11

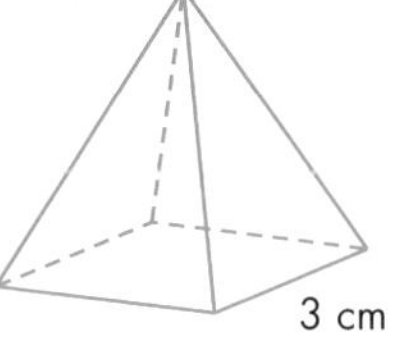

Diagram **not** accurately drawn

The diagram shows a pyramid with a square base.

The length of each side of the base is 3 cm.

The length of each sloping edge is 3 cm.

On the grid of centimetre squares, draw an accurate net of the pyramid. (3)

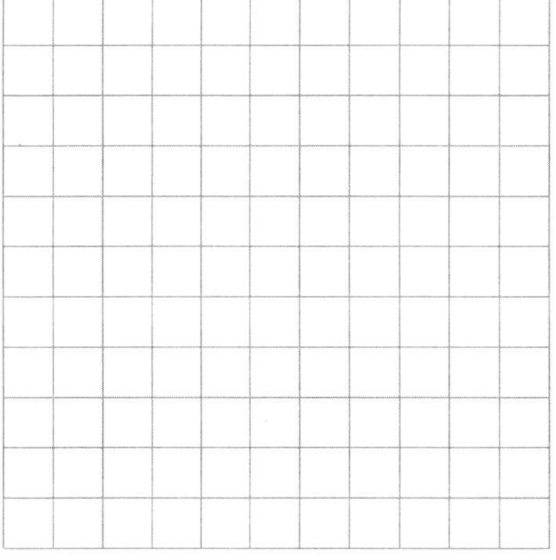

(Total 3 marks)

Edexcel, November 2008, Paper 2 Foundation, Question 20

12 A triangular prism has dimensions as shown.

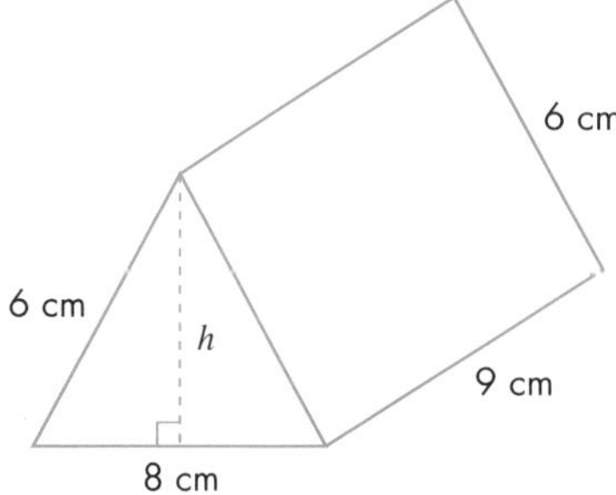

Sketch a net of the prism.

13 The diagram shows a solid shape made from seven 1-centimetre cubes.

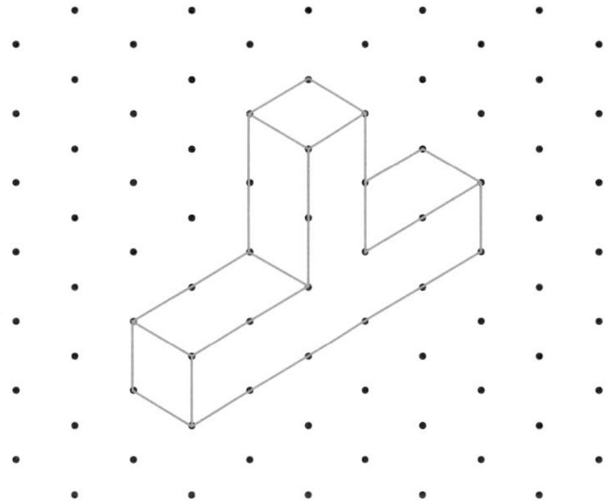

Draw the plan, the front elevation and the side elevation on squared paper.

14 Use isometric paper to copy and complete the drawing of a cuboid 4 cm by 2 cm by 3 cm.

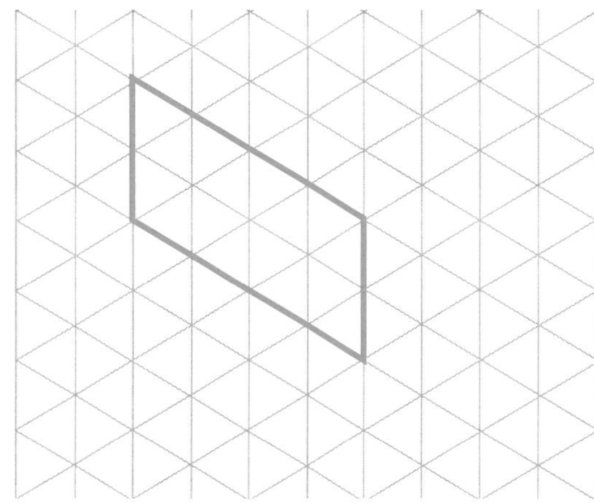

15 The diagram represents a solid made from 5 identical cubes.

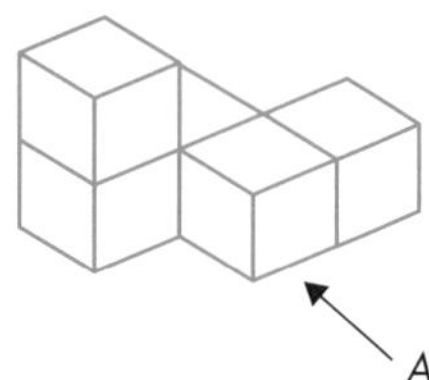

On the grid below, draw the view of the solid from direction *A*. (2)

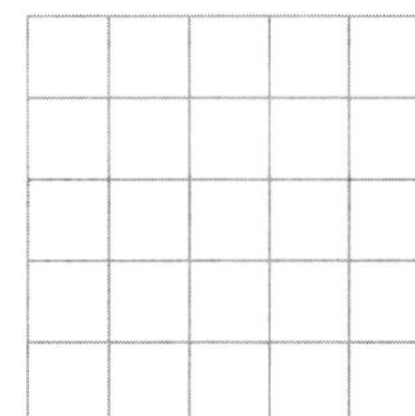

(Total 2 marks)

Edexcel, November 2008, Paper 1 Foundation, Question 21

16 Here are the front elevation, side elevation and the plan of a 3-D shape.

Front elevation

Side elevation

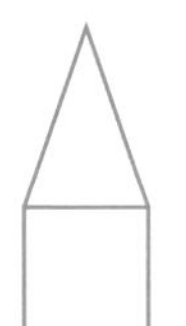

Plan

Draw a sketch of the 3-D shape. (2)

(Total 2 marks)

Edexcel, June 2008, Paper 2 Foundation, Question 21

Worked Examination Questions

1 Draw a net of an open box that has measurements 2 cm by 3 cm by 6 cm.

Step 1: Start with the base. Draw a 3 cm by 6 cm rectangle.

Your measurements must be accurate, so make sure you use a ruler. This step is worth 1 method mark.

Step 2: Now draw the two long sides as 2 cm by 6 cm rectangles on the long sides of the base.

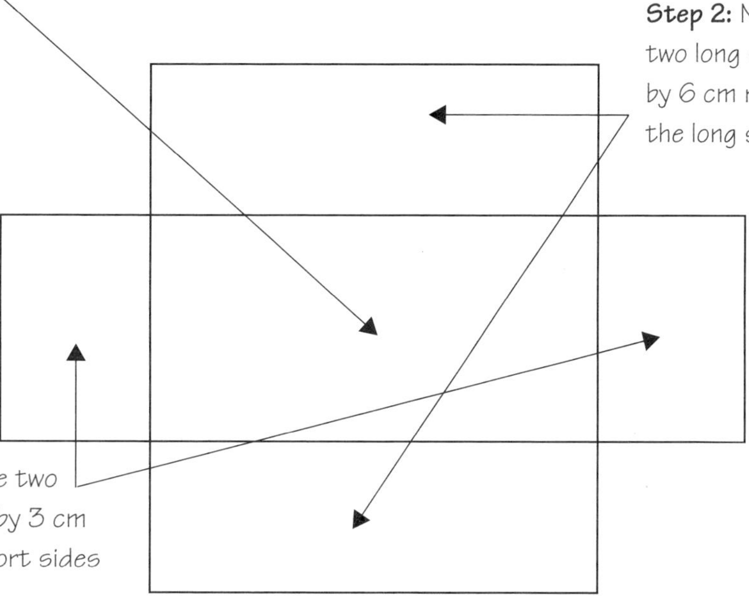

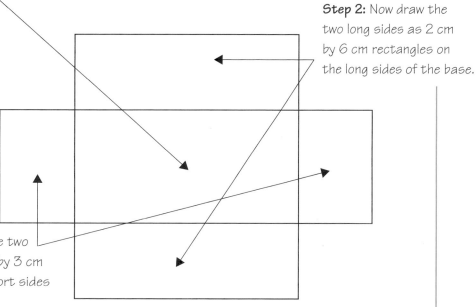

Step 3: Now draw the two short sides as 2 cm by 3 cm rectangles on the short sides of the base.

Drawing these two rectangles accurately is worth 1 mark for accuracy.

Drawing these two rectangles accurately is worth 1 mark for accuracy.

Total: 3 marks

Worked Examination Questions

AU **2** The diagram shows seven **cubes** arranged to make a **3D shape**

Study the diagrams below.

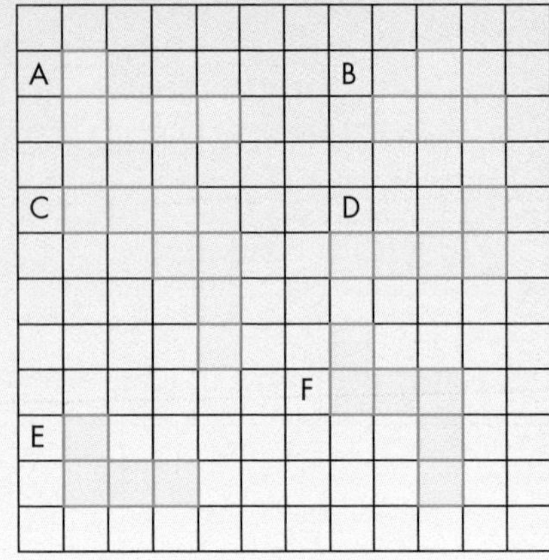

Plan

Front elevation

Side elevation

a Which is the **plan** view?

b Which is the **front elevation** view?

c Which is the **side elevation view**?

a F — You get 1 mark for accuracy for the correct answer.

1 mark

b A — You get 1 mark for accuracy for the correct answer.

1 mark

c E — You get 1 mark for accuracy for the correct answer.

1 mark

Total: 3 marks

Worked Examination Questions

FM **3** Aleks is planning a holiday in Madrid.

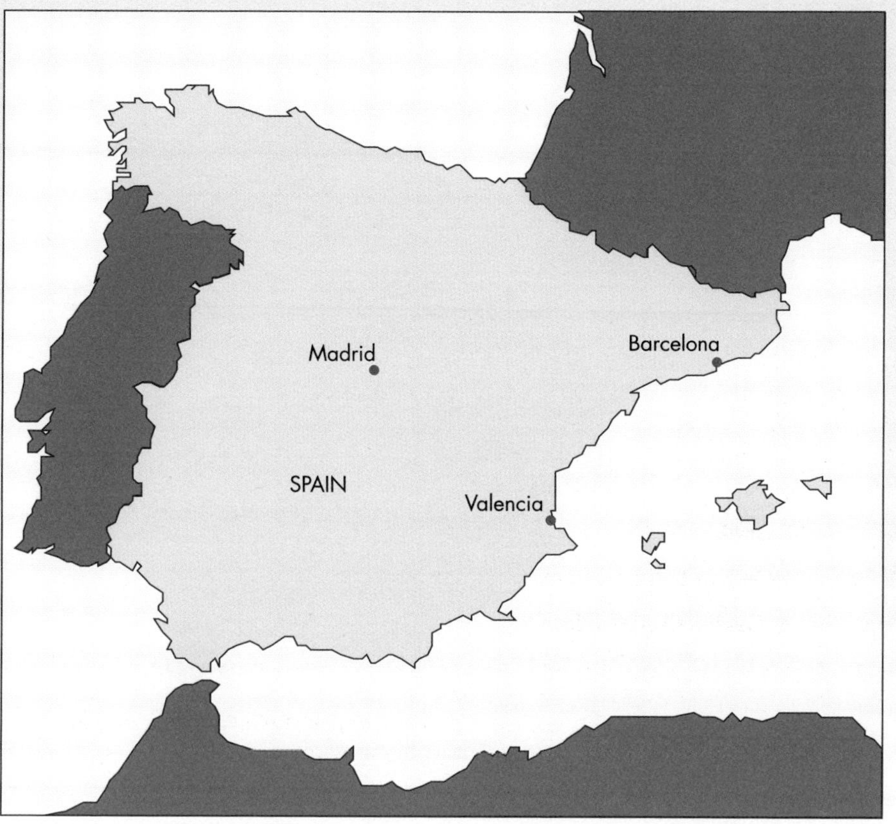

He looks at a map of Spain. The scale of the map is 1 : 10 000 000

Work out the direct distance from Madrid to

a Valencia

b Barcelona

1 cm = 10 000 000 cm = 100 000 m = 100 km

Convert the scale into a suitable format.

According to this, 1 cm represents 100 km

You earn 1 method mark for identifying the key fact that 1 cm = 100 km.

1 mark

a The direct distance from Madrid to Valencia is 310 km

On the map, Madrid to Valencia measures 3.1 cm. So, as 1 cm = 100 km, 3.1 cm must be multiplied by 100 to give the direct distance in kilometres.
You earn 1 mark for accuracy for the correct answer.

1 mark

b The direct distance from Madrid to Barcelona is 460 km

On the map, Madrid to Barcelona measures 4.6 cm. As above, multiply 4.6 cm by 100 to give the direct distance in kilometres.
You earn 1 mark for accuracy for the correct answer.

1 mark

Total: 3 marks

Functional Maths
A place in the sun

Many people buy houses abroad with the idea of renting them out to holidaymakers. They must consider many factors, such as the location, size and nearby attractions, before making their purchase so they can be sure that they have picked a property that will be popular and turn a profit.

Villa Hinojos
Cost: €264 000
Floor space: 110 m^2
Rent per week: £500
Weeks rented per year: 24

Villa Rosa
Cost: €180 000
Floor space: 80 m^2
Rent per week: £350
Weeks rented per year: 25

Villa Cartref
Cost: €189 000
Floor space: 90 m^2
Rent per week: £300
Weeks rented per year: 20

Villa Amapola
Cost: €252 000
Floor space: 105 m^2
Rent per week: £400
Weeks rented per year: 25

Villa Azul
Cost: €237 000
Floor space: 100 m^2
Rent per week: £450
Weeks rented per year: 26

Villa Blanca
Cost: €198 000
Floor space: 72 m^2
Rent per week: £350
Weeks rented per year: 30

Map scale 1 : 300 000
Map not drawn accurately

Your task

Jenny is going to buy a villa in Spain to rent to holidaymakers. She is looking at six properties and must now decide which would make her the most profit.

Investigate which villa would be the best buy and write up your findings as a report to advise Jenny on which property she should buy.

Remember to fully justify your advice using suitable mathematics.

Getting started

Use the scale 1 : 300 000 when working out the real-life distances.

When writing your report you should consider:

- What holidaymakers would look for in their ideal property, including:
  - distance from the airport by road (in kilometres)
  - distance to the coast by road (in kilometres).
- How much each property will cost, including:
  - the cost per square metre (in euros)
  - the cost of the villa (in British pounds)
  - the potential rental income per year (in British pounds).
- Is there anything else that you think Jenny should take into account when choosing her property?
- Think about how to use scales and conversions. For example:
  - What does the scale 2 cm : 100 km, mean?
  - What does the scale 1 : 20, mean?
  - What is £100 in euros, if the conversion rate is £1.00 = €1.50?
  - What is $30 in British pounds, if the conversion rate is $2 = £1?
  - How far is 3500 m in kilometres?

The usefulness of units

Simon Stevin was a Dutch mathematician. He was born in Bruges (now in Belgium) in 1584. He added greatly to the study of several areas of mathematics, including trigonometry and mechanics, the study of motion. He was also influential in the fields of geography, navigation, architecture and musical theory.

Simon Stevin has several other claims to fame.

- He invented a carriage with sails, a little model of which was preserved at Scheveningen until 1802. The carriage itself had been lost long before, but we know that in about 1600 Stevin, with Prince Maurice of Orange and 26 others, used it on the seashore between Scheveningen and Petten. It moved only by the force of the wind and, at its top speed, it was just a bit quicker than horses.
- He had the idea of explaining the tides by the attraction of the moon.

His greatest success, however, was a small pamphlet, first published in Dutch in 1585, in which he put forward the use of a decimal system for measurement throughout. He declared that the universal introduction of decimal coinage, measures and weights was only a question of time.

In the UK, we used imperial units such as, feet, inches, ounces and pounds for many years. Your grandparents will remember them well.

In the later part of the 20th century, the British Government started trying to ensure that we all use metric units here, as much as possible. This is now common practice, apart for the use of the mile.

The metric system is now predominantly used in the UK and most of the world, apart from the USA.

Road signs in the UK give distances in miles, but in Europe distances are given in kilometres.

A brief history of the metric system

1585	Simon Stevin suggested a decimal system for weights and measures.
1790	Thomas Jefferson proposed a decimal-based measurement system for the USA. A vote in the USA congress to replace the UK imperial system by a metric system was lost by just one vote.
1795	The metric system became the official system of measurement in France.
1959	The UK and USA redefined the inch to be 2.54 cm.
1963	The UK redefined the pound to be 0.453 592 37 kilograms.
1985	The UK redefined the gallon to be 4.546 09 litres.
Now	The metric system has been adopted by virtually every country, with the only notable exception being the USA.

It is vital that you understand and can use the metric system, but you should also have some understanding of the older imperial units. This will help you to understand what your grandparents sometimes say and the signs that are still in imperial units in the UK today.

Chapter

Measures: Units

The grades given in this chapter are target grades.

This chapter will show you …

- **G** which units to use when measuring length, weight and capacity
- **F** how to convert from one metric unit to another
- **F** how to convert from one imperial unit to another
- **E** how to convert from imperial units to metric units

Visual overview

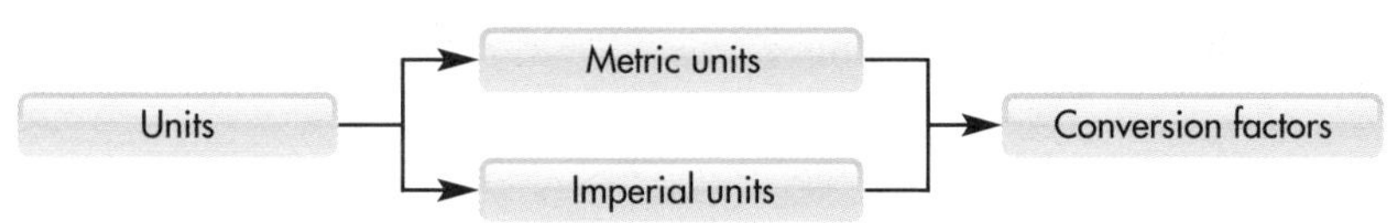

What you should already know

- The basic units used for measuring length, weight and capacity **(KS3 level 4, GCSE grade F)**
- The approximate size of these units **(KS3 level 4, GCSE grade F)**
- How to multiply or divide numbers by 10, 100 or 1000 **(KS3 level 4, GCSE grade F)**

Quick check

1 How many centimetres are there in one metre?

2 How many metres are there in one kilometre?

3 How many grams are there in one kilogram?

4 How many kilograms are there in one tonne?

7.1 Systems of measurement

This section will show you how to:

- decide which units to use when measuring length, weight and capacity

Key words
capacity
imperial
length
metric
volume
weight

There are two systems of measurement currently in use in Britain: the **imperial** system and the **metric** system.

The imperial system is based on traditional units of measurement, many of which were first introduced several hundred years ago. It is gradually being replaced by the metric system, which is used throughout Europe and in many other parts of the world.

The main disadvantage of the imperial system is that it has a lot of awkward conversions, such as 12 inches = 1 foot. The metric system has the advantage that it is based on powers of 10, namely 10, 100, 1000 and so on, so it is much easier to use when you do calculations.

It will be many years before all the units of the imperial system disappear, so you have to know units in both systems.

System	Unit	How to estimate it
	Length	
Metric system	1 metre	A long stride for an average person
	1 kilometre	Two and a half times round a school running track
	1 centimetre	The distance across a fingernail
Imperial system	1 foot	The length of an A4 sheet of paper
	1 yard	From your nose to your fingertips when you stretch out your arm
	1 inch	The length of the top joint of an adult's thumb
	Weight	
Metric system	1 gram	A 1p coin weighs about 4 grams
	1 kilogram	A bag of sugar
	1 tonne	A saloon car
Imperial system	1 pound	A jar full of jam
	1 stone	A bucket three-quarters full of water
	1 ton	A saloon car
	Volume/Capacity	
Metric system	1 litre	A full carton of orange juice
	1 centilitre	A small wine glass is about 10 centilitres
	1 millilitre	A full teaspoon is about 5 millilitres
Imperial system	1 pint	A full bottle of milk
	1 gallon	A half-full bucket of water

Volume and capacity

The term 'capacity' is normally used to refer to the volume of a liquid or a gas.

For example, when referring to the volume of petrol that a car's fuel tank will hold, people may say its capacity is 60 litres or 13 gallons.

In the metric system, there is an equivalence between the units of capacity and volume, as you can see on page 184.

EXAMPLE 1

Choose an appropriate metric unit for each of the following.

a your own height
b the thickness of this book
c the distance from home to school
d your own weight
e the weight of a coin
f the weight of a bus
g a large bottle of lemonade
h a dose of medicine
i a bottle of wine

a metres or centimetres
b millimetres
c kilometres
d kilograms
e grams
f tonnes
g litres
h millilitres
i centilitres

EXERCISE 7A

1 Decide the metric unit you would be most likely to use to measure each of the following.

a The height of your classroom
b The distance from London to Barnsley
c The thickness of your little finger
d The weight of this book
e The amount of water in a fish tank
f The weight of water in a fish tank
g The weight of an aircraft
h A spoonful of medicine
i The amount of wine in a standard bottle
j The length of a football pitch
k The weight of your head teacher
l The amount of water in a bath
m The weight of a mouse
n The amount of tea in a teapot
o The thickness of a piece of wire

F

2 Estimate the approximate metric length, weight or capacity of each of the following.

a This book (both length and weight)

b The length of your school hall

c The capacity of a milk bottle

d A brick (length, width and weight)

e The diameter of a 10p coin, and its weight

f The distance from your school to Manchester

g The weight of a cat

h The amount of water in one raindrop

i The dimensions of the room you are in

j Your own height and weight

FM 3 Bob was asked to put up some decorative bunting from the top of each lamp post in his street. He had three sets of ladders he could use, a two metre, a 3.5 metre and a five metre ladder.

He looked at the lamp posts and estimated that they were about three times his height. He is slightly below average height for an adult male.

Which of the ladders should he use? Give a reason for your choice.

AU 4 The distance from Bournemouth to Basingstoke is shown on a website as 55 miles.

Explain why this unit is used instead of inches, feet or yards.

7.2 Metric units

This section will show you how to:

- convert from one metric unit to another

Key words

centilitre (cl)
centimetre (cm)
gram (g)
kilogram (kg)
kilometre (km)
litre (l)
metre (m)
millilitre (ml)
millimetre (mm)
tonne (t)

You should already know the relationships between these metric units.

Length

10 **millimetres** = 1 **centimetre**

1000 millimetres = 100 centimetres
= 1 **metre**

1000 metres = 1 **kilometre**

Weight

1000 **grams** = 1 **kilogram**

1000 kilograms = 1 **tonne**

Capacity

10 **millilitres** = 1 **centilitre**

1000 millilitres = 100 centilitres
= 1 **litre**

Volume

1000 litres = 1 metre^3

1 millilitre = 1 centimetre^3

Note the equivalence between the units of capacity and volume:

1 litre = 1000 cm^3 which means 1 ml = 1 cm^3

You need to be able to convert from one metric unit to another.

Since the metric system is based on powers of 10, you should be able easily to multiply or divide to change units. Work through the following examples.

EXAMPLE 2

To change *small* units to *larger* units, always *divide*.

Change:

a 732 cm to metres

732 ÷ 100 = 7.32 m

b 410 mm to centimetres

410 ÷ 10 = 41 cm

c 840 mm to metres

840 ÷ 1000 = 0.84 m

d 450 cl to litres

450 ÷ 100 = 4.5 l

EXAMPLE 3

To change *large* units to *smaller* units, always *multiply*.

Change:

a 1.2 m to centimetres

1.2 × 100 = 120 cm

b 0.62 cm to millimetres

0.62 × 10 = 6.2 mm

c 3 m to millimetres

3 × 1000 = 3000 mm

d 75 cl to millilitres

75 × 10 = 750 ml

EXERCISE 7B

F

1 Fill in the gaps, using the information in this section.

a 125 cm = … m
b 82 mm = … cm
c 550 mm = … m
d 2100 m = … km
e 208 cm = … m
f 1240 mm = … m
g 4200 g = … kg
h 5750 kg = … t
i 85 ml = … cl
j 2580 ml = … l
k 340 cl = … l
l 600 kg = … t
m 755 g = … kg
n 800 ml = … l
o 200 cl = … l
p 630 ml = … cl
q 8400 l = … m^3
r 35 ml = … cm^3
s 1035 l = … m^3
t 530 l = … m^3
u 34 km = … m

2 Fill in the gaps, using the information in this section.

a 3.4 m = … mm
b 13.5 cm = … mm
c 0.67 m = … cm
d 7.03 km = … m
e 0.72 cm = … mm
f 0.25 m = … cm
g 0.64 km = … m
h 2.4 l = … ml
i 5.9 l = … cl
j 8.4 cl = … ml
k 5.2 m^3 = … l
l 0.58 kg = … g
m 3.75 t = … kg
n 0.94 cm^3 = … l
o 21.6 l = … cl
p 15.2 kg = … g
q 14 m^3 = … l
r 0.19 cm^3 = … ml

FM 3 Sarif was planning to do some DIY. He wanted to buy two lengths of wood, each 2 m long, and 1.5 cm by 2 cm. He went to the local store where the types of wood were described as:

2000 mm × 15 mm × 20 mm

200 mm × 15 mm × 20 mm

200 mm × 150 mm × 2000 mm

1500 mm × 2000 mm × 20 000 mm

Should he choose any of these? If so, which one?

AU 4 1 litre is equivalent to 1000 millilitres.

Referring to centimetres, explain how you know this.

E

PS 5 How many square millimetres are there in a square kilometre?

HINTS AND TIPS

The answer is not 1 000 000.

7.3 Imperial units

This section will show you how to:

- convert from one imperial unit to another

Key words
foot (ft)
gallon (gal)
inch (in)
mile (m)
ounce (oz)
pint (pt)
pound (lb)
stone (st)
ton (T)
yard (yd)

You need to be familiar with imperial units that are still in daily use. The main ones are:

Length	12 **inches**	=	1 **foot**
	3 feet	=	1 **yard**
	1760 yards	=	1 **mile**
Weight	16 **ounces**	=	1 **pound**
	14 pounds	=	1 **stone**
	2240 pounds	=	1 **ton**
Capacity	8 **pints**	=	1 **gallon**

Examples of the everyday use of imperial measures are:

miles for distances by road

gallons for petrol (in conversation)

feet and inches for people's heights

pints for milk

pounds for the weight of babies (in conversation)

ounces for the weight of food ingredients in a food recipe

EXAMPLE 4

- *To change large units to smaller units, always multiply.*
- *To change small units to larger units, always divide.*

Change:

a 4 feet to inches

$4 \times 12 = 48$ inches

b 5 gallons to pints

$5 \times 8 = 40$ pints

c 36 feet to yards

$36 \div 3 = 12$ yards

d 48 ounces to pounds

$48 \div 16 = 3$ pounds

EXERCISE 7C

E

1 Fill in the gaps, using the information in this section.

a 2 feet = … inches
b 4 yards = … feet
c 2 miles = … yards
d 5 pounds = … ounces
e 4 stone = … pounds
f 3 tons = … pounds
g 5 gallons = … pints
h 4 feet = … inches
i 1 yard = … inches
j 10 yards = … feet
k 4 pounds = … ounces
l 60 inches = … feet
m 5 stone = … pounds
n 36 feet = … yards
o 1 stone = … ounces

2 Fill in the gaps, using the information in this section.

a 8800 yards = … miles
b 15 gallons = … pints
c 1 mile = … feet
d 96 inches = … feet
e 98 pounds = … stones
f 56 pints = … gallons
g 32 ounces = … pounds
h 15 feet = … yards
i 11 200 pounds = … tons
j 1 mile = … inches
k 128 ounces = … pounds
l 72 pints = … gallons
m 140 pounds = … stones
n 15 840 feet = … miles
o 1 ton = … ounces

FM 3 Andrew was asked to do some shopping for his grandmother. She sent him out to get a two-pound bag of sugar from the market. When Andrew got to the market, the only bags of sugar that he saw were:

8-ounce bags, 16-ounce bags, 32-ounce bags and 40-ounce bags

Which bag should he take back for his grandmother?

PS 4 How many square inches are there in a square mile?

D

AU 5 1 kilogram is approximately 2.2 pounds.

Explain how you know that 1 ton is heavier than 1 tonne.

7.4 Conversion factors

This section will show you how to:
- use the approximate conversion factors to change between imperial units and metric units

Key words
conversion factor
imperial
metric

You need to know the approximate conversions between certain **imperial** units and **metric** units.

The **conversion factors** you should be familiar with are given below.

The symbol '≈' means 'is approximately equal to'.

Those you do need to know for your examination are in **bold** type.

Length	1 inch	≈	2.5 centimetres	**Weight**	1 pound	≈	450 grams
	1 foot	≈	**30 centimetres**		**2.2 pounds**	≈	**1 kilogram**
	1 mile	≈	1.6 kilometres				
	5 miles	≈	**8 kilometres**				
Capacity	1 pint	≈	570 millilitres				
	1 gallon	≈	**4.5 litres**				
	$1\frac{3}{4}$ pints	≈	**1 litre**				

EXAMPLE 5

Use the conversion factors above to find the following approximations.

a Change 5 gallons into litres.

$5 \times 4.5 \approx 22.5$ litres

b Change 45 miles into kilometres.

45×1.6 kilometres ≈ 72 kilometres

c Change 5 pounds into kilograms.

$5 \div 2.2 \approx 2.3$ kilograms (rounded to 1 decimal place)

Note: An answer should be rounded when it has several decimal places, since it is only an approximation.

EXERCISE 7D 

F

1 Fill in the gaps to find the approximate conversions for the following. Use the conversion factors on page 191.

a 8 inches = … cm

b 6 kg = … pounds

c 30 miles = … km

d 15 gallons = … litres

e 5 pints = … ml

f 45 litres = … gallons

g 30 cm = … inches

h 80 km = … miles

i 11 pounds = … kg

j 1710 ml = … pints

k 100 miles = … km

l 56 kg = … pounds

m 40 gallons = … litres

n 200 pounds = … kg

o 1 km = … yards

p 1 foot = … cm

q 1 stone = … kg

r 1 yard = … cm

D

FM 2 Which is heavier, a tonne or a ton? Show your working clearly.

FM 3 Which is longer, a metre or a yard? Show your working clearly.

FM 4 The weight of 1 cm^3 of water is about 1 gram.

a What is the weight of 1 litre of water:

i in grams **ii** in kilograms?

b What is the approximate weight of 1 gallon of water:

i in grams **ii** in kilograms?

C

FM 5 While on holiday in France, I saw a sign that said: 'Paris 216 km'. I was travelling on a road that had a speed limit of 80 km/h.

a Approximately how many miles was I from Paris?

b What was the approximate speed limit in miles per hour?

c If I travelled at the top speed all the way, how long would it take me to get to Paris? Give your answer in hours and minutes.

PS 6 While cycling on holiday in France, Tom had to cover a 200-km stretch in one day. He knew that, at home, he averaged 30 mph on the roads.

How long would he expect the journey to take, with no stops?

AU 7 A cowboy's 'ten-gallon' hat could actually hold only 1 gallon of water.

How many cubic inches could a 'ten-gallon' hat hold?

GRADE BOOSTER

- **F** You can convert from one metric unit to another
- **F** You can convert from one imperial unit to another
- **E** You can use the approximate conversion factors to change from imperial units to metric units
- **E** You can solve problems, using conversion factors

What you should know now

- How to convert from one metric unit to another
- How to convert from one imperial unit to another
- How to use conversion factors to change imperial units into metric units
- How to solve problems, using metric units and imperial units

1 Copy and complete this table. Write a sensible unit for each measurement.

	Metric	Imperial
The length of a football pitch	....	Yards
The weight of a newborn baby	....	Pounds
The length of this book	Centimetres	....

2 Two villages are 40 km apart.

a Change 40 km into metres.

b How many miles are the same as 40 km?

3 A school canteen orders 30 litres of milk, but 30 pints of milk are delivered instead. Does the canteen have enough milk?

4 **a** Complete the table by writing a **sensible** metric unit for each measurement.

The first one has been done for you. (3)

The length of the river Nile	6700 kilometres
The height of the world's tallest tree	110 ....
The weight of a chicken's egg	70 ....
The amount of petrol in a full petrol tank of a car	40 ....

b Change 4 metres to centimetres. (1)

c Change 1500 grams to kilograms. (1)

(Total 5 marks)

Edexcel, May 2008, Paper 1 Foundation, Question 13

5

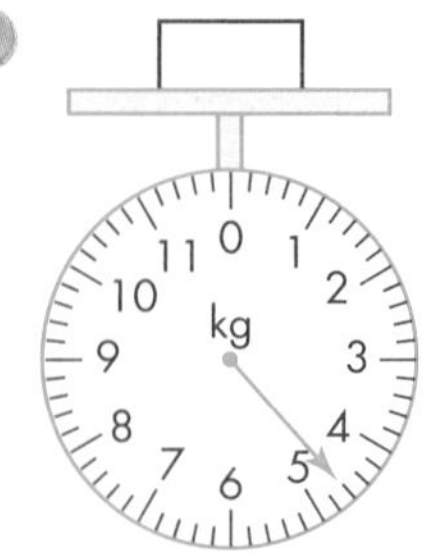

a Write down the weight in kg shown on this scale. (1)

b **i** How many pounds are there in 1 kg?

The weight of a baby is 5 kg.

ii Change 5 kg to pounds. (2)

(Total 3 marks)

Edexcel, November 2008, Paper 2 Foundation, Question 9

6 **a** Write down a sensible metric unit for measuring

i the distance from London to Paris,

ii the amount of water in a swimming pool. (2)

b **i** Change 5 centimetres to millimetres.

ii Change 4000 grams to kilograms. (2)

(Total 4 marks)

Edexcel, November 2008, Paper 1 Foundation, Question 5

7 Brian is driving through France.
The speed limit on the motorway is 130 kilometres per hour.

Change 130 km/h to miles per hour.

8 The diagram below shows the dimensions of a bookcase. The thickness of all the wood used is 30 mm.

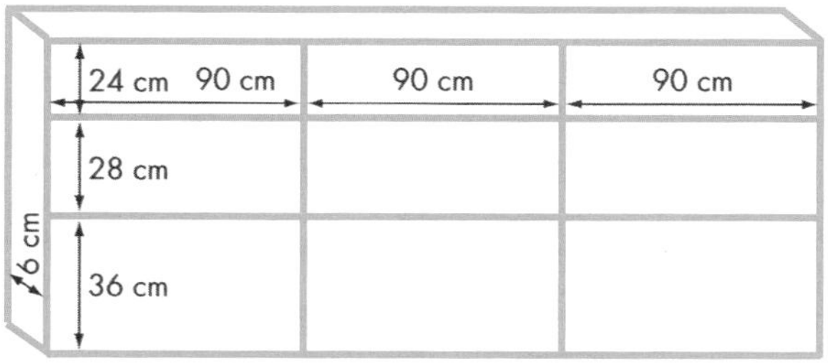

24 cm 90 cm 90 cm 90 cm
28 cm
36 cm
6 cm

a Calculate the height of the bookcase, giving your answer in metres.

b Calculate the length of the bookcase, giving your answer in metres.

9 The distance from Sheffield to Birmingham is shown on a website as 140 kilometres. Explain why this unit is used instead of centimetres.

10 Freyja and Agnes took their grandmother on a self-catering holiday in France. They settled into a cottage and travelled 12 km to the nearest supermarket.

Their grandmother asked them to get 5 lbs of potatoes, 8 oz of butter and 4 pts of milk.

a Their grandmother asked how many miles away the supermarket was. What answer should Freyja and Agnes give her?

b The supermarket only sold goods in metric units. Convert their grandmother's shopping list into metric units.

Worked Examination Questions

PS **1** Beth **weighs** 10 **stone** 5 **pounds** and her **height** is 5 **feet** 4 **inches**.

a Estimate her weight, to the nearest **kilogram**.

b Estimate her height, to the nearest **centimetre**.

a 1 stone = 14 pounds, so she weighs 145 pounds.

2.2 pounds ≈ 1 kg

So, 145 pounds ≈ 145 ÷ 2.2 ≈ 66 kg

3 marks

You get 1 accuracy mark for the answer 145 pounds.

You get 1 method mark for 145 ÷ 2.2 and 1 accuracy mark for 66 kg.

b 1 foot = 12 inches, so her height is 64 inches.

2.5 cm ≈ 1 inch

So, 64 inches ≈ 64 × 2.5 ≈ 160 cm

3 marks

You get 1 accuracy mark for the answer 64 inches.

You get 1 method mark for 64 × 2.5 and 1 accuracy mark for 160 cm.

Total: 6 marks

FM **2** Figure 1 shows a section of a brick wall. Figure 2 shows how layers of bricks are built up.

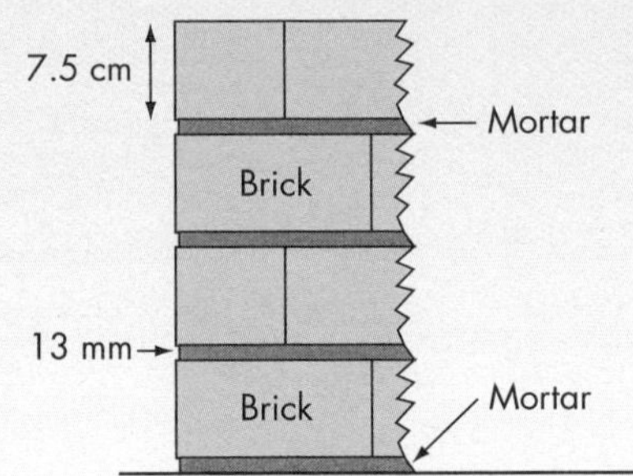

The wall is to be continued until it is at least 1 **metre** high. What is the *smallest* number of *extra* layers of bricks needed?

Height of 1 brick and 1 mortar = 7.5 cm + 1.3 cm = 8.8 cm

The wall needs to be 100 cm tall.

Number of layers of bricks = 100 ÷ 8.8 = 11.36 layers

This means the wall needs to be 12 bricks high.

So number of extra layers of bricks = 12 – 8
= 4 layers of bricks

Total: 4 marks

You get 1 accuracy mark for 8.8 cm.

You get 1 method mark for the calculation 100 ÷ 8.8

You get 1 accuracy mark for rounding up to a whole number of bricks (12).

You get 1 accuracy mark for understanding that 4 layers of bricks are required.

Worked Examination Questions

PS **3** Two towns are 60 km apart.

Driving at 30 mph, how long will it take, in hours and minutes, to drive from one town to the other?

30 mph = 30 × 8 ÷ 5 kph
= 48 kph

You receive 1 method mark for the calculation 30 × 8 ÷ 5 which converts mph into kph
1 accuracy mark for 48 kph.

Time to drive = 60 ÷ 48
= 1.25 hours

You get 1 method mark for 60 (minutes) ÷ 48 (kph) (or the figure you had).
You then get 1 accuracy mark for the figure 1.25.

The 0.25 means one quarter of an hour, which is 15 minutes.

Hence time taken is 1 hour 15 minutes.

You get 1 mark for interpreting 1.25 as 1 hour 15 minutes. If you had left your answer as 1.25 hours you would not get this mark.

Total: 5 marks

AU **4** A café ordered 94 pints of milk, but the milkman delivered 49 litres.

Is the amount delivered more or less than what was ordered?
Show your working.

1 gallon = 4.5 litres = 8 pints

So 1 litre = 8 ÷ 4.5
= 1.78 pints

You get 1 method mark for the calculation to convert litres into pints (8 ÷ 4.5) and 1 mark for working out that 1.78 pints is equal to 4.5 litres.
The 1.78 is rounded. Any correct rounding from the accurate 1.77777, using 1 or more decimal places, will get the mark.
An alternative approximation is 1 litre = 1.75 pints.
If this has been used as the starting point here, then 2 marks would be awarded.

49 litres = 1.78 × 49
= 87.22 pints

You get 1 method mark for 1.78 (or the figure you have) × 49.
1 accuracy mark for any figure between 87 and 89.

So, less has been delivered than was ordered.

This final statement receives 1 mark for independent workings.

Total: 4 marks out of a possible 5

7 Functional Maths
Olympic games

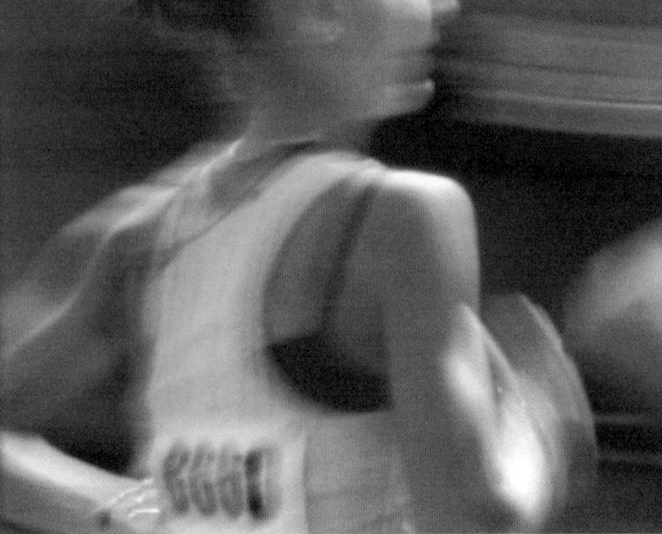

The first Olympic Games were held in Greece in 776 BC, but they were very different to the Games that we know today. In 1896 the first Olympic Games of modern times were held in Athens. Since then, the Games have gradually evolved, giving us our current system of international Olympic Games.

The Olympic Games show us that the way we measure things, from distance to weight, have changed over time. For example, over the centuries many countries have moved from using the imperial system of measurements to the metric system. This has meant that in order to maintain world records and make comparisons, measurements must be converted.

Your task

On the right are results from Olympic Games in two different years.

Write a report for team Great Britain, comparing the two sets of data. In your report you should:

- Show who was the best athlete overall in each category
- State any assumptions that you have made when completing your calculations, comparing the data and making statements.

Getting started

Use these questions to remind yourself of the key metric and imperial conversions.

1 Name three metric units of length.
2 Name three imperial units of length.
3 State the metric-to-metric conversions for length and use these in three conversions.
4 State the imperial-to-imperial conversion for length and use these in three conversions.
5 State the conversion facts that you know for metric-to-imperial measurements of length.

Extension

Design a stadium for the next Olympic Games.

- Plan the dimensions of your stadium using imperial and metric units, and draw it to scale.
- You will need to research past Olympic stadiums in order to find out the approximate sizes and requirements of an Olympic Stadium.

Year 1

Event	Winner (Women)	Result
100-yard sprint	Kathryn Ball	10.6 s
220-yard race	Kathryn Ball	25.2 s
880-yard race	Dianne Edwards	2 min 10 s
80-yards hurdle	Cho Ming	15.8 s
440-yards hurdle	Alejandra Lopez	58.3 s
High jump	Janet Guggiani	4 ft 5 in
Long jump	Barbara Charlton	5 yd 1 ft 8 in
Discus	Ife Adebayor	26 yd 2 ft 3 in
Javelin	Paula Ivan	25 yd 2 ft 11 in

Year 2

Event	Winner (Women)	Result
100-metre sprint	Aneta Jarzebska	13.1 s
200-metre race	Karim Djebli	28.3 s
800-metre race	Marie Auvergne	2 min 25 s
80-metre hurdle	Li Du	14.3 s
300-metre hurdle	Gabriela Lopez	56.7 s
High jump	Rachel Evans	1.45 m
Long jump	Brittany Banks	5.03 m
Discus	Ola Kubot	24.31 m
Javelin	Naveen Challa	23.72 m

Why this chapter matters

If you look carefully, you will be able to spot symmetry all around you. It is present in the natural world and in objects made by man. But, does it have a purpose and why do we need it?

Symmetry in nature, art and literature

Symmetry is everywhere you look in nature. Plants and animals have symmetrical body shapes and patterns. For example, if you divide a leaf in half, you will see that one half is the same shape as the other half.

Where is the symmetry in this butterfly, star fish and peacock? What effect does this symmetry have?

Pegasus.

This painting by a Dutch artist called M.C. Escher (1898–1972) uses line symmetry and rotational symmetry. Why do you think Escher used symmetry in his paintings?

Here is an extract from a poem called *The Tiger*, by William Blake (1757–1827).

Tiger, tiger, burning bright
In the forests of the night,
What immortal hand or eye
Could frame thy fearful symmetry?

Can you identify symmetry in the face of the tiger? And how about in the words of the poem?

The pattern of stripes is unique to each tiger and can be found on their skin as well as their fur. What purpose do these symmetrical stripes serve?

Symmetry in structures

St Peter's Basilica, Rome.

St Peter's Basilica, in the Vatican City in Rome, was started in 1506 and completed in 1626. It is a very symmetrical structure – see if you can identify all the symmetry that is present.

Why do you think that the designers of this building used symmetry?

These examples of natural and manmade symmetry hint at the place of symmetry in the world: it occurs naturally to give animals their uniform patterning, it creates aesthetically pleasing paintings and it can help to give structural stability. Now think about where symmetry occurs in your own life – how important is it to you?

Chapter

Geometry: Symmetry

1 Lines of symmetry

2 Rotational symmetry

The grades given in this chapter are target grades.

This chapter will show you …

G how to draw the lines of symmetry on a 2D shape

F how to find the order of rotational symmetry for a 2D shape

Visual overview

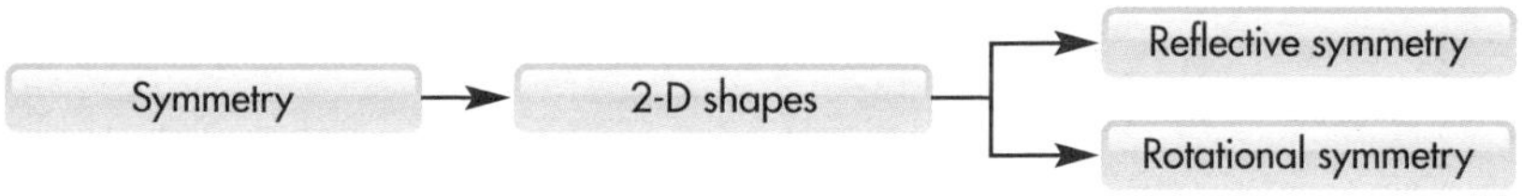

Symmetry → 2-D shapes → Reflective symmetry / Rotational symmetry

What you should already know

- A triangle is a 2D shape with three straight sides.
  (KS3 level 3, GCSE grade G)
- A quadrilateral is a 2D shape with four straight sides.
  (KS3 level 3, GCSE grade G)

Quick check

Write down the names of these 2D shapes.

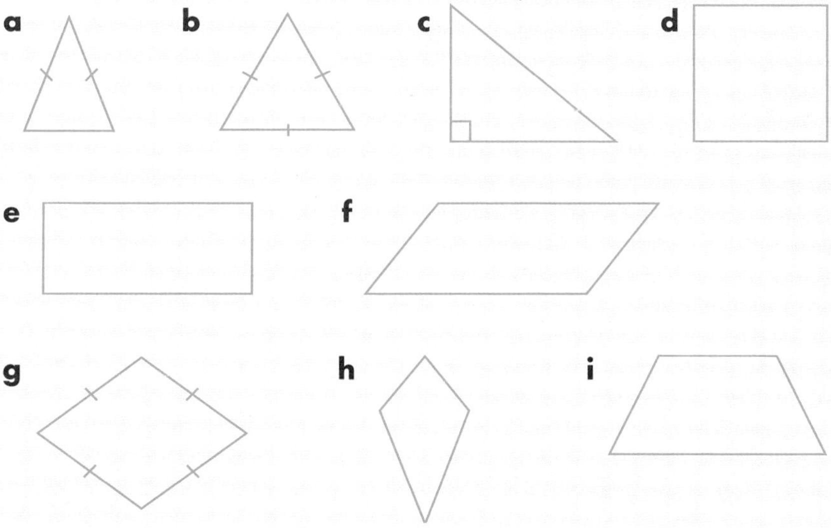

a **b** **c** **d**

e **f**

g **h** **i**

8.1 Lines of symmetry

This section will show you how to:

- draw the lines of symmetry on a 2D shape
- recognise shapes with reflective symmetry

Key words
line of symmetry
mirror line
symmetry

ACTIVITY

Mirror writing

You need a plane mirror and some plain or squared paper.

You probably know that certain styles of some upright capital letters have one or more lines of symmetry. For example, the upright A given below has one line of symmetry (shown here as a dashed line).

Draw a large A on your paper and put the mirror along the line of symmetry.

What do you notice when you look in the mirror?

Upright capital letters such as **A**, **O** and **M** have a vertical line of symmetry. Can you find any others?

Other upright capital letters (**E**, for example) have a horizontal line of symmetry. Can you find any others?

E

Now try to form words that have a vertical or a horizontal line of symmetry.

Here are two examples:

Make a display of all the different words you have found.

Many 2D shapes have one or more lines of **symmetry**.

A **line of symmetry** is a line that can be drawn through a shape so that what can be seen on one side of the line is the mirror image of what is on the other side. This is why a line of symmetry is sometimes called a **mirror line**.

It is also the line along which a shape can be folded exactly onto itself.

> **HINTS AND TIPS**
>
> **Remember:** you can use tracing paper to check for symmetry. For line symmetry, use it to find the mirror line.
> To check for rotational symmetry, trace the shape and turn your tracing around, over the shape.

Finding lines of symmetry

In an examination, you cannot use a mirror to find lines of symmetry but it is just as easy to use tracing paper, which is always available in any mathematics examination.

For example, to find the lines of symmetry for a rectangle, follow these steps.

1 Trace the rectangle.

2 Draw a line on the tracing paper where you think there is a line of symmetry.

3 Fold the tracing paper along this line. If the parts match, you have found a line of symmetry. If they do not match, try a line in another position.

4 Next, find out whether this is also a line of symmetry. You will find that it is.

5 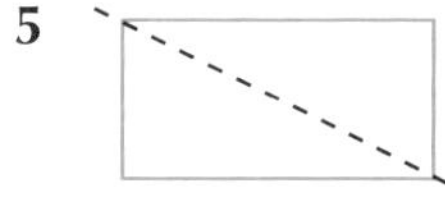

Now see whether this is a line of symmetry. You will find that it is *not* a line of symmetry.

6 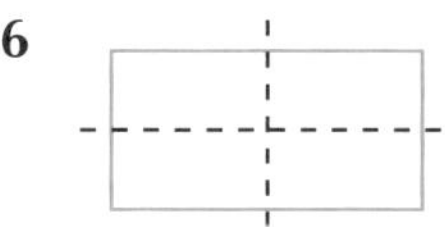

Your completed diagram should look like this. It shows that a rectangle has *two* lines of symmetry.

EXAMPLE 1

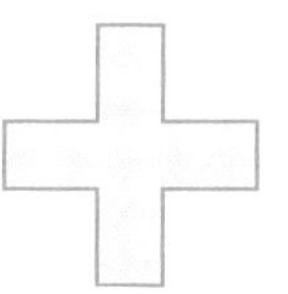

Find the number of lines of symmetry for this cross.

First, follow steps 1 to 4, which give the vertical and horizontal lines of symmetry.

Then, search for any other lines of symmetry in the same way.

There are two more, as the diagram shows.

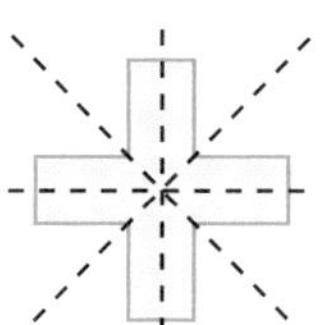

So, this cross has a total of four lines of symmetry.

EXERCISE 8A

1 Copy these shapes and draw on the lines of symmetry for each one. If it will help you, use tracing paper or a mirror to check your results.

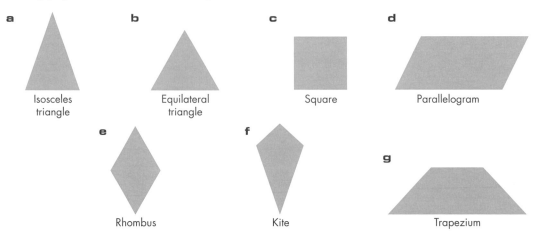

a Isosceles triangle

b Equilateral triangle

c Square

d Parallelogram

e Rhombus

f Kite

g Trapezium

2 How many shapes with lines of symmetry can you find in this drawing of a temple?

Copy each one you find and draw on the lines of symmetry.

3 **a** Find the number of lines of symmetry for each of these regular polygons.

i Regular pentagon

ii

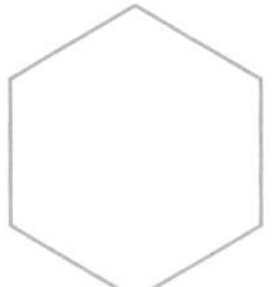

Regular hexagon

iii

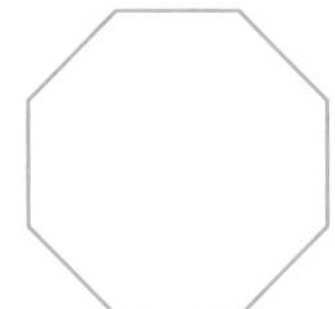

Regular octagon

b How many lines of symmetry do you think a regular decagon has? (A decagon is a ten-sided polygon.)

4 Copy these star shapes and draw in all the lines of symmetry for each one.

a

b

c

5 Copy these patterns and draw in all the lines of symmetry for each one.

a 

b 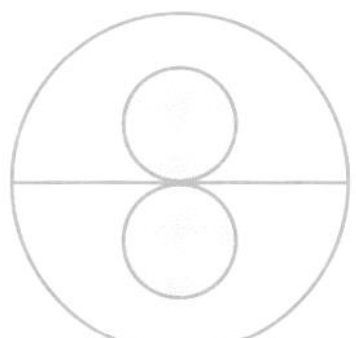

c 

d 

e 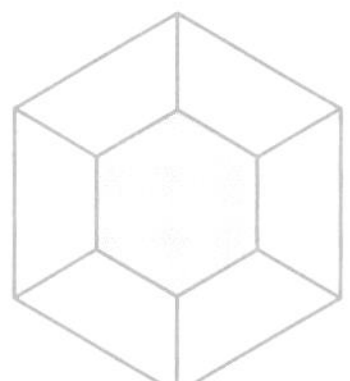

f 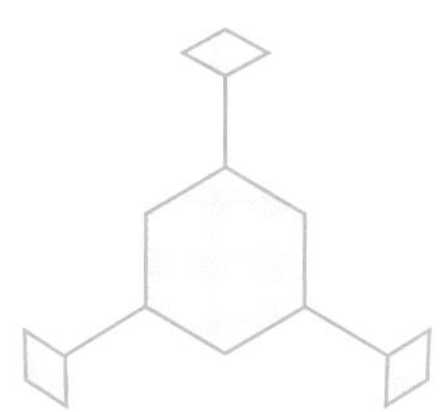

6 Write down the number of lines of symmetry for each of these flags.

Austria

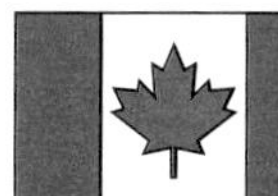
Canada

Iceland

Switzerland

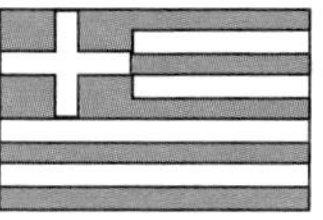
Greece

7 **a** These road signs all have lines of symmetry. Copy them and draw on the lines of symmetry for each one.

b Draw sketches of other common signs that also have lines of symmetry. State the number of lines of symmetry in each case.

8 The animal and plant kingdoms are full of symmetry. Four examples are given below. State the number of lines of symmetry for each one.

a 

b 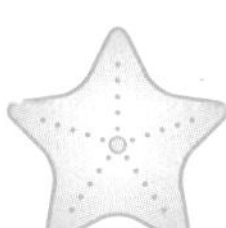

c 

d 

Can you find other examples? Find suitable pictures, copy them and state the number of lines of symmetry each one has.

9 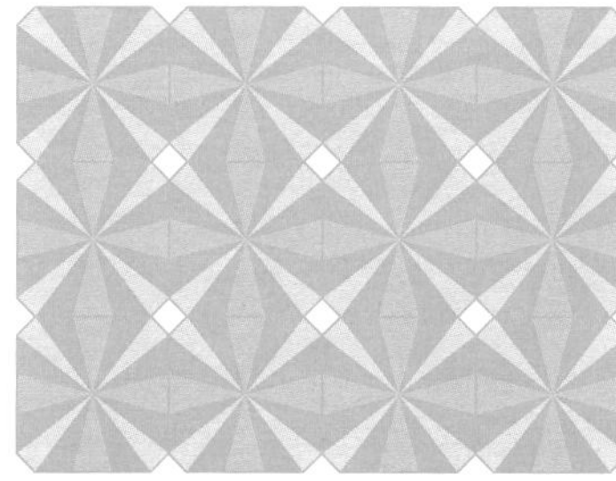

This decorative wallpaper pattern is made by repeating shapes that have lines of symmetry. By using squared or isometric paper, try to make a similar pattern of your own.

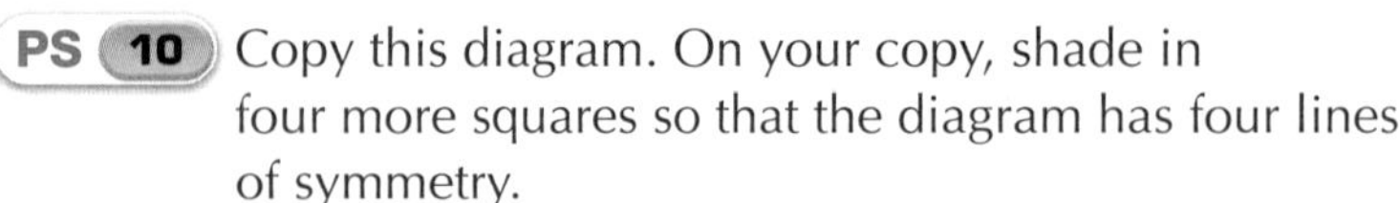

PS 10 Copy this diagram. On your copy, shade in four more squares so that the diagram has four lines of symmetry.

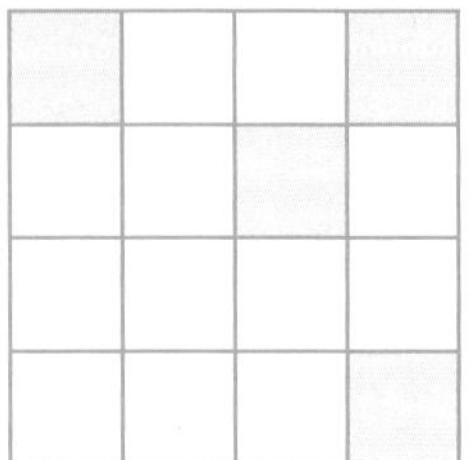

AU 11 Billy says that all triangles have either none or one or two or three lines of symmetry. Is Billy correct? Explain your answer.

8.2 Rotational symmetry

This section will show you how to:

- find the order of rotational symmetry for a 2D shape
- recognise shapes with rotational symmetry

Key words

order of rotational symmetry
rotational symmetry

A 2D shape has **rotational symmetry** if it can be rotated about a point to look exactly the same in a new position.

The **order of rotational symmetry** is the number of different positions in which the shape looks the same when it is rotated about the point.

The easiest way to find the order of rotational symmetry for any shape is to trace it and count the number of times that the shape stays the same as you turn the tracing paper through one complete turn.

EXAMPLE 2

Find the order of rotational symmetry for this shape.

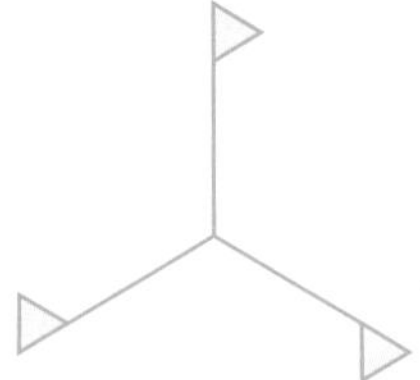

First, hold the tracing paper on top of the shape and trace the shape. Then rotate the tracing paper and count the number of times the tracing matches the original shape in one complete turn.

You will find three different positions.

So, the order of rotational symmetry for the shape is 3.

EXERCISE 8B

1 Copy these shapes and write below each one the order of rotational symmetry. If it will help you, use tracing paper.

a 
Square

b 
Rectangle

c 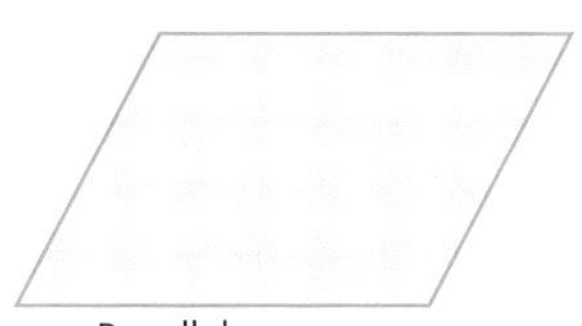
Parallelogram

d 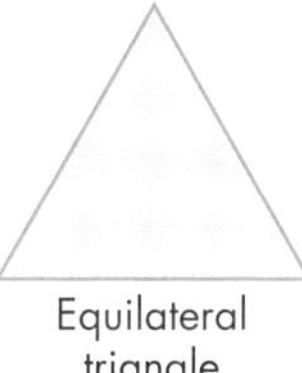
Equilateral triangle

e 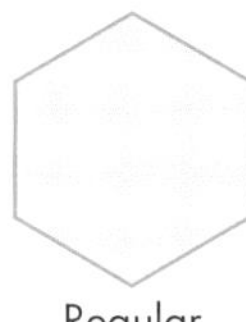
Regular hexagon

HINTS AND TIPS

Remember: a shape with rotational symmetry of order 1 has no rotational symmetry.

2 Find the order of rotational symmetry for each of these shapes.

a  **b**  **c**  **d**  **e** 

3 The following are Greek capital letters. Write down the order of rotational symmetry for each one.

a  Φ **b** H **c** Z **d** Θ **e** Ξ

4 Copy these shapes on tracing paper and find the order of rotational symmetry for each one.

a 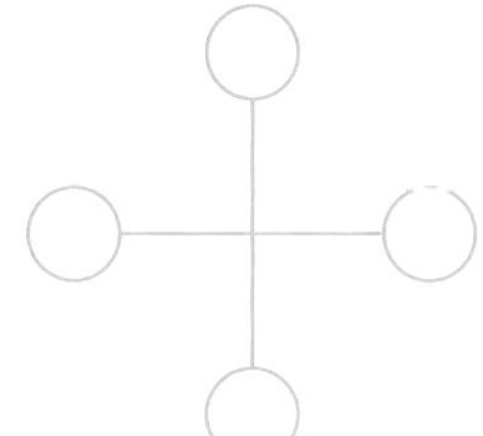 **b**  **c** 

d 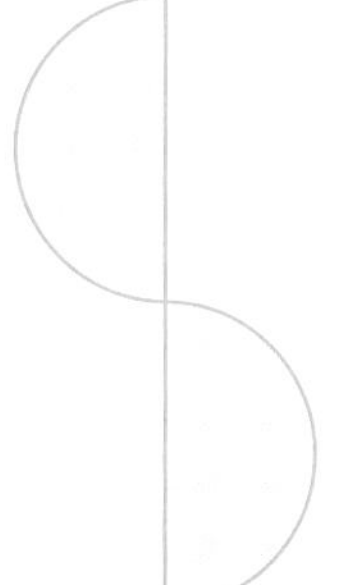 **e** 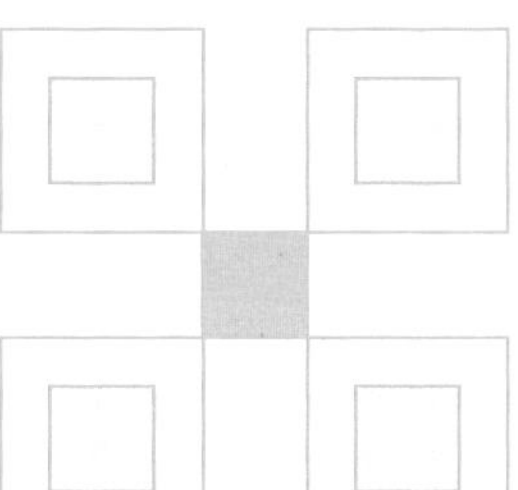 **f** 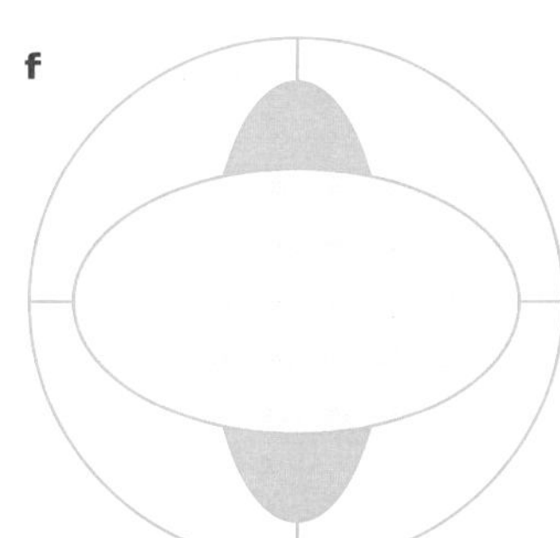

F

5 The upright capital letter A fits exactly onto itself only *once*. So, its order of rotational symmetry is 1. This means that it has *no* rotational symmetry. Write down all the upright capital letters of the alphabet that have rotational symmetry of order 1.

6 Obtain a pack of playing cards or a set of dominoes. Which cards or dominoes have rotational symmetry? Can you find any patterns? Write down everything you discover about the symmetry of the cards or dominoes.

7 Here is an Islamic star pattern.

Inside the star there are two patterns that have rotational symmetry.

a What is the order of rotational symmetry of the whole star?

b What is the order of rotational symmetry of the two patterns inside the star?

FM 8 Design a logo that has rotational symmetry of order 2 to advertise a new brand of soap.

PS 9 Copy the grid on the right. On your copy, shade in four squares so that the shape has rotational symmetry of order 2.

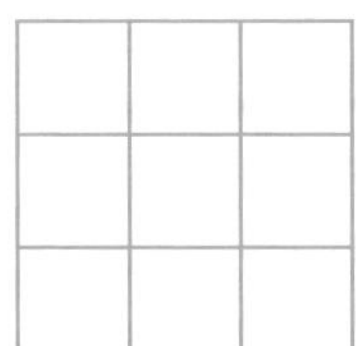

AU 10 Copy the table below. On your copy, write the letter for each shape in the correct box. The first one has been done for you.

		Number of lines of symmetry			
		0	1	2	3
Order of rotational symmetry	1		A		
	2				
	3				

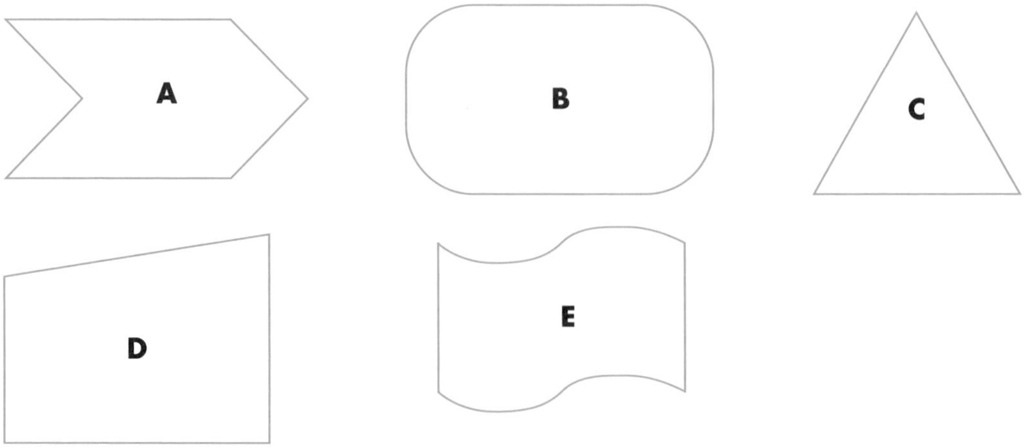

A

B

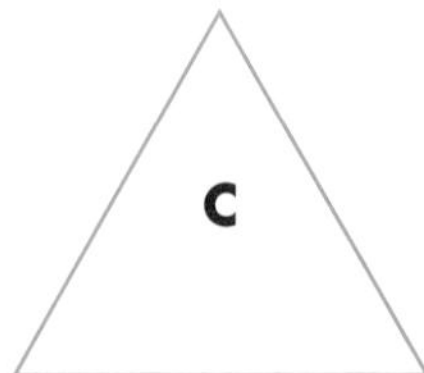

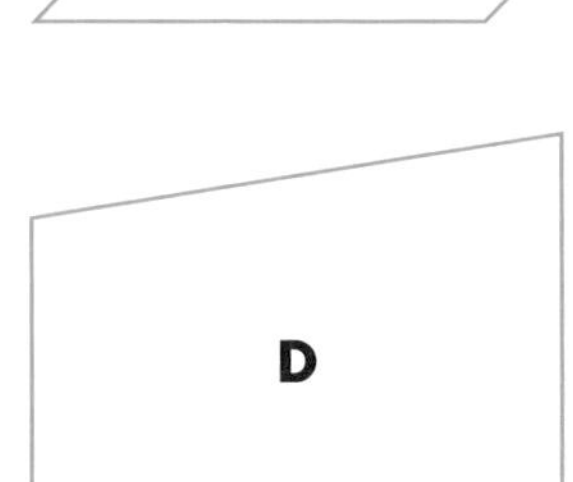

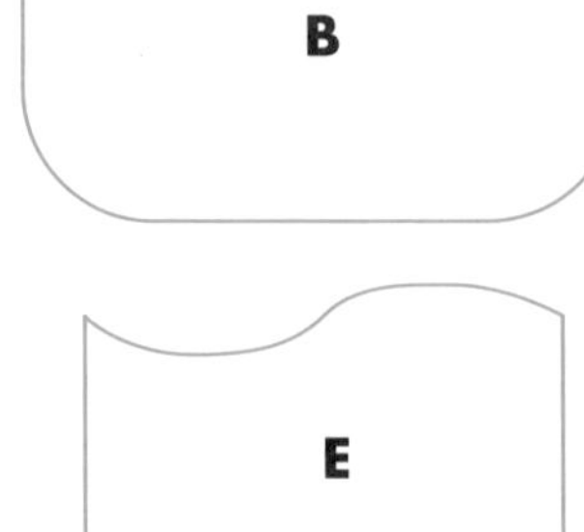

ACTIVITY

Pentomino patterns

Pentominoes are shapes made with five squares that touch edge-to-edge.

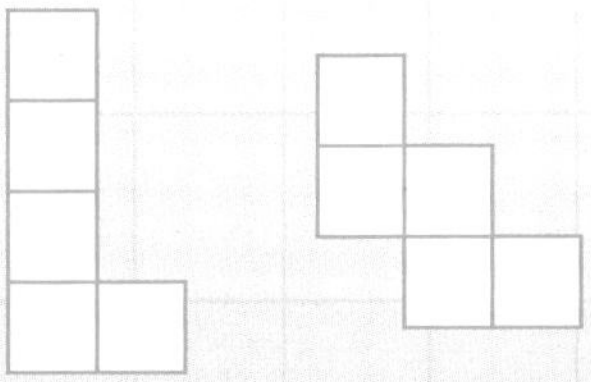

There are 12 possible pentomino shapes.

Two of them are shown here.

Find the other 10.

When you have found them all, investigate the line symmetry and rotational symmetry for each pentomino.

GRADE BOOSTER

- **G** You can draw lines of symmetry on basic 2D shapes.
- **F** You can draw lines of symmetry on more complex 2D shapes.
- **F** You can find the order of rotational symmetry for more complex 2D shapes.

What you should know now

- How to recognise lines of symmetry and draw them on 2D shapes
- How to recognise whether a 2D shape has rotational symmetry and find its order of rotational symmetry

1 The diagram shows a pentagon. It has one line of symmetry.

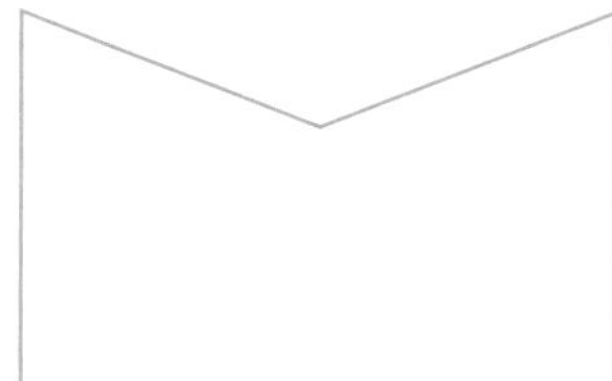

Copy the diagram and draw the line of symmetry.

2

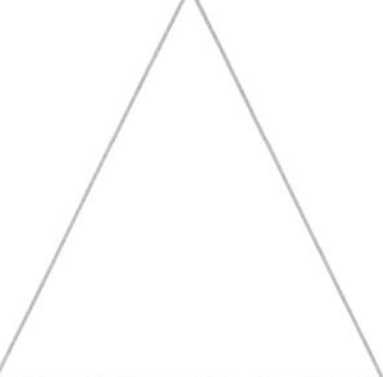

Draw the line of symmetry on this triangle. (1)

(Total 1 mark)

Edexcel, May 2008, Paper 12 Foundation, Question 7(b)

3 Here are four shapes.

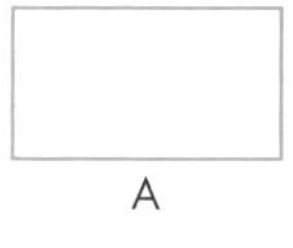

A

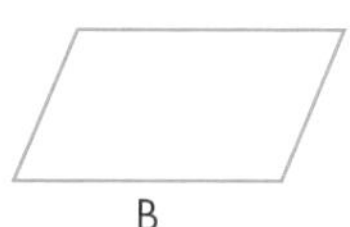

B

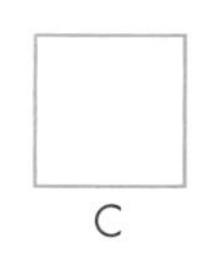

C

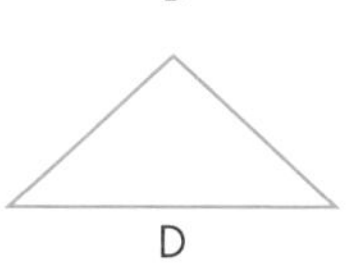

D

Write down the letter of the shape which has

i exactly **one** line of symmetry, (1)

ii **no** lines of symmetry, (1)

iii exactly **two** lines of symmetry. (1)

(Total 3 marks)

Edexcel, November 2008, Paper 13 Foundation, Question 2

4 The diagram shows a slab in the shape of a pentagon. It has one line of symmetry.

Copy the diagram and draw the line of symmetry.

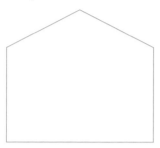

5 **a** Copy this rectangle and draw the lines of symmetry on it.

b What is the order of rotational symmetry of a rectangle?

6 Here is a list of 8 numbers.

11 16 18 36

68 69 82 88

From these numbers, write down a number which has

a exactly **one** line of symmetry, (1)

b 2 lines of symmetry **and** rotational symmetry of order 2, (1)

c rotational symmetry of order 2 but **no** lines of symmetry. (1)

(Total 3 marks)

Edexcel, June 2005, Paper 1 Foundation, Question 5

7 **a** On the diagram below, shade **one** square so that the shape has exactly one line of symmetry. (1)

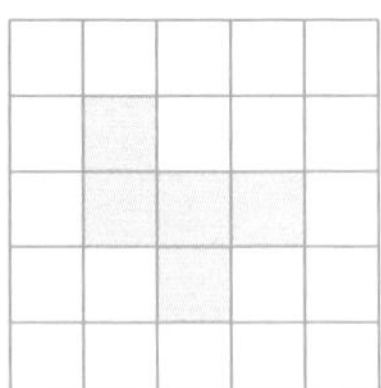

b On the diagram below, shade **one** square so that the shape has rotational symmetry of order 2. (1)

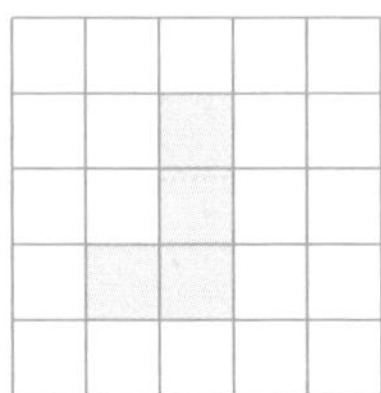

(Total 2 marks)

Edexcel , November 2008, Paper 1 Foundation, Question 10

8 This shape is made from three equilateral triangles and a regular hexagon.

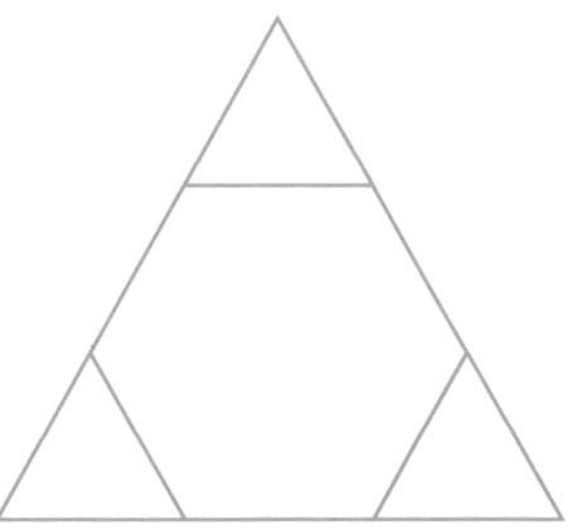

a Write down the order of rotational symmetry of the shape.

b Copy the shape and draw all the lines of symmetry on it.

9 **a** This is a diagram of a wall tile. Copy the diagram and draw in the lines of symmetry.

b Four of these tiles are used to make a pattern.

i Copy and complete this diagram to show the final pattern.

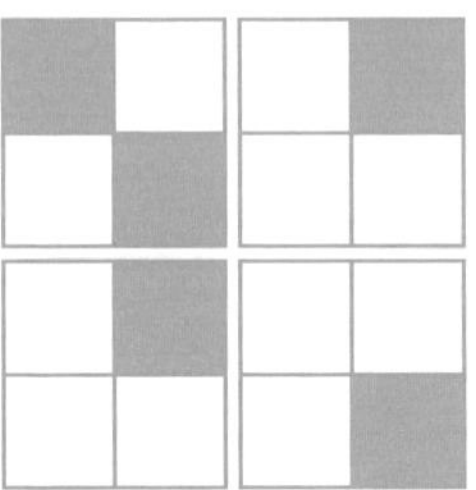

ii How many lines of symmetry does your final pattern have?

Worked Examination Questions

M A T H S

1 Which of the letters above has
  a **line symmetry**
  b **rotational symmetry** of order 2?

a M A T H

2 marks

These letters have line symmetry as shown. 2 marks are available for finding all five lines of symmetry. You would lose a mark if you left one out.

b H S

You need to get both letters for this mark.

1 mark   **Total:** 3 marks

2 A pattern has **rotational symmetry of order 4** and no line symmetry.
Part of the pattern is shown on the right. Complete the pattern.

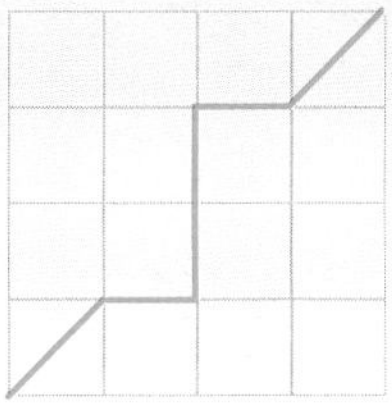

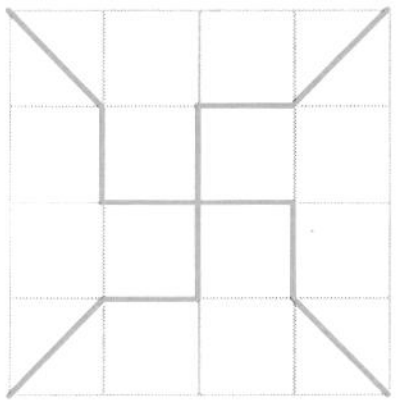

Trace the part of the pattern and rotate it about the centre of the grid three times through 90° to form the pattern.
You get 2 marks for completing the pattern correctly.
You would lose a mark if you made an error in the pattern.

Total: 2 marks

3 Add two squares to the diagram below so that it has **rotational symmetry of order 2**.

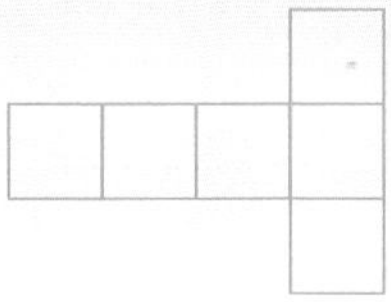

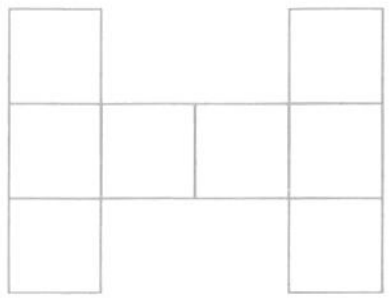

You can use tracing paper to check your answer.
You get 1 mark for the correct pattern.

Total: 1 mark

PS **4** Copy the diagram. Shade in more squares so that the pattern has **rotational symmetry of order 4**.

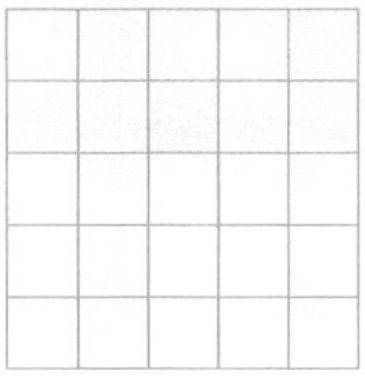

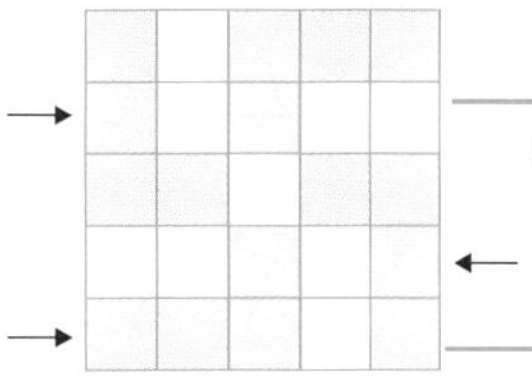

You can use tracing paper to check your answer.

You get 1 mark for each correct section.

Total: 3 marks

8

Problem solving
Finding the symmetry

Symmetrical objects are all around us: in our homes, at school and work, and in the 'great outdoors'. In this task you are going to investigate two types of symmetry: line symmetry and rotational symmetry.

Getting started

To start with, think about the following questions.

- Think of five objects in your room at home that have line symmetry. How many **lines** of symmetry does each object have?
- Look around your classroom and find two objects that have rotational symmetry. What is their **order of rotational** symmetry?
- Where does symmetry naturally occur? Name some plants and animals that have symmetry and give the number of **lines** of symmetry and/or the **order of rotational** symmetry found on each.

Your task

Here are some examples of symmetry in modern art and ornate metal railings. With a partner, discuss the symmetry that you can see in each picture.

Now, design your own modern art piece or metal railings that have line symmetry, rotational symmetry, or both. You must explain the symmetry in your design to your partner.

Symmetry in art

Stephen Pitts – Triangular Colour Wheel

Shana McCormick – Symmetrical Unison

Symmetry in railings

Why this chapter matters

Line graphs are used in many media, including newspapers and the textbooks of most of the subjects that you learn in school.

Graphs show the relationship between two variables. Often one of these variables is time and the graph shows how the other variable changes over time.

For example, this graph on the right shows how the exchange rate between the dollar and the pound changed over five months in 2009.

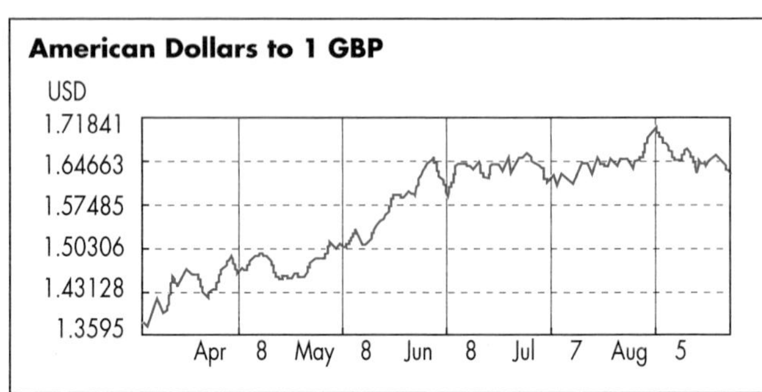

The earliest line graphs, such as the one shown on the right, appeared in the book *A Commercial and Political Atlas*, written in 1786 by William Playfair, who also used bar charts and pie charts for the first time. Playfair argued that charts and graphs communicated information to an audience better than tables of data. Do you think this is true?

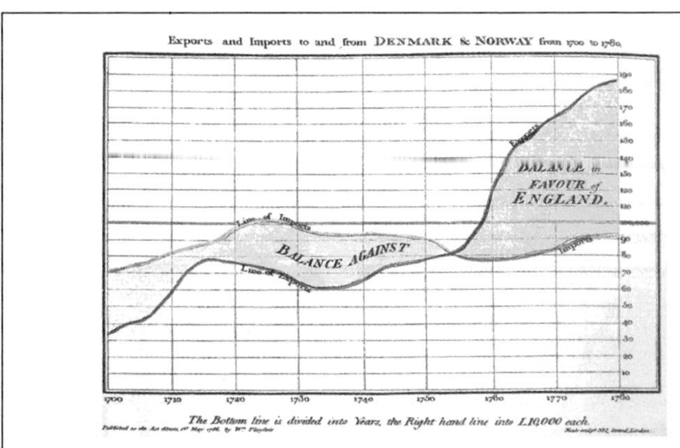

A line graph showing imports and exports to and from Denmark and Norway by William Playfair.

The graph below right shows all the data from a racing car going round a circuit. Engineers can use this to fine-tune parts of the car to give the best performance. It perfectly illustrates that graphs give a visual representation of how variables change and can be used to compare data in a way that looking at lists of data cannot.

Think about instances in school and everyday life where a line graph would help you to communicate information more effectively.

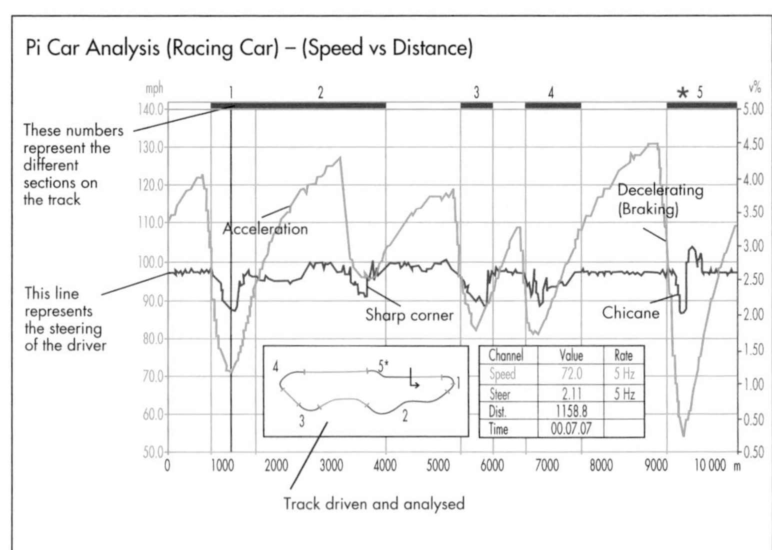

Chapter

Algebra: Graphs

The grades given in this chapter are target grades.

This chapter will show you …

F how to read information from a conversion graph

E to D how to read information from a travel graph

E to D how to draw a straight-line graph from its equation

Visual overview

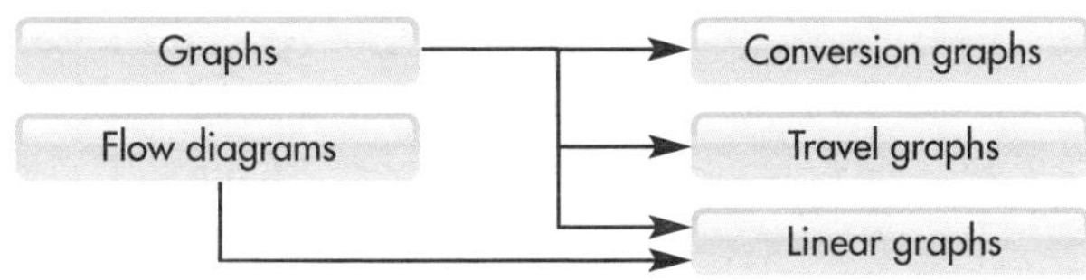

What you should already know

- How to plot coordinates in the first quadrant (KS3 level 3, GCSE grade G)
- How speed, distance and time are related (from Chapter 9) (KS3 level 6, GCSE grade D)
- How to substitute numbers into a formula (from Chapter 7) (KS3 level 5, GCSE grade E)
- How to use a flow diagram to set up an expression (from Chapter 13) (KS3 level 5, GCSE grade E)
- How to read and estimate from scales (KS3 level 5, GCSE grade E)

Quick check

Write down the coordinates of the following points.

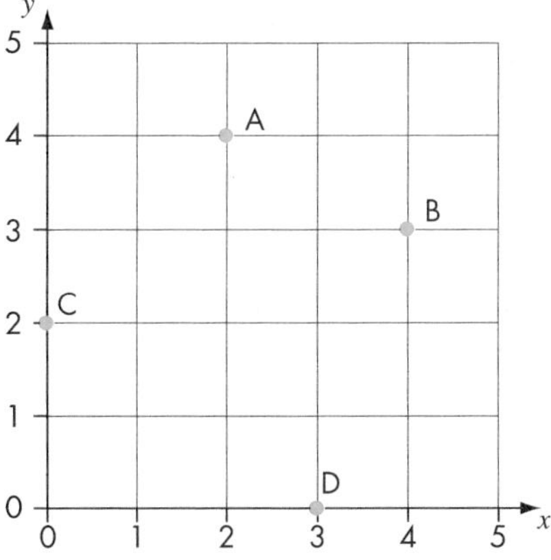

9.1 Conversion graphs

This section will show you how to:

- convert from one unit to another unit by using a graph

Key words

conversion graph
estimate
scales

Look at Examples 1 and 2, and make sure that you can understand the conversions. You need to be able to read these types of graph by finding a value on one axis and following it through to the other axis. Make sure you understand the **scales** on the axes to help you **estimate** the answers.

EXAMPLE 1

This is a **conversion graph** between litres and gallons.

a How many litres are there in 5 gallons?

b How many gallons are there in 15 litres?

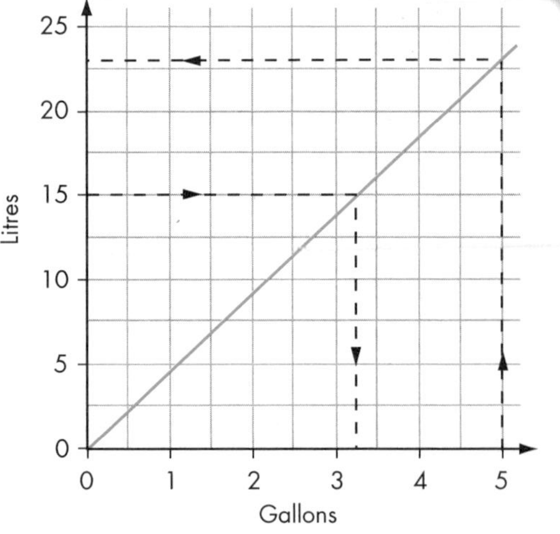

From the graph you can see that:

a 5 gallons are approximately equivalent to 23 litres.

b 15 litres are approximately equivalent to $3\frac{1}{4}$ gallons.

EXAMPLE 2

This is a graph of the charges made for units of electricity used in the home.

a How much will a customer who uses 500 units of electricity be charged?

b How many units of electricity will a customer who is charged £20 have used?

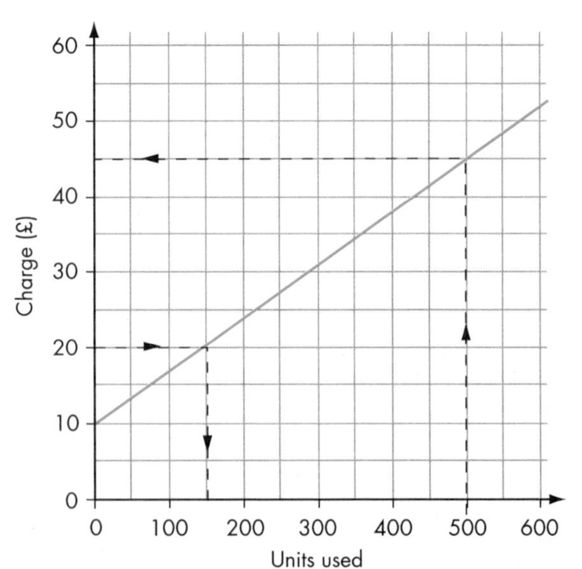

From the graph you can see that:

a A customer who uses 500 units of electricity will be charged £45.

b A customer who is charged £20 will have used about 150 units.

EXERCISE 9A

FM 1 This is a conversion graph between kilograms (kg) and pounds (lb).

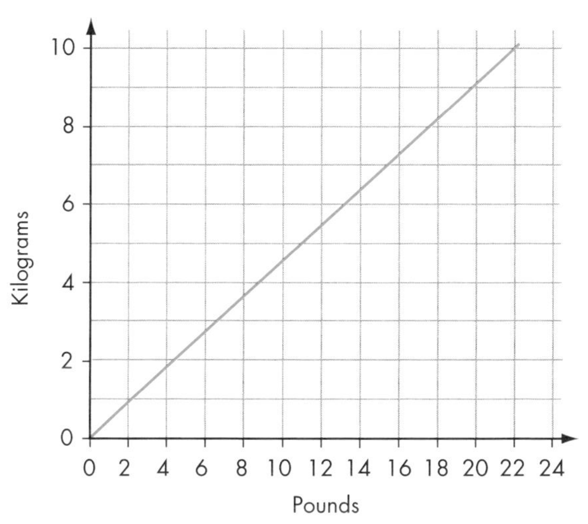

a Use the graph to make an approximate conversion of:

i 18 lb to kilograms

ii 5 lb to kilograms

iii 4 kg to pounds

iv 10 kg to pounds.

b Approximately how many pounds are equivalent to 1 kg?

c Explain how you could use the graph to convert 48 lb to kilograms.

FM 2 This is a conversion graph between inches (in) and centimetres (cm).

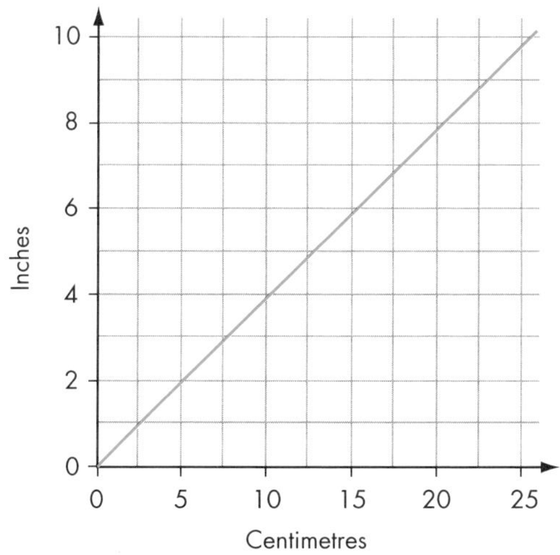

a Use the graph to make an approximate conversion of:

i 4 in to centimetres

ii 9 in to centimetres

iii 5 cm to inches

iv 22 cm to inches.

b Approximately how many centimetres are equivalent to 1 in?

c Explain how you could use the graph to convert 18 into centimetres.

FM 3 This graph was produced to show the approximate equivalence of the British pound (£) to the Singapore dollar ($).

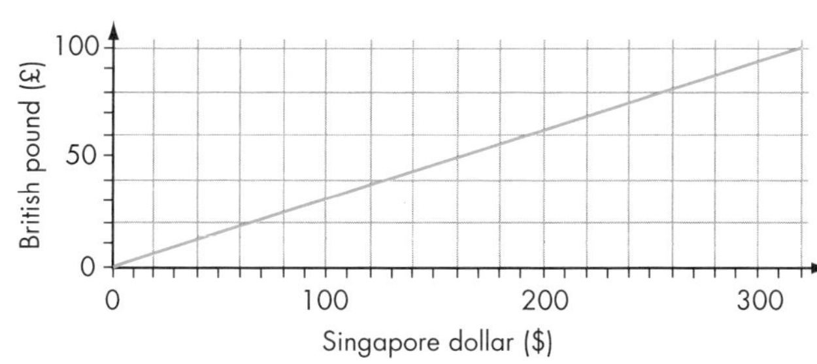

a Use the graph to make an approximate conversion of:

i £100 to Singapore dollars

ii £30 to Singapore dollars

iii $150 to British pounds

iv $250 to British pounds.

b Approximately how many Singapore dollars are equivalent to £1?

AU c What would happen to the conversion line on the graph if the pound became weaker against the Singapore dollar?

F

F

4 A hire firm hired out industrial blow heaters. They used the following graph to approximate what the charges would be.

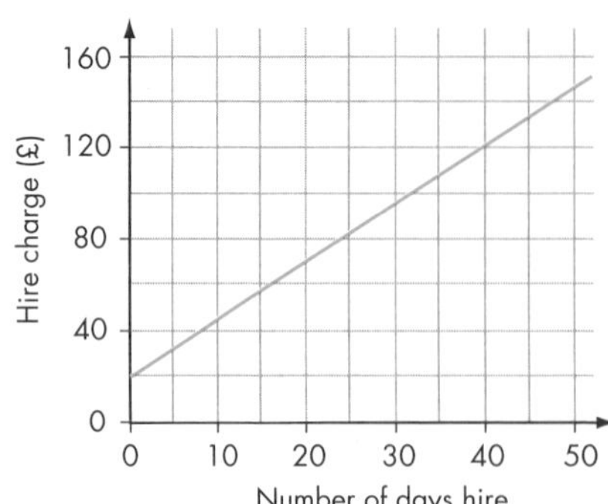

a Use the graph to find the approximate charge for hiring a heater for:

i 40 days

ii 25 days.

b Use the graph to find out how many days' hire you would get for a cost of:

i £100

ii £140.

5 A conference centre had the following chart on the office wall so that the staff could see the approximate cost of a conference, based on the number of people attending it.

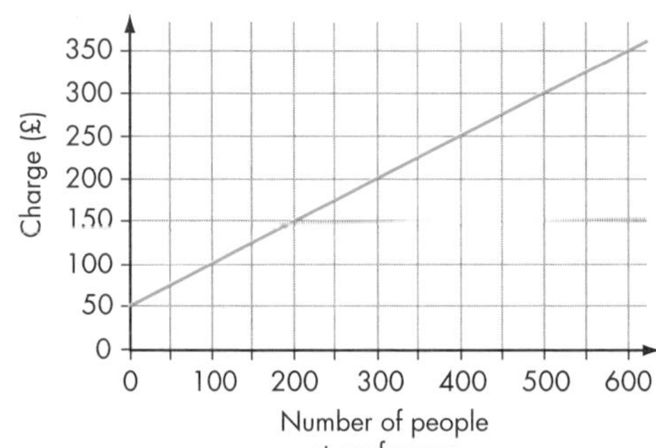

a Use the graph to find the approximate charge for:

i 100 people

ii 550 people.

b Use the graph to estimate how many people can attend a conference at the centre for a cost of:

i £300

ii £175.

6 At a small shop, the manager marked all goods at the pre-VAT prices and the sales assistant had to use the following chart to convert these marked prices to selling prices.

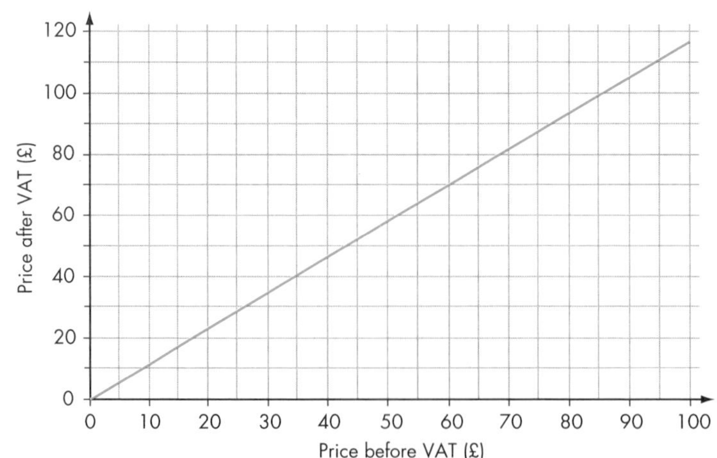

a Use the chart to find the selling price of goods marked:

i £60

ii £25.

b What was the marked price if you bought something for:

i £100

ii £45?

7 Granny McAllister still finds it hard to think in degrees Celsius. So she always uses the following conversion graph to help her to understand the weather forecast.

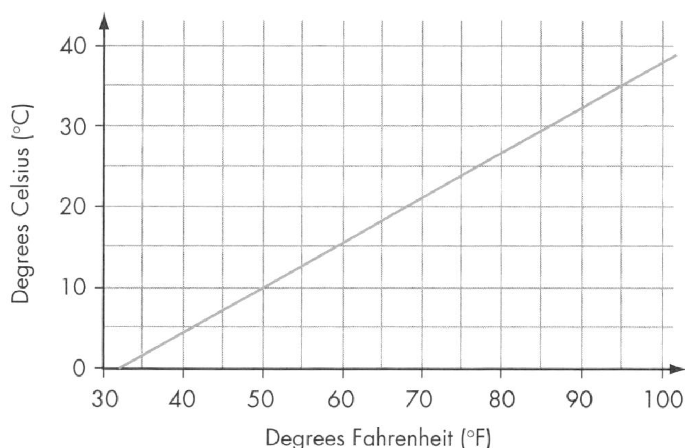

a Use the graph to make an approximate conversion of:

i 35 °C to Fahrenheit

ii 20 °C to Fahrenheit

iii 50 °F to Celsius

iv 90 °F to Celsius.

b Water freezes at 0 °C. What temperature is this in Fahrenheit?

FM **8** Tea is sold at a school fete between 1.00 pm and 2.30 pm. The numbers of cups of tea that had been sold were noted at half-hour intervals.

Time	1.00	1.30	2.00	2.30	3.00	3.30
No. of cups of tea sold	0	24	48	72	96	120

a Draw a graph to illustrate this information. Use a scale from 1 to 4 hours on the horizontal time axis, and from 1 to 120 on the vertical axis for numbers of cups of tea sold.

b Use your graph to estimate when the 60th cup of tea was sold.

FM **9** I lost my fuel bill, but while talking to my friends I found out that:

Bill, who had used 850 units, was charged £57.50
Wendy, who had used 320 units, was charged £31
Rhanni, who had used 540 units, was charged £42.

a Plot the given information and draw a straight-line graph. Use a scale from 0 to 900 on the horizontal units axis, and from £0 to £60 on the vertical cost axis.

b Use your graph to find what I will be charged for 700 units.

E

AU 10 The graph shows the number of passengers arriving in Exeter each day on a particular train that is due at 0815 in December 2009.

The first of December was a Tuesday. Sundays are marked with red lines.

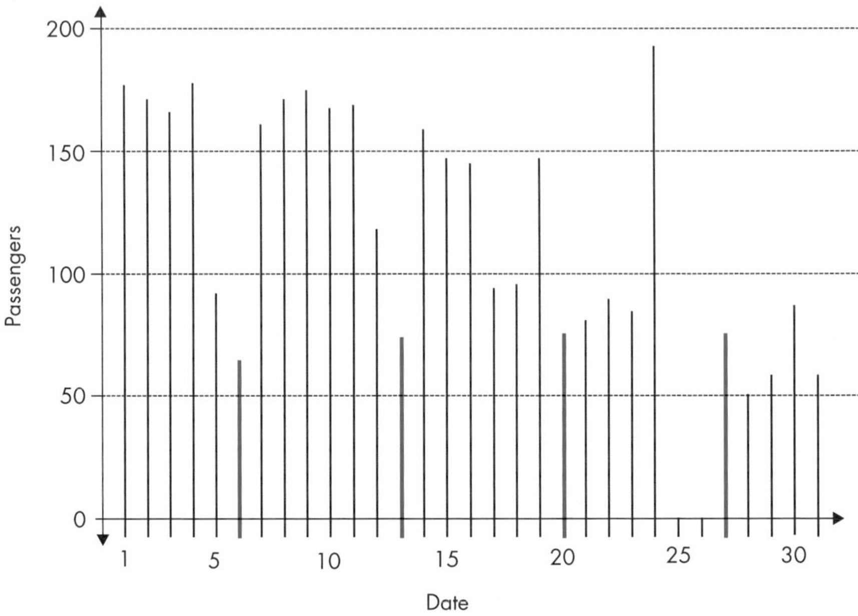

a No passengers used the train on the 25th and 26th. Why was this?

b One of the schools in Exeter closed for Christmas on 16th December. What evidence is there on the graph to support this?

c There was a steady increase in passengers using the train on Saturday through the month and a large number using the train on the 24th. What reason can you give for this?

C

PS AU 11 Leon is travelling from Paris to Calais.

When he sets off he has enough fuel to drive about 100 miles.

Sometime into the journey he sees this sign.

Calais 125 km ⇨
⇦ Paris 75 km

Does he have enough fuel to get to Calais or will he have to visit a petrol station?

Justify your answer.

PS 12 Two taxi companies use these rules for calculating fares:

CabCo: £2.50 basic charge and £0.75 per kilometre

YellaCabs: £2.00 basic charge and £0.80 per kilometre

This map shows the distances, in kilometres, that three friends, Anya (A), Bettina (B) and Calista (C) live from a restaurant (R) and from each other.

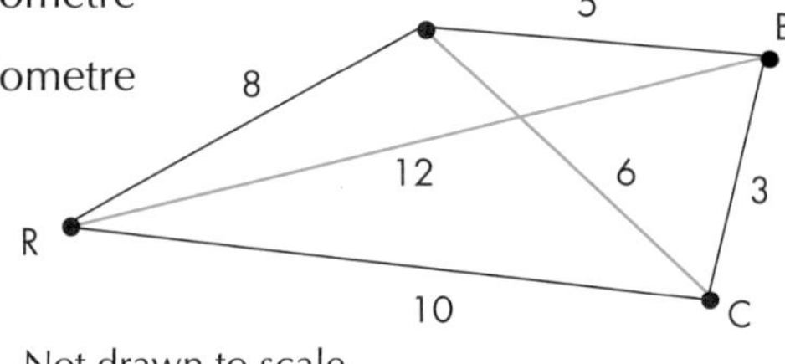

Not drawn to scale

C

You may find a copy of the grid below useful in answering the following question.

a If they each take an individual cab home, which company should they each choose?

b Work out the cheapest way they can travel home if two, or all three, share a cab.

HINTS AND TIPS

Draw a graph for both companies on the grid. Use this to work out the costs of the journeys.

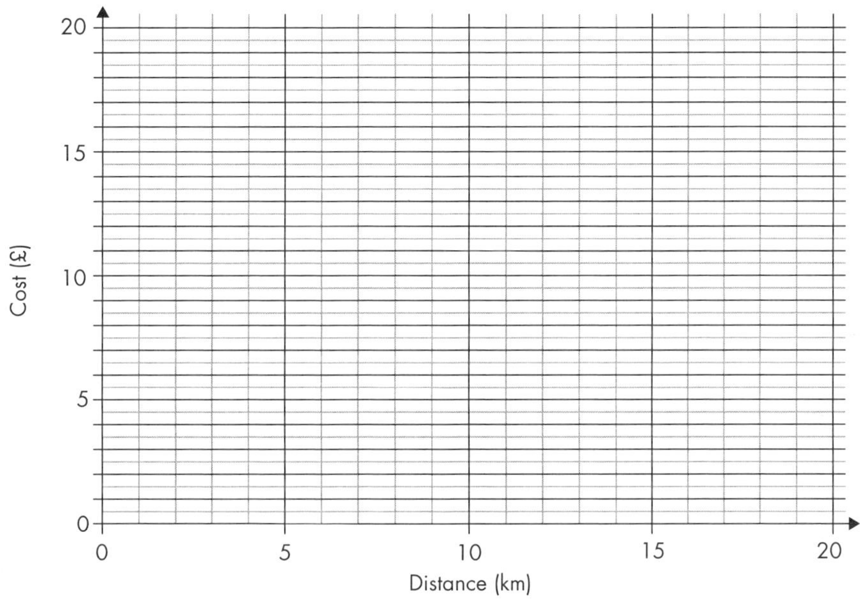

9.2 Travel graphs

This section will show you how to:

- read information from a travel graph
- find an average speed from a travel graph

Key words

average speed
distance–time graph
travel graph

As the name suggests, a **travel graph** gives information about how someone or something has travelled over a given time period. It is also called a **distance–time graph**.

A travel graph is read in a similar way to the conversion graphs you have just done. But you can also find the **average speed** from a distance–time graph by using the formula:

$$\text{average speed} = \frac{\text{total distance travelled}}{\text{total time taken}}$$

EXAMPLE 3

The distance–time graph below represents a car journey from Barnsley to Nottingham, a distance of 50 km, and back again.

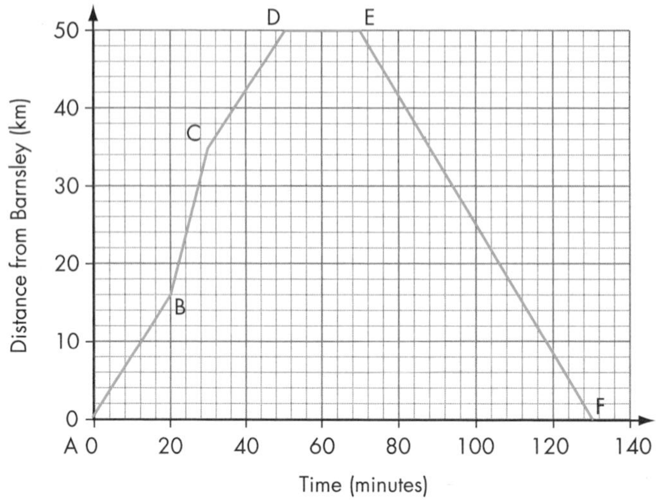

a What can you say about points B, C and D?

b What can you say about the journey from D to F?

c Work out the average speed for each of the five stages of the journey.

From the graph:

a B: After 20 minutes the car was 16 km away from Barnsley.

C: After 30 minutes the car was 35 km away from Barnsley.

D: After 50 minutes the car was 50 km away from Barnsley, so at Nottingham.

b D–F: The car stayed at Nottingham for 20 minutes, and then took 60 minutes for the return journey.

c The average speeds over the five stages of the journey are worked out as follows.

A to B represents 16 km in 20 minutes.

20 minutes is $\frac{1}{3}$ of an hour, so we need to multiply by 3 to give distance/hour. Multiplying both numbers by 3 gives 48 km in 60 minutes, which is 48 km/h.

B to C represents 19 km in 10 minutes.

Multiplying both numbers by 6 gives 114 km in 60 minutes, which is 114 km/h.

C to D represents 15 km in 20 minutes.

Multiplying both numbers by 3 gives 45 km in 60 minutes, which is 45 km/h.

D to E represents a stop: no further distance travelled.

E to F represents the return journey of 50 km in 60 minutes, which is 50 km/h.

So, the return journey was at an average speed of 50 km/h.

You always work out the distance travelled in 1 hour to get the speed in kilometres per hour (km/h) or miles per hour (mph or miles/h).

D

EXERCISE 9B

FM 1 Paul was travelling in his car to a meeting. This distance–time graph illustrates his journey.

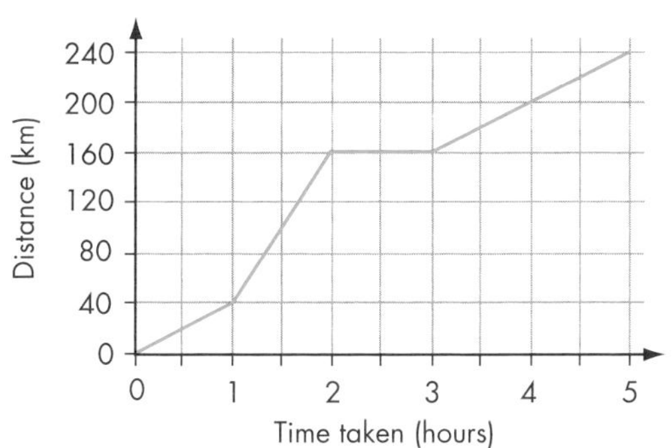

HINTS AND TIPS

Read the question carefully. Paul set off at 7 o'clock in the morning and the graph shows the time after this.

a How long after he set off did he

i stop for his break

ii set off after his break

iii get to his meeting place?

b At what average speed was he travelling:

i over the first hour

ii over the second hour

iii for the last part of his journey?

c The meeting was scheduled to start at 10.30 am.

What is the latest time he should have left home?

HINTS AND TIPS

If part of a journey takes 30 minutes, for example, just double the distance to get the average speed per hour.

2 A small bus set off from Leeds to pick up Mike and his family. It then went on to pick up Mike's parents and grandparents. It then travelled further, dropping them all off at a hotel. The bus then went on a further 10 km to pick up another party and took them back to Leeds. This distance–time graph illustrates the journey.

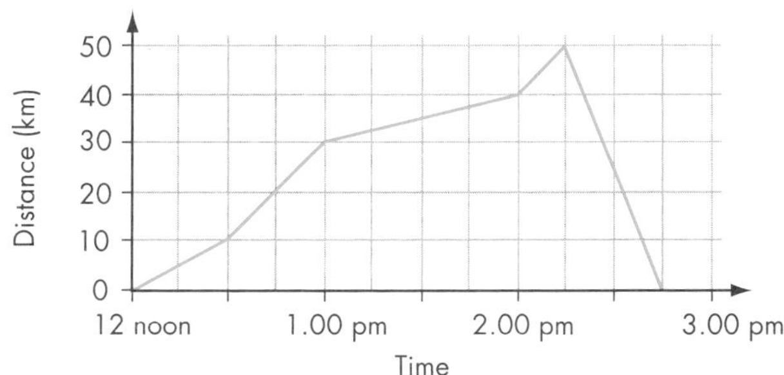

a How far from Leeds did Mike's parents and grandparents live?

b How far from Leeds is the hotel at which they all stayed?

c What was the average speed of the bus on its way back to Leeds?

D

3 James was travelling to Cornwall on his holidays. This distance–time graph illustrates his journey.

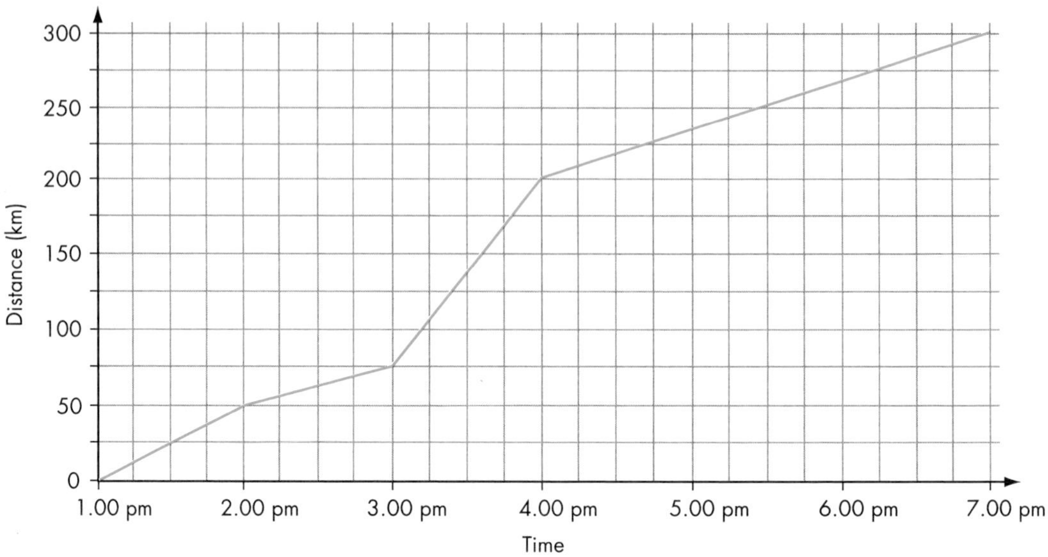

a His greatest speed was on the motorway.

i How far did he travel along the motorway?

ii What was his average speed on the motorway?

b **i** When did he travel most slowly?

ii What was his lowest average speed?

> **HINTS AND TIPS**
>
> **Remember** that the graph is made up of straight lines, as it shows average speed for each section of the journey. In reality, speed is rarely constant – except sometimes on motorways.

PS **4** Azam and Jafar were having a race. The distance–time graph below illustrates the distances covered.

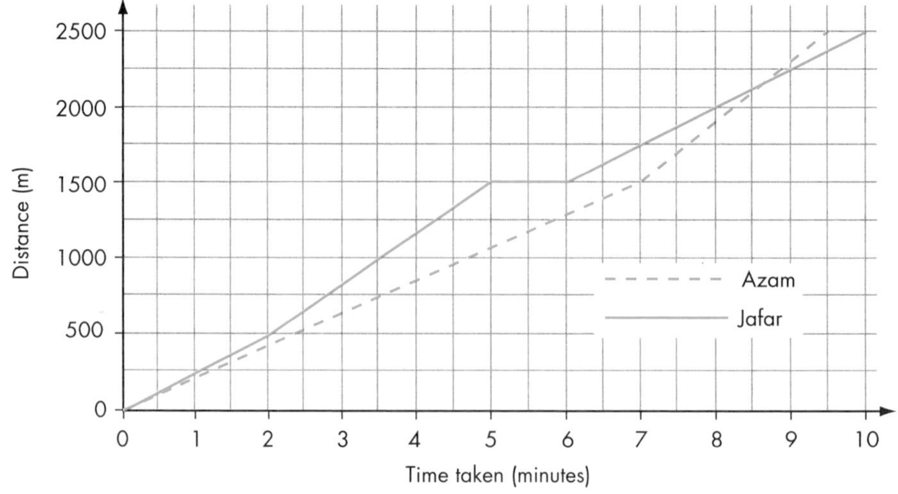

Write a commentary to describe the race.

D

FM 5 Three friends, Patrick, Araf and Sean, ran a 1000 metres race. The race is illustrated on the distance–time graph below.

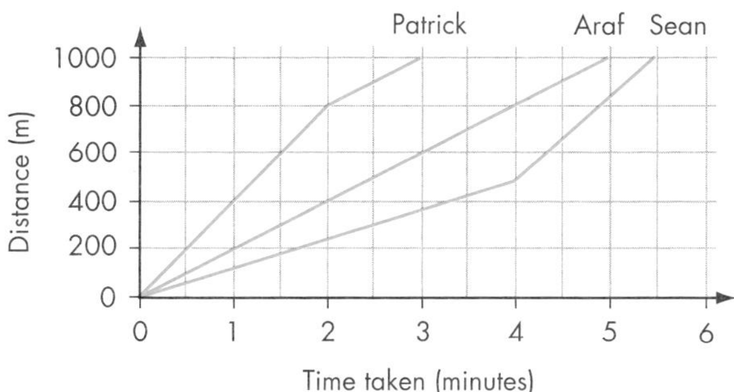

The school newspaper gave the following report of Patrick's race:

'Patrick took an early lead, running the first 800 metres in 2 minutes. He then slowed down a lot and ran the last 200 metres in 1 minute, to finish first in a total time of 3 minutes.'

a Describe the races of Araf and Sean in a similar way.

b **i** What is the average speed of Patrick in kilometres per hour?

ii What is the average speed of Araf in kilometres per hour?

iii What is the average speed of Sean in kilometres per hour?

AU 6 A walker sets off at 9.00 am from point P to walk along a trail at a steady pace of 6 km per hour.

90 minutes later, a cyclist sets off from P on the same trail at a steady pace of 15 km per hour.

At what time did the cyclist overtake the walker?

You may use a graph to help you solve this question.

HINTS AND TIPS

This question can be done by many methods, but drawing a distance–time graph is the easiest.

D

PS **7** Three school friends all set off from school at the same time, 3.45 pm. They all lived 12 km away from the school. The distance–time graph below illustrates their journeys.

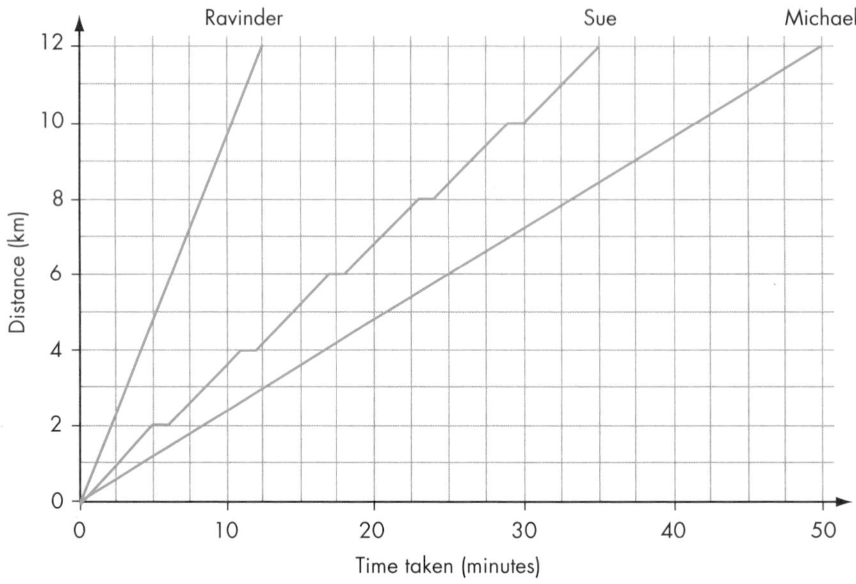

One of them went by bus, one cycled and one was taken by car.

a **i** Explain how you know that Sue used the bus.

ii Who went by car?

b At what time did each friend get home?

c **i** When the bus was moving, it covered 2 kilometres in 5 minutes. What is this speed in kilometres per hour?

ii Overall, the bus covered 12 kilometres in 35 minutes. What is this speed in kilometres per hour?

iii How many stops did the bus make before Sue got home?

9.3 Flow diagrams and graphs

This section will show you how to:
- find the equations of horizontal and vertical lines
- use flow diagrams to draw graphs

Key words
equation of a line
flow diagram
function
input value
line segment
negative coordinates
output value
x-value
y-value

Plotting negative coordinates

A set of axes can form four sectors called quadrants, but so far, all the points you have read or plotted on graphs have been coordinates in the first quadrant. The grid below shows you how to read and plot coordinates in all four quadrants and how to find the equations of vertical and horizontal lines. This involves using **negative coordinates**.

The coordinates of a point are given in the form (x, y), where x is the number along the x-axis and y is the number up the y-axis.

The coordinates of the four points on the grid are:

A(2, 3) B(–1, 2) C(–3, –4) D(1, –3)

The x-coordinate of all the points on line X are 3. So you can say the **equation of line** X is $x = 3$.

The y-coordinate of all the points on line Y are –2. So you can say the equation of line Y is $y = -2$.

Note: The equation of the x-axis is $y = 0$ and the equation of the y-axis is $x = 0$.

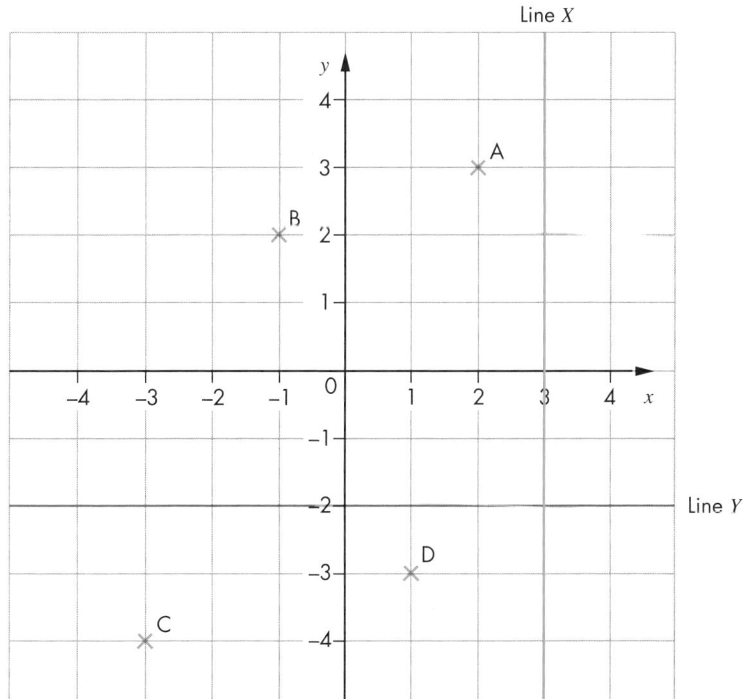

Flow diagrams

One way of drawing a graph is to obtain a set of coordinates from an equation by means of a **flow diagram**. These coordinates are then plotted and the graph is drawn.

In its simplest form, a flow diagram consists of a single box, which may be thought of as containing a mathematical operation, called a **function**. A set of numbers fed into one side of the box is changed by the operation into another set, which comes out from the opposite side of the box. For example, the box shown below represents the operation of multiplying by 3.

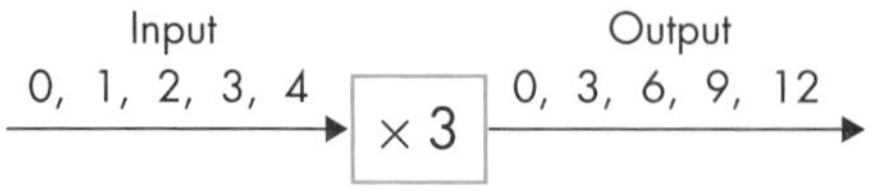

Input: 0, 1, 2, 3, 4 → [× 3] → Output: 0, 3, 6, 9, 12

The numbers that are fed into the box are called **input values** and the numbers that come out are called **output values**.

The input and output values can be arranged in a table.

x	0	1	2	3	4
y	0	3	6	9	12

The input values are called ***x*-values** and the output values are called ***y*-values**. These form a set of coordinates that can be *plotted on a graph*. In this case, the coordinates are (0, 0), (1, 3), (2, 6), (3, 9) and (4, 12).

Most functions consist of more than one operation, so the flow diagrams consist of more than one box. In such cases, you need to match the *first* input values to the *last* output values. The values produced in the middle operations are just working numbers and can be missed out.

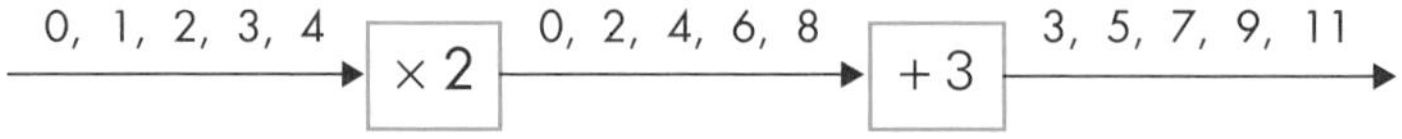

0, 1, 2, 3, 4 → [× 2] → 0, 2, 4, 6, 8 → [+ 3] → 3, 5, 7, 9, 11

So, for the two-box flow diagram the table looks like this.

x	0	1	2	3	4
y	3	5	7	9	11

This gives the coordinates (0, 3), (1, 5), (2, 7), (3, 9) and (4, 11).

The two flow diagrams above represent respectively the equation $y = 3x$ and the equation $y = 2x + 3$, as shown below.

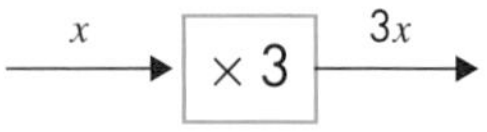

x → [× 3] → $3x$

$y = 3x$

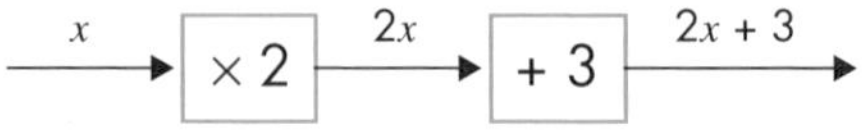

x → [× 2] → $2x$ → [+ 3] → $2x + 3$

$y = 2x + 3$

It is now an easy step to plot the coordinates for each equation on a set of axes, to produce the graphs of $y = 3x$ and $y = 2x + 3$, as shown below.

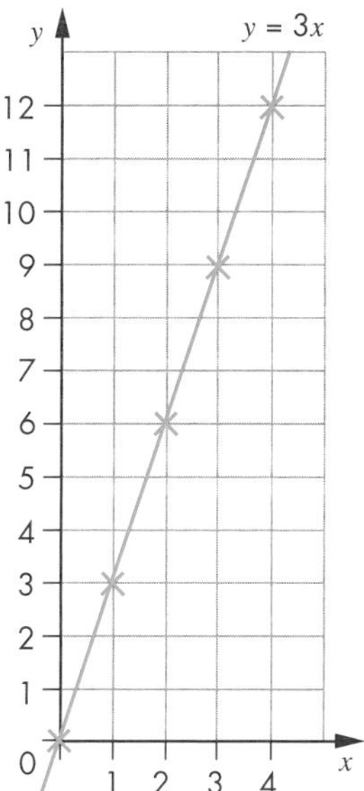

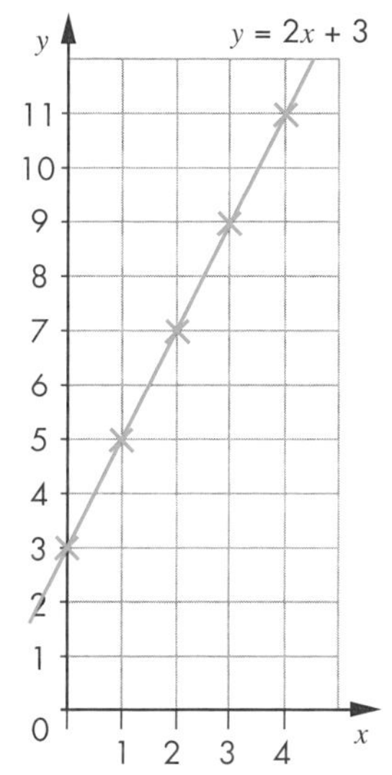

Remember:
Always label graphs.

Note: The line drawn is a **line segment**, as it is only part of an infinitely long line. You will not be penalised if you extend the line beyond the given range of values.

One of the practical problems in graph work is deciding the range of values for the axes. In examinations this is not usually a problem as the axes are drawn for you. Throughout this section, diagrams like the one below will show you the range for your axes for each question. These diagrams are not necessarily drawn to scale.

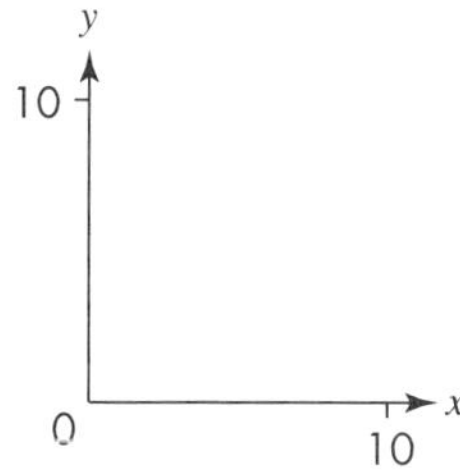

This particular diagram means draw the x-axis (horizontal axis) from 0 to 10 and the y-axis (vertical axis) from 0 to 10. You can use any type of graph or squared paper to draw your axes.

Note that the *scale* on each axis need *not always be the same.*

EXAMPLE 4

Use the flow diagram below to draw the graph of $y = 4x - 1$.

x —0, 1, 2, 3, 4→ [×4] —?, ?, ?, ?, ?→ [− 1] —?, ?, ?, ?, ?→ y

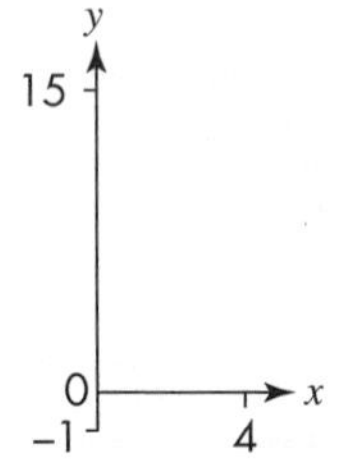

Now enter the values in a table.

x	0	1	2	3	4
y					

The table becomes:

x	0	1	2	3	4
y	−1	3	7	11	15

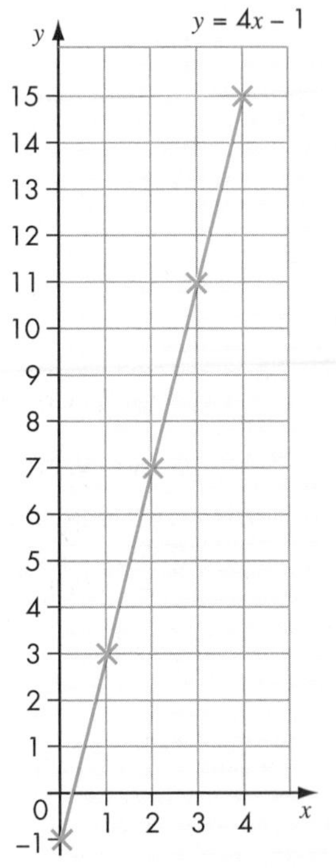

So, the coordinates are:

(0, −1), (1, 3), (2, 7), (3, 11), (4, 15)

Plot these points and join them up to obtain the graph shown on the right.

This is the graph of $y = 4x - 1$.

Always label your graphs. In an examination, you may need to draw more than one graph on the same axes. If you do not label your graphs you may lose marks.

EXERCISE 9C

1

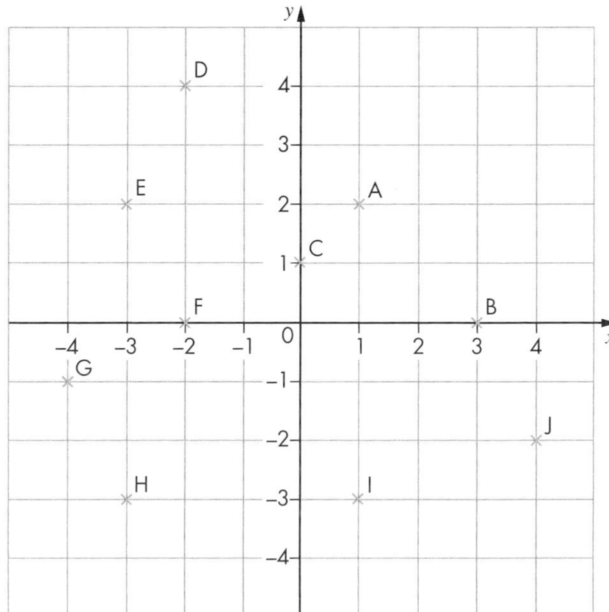

a Write down the coordinates of all the points A to J on the grid.

b Write down the coordinates of the midpoint of the line joining:

i A and B **ii** H and I **iii** D and J.

c Write down the equations of the lines labelled 1 to 4 on the grid.

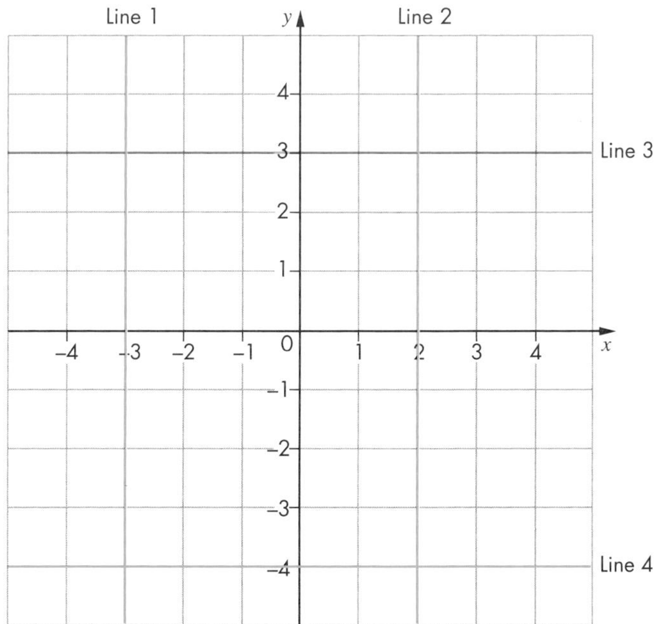

d Write down the equation of the line that is exactly halfway between:

i line 1 and line 2 **ii** line 3 and line 4.

2 Draw the graph of $y = x + 2$.

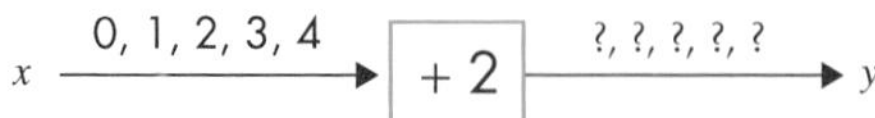

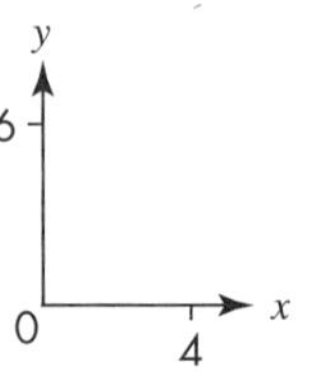

x	0	1	2	3	4
y					

3 Draw the graph of $y = 2x - 2$.

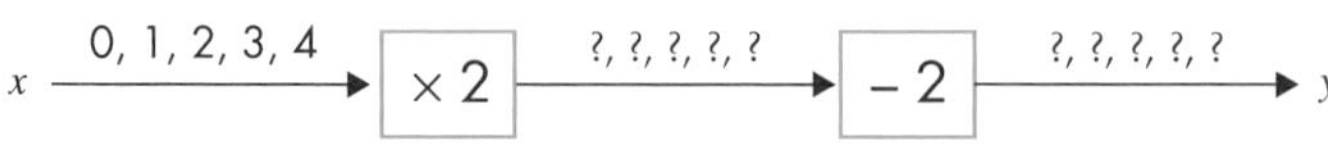

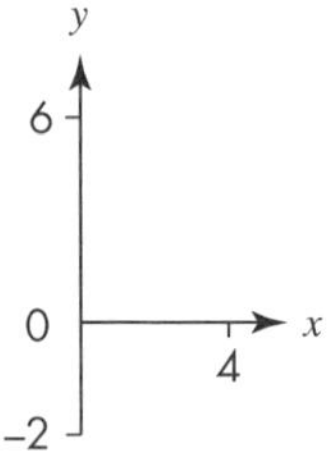

x	0	1	2	3	4
y					

4 Draw the graph of $y = \frac{x}{3} + 1$.

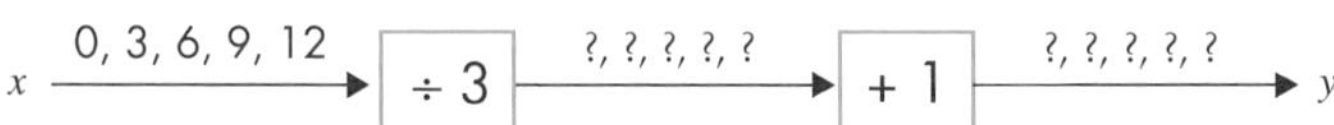

x	0	3	6	9	12
y					

HINTS AND TIPS

If the x-value is divided by a number, then choose multiples of that number as input values. It makes calculations and plotting points much easier.

5 Draw the graph of $y = \frac{x}{2} - 4$.

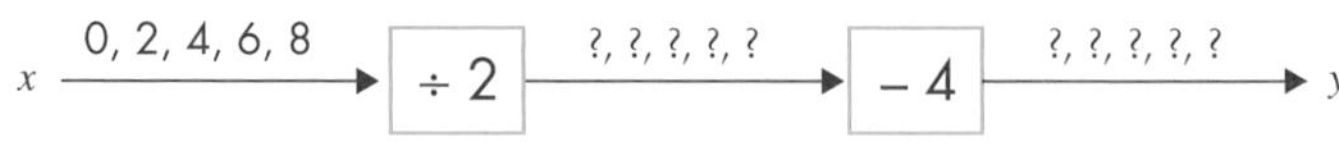

x	0	2	4	6	8
y					

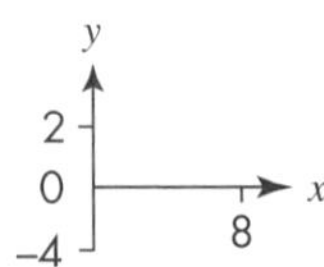

6 **a** Draw the graphs of $y = 2x$ and $y = x + 6$ on the same grid.

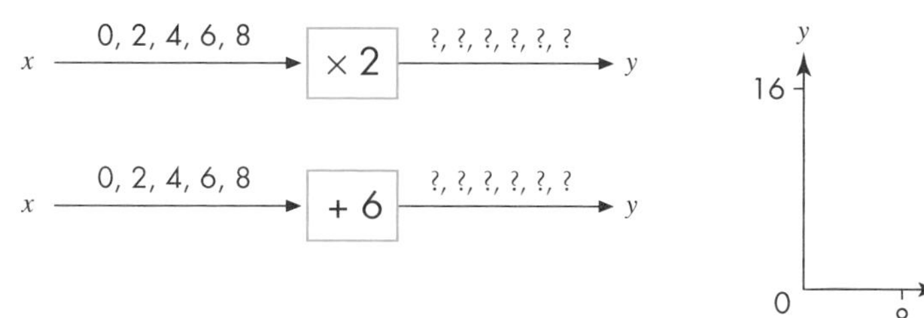

b At which point do the lines intersect?

7 **a** Draw the graphs of $y = x - 3$ and $y = 2x - 6$ on the same grid.

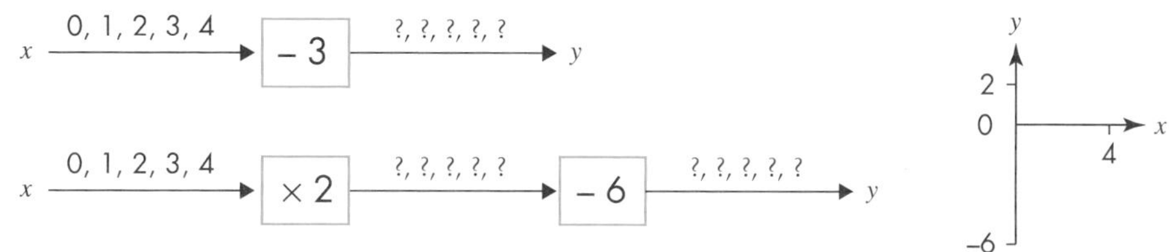

b At which point do the lines intersect?

8 Draw the graph of $y = 5x - 1$. Choose your own inputs and axes.

FM 9 A tea shop sells two types of afternoon tea: a cream tea which costs £3.50 and a high tea which costs £5.00.

To work out the cost of different combinations of teas, they use a wall chart or a flow diagram.

a Use the flow chart to work out the cost of three cream teas and two high teas.

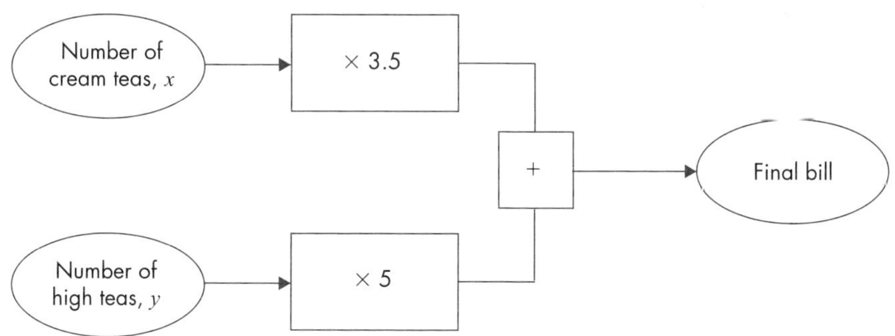

D

b The wall chart is partially filled in.

Use the flow diagram to complete the chart.

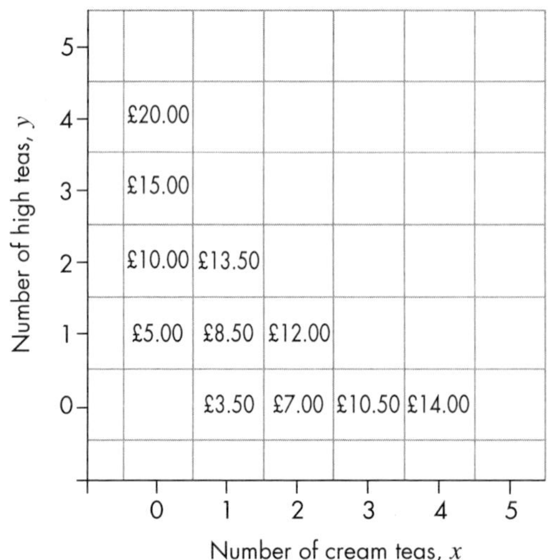

c A party paid £30.50. Can you say for sure what their order was?

AU 10 A teacher reads out the following 'think of a number' problem:

'I am thinking of a number: I multiply it by 3 and add 1.'

a Represent this using a flow diagram.

b If the input is x and the output is y, write down a relationship between x and y.

c Draw a graph for x-values from 0 to 5.

d Explain how you could use the graph to find the number the teacher thought of if the final answer was 13.

PS 11 This flow diagram connects two variables X and Y.

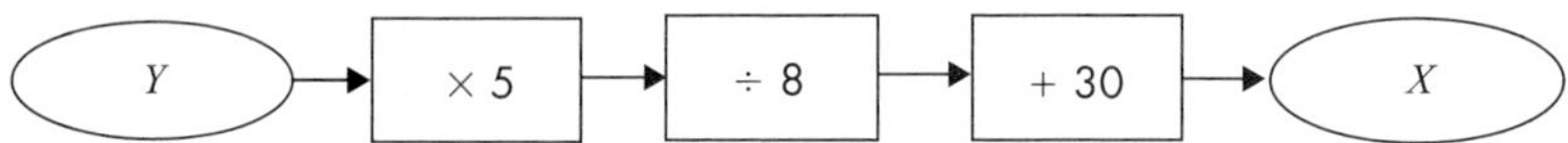

This graph connects the variable Y and X.

Fill in the missing values on the Y-axis.

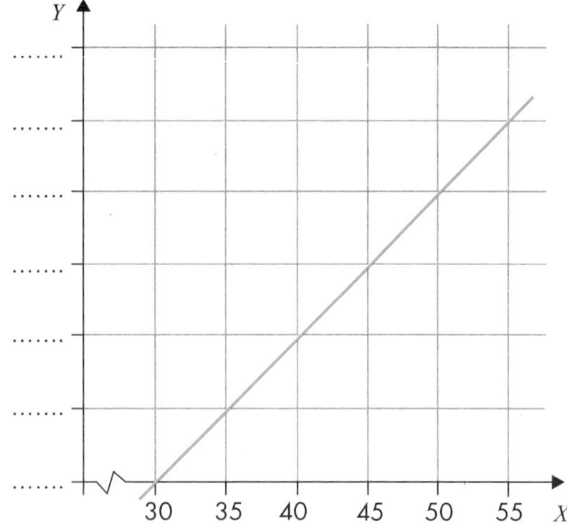

9.4 Linear graphs

This section will show you how to:

- draw linear graphs without using flow diagrams
- find the gradient of a straight line
- use the gradient to draw a straight line

Key words
gradient
linear graphs
slope

This chapter is concerned with drawing straight-line graphs. These graphs are usually referred to as **linear graphs**.

The minimum number of points needed to draw a linear graph is two but it is better to plot three or more because that gives at least one point to act as a check. There is no rule about how many points to plot but here are some tips for drawing graphs.

- Use a sharp pencil and mark each point with an accurate cross.
- Position your eyes directly over the graph. If you look from the side, you will not be able to line up your ruler accurately.

Drawing graphs by finding points

This method is a bit quicker and does not need flow diagrams. However, if you prefer flow diagrams, use them.

Follow through Example 5 to see how this method works.

EXAMPLE 5

Draw the graph of $y = 4x - 5$ for values of x from 0 to 5. This is usually written as $0 \leqslant x \leqslant 5$.

Choose three values for x: these should be the highest and lowest x-values and one in between.

Work out the y-values by substituting the x-values into the equation.

Keep a record of your calculations in a table, as shown below.

x	0	3	5
y			

When $x = 0$, $y = 4(0) - 5 = -5$
This gives the point (0, –5).

When $x = 3$, $y = 4(3) - 5 = 7$
This gives the point (3, 7).

When $x = 5$, $y = 4(5) - 5 = 15$
This gives the point (5, 15).

Hence your table is:

x	0	3	5
y	–5	7	15

You now have to decide the extent (range) of the axes. You can find this out by looking at the coordinates that you have so far.

The smallest x-value is 0, the largest is 5.
The smallest y-value is –5, the largest is 15.

Now draw the axes, plot the points and complete the graph.

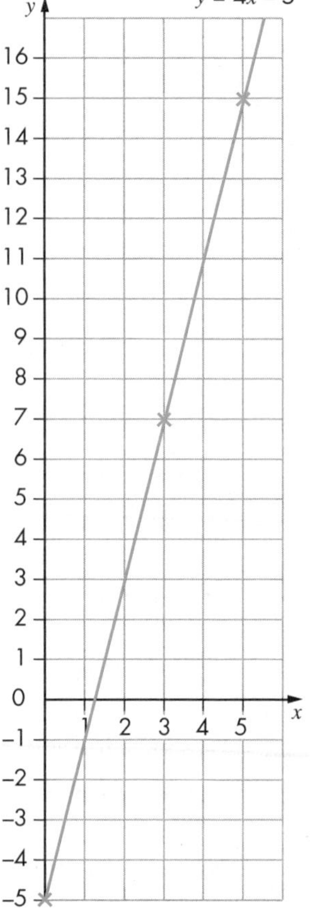

It is nearly always a good idea to choose 0 as one of the x-values. In an examination, the range for the x-values will usually be given and the axes will already be drawn.

EXERCISE 9D

Read through these hints before drawing the following linear graphs.

- Use the highest and lowest values of x given in the range.
- Do not pick x-values that are too close together, such as 1 and 2. Try to space them out so that you can draw a more accurate graph.
- Always label your graph with its equation. This is particularly important when you are drawing two graphs on the same set of axes.
- If you want to use a flow diagram, use one.
- Create a table of values. You will often have to complete these in your examinations.

D

1 Draw the graph of $y = 3x + 4$ for x-values from 0 to 5 ($0 \leqslant x \leqslant 5$).

2 Draw the graph of $y = 2x - 5$ for $0 \leqslant x \leqslant 5$.

HINTS AND TIPS

Complete the table of values first, then you will know the extent of the y-axis.

3 Draw the graph of $y = \frac{x}{2} - 3$ for $0 \leqslant x \leqslant 10$.

4 Draw the graph of $y = 3x + 5$ for $-3 \leqslant x \leqslant 3$.

5 Draw the graph of $y = \frac{x}{3} + 4$ for $-6 \leqslant x \leqslant 6$.

6 **a** On the same set of axes, draw the graphs of $y = 3x - 2$ and $y = 2x + 1$ for $0 \leqslant x \leqslant 5$.

b At which point do the two lines intersect?

7 **a** On the same axes, draw the graphs of $y = 4x - 5$ and $y = 2x + 3$ for $0 \leqslant x \leqslant 5$.

b At which point do the two lines intersect?

8 **a** On the same axes, draw the graphs of $y = \frac{x}{3} - 1$ and $y = \frac{x}{2} - 2$ for $0 \leqslant x \leqslant 12$.

b At which point do the two lines intersect?

9 **a** On the same axes, draw the graphs of $y = 3x + 1$ and $y = 3x - 2$ for $0 \leqslant x \leqslant 4$.

b Do the two lines intersect? If not, why not?

10 **a** Copy and complete the table to draw the graph of $x + y = 5$ for $0 \leqslant x \leqslant 5$.

x	0	1	2	3	4	5
y	5		3		1	

b Now draw the graph of $x + y = 7$ for $0 \leqslant x \leqslant 7$.

11 Ian the electrician used this formula to work out how much to charge for a job:

$C = 25 + 30H$

where C is the charge and H is how long the job takes.

John the electrician uses this formula:

$C = 35 + 27.5H$

a On a copy of the grid, draw lines to represent these formulae.

FM **b** For what length of job do Ian and John charge the same amount?

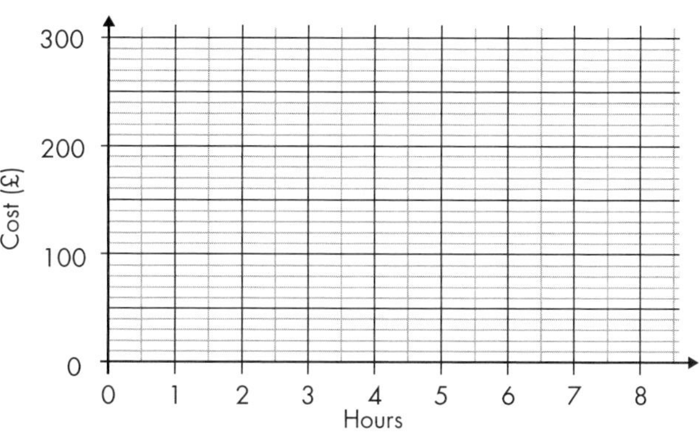

AU **12** **a** Draw the graphs $y = 4$, $y = x$ and $x = 1$ on a copy of the grid on the right.

b What is the area of the triangle formed by the three lines?

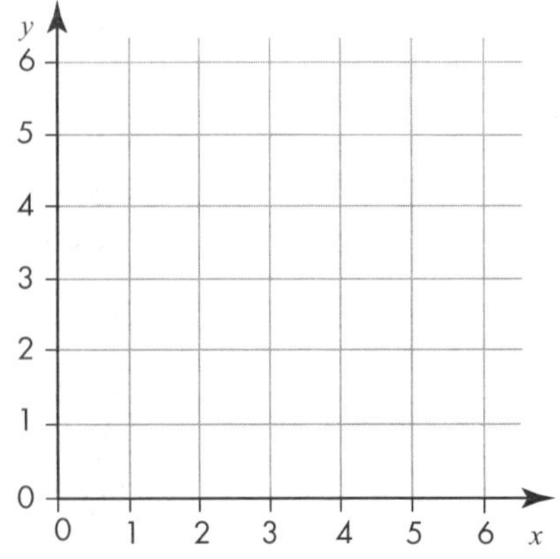

PS **13** The two graphs below show y against x and y against z.

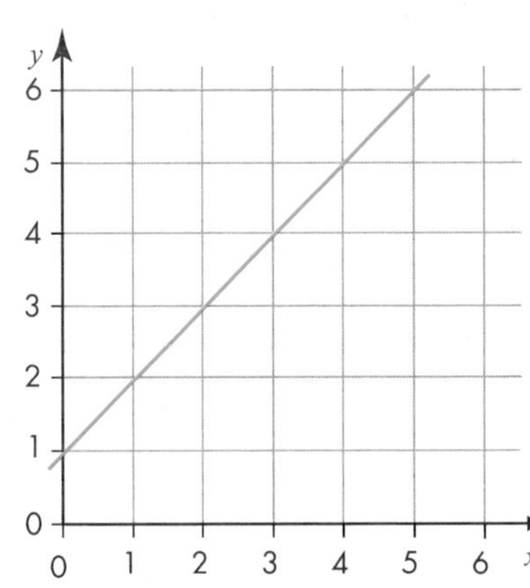

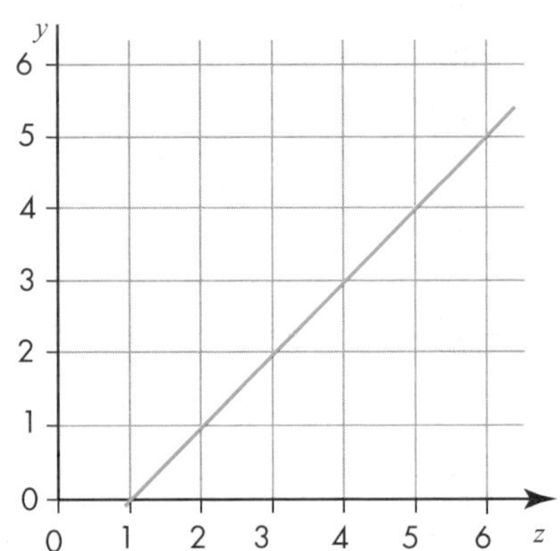

On a copy of the blank grid, show the graph of x against z.

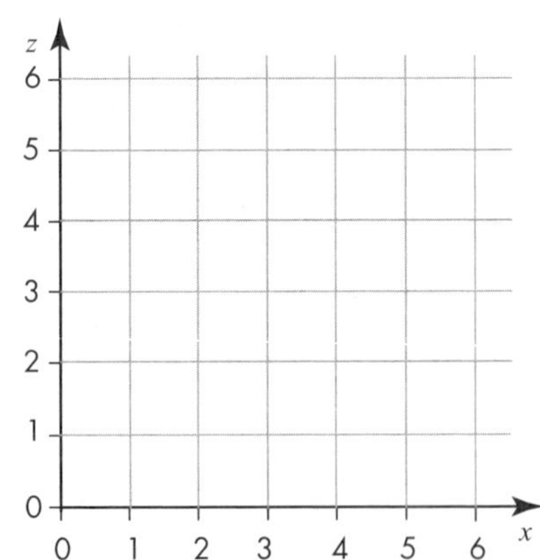

HINTS AND TIPS

When $x = 1$, $y = 2$ and when $y = 2$, $z = 3$.
So, plot coordinates for $x = 1$, $z = 3$.

Gradient

The **slope** of a line is called its **gradient**. The steeper the slope of the line, the larger the value of the gradient.

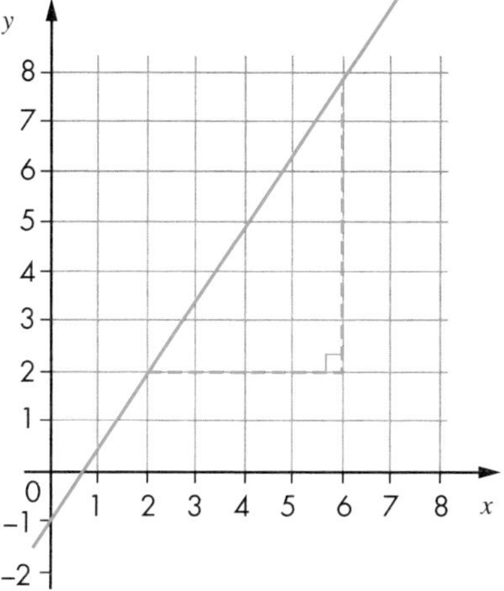

The gradient of the line shown here can be measured by drawing, as large as possible, a right-angled triangle which has part of the line as its hypotenuse (sloping side). The gradient is then given by:

$$\text{gradient} = \frac{\text{distance measured up}}{\text{distance measured along}}$$

$$= \frac{\text{difference on } y\text{-axis}}{\text{difference on } x\text{-axis}}$$

For example, to measure the steepness of the line in the first diagram, below, you first draw a right-angled triangle that has part of this line as its hypotenuse. It does not matter where you draw the triangle but it makes the calculations much easier if you choose a sensible place. This usually means using existing grid lines, so that you avoid fractional values. See the second and third diagrams below.

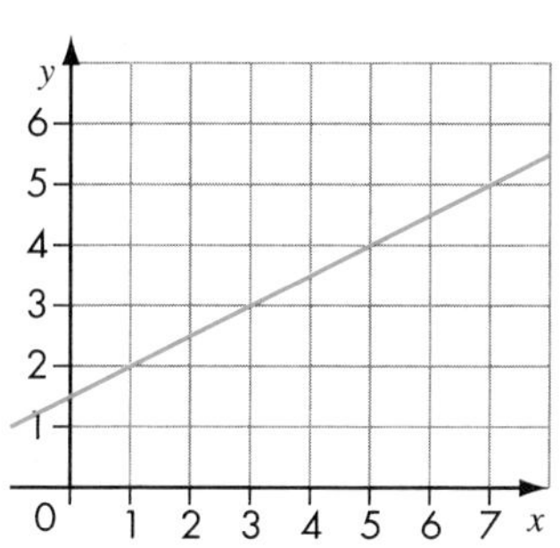

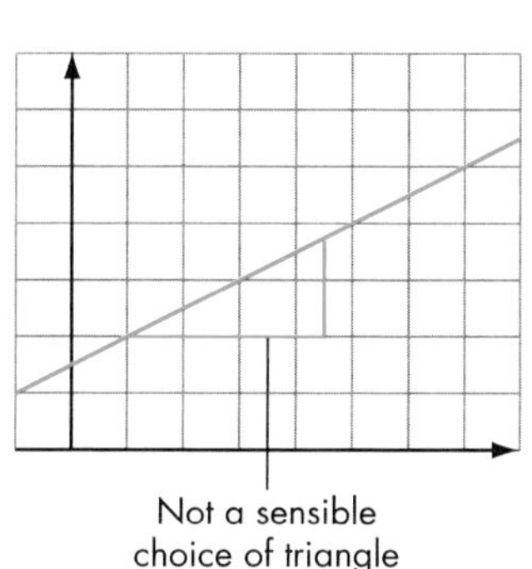

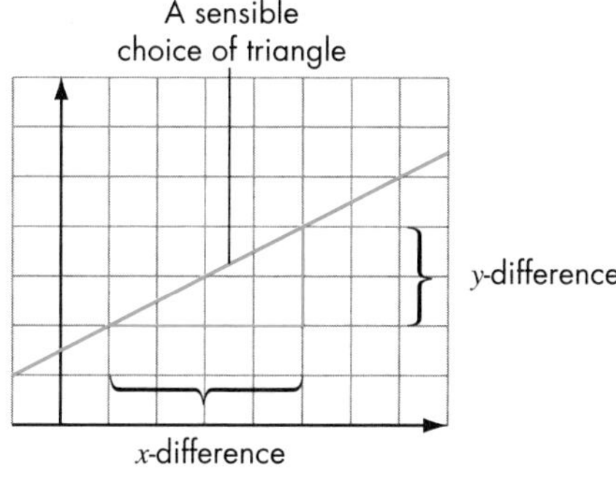

After you have drawn the triangle, measure (or count) how many squares there are on the vertical side. This is the difference between the y-coordinates. In this case, it is 2.

Then measure (or count) how many squares there are on the horizontal side. This is the difference between the x-coordinates. In the case above, this is 4.

To work out the gradient, you make the following calculation.

$$\text{gradient} = \frac{\text{difference of the } y\text{-coordinates}}{\text{difference of the } x\text{-coordinates}}$$

$$= \frac{2}{4} = \frac{1}{2} \text{ or } 0.5$$

Note that the value of the gradient is not affected by where the triangle is drawn. As you are calculating the ratio of two sides of the triangle, the gradient will always be the same wherever you draw the triangle.

Remember: Take care when finding the differences between the coordinates of the two points. Choose one point as the first and the other as the second, and subtract in the *same order* each time to find the difference. When a line slopes *down from right to left* (/) the gradient is always positive, but when a line slopes *down from left to right* (\) the gradient is always negative, so you must make sure there is a minus sign in front of the fraction.

EXAMPLE 6

Find the gradient of each of these lines.

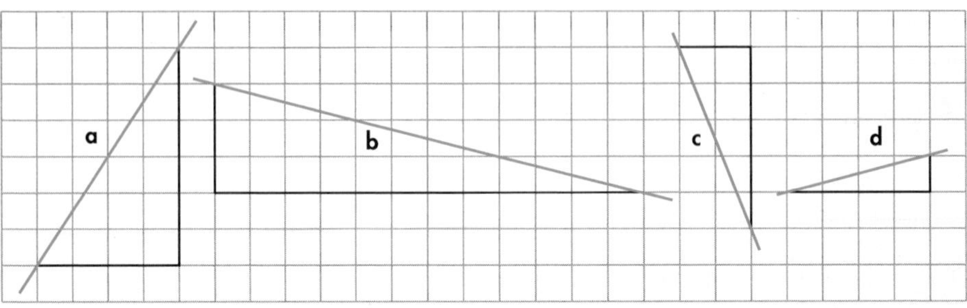

In each case, a sensible choice of triangle has already been made.

a y-difference = 6, x-difference = 4 Gradient $= 6 \div 4 = \frac{3}{2} = 1.5$

b y-difference = 3, x-difference = 12 Line slopes down from left to right,

so gradient $= -(3 \div 12) = -\frac{1}{4} = -0.25$

c y-difference = 5, x-difference = 2 Line slopes down from left to right,

so gradient $= -(5 \div 2) = -\frac{5}{2} = -2.5$

d y-difference = 1, x-difference = 4 Gradient $= 1 \div 4 = \frac{1}{4} = 0.25$

Drawing a line with a certain gradient

To draw a line with a certain gradient, you need to 'reverse' the process described above. Use the given gradient to draw the right-angled triangle first. For example, take a gradient of 2.

Start at a convenient point (A in the diagrams opposite). A gradient of 2 means for an x-step of 1 the y-step must be 2 (because 2 is the fraction $\frac{2}{1}$). So, move one square across and two squares up, and mark a dot.

Repeat this as many times as you like and draw the line. You can also move one square back and two squares down, which gives the same gradient, as the third diagram shows.

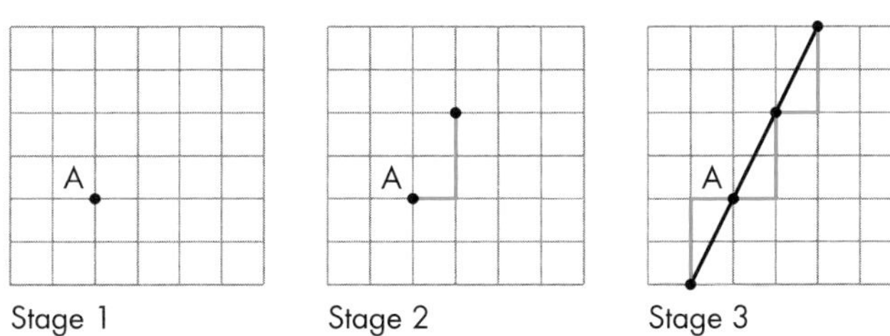

Remember: For a positive gradient you move across (left to right) and then *up*. For a negative gradient you move across (left to right) and then *down*.

EXAMPLE 7

Draw lines with these gradients. **a** $\frac{1}{3}$ **b** -3 **c** $-\frac{1}{4}$

a This is a fractional gradient which has a y-step of 1 and an x-step of 3. Move three squares across and one square up every time.

b This is a negative gradient, so for every one square across, move three squares down.

c This is also a negative gradient and it is a fraction. So for every four squares across, move one square down.

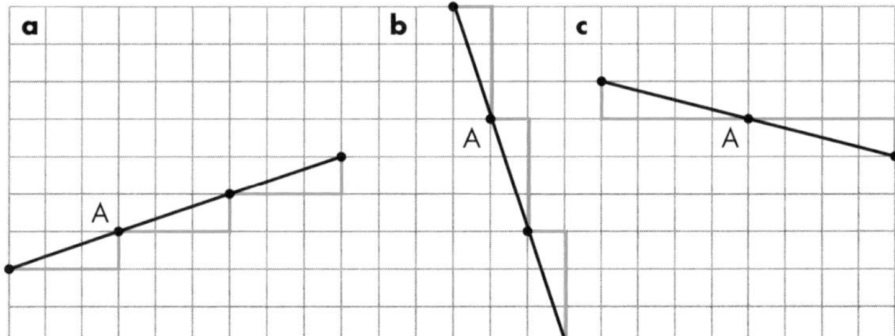

EXERCISE 9E

D

FM **1** Ravi was ill in hospital.
This is his temperature chart for the two weeks he was in hospital.

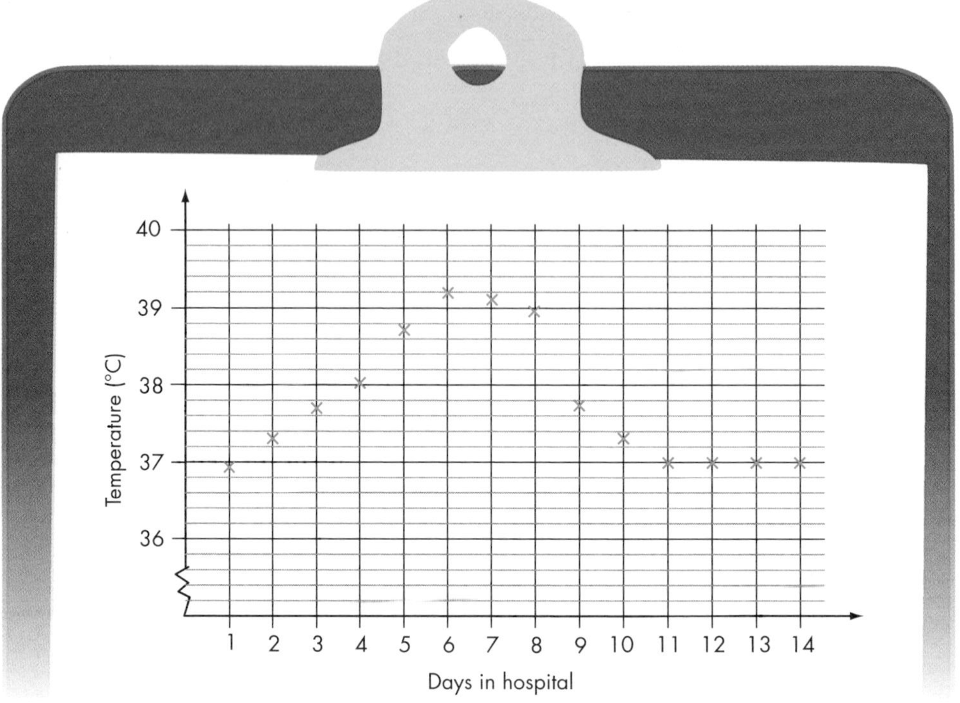

a What was Ravi's highest temperature?

b Between which days did Ravi's temperature increase the most? Explain how you can tell.

c Between which days did Ravi's temperature fall the most? Explain how you can tell.

d When Ravi's temperature went over 38.5 °C he was put on an antibiotic drip.

  i On what day did Ravi go on the drip?

  ii How many days did it take for the antibiotics to work before Ravi's temperature started to come down?

e Once a patient's temperature returns to normal for four days, they are allowed home. What is the normal body temperature?

C

2 Find the gradient of each of these lines.

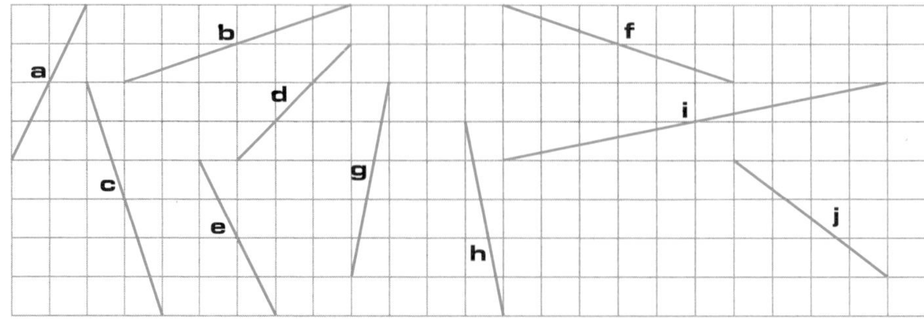

3 Find the gradient of each of these lines. What is special about these lines?

a

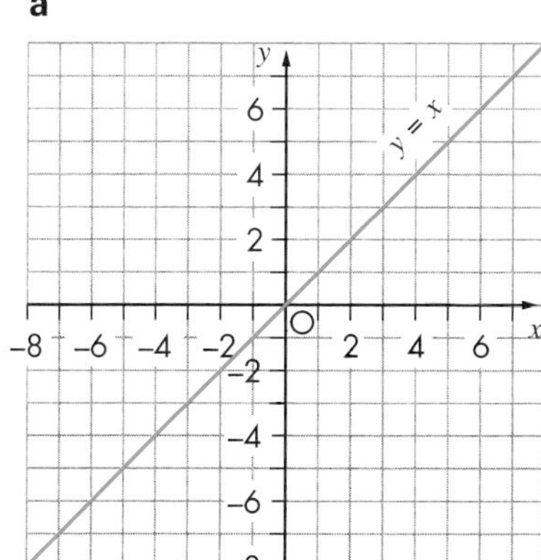

b

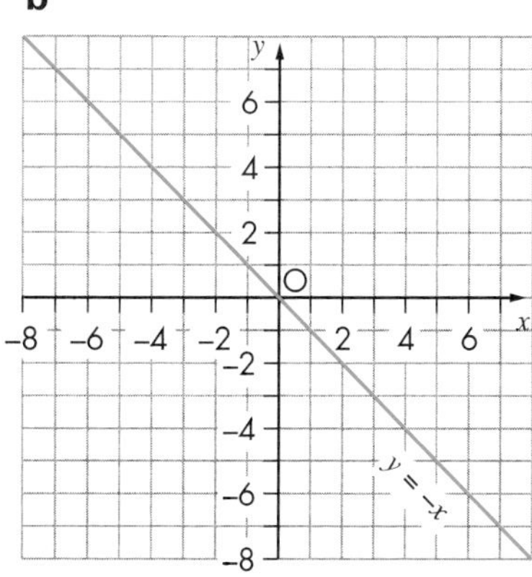

4 Draw lines with these gradients.

a 4 **b** $\frac{2}{3}$ **c** –2 **d** $-\frac{4}{5}$ **e** 6 **f** –6

AU 5 Students in a class were asked to predict the y-value for an x-value of 10 for this line.

Rob says 'The gradient is 1, so the line is $y = x + 2$. When $x = 10$, $y = 12$.'

Rob is wrong.

Explain why and work out the correct y-value.

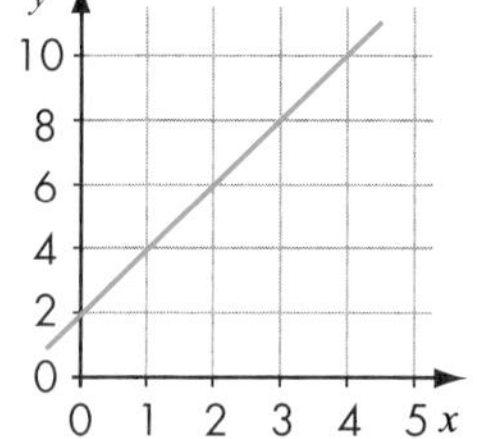

FM PS 6 The Health and Safety regulations for vent pipes from gas appliances state that the minimum height depends on the pitch (gradient) of the roof.

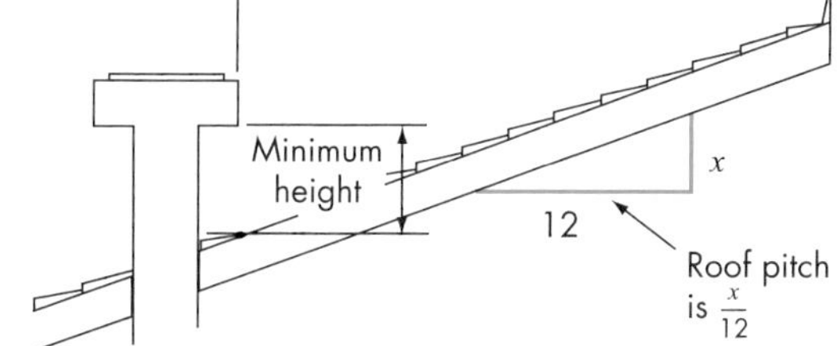

This is the rule:

Minimum height = 1 metre or twice the pitch in metres, whichever is greatest

a What is the minimum height of a roof with a pitch of 2?

b What is the minimum height of a roof with a pitch of 0.5?

c What is the minimum height for these two roofs?

i

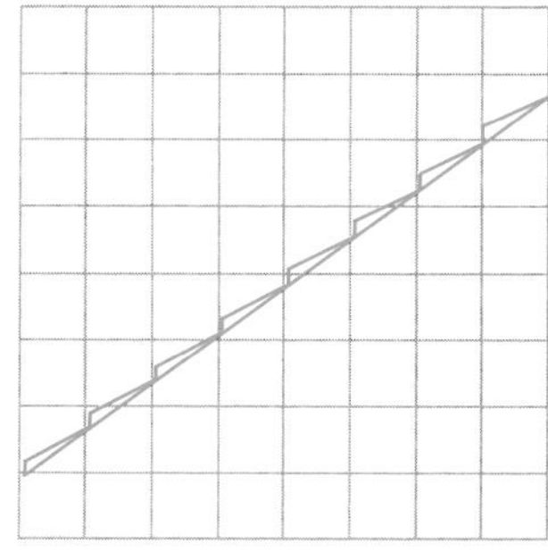

ii

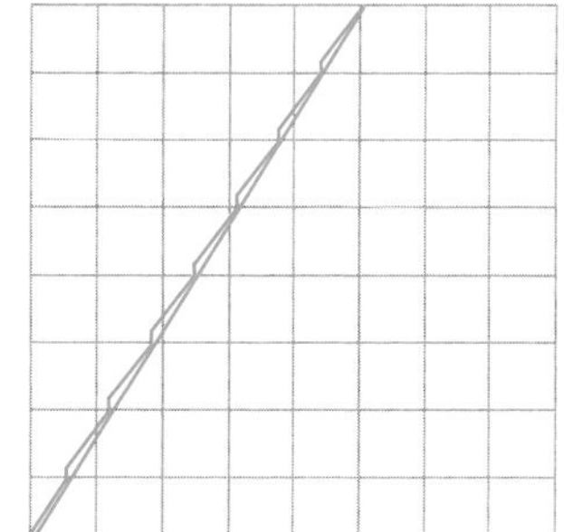

GRADE BOOSTER

- **F** You can read off values from a conversion graph
- **E** You can plot points in all four quadrants
- **E** You can read off distances and times from a travel graph
- **E** You can draw a linear graph given a table of values to complete
- **D** You can find an average speed from a travel graph
- **D** You can draw a linear graph without being given a table of values

What you should know now

- How to use conversion graphs
- How to use travel graphs to find distances, times and speeds
- How to draw a linear graph

1 This conversion graph can be used to change between metres and feet.

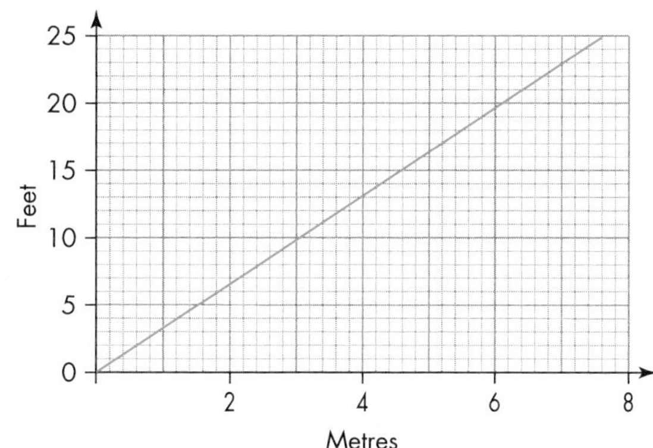

a Use the conversion graph to change 6 metres to feet. (1)

b Use the conversion graph to change 8 feet to metres. (1)

Robert jumps 4 metres.

James jumps 12 feet.

c **i** Who jumps furthest, Robert or James?

ii How did you get your answer? (2)

(Total 4 marks)

Edexcel, November 2008, Paper 13 Foundation, Question 10

2

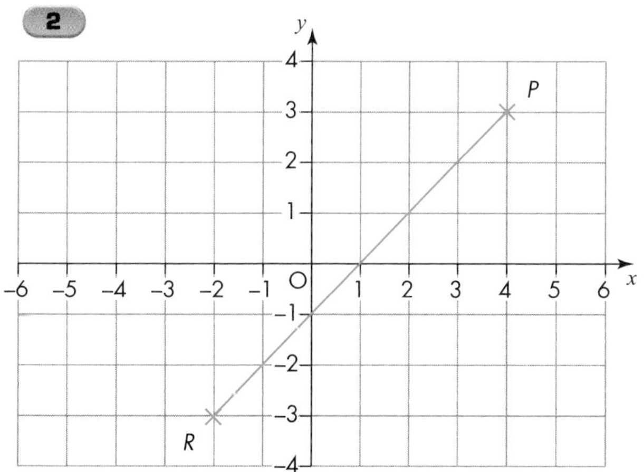

a Write down the coordinates of the point *P*. (1)

b On the grid, mark the point (–3, 1) with a cross (x). (1)

c Write down the coordinates of the midpoint of the line *PR*. (2)

(Total 4 marks)

Edexcel, June 2007, Paper 10 Foundation, Question 3

3

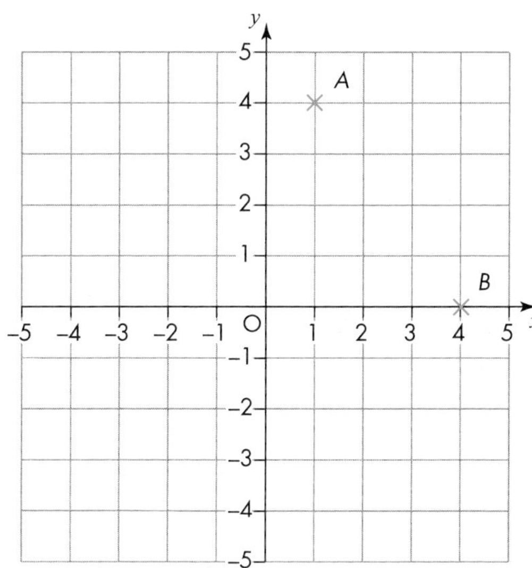

a **i** Write down the coordinates of the point *A*.

ii Write down the coordinates of the point *B*. (2)

b **i** On the grid, plot the point (3, 2). Label this point *P*.

ii On the grid, plot the point (–4, 3). Label this point *Q*. (2)

(Total 4 marks)

Edexcel, May 2008, Paper 1 Foundation, Question 7

4 **a** Complete the table of values for $y = 2x - 3$

x	–1	0	1	2	3
y	–5		–1		3

b On a grid, draw the graph of $y = 2x - 3$ for values of x from –1 to +3. Take the x-axis from –1 to +3 and the y-axis from –5 to 3.

c Find the coordinates of the point where the line $y = 2x - 3$ crosses the line $y = -2$.

5 **a** Complete the table of values for $y = 3x - 1$

x	–2	–1	0	1	2
y		–4		2	

(2)

b On the grid, draw the graph of $y = 3x - 1$

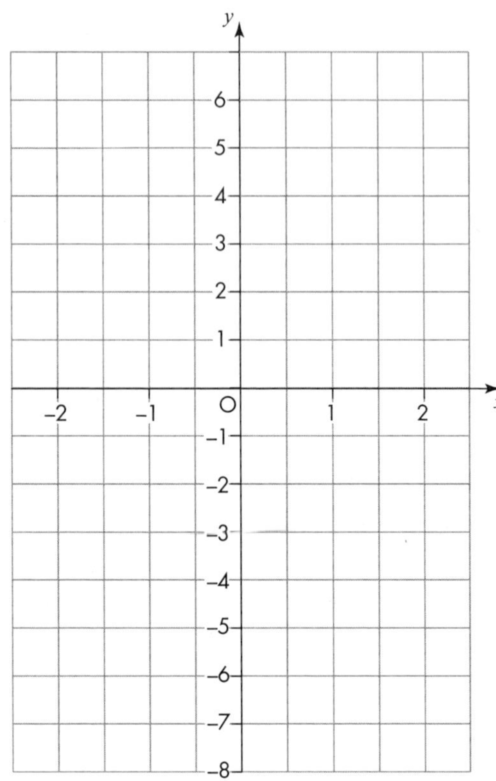

(2)

(Total 4 marks)

Edexcel, November 2008, Paper 10 Foundation, Question 8

6 **a** Complete the table of values for $y = 3x + 1$

x	–3	–2	–1	0	1	2
y	–8		–2			

(2)

b On the grid, draw the graph of $y = 3x + 1$

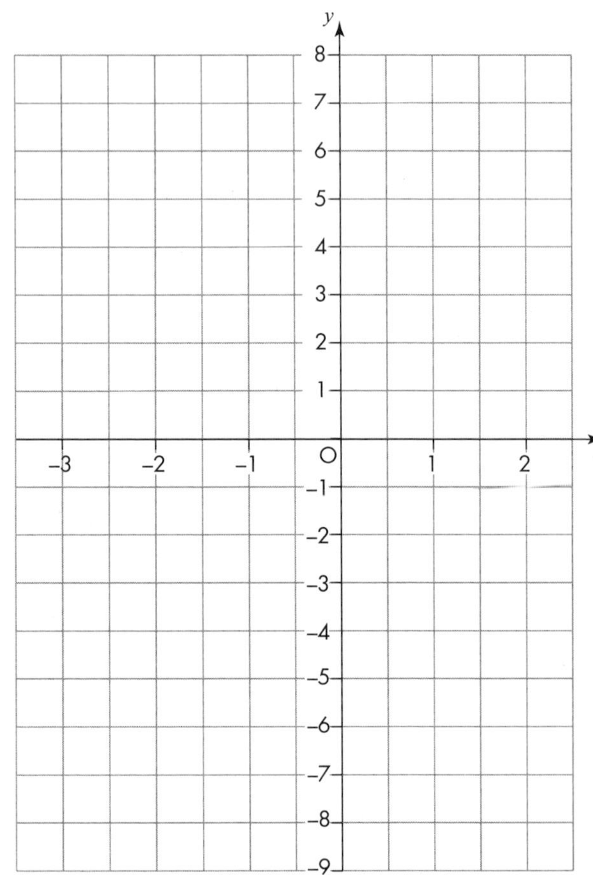

(2)

(Total 4 marks)

Edexcel, May 2008, Paper 1 Foundation, Question 18

7 **a** Draw a set of axes on a grid and label the x-axis from –4 to +4 and the y-axis from –8 to +10. On this grid, draw and label the lines $y = -5$ and $y = 2x + 1$.

b Write down the coordinates of the point where the lines $y = -5$ and $y = 2x + 1$ cross.

8

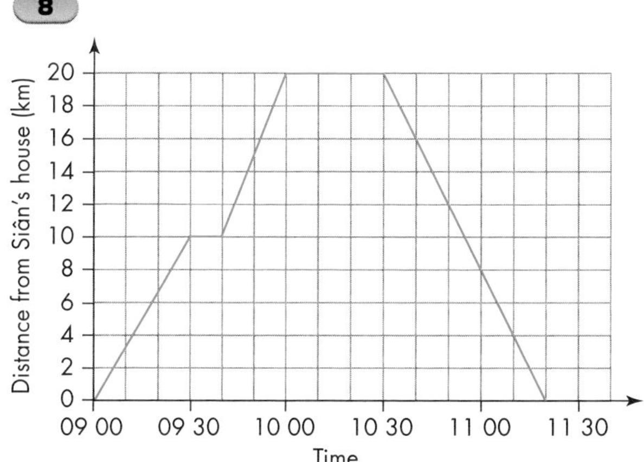

a How far is Siân from her house at 09 30? (1)

The library is 20 km from Siân's house.

b **i** At what time did Siân arrive at the library?

ii How long did Siân spend at the library? (2)

Siân left the library at 10 30 to travel back to her house.

c At what time did Siân arrive back at her house? (1)

(Total 4 marks)

Edexcel, May 2008, Paper 12 Foundation, Question 10

9 Pete visited his friend and then returned home.

The travel graph shows some information about Pete's journey.

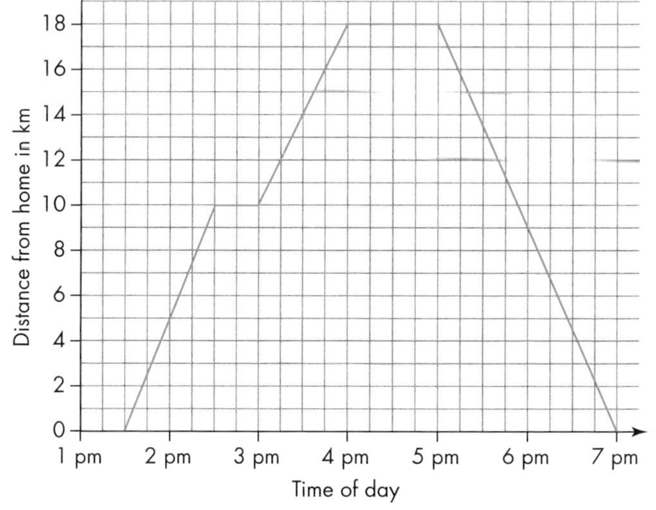

a Write down the time that Pete started his journey. (1)

At 2.30 pm Pete stopped for a rest.

b **i** Find his distance from home when he stopped for this rest.

ii How many minutes was this rest? (2)

Pete stayed with his friend for one hour.
He then returned home.

c Work out the total distance travelled by Pete on this journey. (2)

(Total 5 marks)

Edexcel, November 2008, Paper 12 Foundation, Question 8

10 Here are six temperature/time graphs.

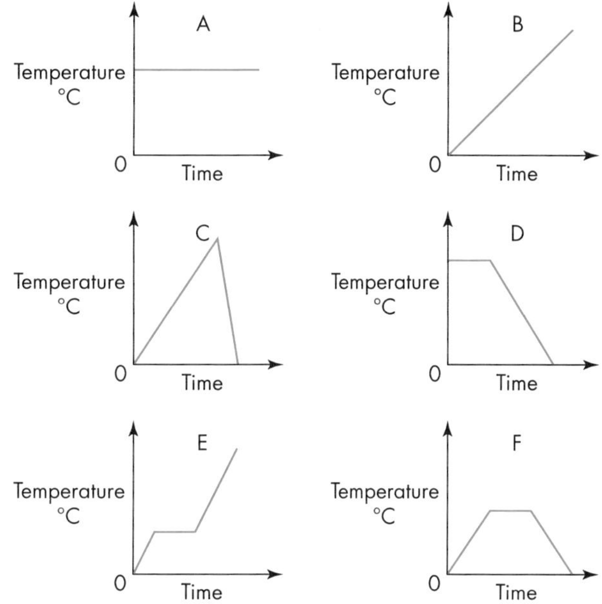

Each sentence in the table describes one of the graphs. Write the letter of the correct graph next to each sentence.

The first one has been done for you. (3)

The temperature starts at 0°C and keeps rising.	**B**
The temperature stays the same for a time and then falls.	
The temperature rises and then falls quickly.	
The temperature is always the same.	
The temperature rises, stays the same for a time and then falls.	
The temperature rises, stays the same for a time and then rises again.	

(Total 3 marks)

Edexcel, November 2008, Paper 12 Foundation, Question 11

Worked Examination Questions

FM **1** The distance–time graph shows the journey of a train between two stations. The stations are 6 kilometres apart.

a During the journey the train stopped at a signal. For how long was the train stopped?

b What was the average speed of the train for the whole journey? Give your answer in kilometres per hour.

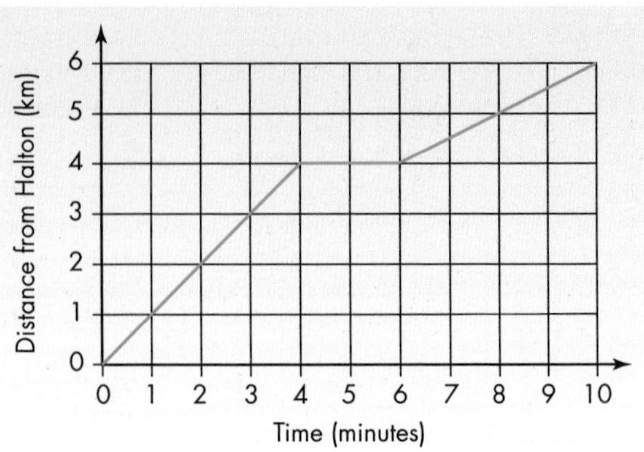

a The train stopped for 2 minutes (where the line is horizontal).

You get 1 mark for the correct answer.

b The train travels 6 km in 10 minutes. This is 36 km in 60 minutes (multiply both numbers by 6). So the average speed is 36 km/h.

You will get 1 method mark for writing down any distance and an equivalent time and 1 accuracy mark for the answer.

Total: 3 marks

FM **2** The graph shows the increase in rail fares as a percentage of the fares in 1997 since the railways were privatised in 1997.

a In what year did first class fares double in price from 1997?

b Approximately how much would a regulated standard class fare that cost £20 in 1997 cost in 2002?

c Approximately how much would a first class fare that cost £100 in 1997 cost in 2010?

d During which period did first class fares rise the most? How can you tell?

e An unregulated standard class fare cost £35 in 2008. Approximately how much would this fare have been in 1997?

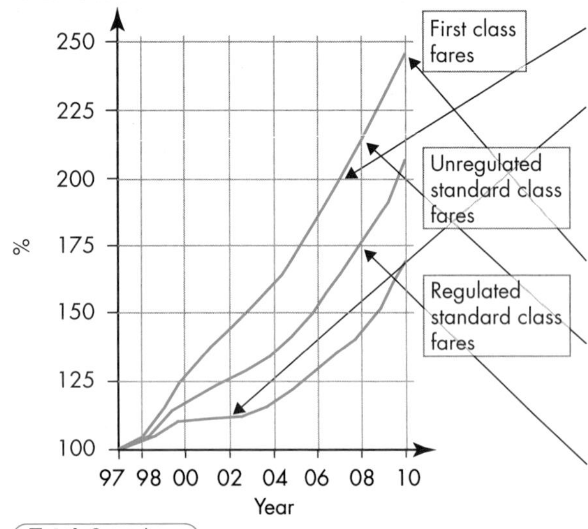

a 2007. Read when the fares were 200%.

This is worth 1 mark.

b £22.00. Read the percentage increase in 2002 and multiply the fare by this figure (about 10%).

This is worth 1 method mark and 1 mark for accuracy.

c £223. Read the percentage increase in 2010.

The correct answer is worth 1 mark.

d 2005–2010: this is the steepest part of the graph.

The correct answer is worth 1 mark.

e £20. Unregulated fares have gone up by 75%. 75% of £20 is £15.

The correct answer is worth 1 mark.

Total: 6 marks

Worked Examination Questions

PS **3** This graph shows the conversion between two variables x and y.

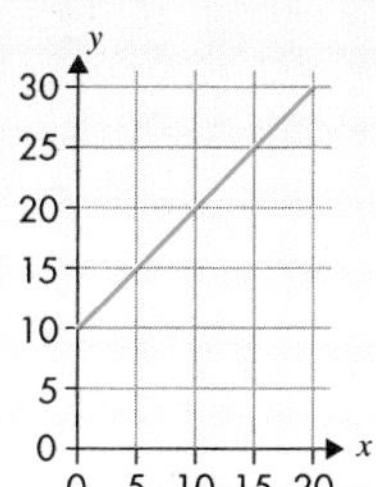

This graph shows the conversions between two variables y and z.

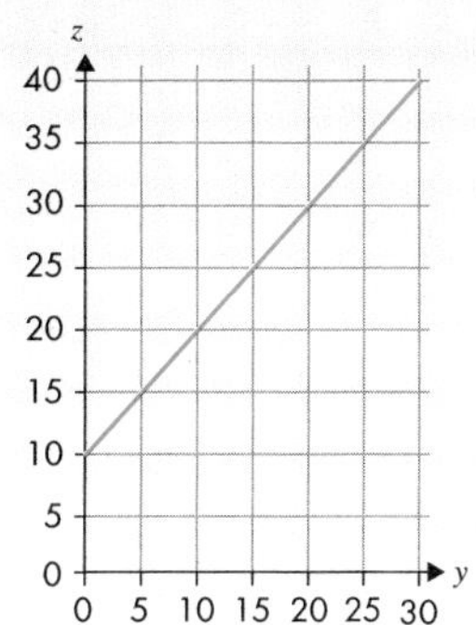

On the graph below, draw the conversion between x and z.

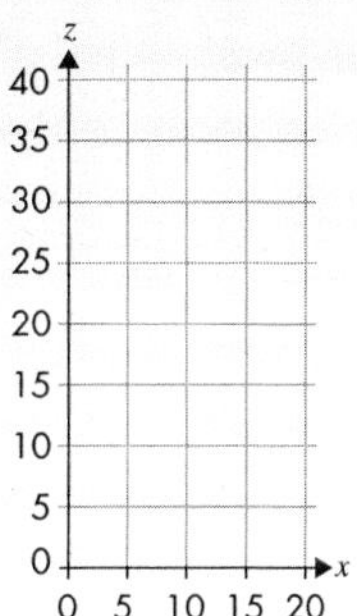

When $x = 0, y = 10$ and when $y = 10, z = 20$

When $x = 20, y = 30$ and when $y = 30, z = 40$

Connect at least two x-values to y-values. The obvious choices are 0 and 20 as these are the limits for x.
This is worth 1 method mark and 1 accuracy mark.

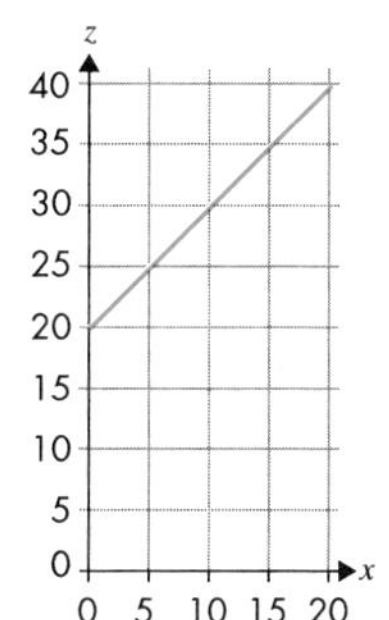

Draw a line joining the pairs of x- and z-values, i.e. (0, 20) to (20, 40).
This is worth 1 accuracy mark.

Total: 3 marks

AU **4** Three lines are shown:

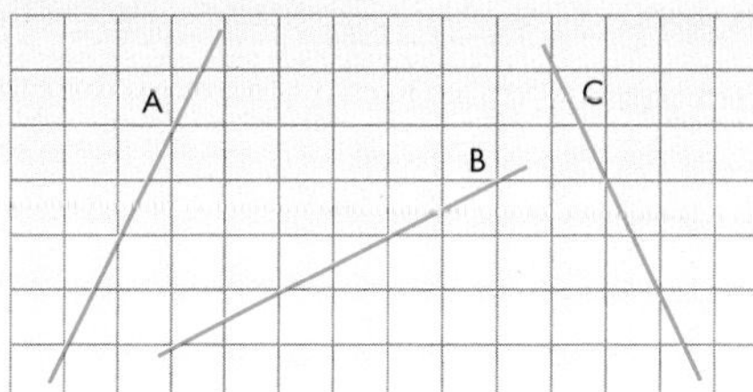

a Write down one thing that is the same about the gradients of line A and B.

b Write down one thing that is the same about the gradients of line A and C.

c Write down one thing that is different about the gradients of line B and C.

Gradient of A is 2, B is $\frac{1}{2}$ and C is -2

First work out gradients for 1 mark.

a They are both positive.

b They both have values of 2.

c One is positive and one is negative.

Write down something obvious. This question is testing your understanding of gradients so you could have said anything valid.
You get 1 mark for a statement about line B and 1 mark for a statement about line C.

Total: 4 marks

Functional Maths
Planning a motorbike trip to France

A group of friends are going on a holiday to France. They have decided to go on motorbikes and have asked you to join them as a pillion passenger. The ferry will take your group to Boulogne and the destination is Perpignan.
Planning your motorbike trip will involve a range of mathematics, much of which can be represented on graphs.

Your task

The following task will require you to work in groups of 2–3.

Using all the information that you gather from these pages and your own knowledge, investigate the key mathematical elements of your motorbike trip.

You must draw at least one conversion graph and try to use as many different mathematical methods as possible.

Getting started

Start by thinking about the mathematics that you use when you go on holiday. Here are a few questions to get you going.

- What information do you need before you travel?
- How many euros (€) are there in one British pound (£)?
- What differences might you find when you travel abroad?
- What are the differences between metric and imperial units of measure? (It may help to list the conversion facts that you know.)
- After approximately how long would you need to stop to rest when travelling?

Handy hints

There are a number of measures and units in France that will need converting when you get there. Two of the most noticeable are:

- **Currency:** in France the currency is in euros
- **Distances**: in France, distances are measured in kilometres, and hence speeds are in kilometres per hour.

1 Euro = £0.90

1 gallon ≈ 4 litres

1 mile ≈ 1.6 km

A motorbike fuel tank holds 3 gallons and travels about 45 miles per gallon.

The cost of petrol in France is 1.2 per litre.

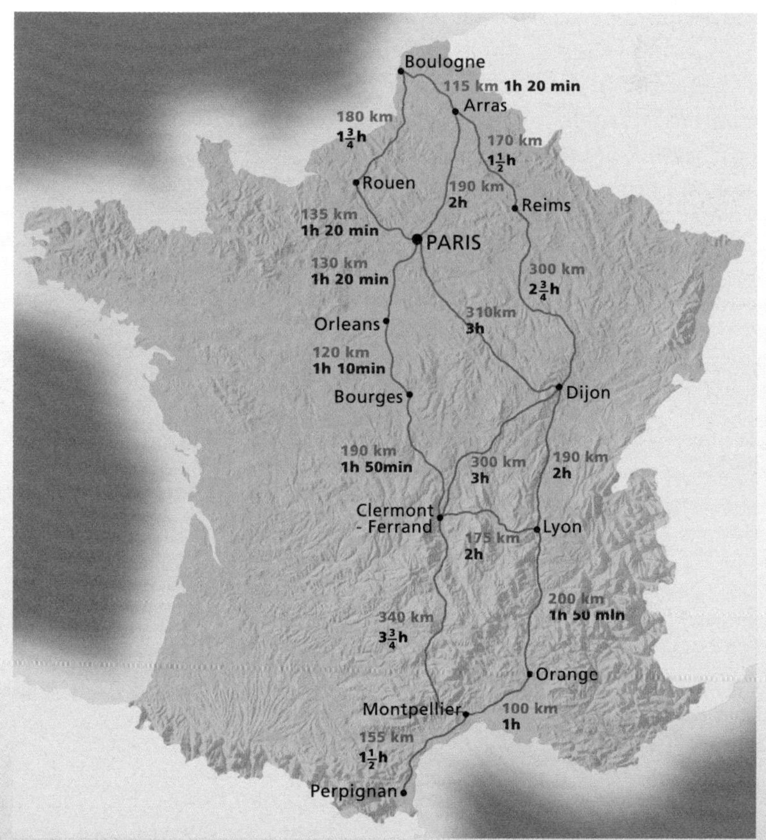

Why this chapter matters

It is essential that we understand angles. They help us to construct everything, from a building to a table. So, angles literally shape our world.

Ancient measurement of angles

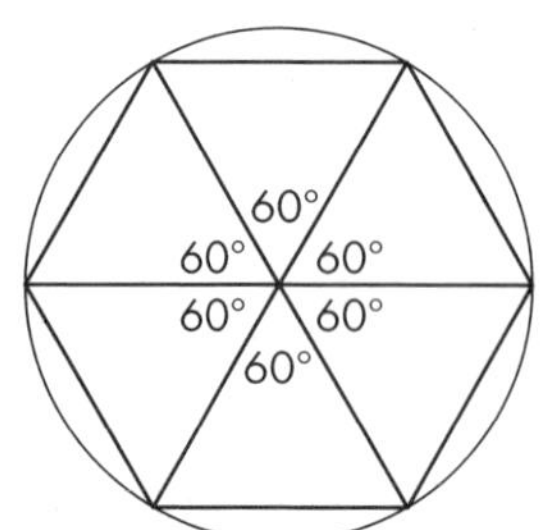

Ancient civilisations used **right angles** in surveying and in constructing buildings, however, not everything can be measured in right angles. There is a need for a smaller, more useful unit. The ancient Babylonians chose a unit angle that led to the development of the **degree**, which is what we still use now.

Most historians think that the ancient Babylonians believed that the 'circle' of the year consisted of 360 days. Mathematics historians also generally believe that the ancient Babylonians knew that the side of a **regular hexagon** inscribed in a circle is equal to the **radius** of the circle. This may have led to the division of the full circle (360 'days') into six equal parts, each part consisting of 60 'days', as shown opposite. They divided one angle of an **equilateral triangle** into 60 equal parts, now called degrees, then further subdivided a degree into 60 equal parts, called **minutes**, and a minute into 60 equal parts, called **seconds**.

The divisions 'minutes' and 'second' are also used in time-keeping.

Modern theodolite.

Modern measurement of angles

Modern surveyors use a **theodolite** for measuring angles. A modern theodolite comprises a movable telescope mounted within horizontal and a vertical axis. When the telescope is pointed at a desired object, the angle of each of these axes can be measured with great precision, typically on the scale of **arcseconds**. (There are 3600 arcseconds in 1°.)

This can be used for measuring both horizontal and vertical angles. It is a key tool in surveying and engineering work, particularly on inaccessible ground, but theodolites have been adapted for other specialised purposes in fields such as meteorology and rocket-launch technology.

Chapter

Geometry and measures: Angles

The grades given in this chapter are target grades.

1 Measuring and drawing angles
2 Angle facts
3 Angles in a triangle
4 Angles in a polygon
5 Regular polygons
6 Parallel lines
7 Special quadrilaterals
8 Bearings

This chapter will show you ...

- F how to measure and draw angles
- F how to find angles on a line and at a point
- E to D how to find angles in a triangle and in any polygon
- E to D how to use bearings
- D how to calculate angles in parallel lines
- C how to calculate interior and exterior angles in polygons

Visual overview

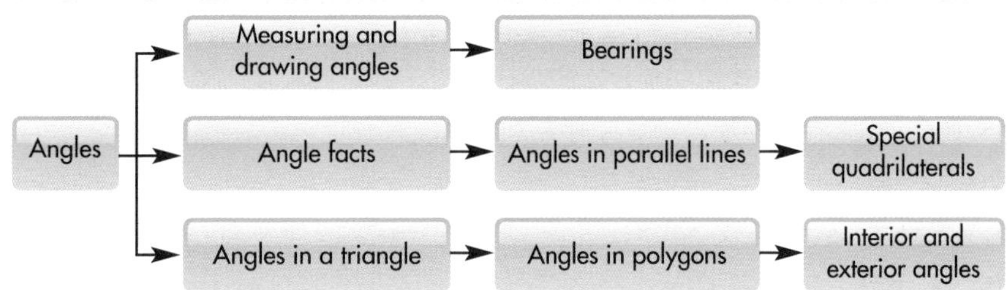

What you should already know

- How to use a protractor to measure an acute angle (KS3 level 5, GCSE grade F)
- The meaning of the terms 'acute', 'obtuse', 'reflex', 'right' and how to use these terms to describe angles (KS3 level 5, GCSE grade G)
- That a polygon is a 2D shape with any number of straight sides (KS3 level 5, GCSE grade G)
- That a diagonal is a line joining two vertices of a polygon (KS3 level 5, GCSE grade G)
- The meaning of the terms 'parallel lines' and 'perpendicular lines' (KS3 level 5, GCSE grade G)

Quick check

State whether these angles are acute, obtuse or reflex.

1 135°   2 68°   3 202°   4 98°   5 315°

10.1 Measuring and drawing angles

This section will show you how to:

- measure and draw an angle of any size

Key words
acute angle
obtuse angle
protractor
reflex angle

When you are using a **protractor**, it is important that you:

- place the centre of the protractor *exactly* on the corner (vertex) of the angle
- lay the base-line of the protractor *exactly* along one side of the angle.

You must follow these two steps to obtain an accurate value for the angle you are measuring.

You should already have discovered how easy it is to measure **acute angles** and **obtuse angles**, using the common semicircular protractor.

EXAMPLE 1

Measure the angles ABC, DEF and GHI in the diagrams below.

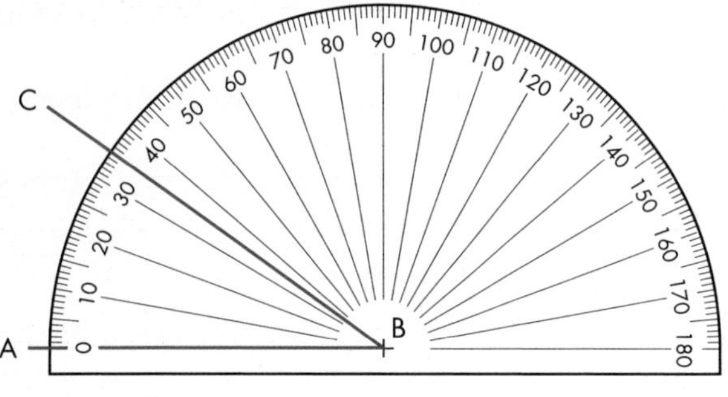

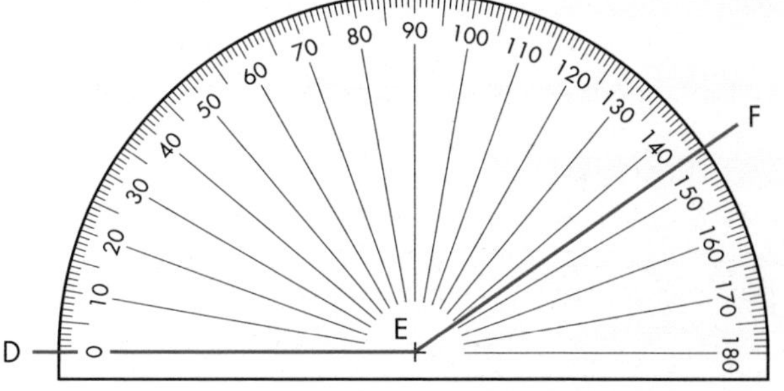

Acute angle ABC is 35° and obtuse angle DEF is 145°.

To measure **reflex angles**, such as angle GHI, it is easier to use a circular protractor if you have one.

Note the notation for angles.

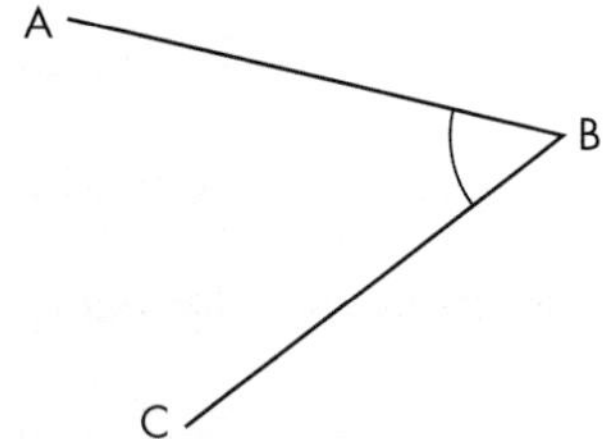

Angle ABC, or ∠ABC, means the angle at B between the lines AB and BC.

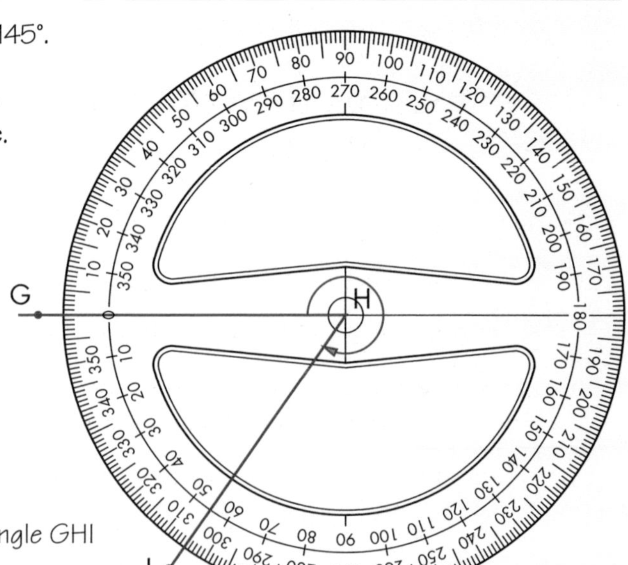

Reflex angle GHI is 305°.

EXERCISE 10A

1 Use a protractor to measure the size of each marked angle.

a

b

c

d

e

f

g

h

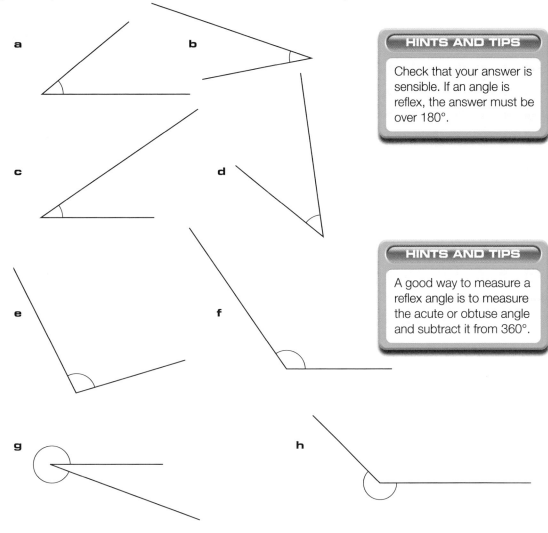

> **HINTS AND TIPS**
>
> Check that your answer is sensible. If an angle is reflex, the answer must be over 180°.

> **HINTS AND TIPS**
>
> A good way to measure a reflex angle is to measure the acute or obtuse angle and subtract it from 360°.

2 Use a protractor to draw angles of the following sizes.

a 30° **b** 60° **c** 90° **d** 10° **e** 20° **f** 45° **g** 75°

3 **a** **i** Draw any three acute angles.

ii Estimate their sizes. Record your results.

iii Measure the angles. Record your results.

iv Work out the difference between your estimate and your measurement for each angle. Add all the differences together. This is your total error.

b Repeat parts **i** to **iv** of part **a** for three obtuse angles.

c Repeat parts **i** to **iv** of part **a** for three reflex angles.

d Which type of angle are you most accurate with, and which type are you least accurate with?

F

FM 4 It is only safe to climb this ladder if the angle between the ground and the ladder is between 72° and 78°.

Is it safe for Oliver to climb the ladder?

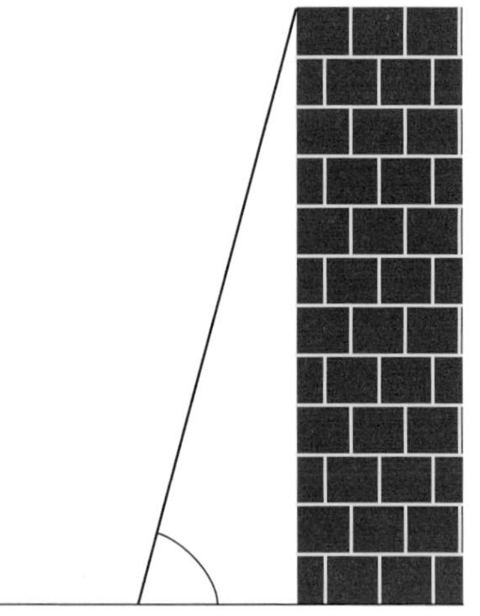

PS 5 An obtuse angle is 10° more than an acute angle.
Write down a possible value for the size of the obtuse angle.

AU 6 Which angle is the odd one out?

Give a reason for your answer.

a

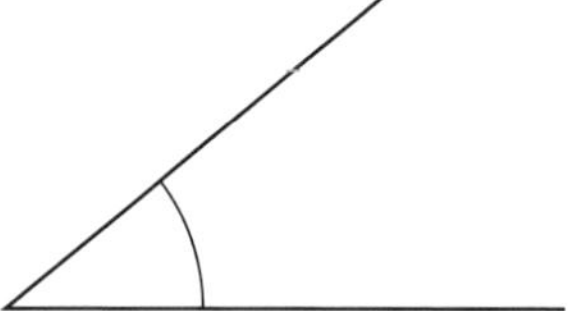

b

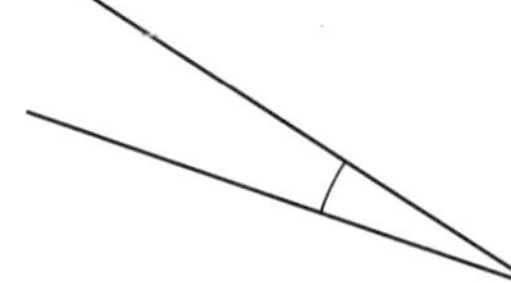

c

d

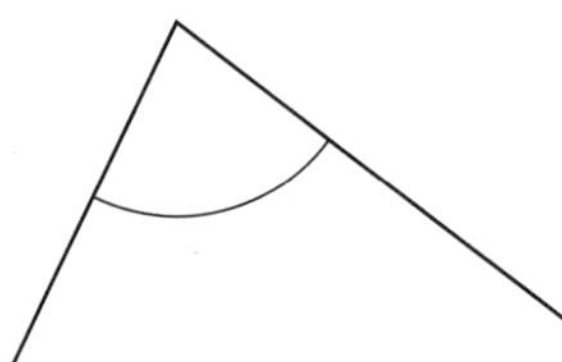

E

7 Use a ruler and a protractor to draw these triangles accurately. Then measure the unmarked angle in each one.

a

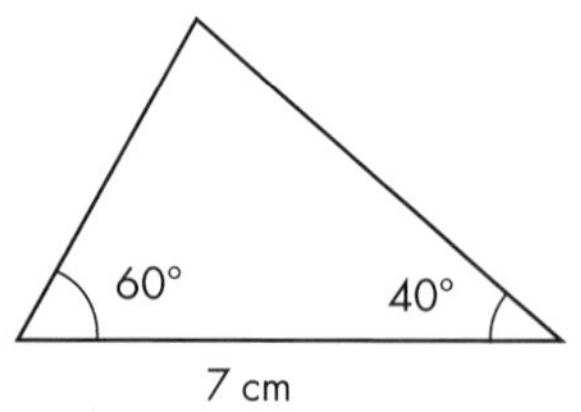

b

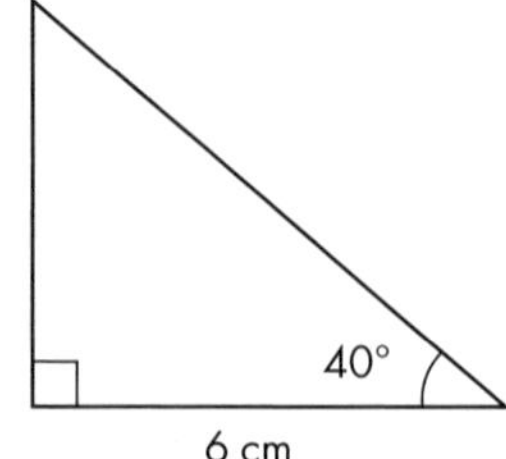

c

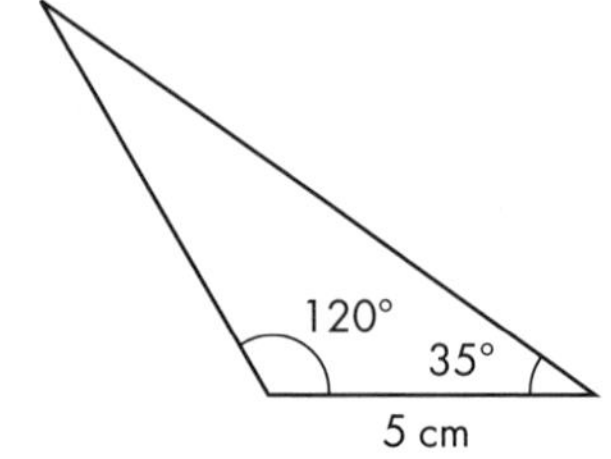

10.2 Angle facts

This section will show you how to:

- calculate angles on a straight line and angles around a point and use opposite angles

Key words
angles around a point
angles on a straight line
opposite angles

Angles on a line

The **angles on a straight line** add up to 180°.

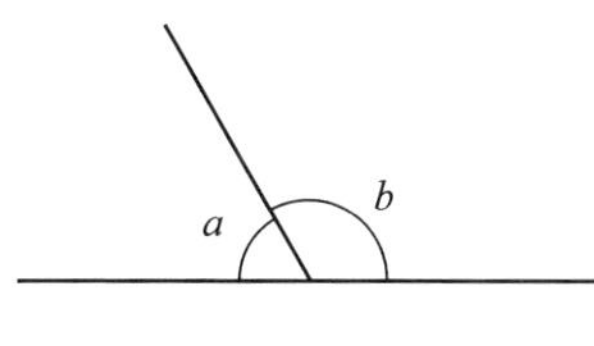

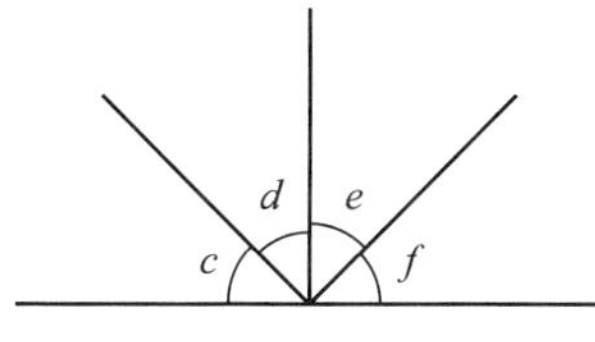

$a + b = 180°$ $\qquad$ $c + d + e + f = 180°$

Draw an example for yourself (and measure a and b) to show that the statement is true.

Angles around a point

The sum of the **angles around a point** is 360°. For example:

$$a + b + c + d + e = 360°$$

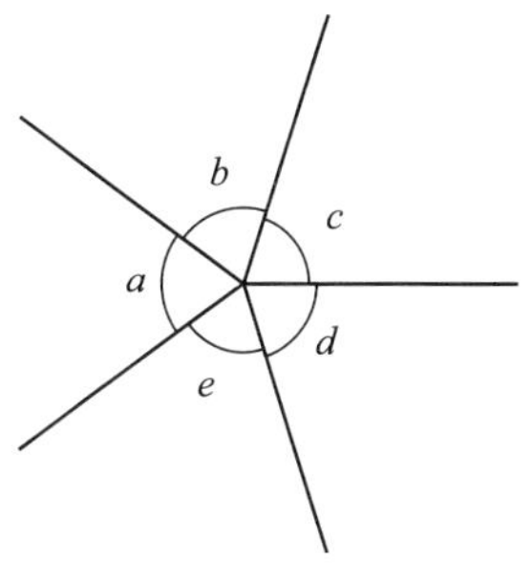

Again, check this for yourself by drawing an example and measuring the angles.

EXAMPLE 2

Find the size of angle x in the diagram.

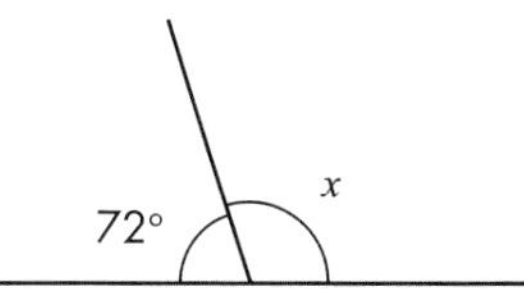

Angles on a straight line add up to 180°.

$x + 72° = 180°$

So, $x = 180° - 72°$

$x = 108°$

Sometimes equations can be used to solve angle problems.

EXAMPLE 3

Find the value of x in the diagram.

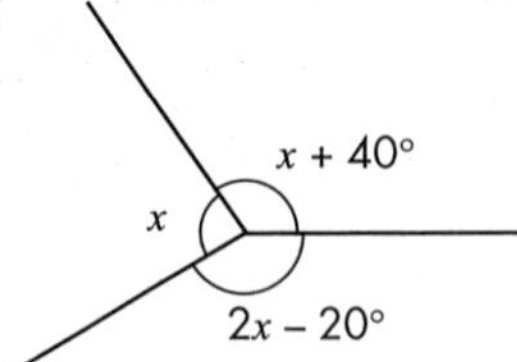

These angles are around a point, so they must add up to 360°.

Therefore, $x + x + 40° + 2x - 20° = 360°$

$$4x + 20° = 360°$$

$$4x = 340°$$

$$x = 85°$$

Opposite angles

Opposite angles are equal.

So $a = c$ and $b = d$.

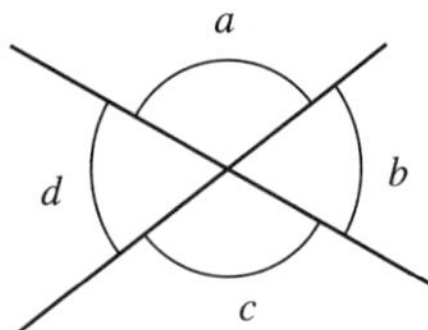

Sometimes opposite angles are called **vertically opposite angles**.

EXAMPLE 4

Find the value of x in the diagram.

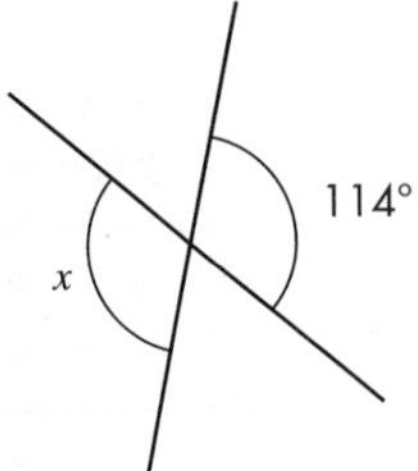

The two angles are opposite, so $x = 114°$.

EXERCISE 10B

Calculate the size of the angle marked x in each of these examples.

1 **a**

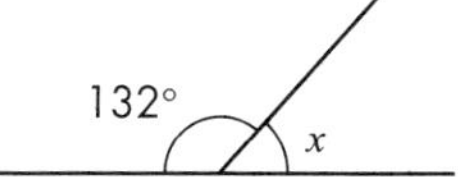

b

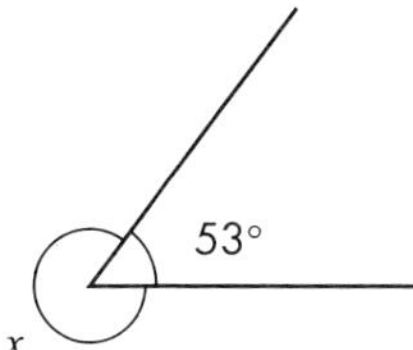

c

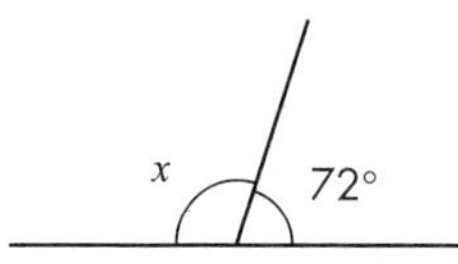

d

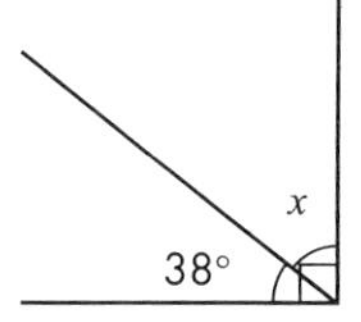

e

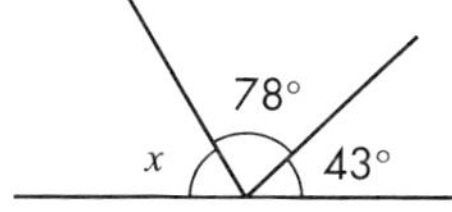

f

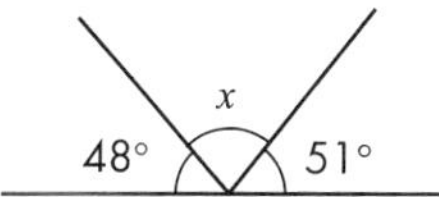

g

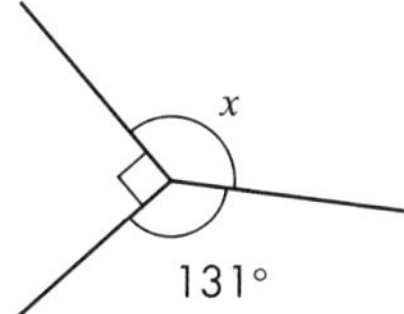

h

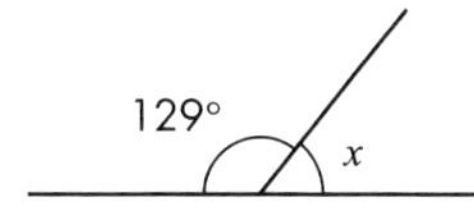

i

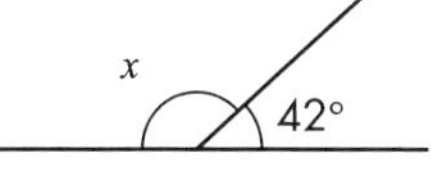

j

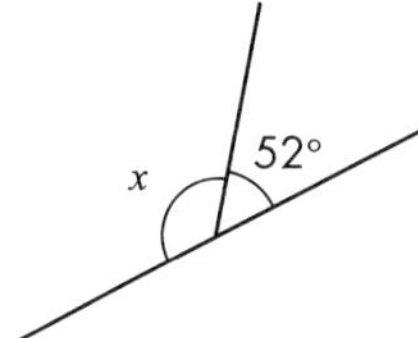

k

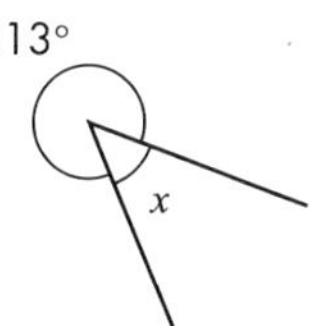

l

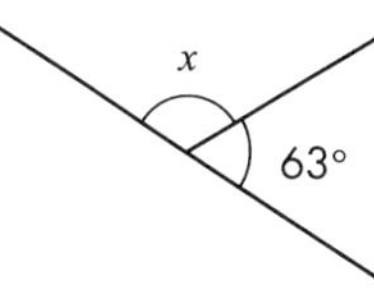

m

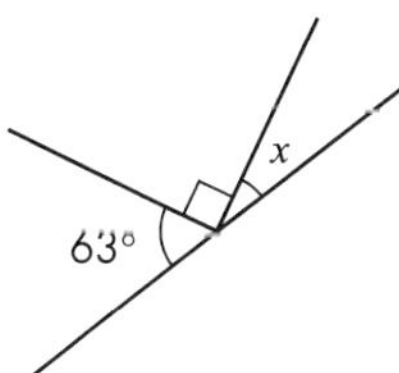

n

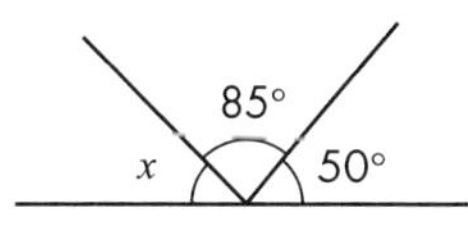

o

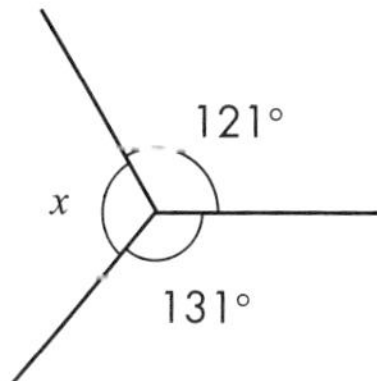

p

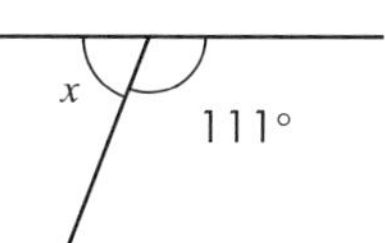

q

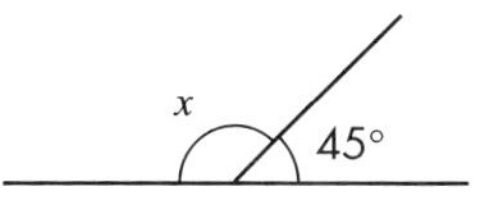

r

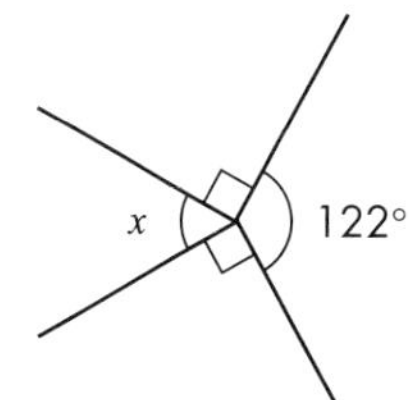

F

s

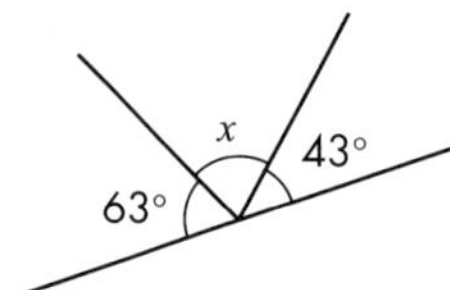

t

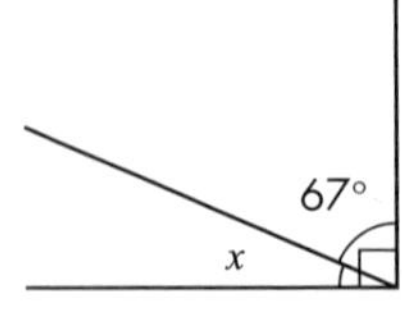

u

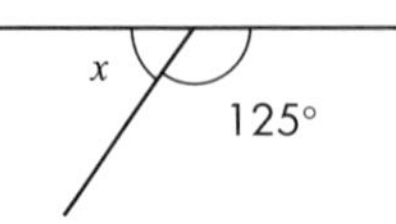

v

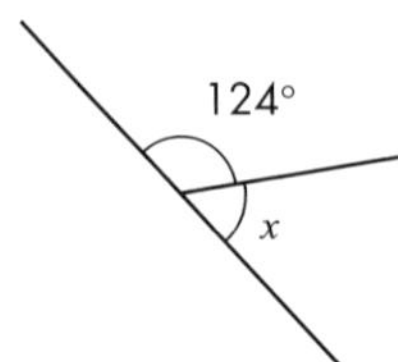

> **HINTS AND TIPS**
>
> Never measure angles in questions like these. Diagrams in examinations are not drawn accurately. Always calculate angles unless you are told to measure them.

2 Write down the value of x in each of these diagrams.

a

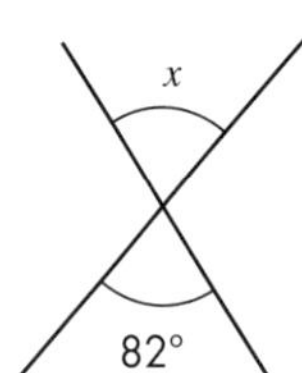

b

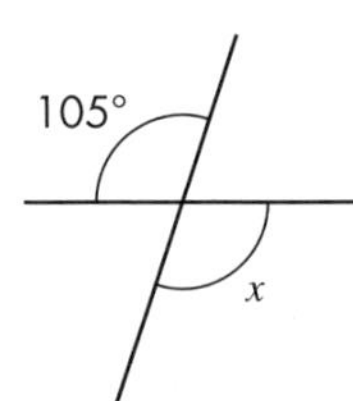

c

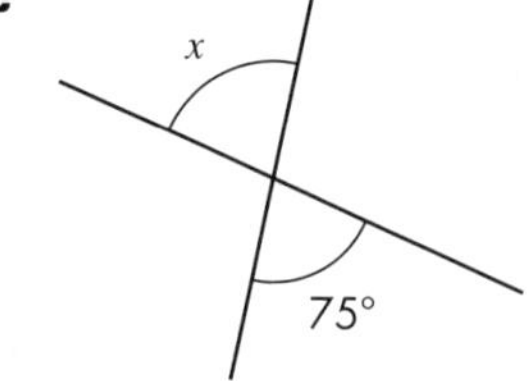

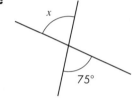

AU 3 In the diagram, angle ABD is 45° and angle CBD is 125°.

Decide whether ABC is a straight line. Write down how you decided.

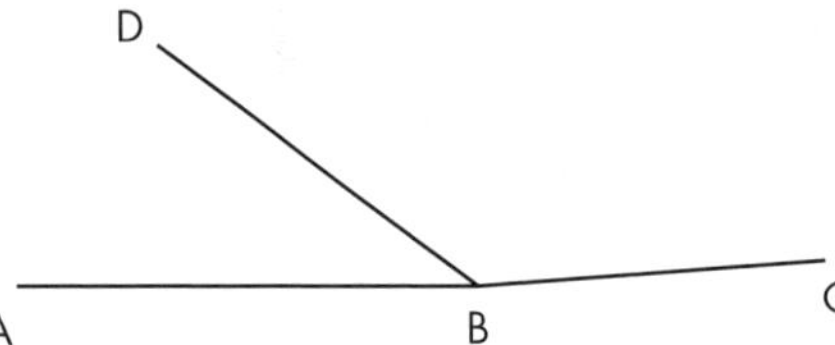

E

4 Calculate the value of x in each of these examples.

a

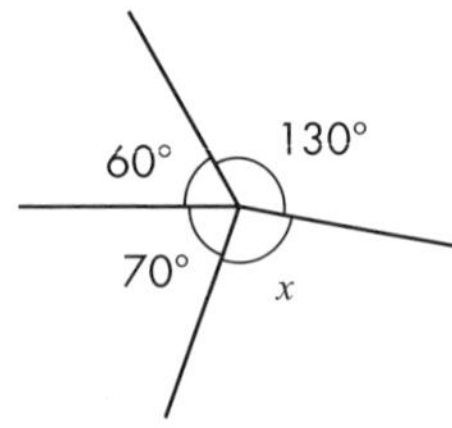

b

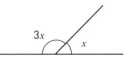

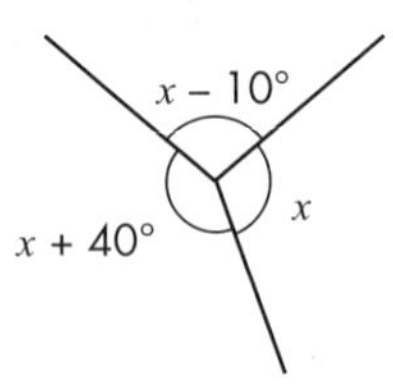

c

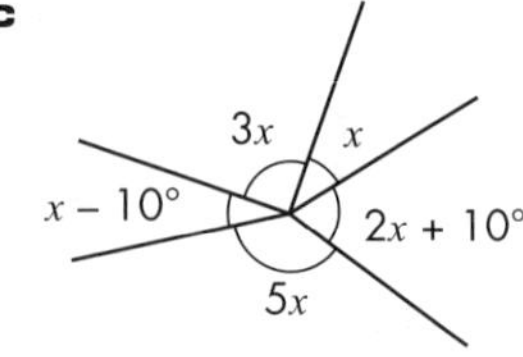

5 Calculate the value of x in each of these examples.

a

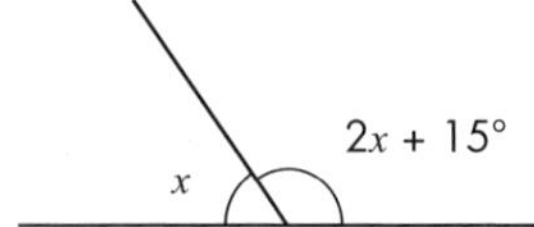

b

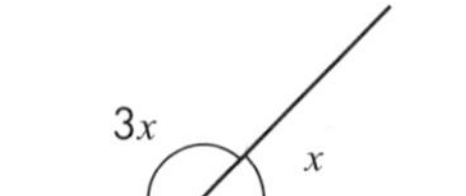

c

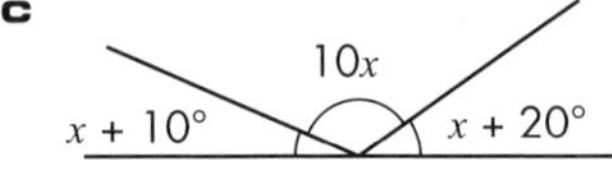

E

6 Calculate the value of x first and then calculate the value of y in each of these examples.

a

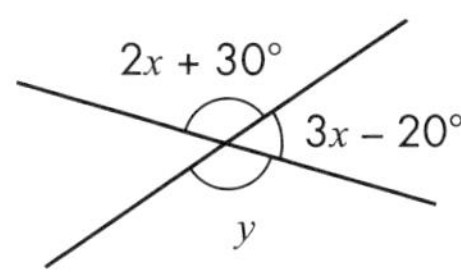

b

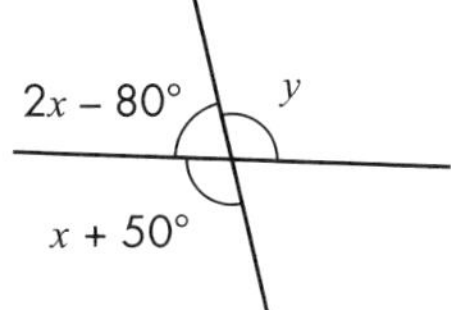

c

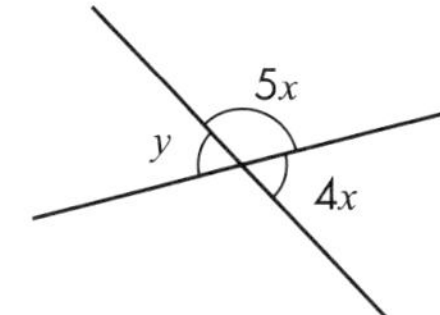

AU 7 Ella has a collection of tiles. They are all equilateral triangles and are all the same size.

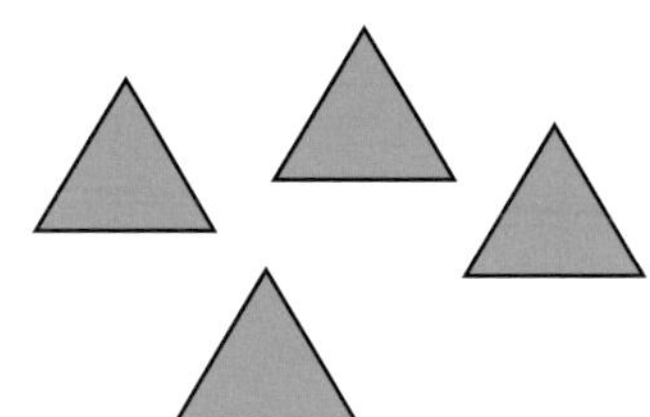

She says that six of the tiles will fit together and leave no gaps.

Explain why Ella is correct.

HINTS AND TIPS

All the angles in an equilateral triangle are 60°.

PS 8 Work out the value of y in the diagram.

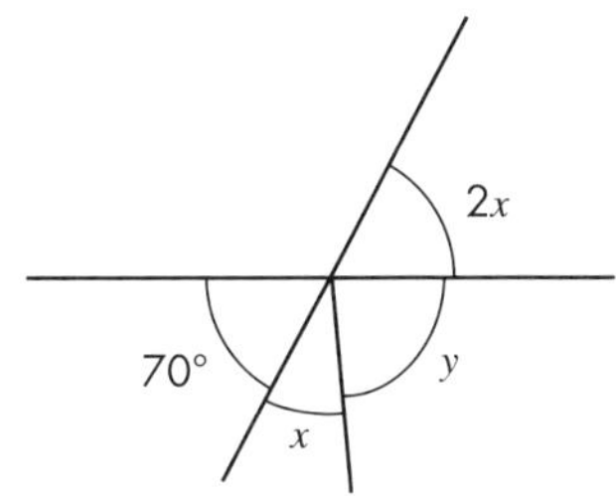

10.3 Angles in a triangle

This section will show you how to:

- calculate the size of angles in a triangle

Key words

angles in a triangle
equilateral triangle
exterior angle
interior angle
isosceles triangle
right-angled triangle

ACTIVITY

Angles in a triangle

You need a protractor.

Draw a triangle. Label the corners (vertices) A, B and C.

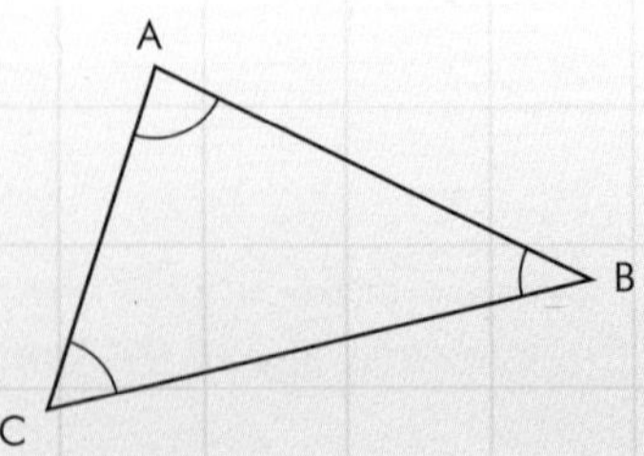

Use a ruler and make sure that the corners of your triangle form proper angles.

Like this. Not like this … … or this.

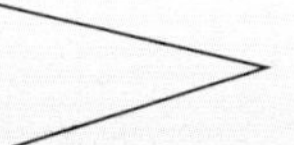

Measure each angle, A, B and C.

Write them down and add them up:

Angle A = ……°

Angle B = ……°

Angle C = ……°

Total =

Repeat this for five more triangles, including at least one with an obtuse angle.

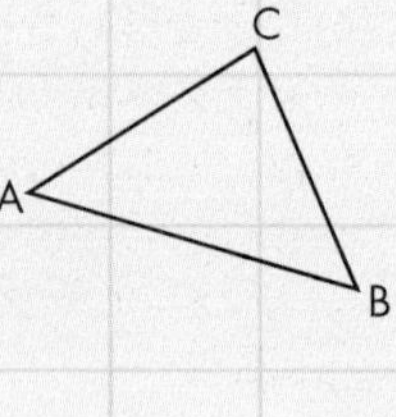

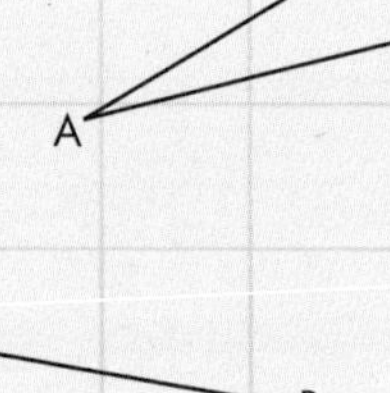

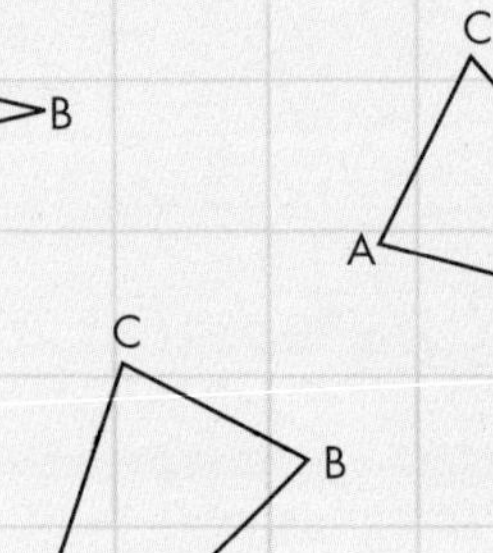

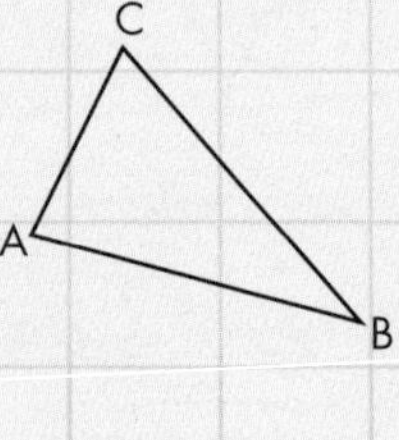

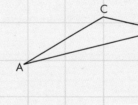

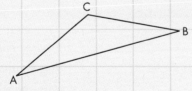

What conclusion can you draw about the sum of the angles in a triangle?

Remember:

You will not be able to measure with total accuracy.

You should have discovered that the three **angles in a triangle** add up to 180°.

$a + b + c = 180°$

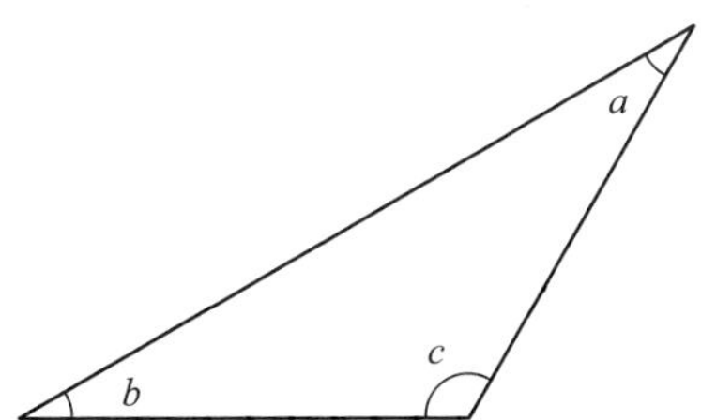

EXAMPLE 5

Calculate the size of angle a in the triangle below.

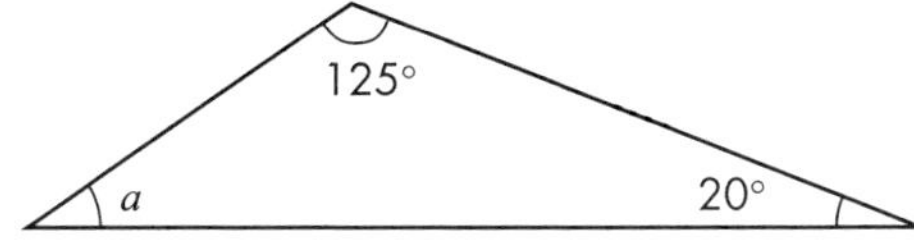

Angles in a triangle add up to 180°

Therefore, $a + 20° + 125° = 180°$

$a + 145° = 180°$

So $a = 35°$

Special triangles

Equilateral triangle

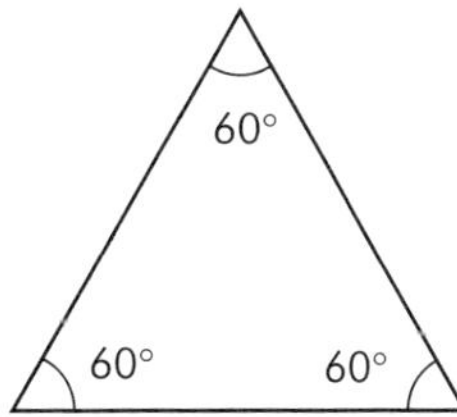

An **equilateral triangle** is a triangle with all its sides equal. Therefore, all three **interior angles** are 60°.

Isosceles triangle

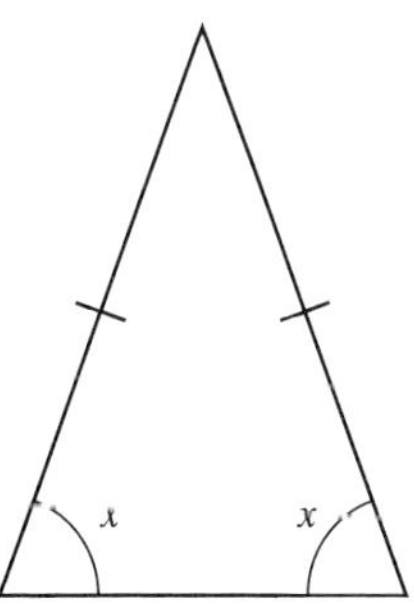

An **isosceles triangle** is a triangle with two equal sides and, therefore, with two equal interior angles (at the foot of the equal sides).

Notice how to mark the equal sides and equal angles.

Right-angled triangle

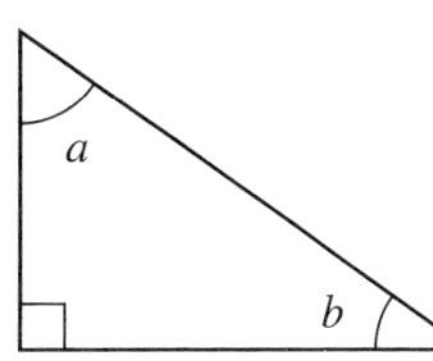

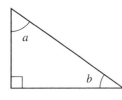

A **right-angled triangle** has an interior angle of 90°.
$a + b = 90°$

EXERCISE 10C

1 Find the size of the angle marked with a letter in each of these triangles.

a

60°
50°
a

b

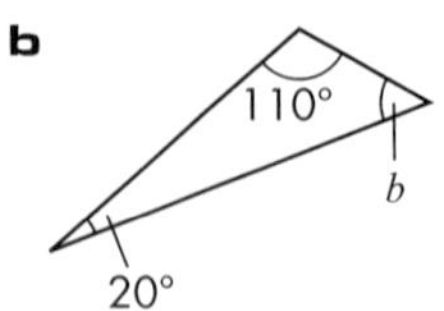

c

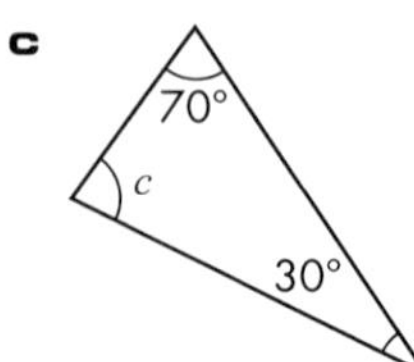

d

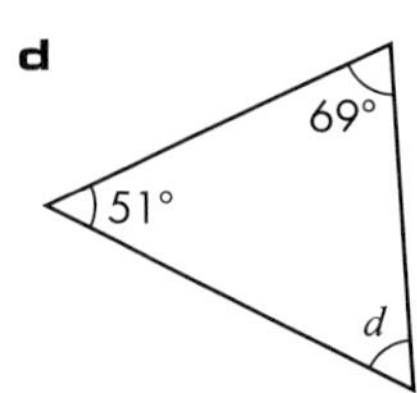

e

67°
e
38°

f

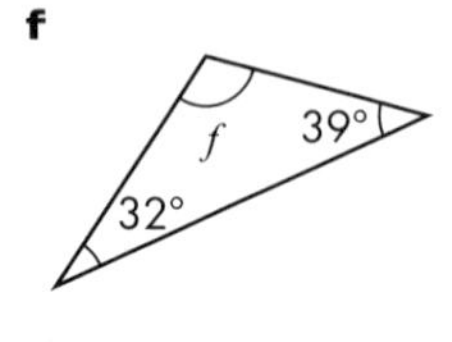

g

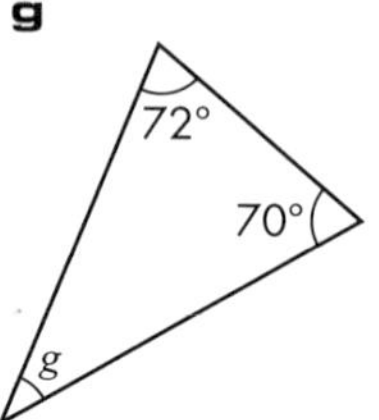

h

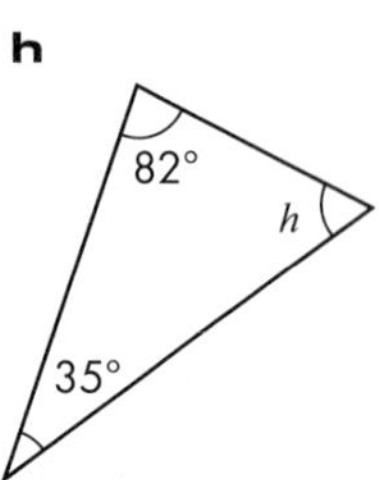

2 Do any of these sets of angles form the three angles of a triangle? Explain your answer.

a 35°, 75°, 80°
b 50°, 60°, 70°
c 55°, 55°, 60°
d 60°, 60°, 60°
e 35°, 35°, 110°
f 102°, 38°, 30°

3 Two interior angles of a triangle are given in each case. Find the third one indicated by a letter.

a 20°, 80°, a
b 52°, 61°, b
c 80°, 80°, c
d 25°, 112°, d
e 120°, 50°, e
f 122°, 57°, f

4 In the triangle on the right, all the interior angles are the same.

a What is the size of each angle?
b What is the name of a special triangle like this?
c What is special about the sides of this triangle?

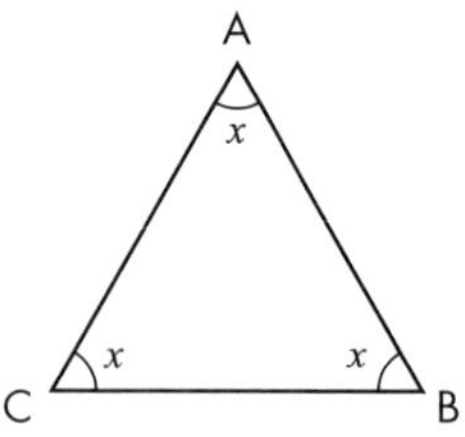

5 In the triangle on the right, two of the angles are the same.

a Work out the size of the lettered angles.
b What is the name of a special triangle like this?
c What is special about the sides AC and AB of this triangle?

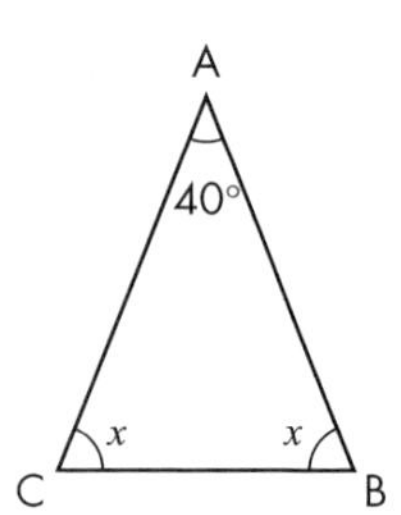

E

AU **6** In the triangle on the right, the angles at B and C are the same. Write down the size of the lettered angles.

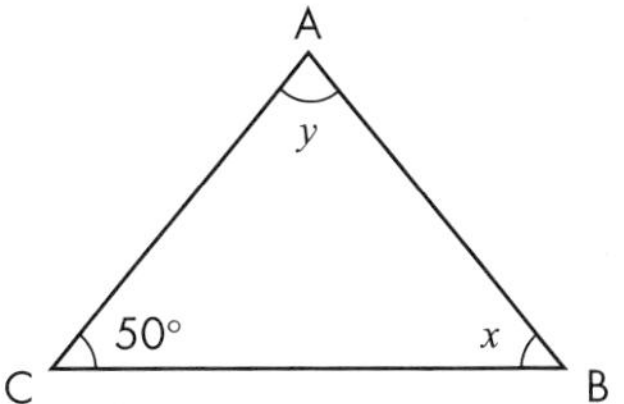

7 Find the size of the **exterior angle** marked with a letter in each of these diagrams.

a

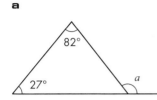

b

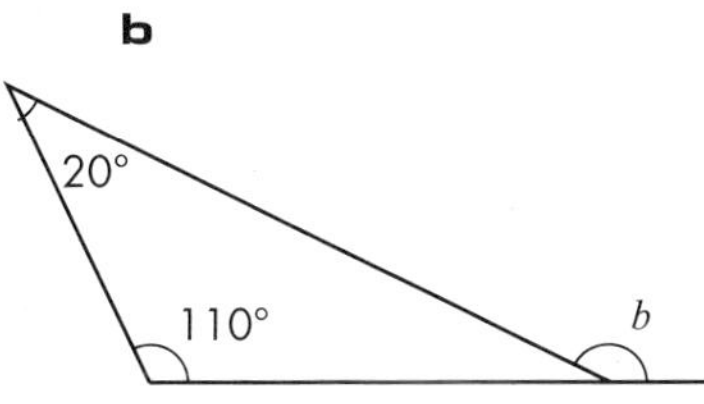

c

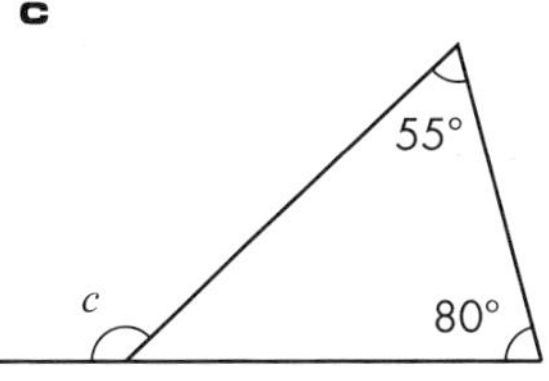

FM **8** A town planner has drawn this diagram to show three paths in a park but they have missed out the angle marked x.

Work out the value of x.

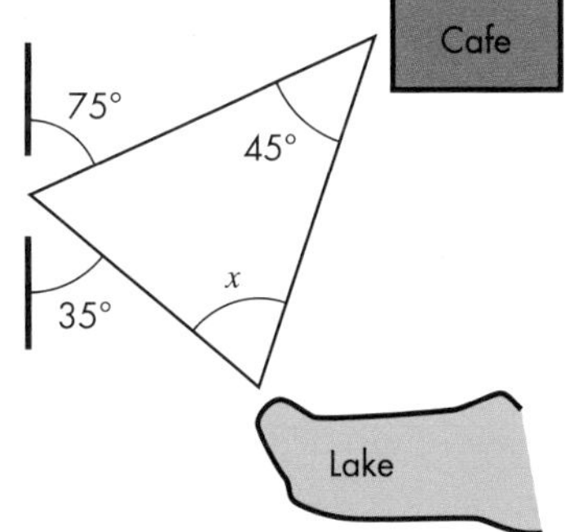

PS **9** What is the special name for triangle DEF?

Show all your working to explain your answer.

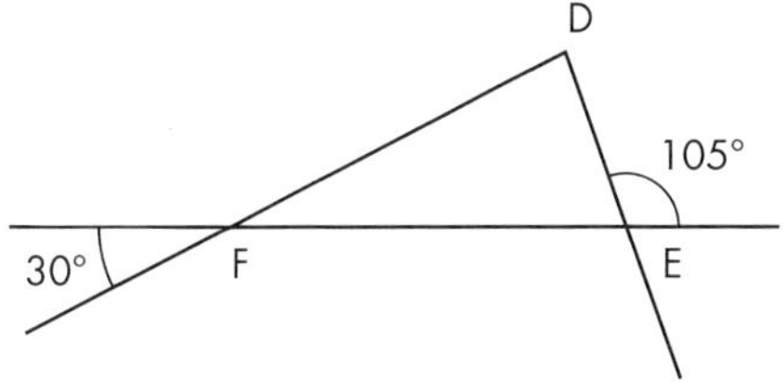

AU **10** The diagram shows three intersecting straight lines.

Work out the values of a, b and c.

Give reasons for your answers.

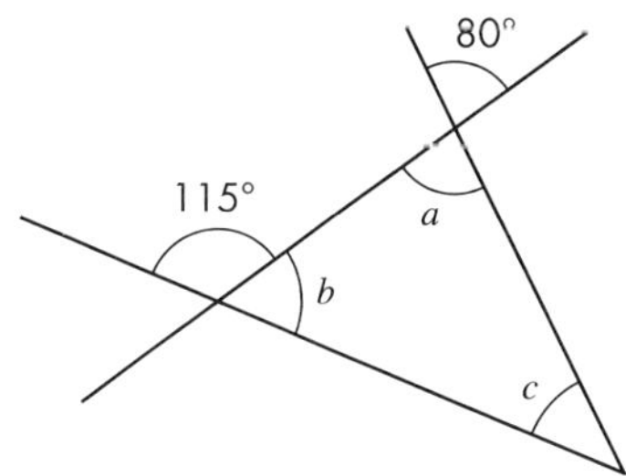

D

11 By using algebra, show that $x = a + b$.

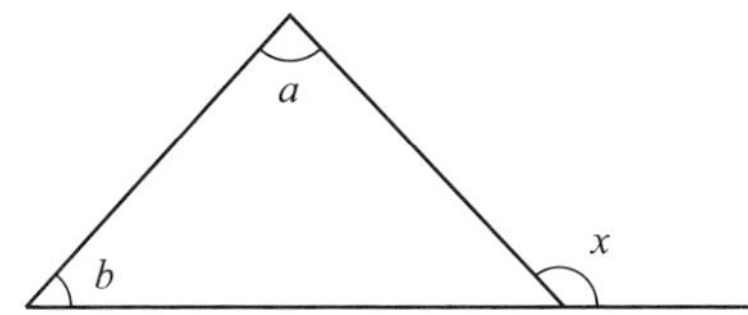

10.4 Angles in a polygon

This section will show you how to:

- calculate the sum of the interior angles in a polygon

Key words
decagon
heptagon
hexagon
interior angle
nonagon
octagon
pentagon
polygon
quadrilateral

ACTIVITY

Angle sums from triangles

Draw a **quadrilateral** (a four-sided shape).
Draw in a diagonal to make it into two triangles.

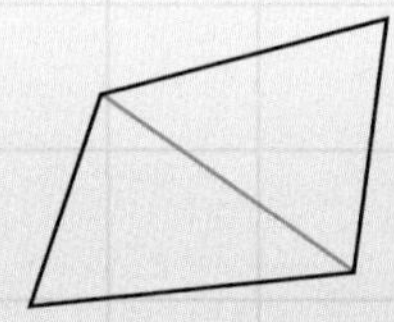

You should be able to copy and complete this statement:

The sum of the angles in a quadrilateral is equal to the sum of the angles in …… triangles, which is …… × 180° = ……°.

Now draw a **pentagon** (a five-sided shape).

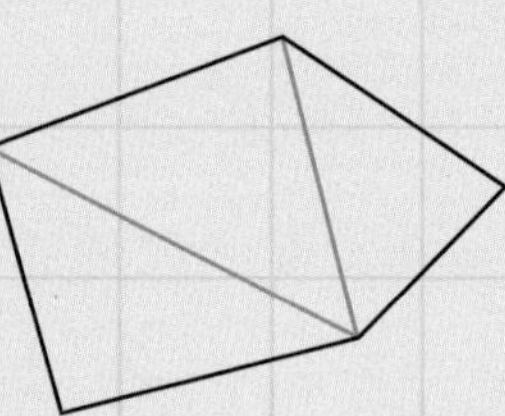

Draw in the diagonals to make it into three triangles.

You should be able to copy and complete this statement:

The sum of the angles in a pentagon is equal to the sum of the angles in …… triangles, which is …… × 180° = ……°.

Next, draw a **hexagon** (a six-sided shape).

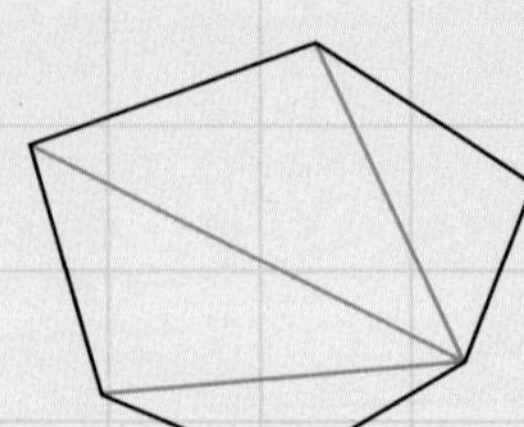

Draw in the diagonals to make it into four triangles.

You should be able to copy and complete this statement:

The sum of the angles in a hexagon is equal to the sum of the angles in …… triangles, which is …… × 180° = ……°.

Now, complete the table below. Use the number pattern to carry on the angle sum up to a **decagon** (ten-sided shape).

Shape	Number of sides	Triangles	Angle sum
triangle	3	1	180°
quadrilateral	4	2	
pentagon	5	3	
hexagon	6	4	
heptagon	7		
octagon	8		
nonagon	9		
decagon	10		

If you have spotted the number pattern, you should be able to copy and complete this statement:

The number of triangles in a 20-sided shape is ……, so the sum of the angles in a 20-sided shape is …… × 180° = ……°.

So for an n-sided **polygon**, the sum of the **interior angles** is $180(n - 2)°$.

EXAMPLE 6

Calculate the size of angle a in the quadrilateral below.

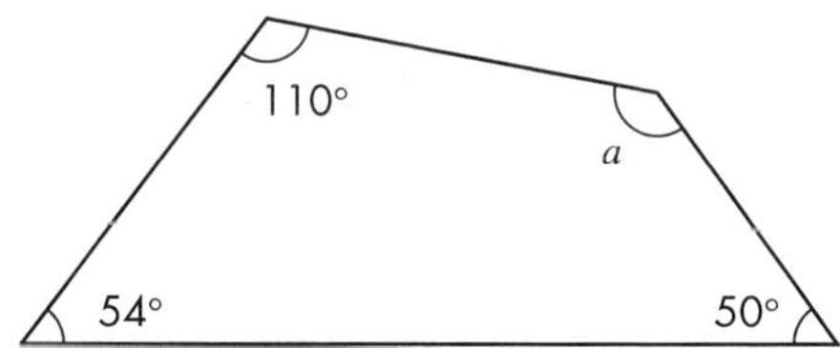

Angles in a quadrilateral add up to 360°.

Therefore, $a + 50° + 54° + 110° = 360°$

$a + 214° = 360°$

So, $a = 146°$

EXERCISE 10D

E

1 Find the size of the angle marked with a letter in each of these quadrilaterals.

a

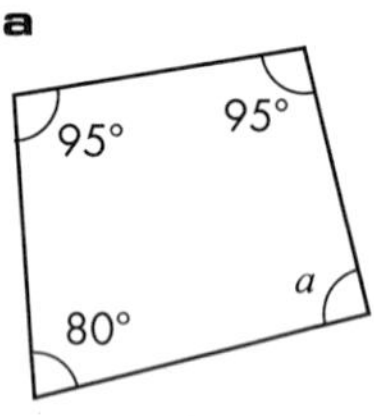

b

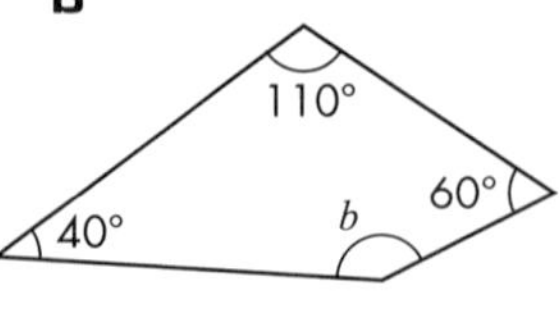

c

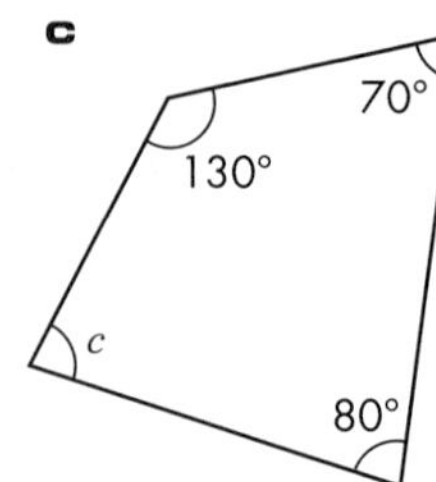

d

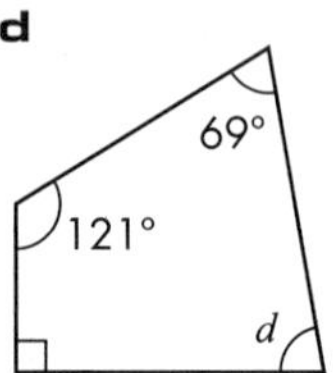

e

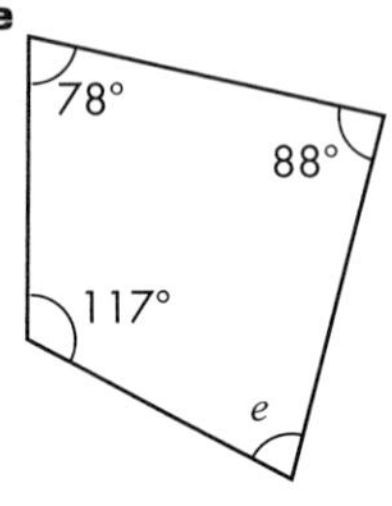

f

f 49° 49° 131°

g

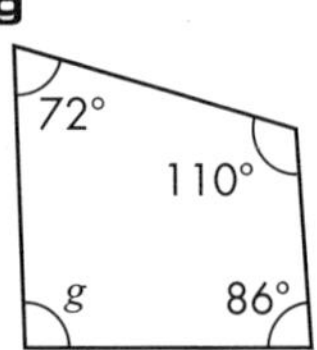

h

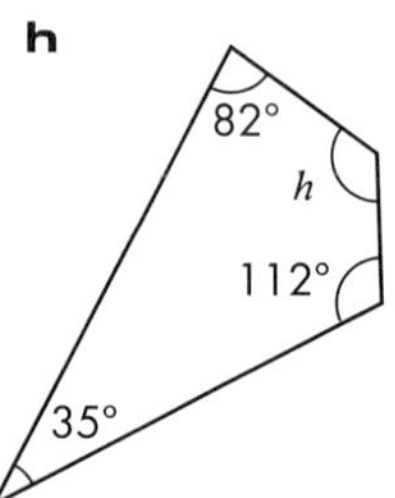

> **HINTS AND TIPS**
>
> **Remember**, the sum of the interior angles of a quadrilateral is 360°.

2 Do any of these sets of angles form the four interior angles of a quadrilateral? Explain your answer.

a 135°, 75°, 60°, 80°
b 150°, 60°, 80°, 70°
c 85°, 85°, 120°, 60°
d 80°, 90°, 90°, 110°
e 95°, 95°, 60°, 110°
f 102°, 138°, 90°, 30°

3 Three interior angles of a quadrilateral are given. Find the fourth one indicated by a letter.

a 120°, 80°, 60°, a
b 102°, 101°, 90°, b
c 80°, 80°, 80°, c
d 125°, 112°, 83°, d
e 120°, 150°, 50°, e
f 122°, 157°, 80°, f

4 In the quadrilateral on the right, all the angles are the same.

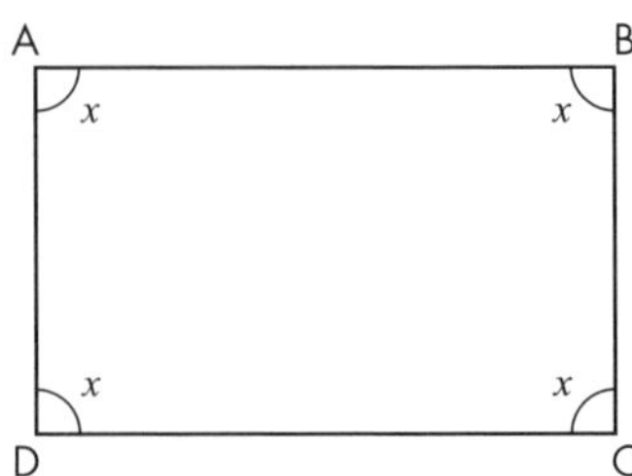

a What is each angle?

b What is the name of a special quadrilateral like this?

c Is there another quadrilateral with all the angles the same? What is it called?

D

5 Work out the size of the angle marked with a letter in each of the polygons below. You may find the table you completed on page 269 useful.

a

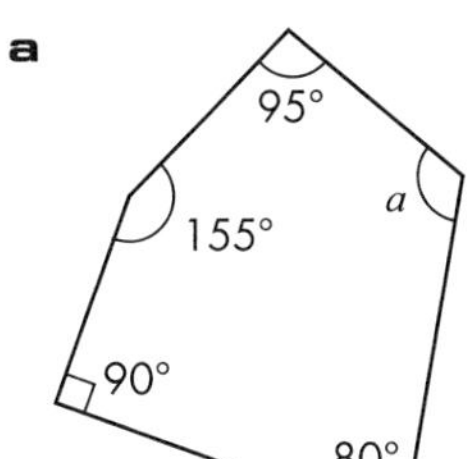

b

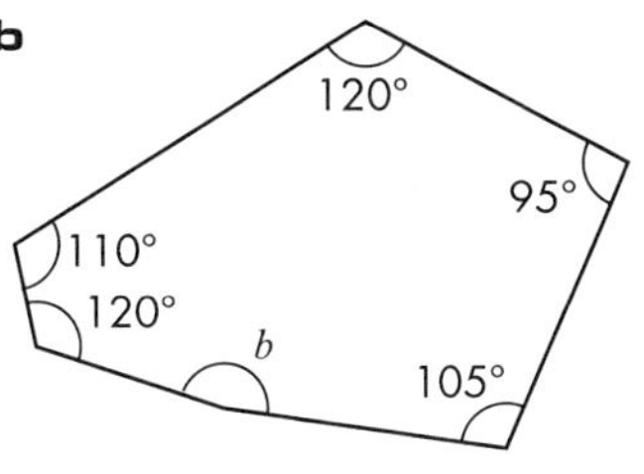

c

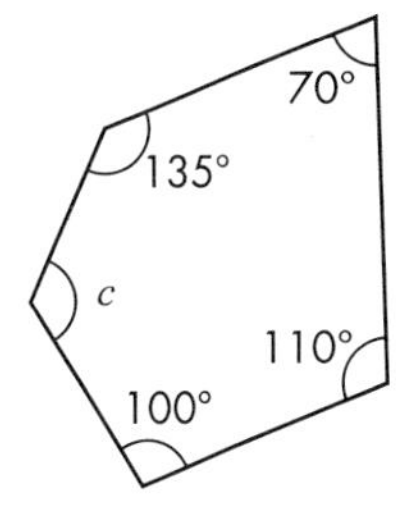

d

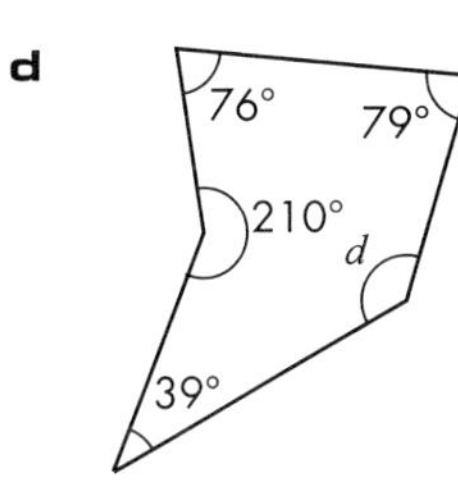

e

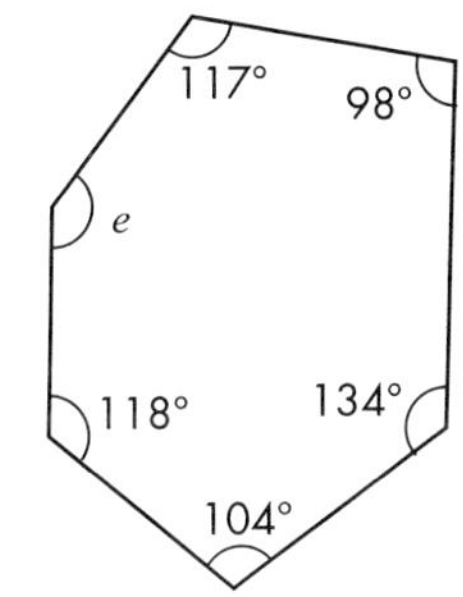

f

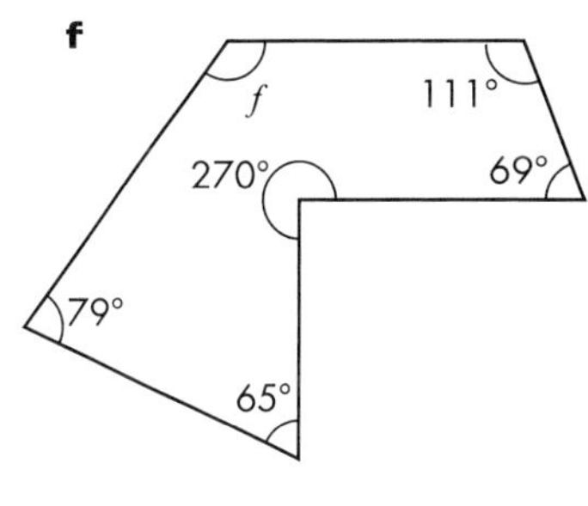

g

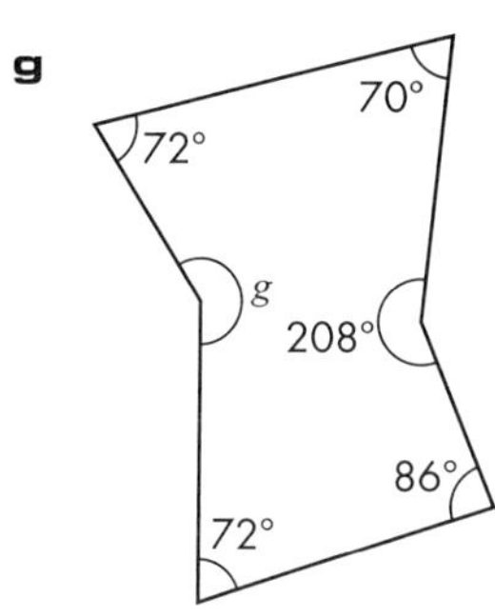

h

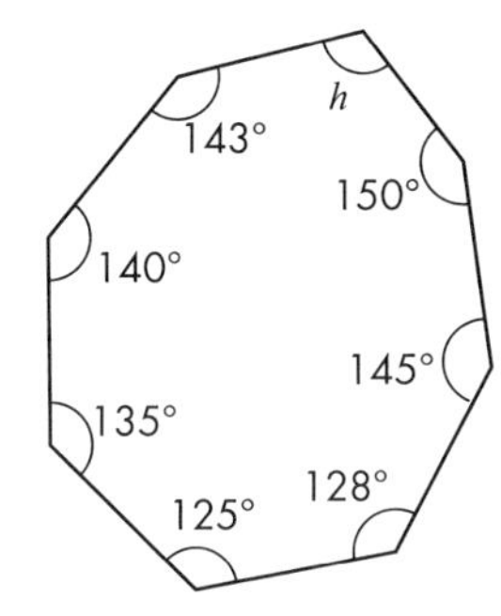

HINTS AND TIPS

Remember, the sum of the interior angles of an n-sided polygon is $180(n - 2)°$.

FM 6 Anna is drawing this logo for a school magazine.

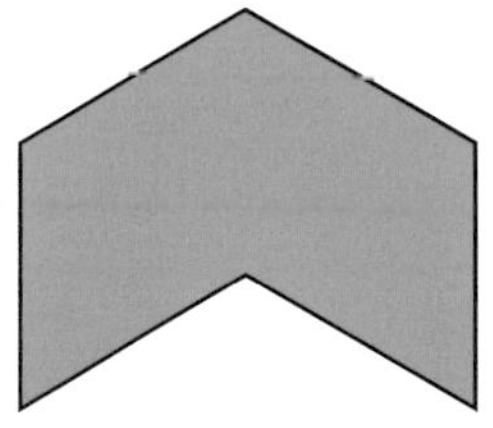

HINTS AND TIPS

First draw the four equilateral triangles on the diagram.

It is made up of four equilateral triangles that are all the same size.

She needs to know the sizes of the six angles so that she can draw it accurately.

What are the sizes of the six angles?

D

PS 7 This quadrilateral is made from two isosceles triangles.
They are both the same size.

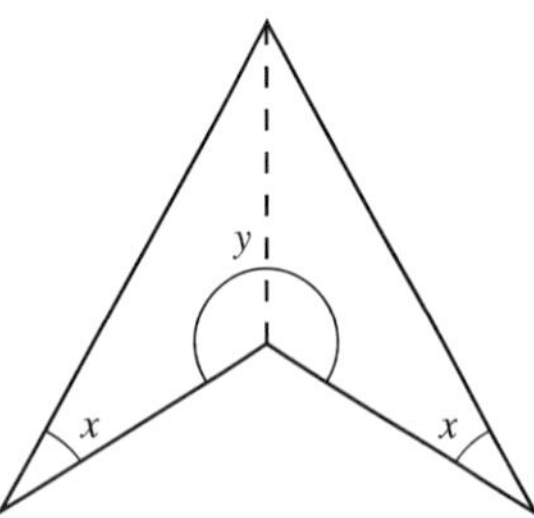

Find the value of y in terms of x.

AU 8 The diagram shows the four angles in a quadrilateral.

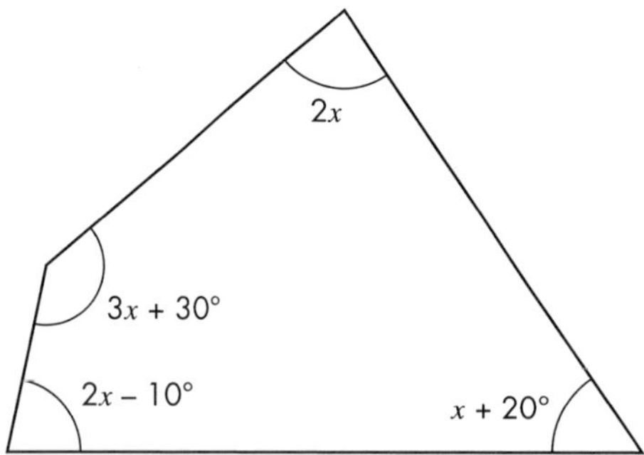

a Write down an equation in terms of x.

b Solve the equation and find the size of the smallest angle in the quadrilateral.

10.5 Regular polygons

This section will show you how to:

- calculate the exterior angles and the interior angles of a regular polygon

Key words
exterior angle
interior angle
regular polygon

ACTIVITY

Regular polygons

To do this activity, you will need to have done the one on pages 268–269.

You will also need a calculator.

Below are five **regular polygons**.

Square
4 sides

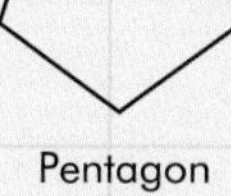
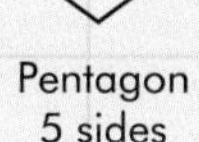

Pentagon
5 sides

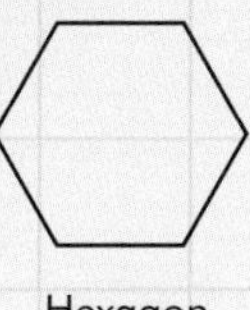

Hexagon
6 sides

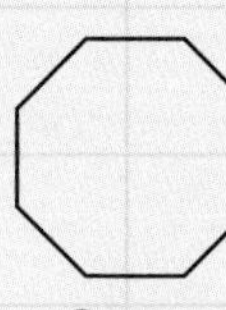

Octagon
8 sides

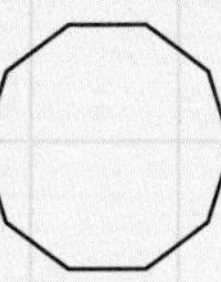

Decagon
10 sides

A polygon is regular if all its **interior angles** are equal and all its sides have the same length.

A square is a regular four-sided shape that has an angle sum of 360°.

So, each angle is 360° ÷ 4 = 90°.

A regular pentagon has an angle sum of 540°.

So, each angle is 540° ÷ 5 = 108°.

Copy and complete the table below.

Shape	Number of sides	Angle sum	Each angle
square	4	360°	90°
pentagon	5	540°	108°
hexagon	6	720°	
octagon	8		
nonagon	9		
decagon	10		

Interior and exterior angles

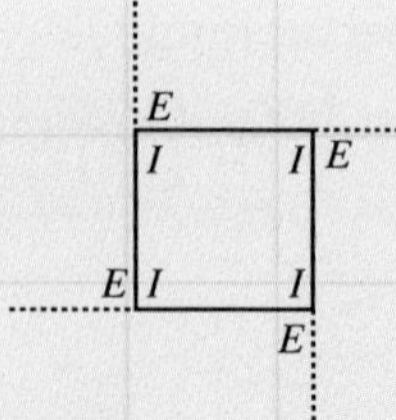

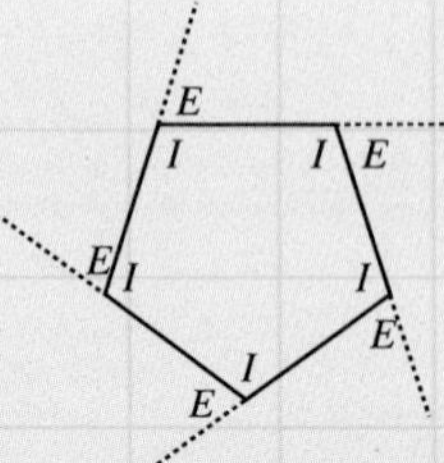

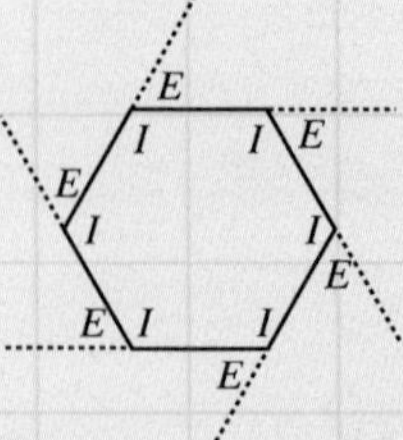

Look at these three regular polygons. At each vertex of each regular polygon, there is an interior angle, I, and an **exterior angle**, E. Notice that: $I + E = 180°$.

Clearly, the exterior angles of a square are each 90°. So, the sum of the exterior angles of a square is $4 \times 90° = 360°$.

You can calculate the exterior angle of a regular pentagon as follows. From the table on the previous page, you know that the interior angle of a regular pentagon is 108°.

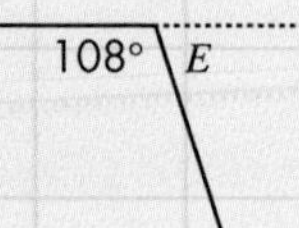

So, the exterior angle is $180° - 108° = 72°$.

Therefore, the sum of the exterior angles is $5 \times 72° = 360°$.

Now copy and complete the table below for regular polygons.

Regular polygon	Number of sides	Interior angle	Exterior angle	Sum of exterior angles
square	4	90°	90°	$4 \times 90° = 360°$
pentagon	5	108°	72°	$5 \times 72° =$
hexagon	6	120°		
octagon	8			
nonagon	9			
decagon	10			

From the table, you can see that the sum of the exterior angles is always 360°.

You can use this information to find the exterior angle and the interior angle for any regular polygon.

For an n-sided regular polygon, the exterior angle is given by $E = \dfrac{360°}{n}$

and the interior angle is given by $I = 180° - E$.

EXAMPLE 7

Calculate the size of the exterior and interior angle for a regular 12-sided polygon (a regular dodecagon).

$$E = \frac{360°}{12} = 30° \quad \text{and} \quad I = 180° - 30° = 150°$$

EXERCISE 10E

1 Each diagram shows an interior angle of a regular polygon. For each polygon, answer the following.

i What is its exterior angle?

ii How many sides does it have?

iii What is the sum of its interior angles?

a

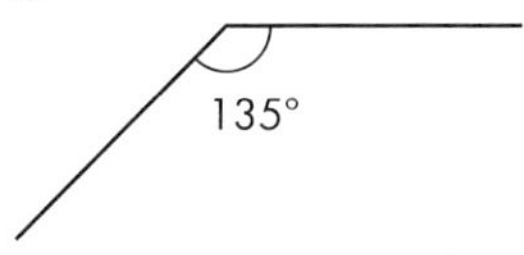

b

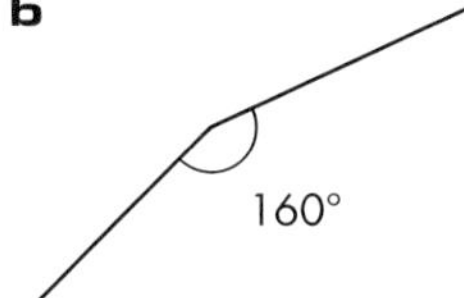

c

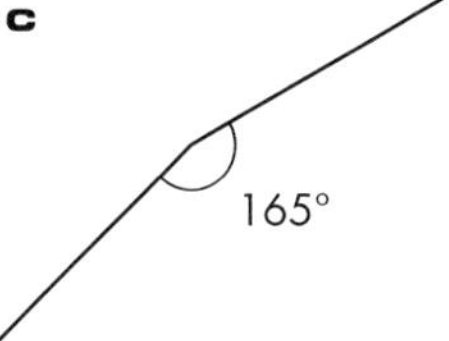

d

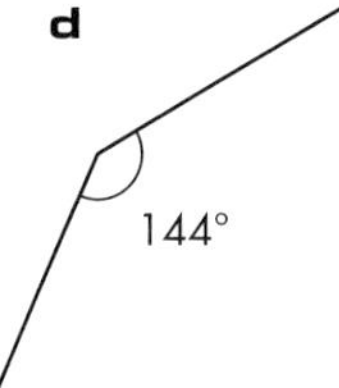

2 Each diagram shows an exterior angle of a regular polygon. For each polygon, answer the following.

i What is its interior angle?

ii How many sides does it have?

iii What is the sum of its interior angles?

> **HINTS AND TIPS**
>
> **Remember** that the angle sum is calculated as (number of sides – 2) × 180°.

a

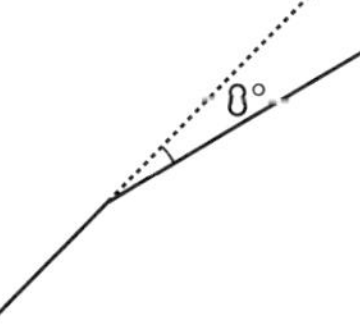

b

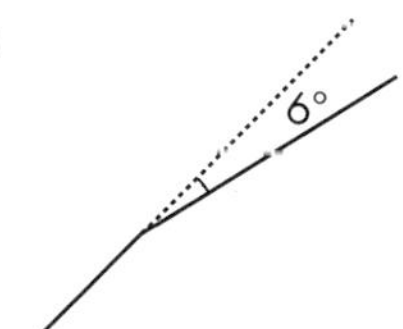

c

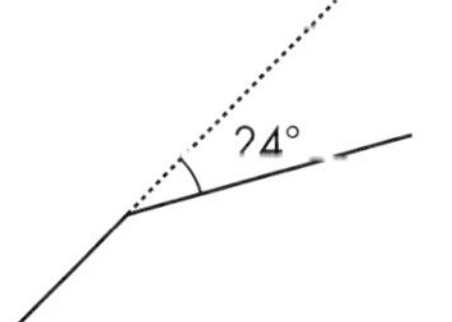

d

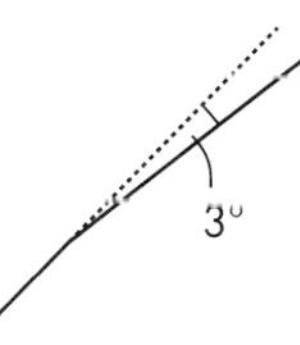

3 Each of these cannot be the interior angle of a regular polygon. Explain why.

a

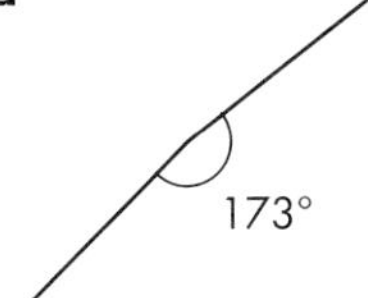

b

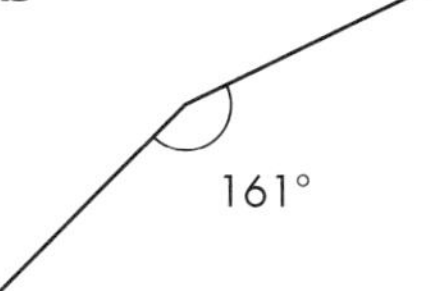

c

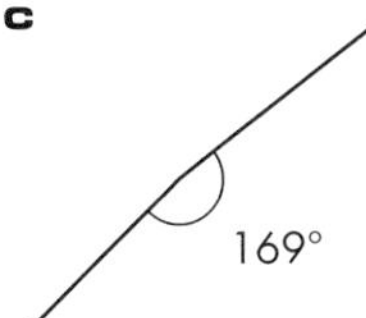

d

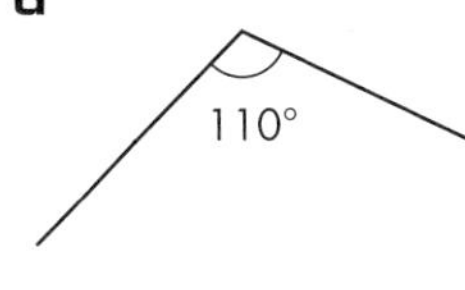

C

4 Each of these cannot be the exterior angle of a regular polygon. Explain why.

a 7° **b** 26° **c** 44° **d** 13°

5 Draw a sketch of a regular octagon and join each vertex to the centre.

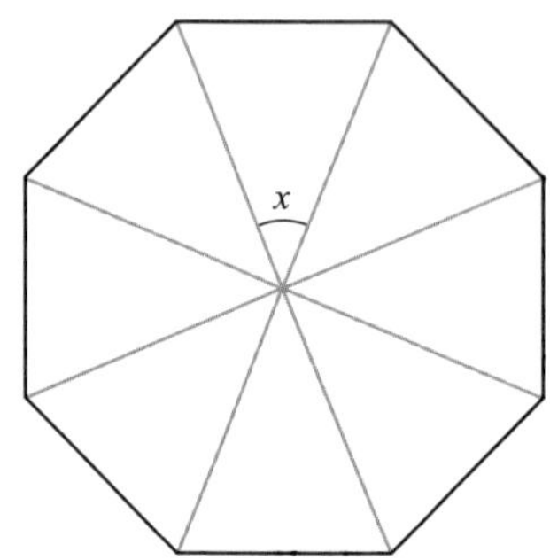

Calculate the value of the angle at the centre (marked x).

What connection does this have with the exterior angle?

Is this true for all regular polygons?

FM 6 A joiner is making tables so that the shape of each one is half a regular octagon, as shown in the diagram.

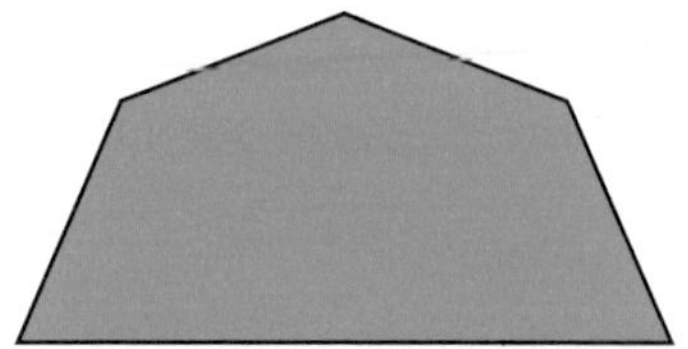

He needs to know the size of each angle on the table top.

What are the sizes of the angles?

PS 7 This star shape has ten sides that are equal in length.

Each reflex interior angle is 200°.

Work out the size of each acute interior angle.

> **HINTS AND TIPS**
>
> Find the sum of the interior angles of a decagon first.

AU 8 The diagram shows part of a regular polygon.

Each interior angle is 144°.

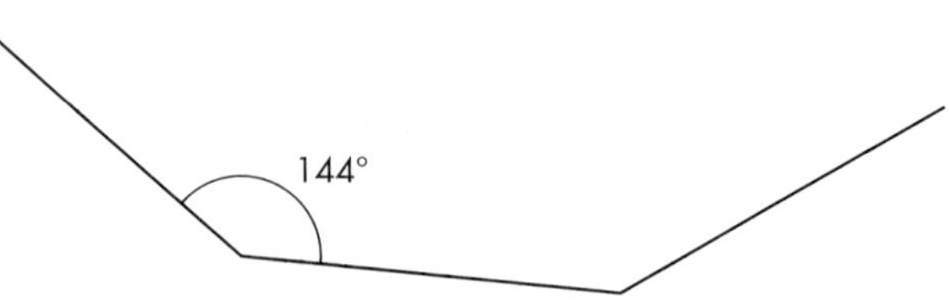

a What is the size of each exterior angle of the polygon?

b How many sides does the polygon have?

10.6 Parallel lines

This section will show you how to:

- find angles in parallel lines

Key words
allied angles
alternate angles
corresponding angles
interior angles

ACTIVITY

Angles in parallel lines

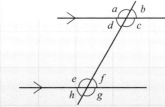

You need tracing paper or a protractor.

Draw two parallel lines about 5 cm apart and a third line that crosses both of them.

The arrowheads indicate that the lines are parallel and the line that crosses the parallel lines is called a *transversal*.

Notice that eight angles are formed. Label these a, b, c, d, e, f, g and h.

Measure or trace angle a. Find all the angles on the diagram that are the same size as angle a.

Measure or trace angle b. Find all the angles on the diagram that are the same size as angle b.

What is the sum of $a + b$?

Find all the pairs of angles on the diagram that add up to 180°.

Angles like these

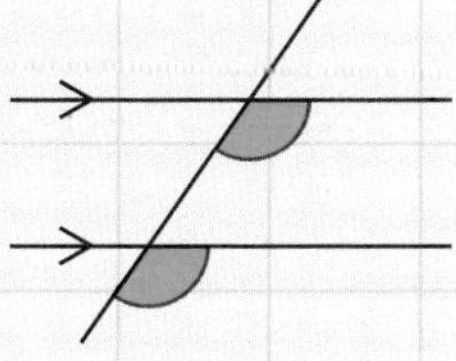

are called **corresponding angles** (Look for the letter F).

Corresponding angles are equal.

Angles like these

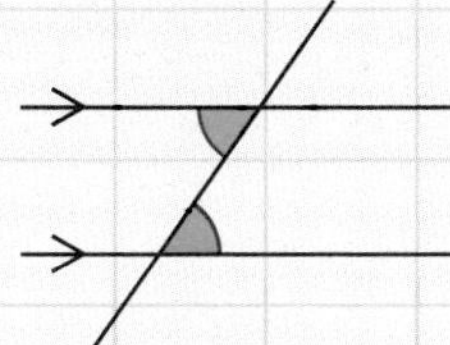

are called **alternate angles** (Look for the letter Z).

Alternate angles are equal.

Angles like these are called **allied angles** or **interior angles** (Look for the letter C).

Allied angles add to 180°.

Copy and complete these statements to make them true.

1 Angles *h* and ...... are corresponding angles.

2 Angles *d* and ...... are alternate angles.

3 Angles *e* and ...... are allied angles.

4 Angles *b* and ...... are corresponding angles.

5 Angles *c* and ...... are allied angles.

6 Angles *c* and ...... are alternate angles.

Note that in examinations you should use the correct terms for types of angles. Do *not* call them F, Z or C angles as you will lose marks for quality of written communication.

EXAMPLE 8

State the size of each of the lettered angles in the diagram.

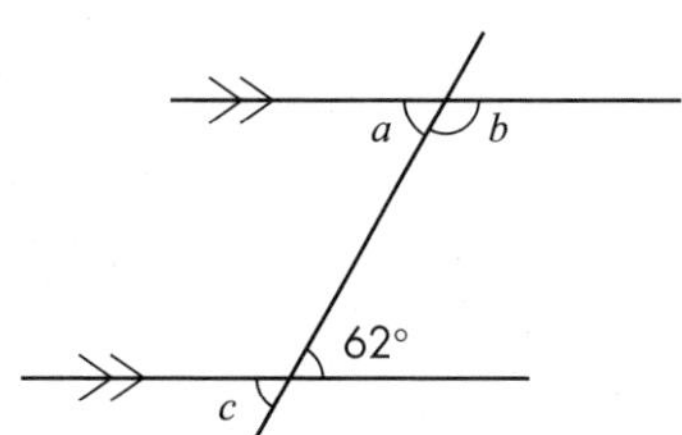

a = 62° (alternate angle)

b = 118° (allied angle or angles on a line)

c = 62° (vertically opposite angle)

EXERCISE 10F

1 State the sizes of the lettered angles in each diagram.

a

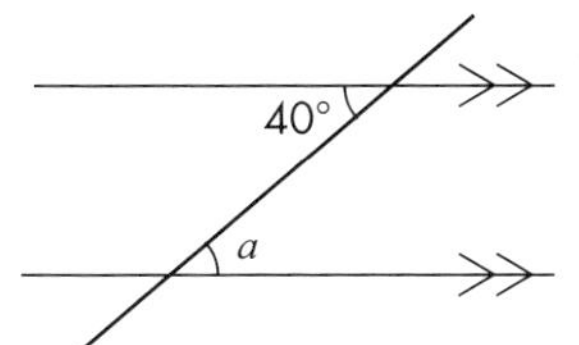

b

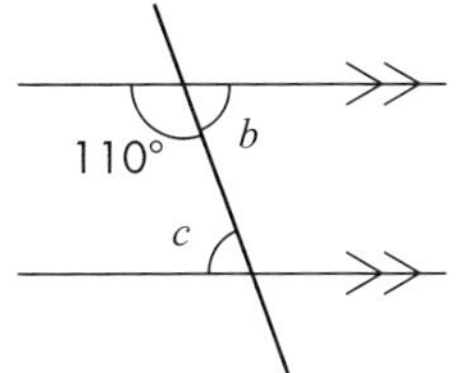

c

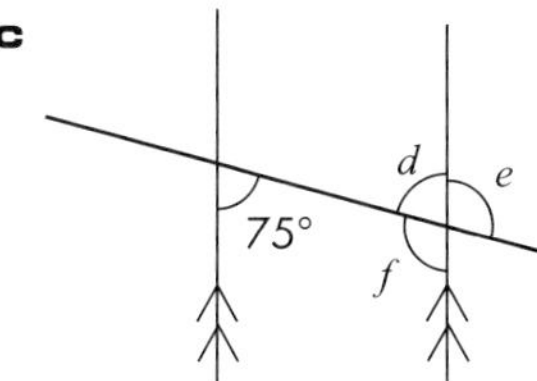

d

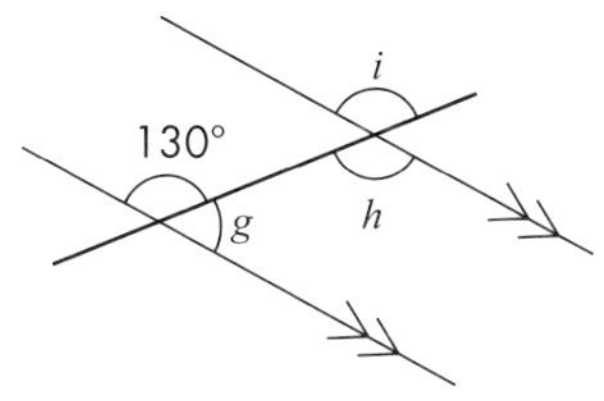

e

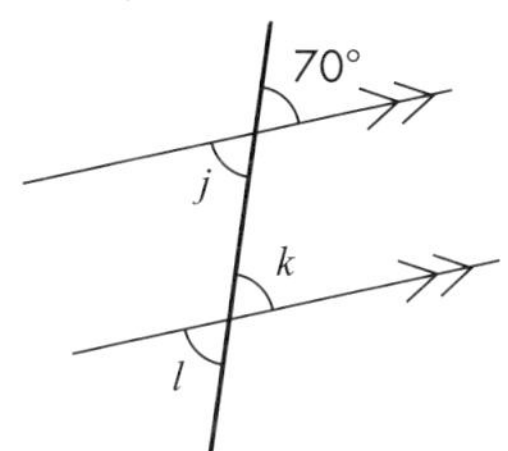

f

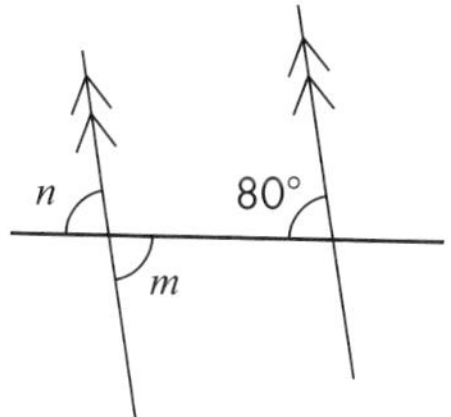

2 State the sizes of the lettered angles in each diagram.

a

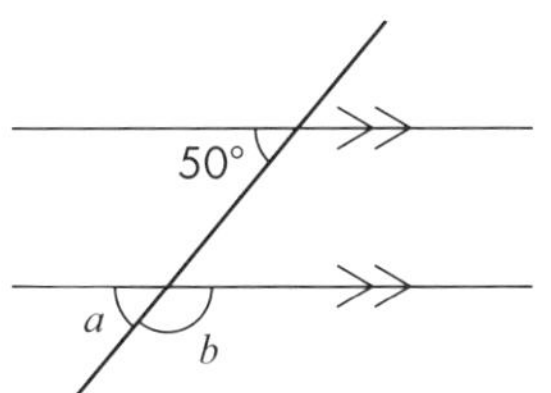

b

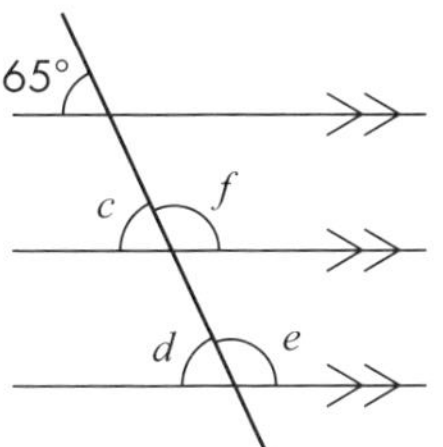

c

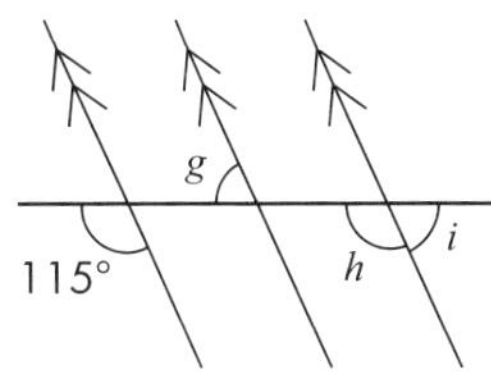

d

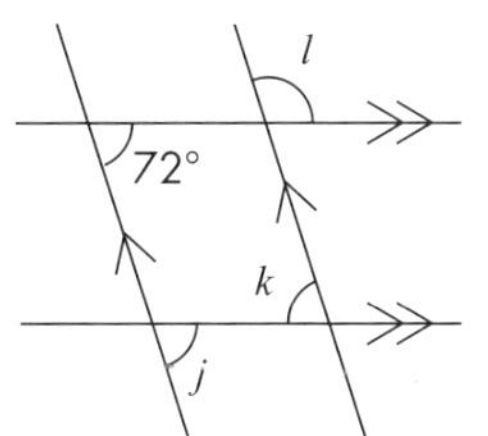

e

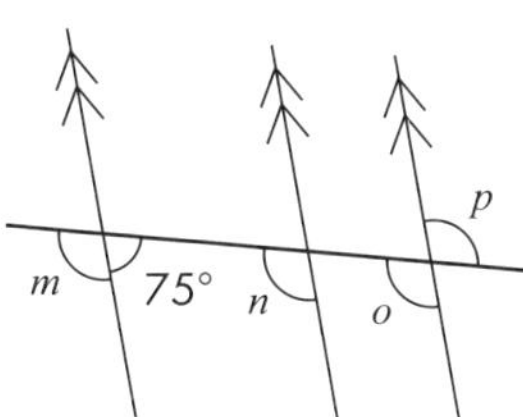

f

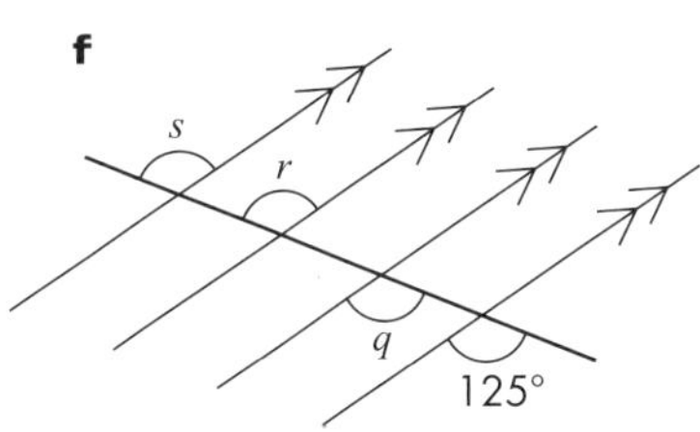

3 State the sizes of the lettered angles in these diagrams.

a

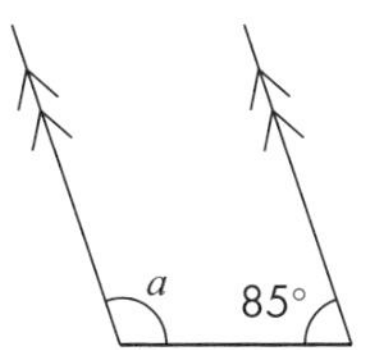

b

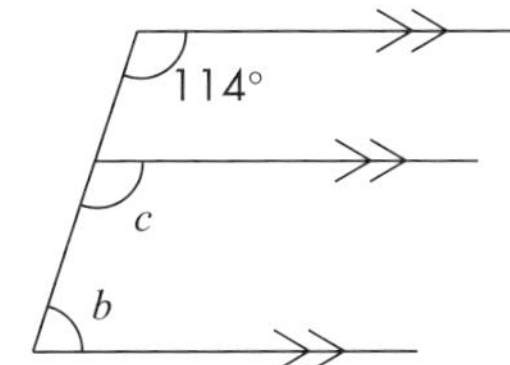

D

D

4 Calculate the values of x and y in these diagrams.

a y, $2x$, $x + 30°$

b $3x$, $2x + 25°$, y

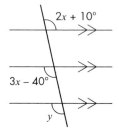

c $3x + 10°$, $4x - 20°$, y

5 Calculate the values of x and y in these diagrams.

a $2x + 10°$, $3x - 40°$, y

b

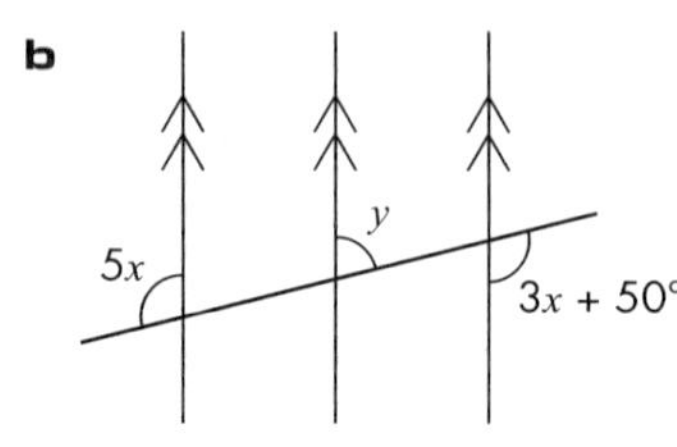

c

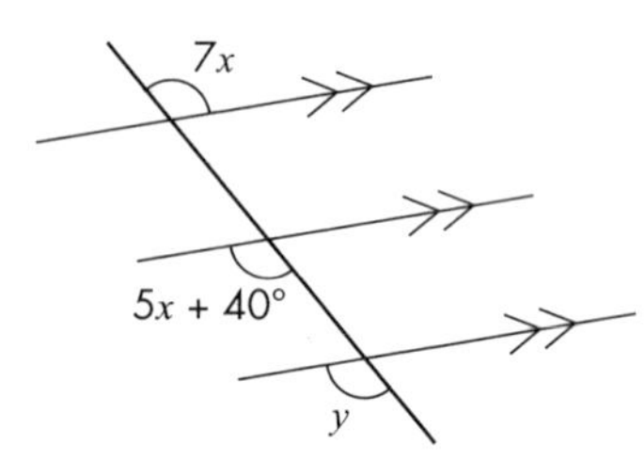

FM 6 A company makes signs in the shape of a chevron.

This is one of their signs. It has one line of symmetry.

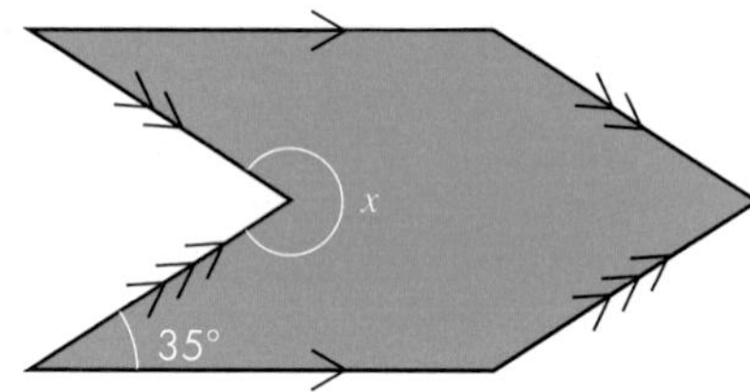

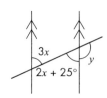

HINTS AND TIPS

Draw the line of symmetry on the shape first.

The designer for the company needs to know the size of the angle marked x on the diagram.

Work out the size of angle x.

PS 7 In the diagram, AE is parallel to BD.

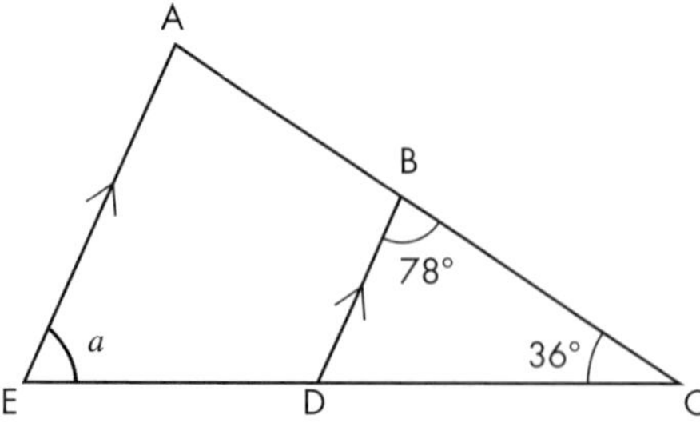

Work out the size of angle a.

Give clear reasons as to how you obtained your answers.

AU **8** Lizzie is writing out a solution to this question.

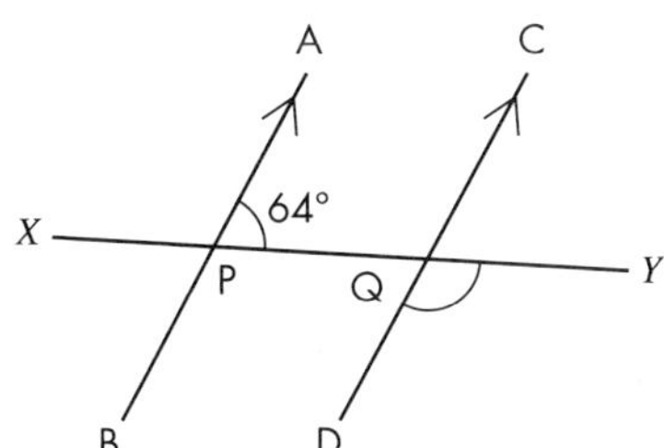

The line XY crosses the parallel lines AB and CD at P and Q.

Work out the size of angle DQY.

Give reasons for your answer.

This is her solution.

Angle PQD = 64° (corresponding angles)

So angle DQY = 124° (angles on a line = 190°)

Lizzie has made a number of errors in her solution.

Write out a correct solution for the question.

D

9 Use the diagram to prove that the three angles in a triangle add up to 180°.

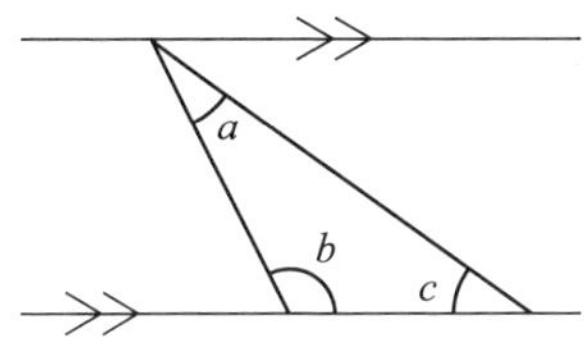

10 Prove that $p + q + r = 180°$.

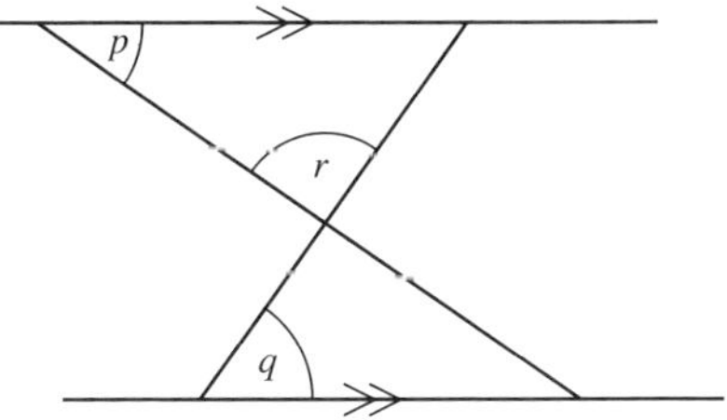

C

10.7 Special quadrilaterals

This section will show you how to:
- use angle properties in quadrilaterals

Key words
kite
parallelogram
rhombus
trapezium

You should know the names of the following quadrilaterals, be familiar with their angle properties and know how to use the three-letter notation to describe any angle.

Parallelogram

- A **parallelogram** has opposite sides parallel.
- Its opposite sides are equal.
- Its diagonals bisect each other.
- Its opposite angles are equal. That is:

  angle BAD = angle BCD
  angle ABC = angle ADC

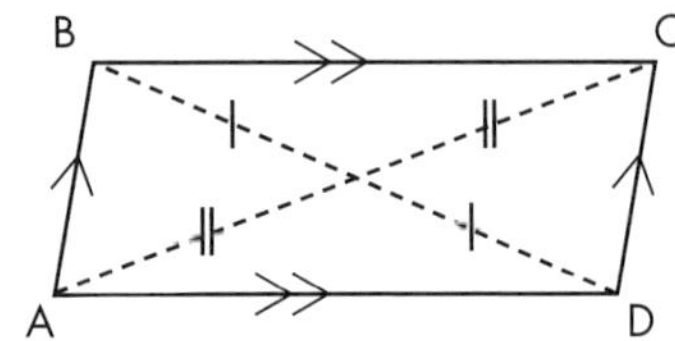

Rhombus

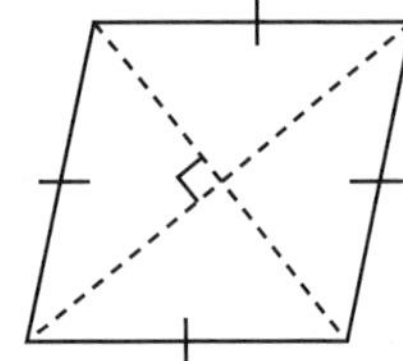

- A **rhombus** is a parallelogram with all its sides equal.
- Its diagonals bisect each other at right angles.
- Its diagonals also bisect the angles.

Kite

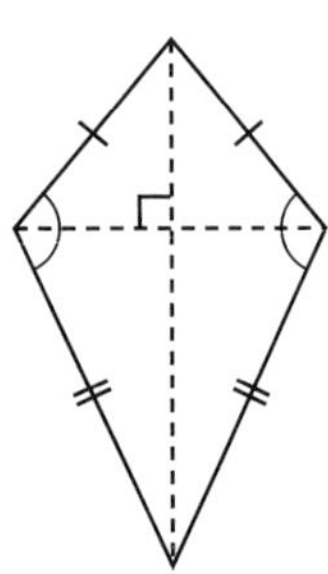

- A **kite** is a quadrilateral with two pairs of equal adjacent sides.
- Its longer diagonal bisects its shorter diagonal at right angles.
- The opposite angles between the sides of different lengths are equal.

Trapezium

- A **trapezium** has two parallel sides.
- The sum of the interior angles at the ends of each non-parallel side is 180°. That is:

  angle BAD + angle ADC = 180°
  angle ABC + angle BCD = 180°

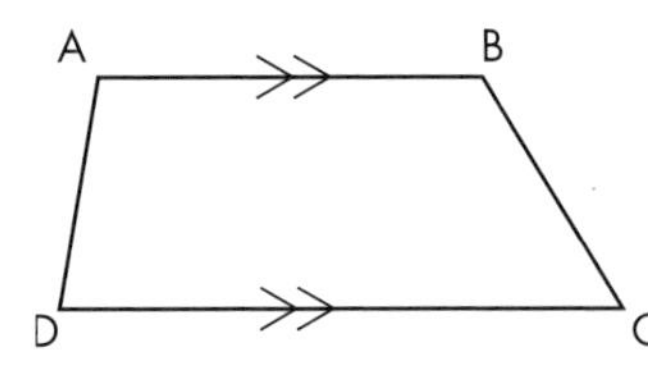

EXERCISE 10G

1 For each of the trapeziums, calculate the sizes of the lettered angles.

a

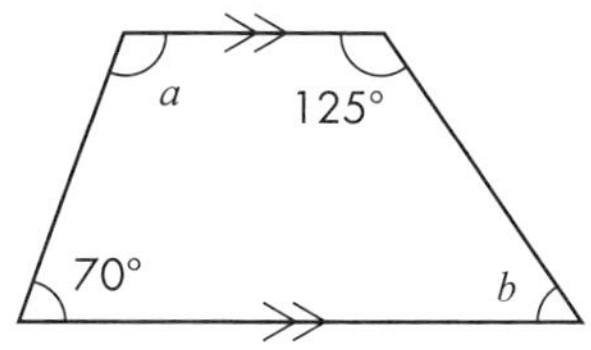

b

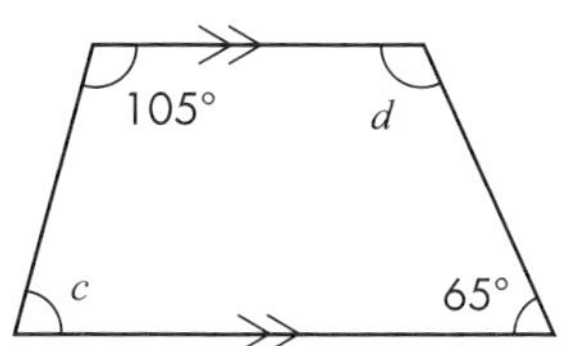

c

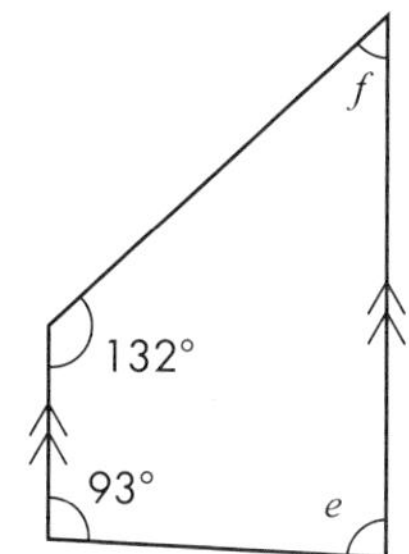

2 For each of these parallelograms, calculate the sizes of the lettered angles.

a

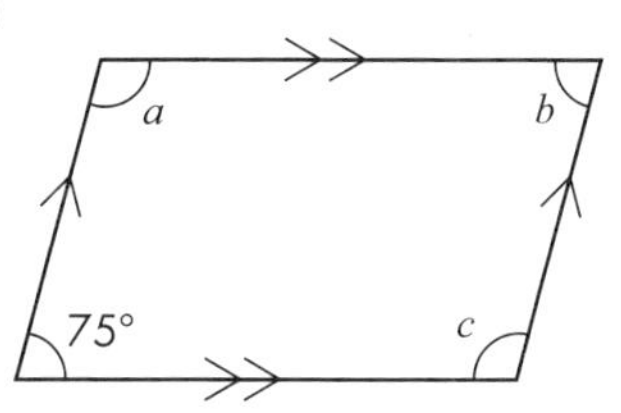

b

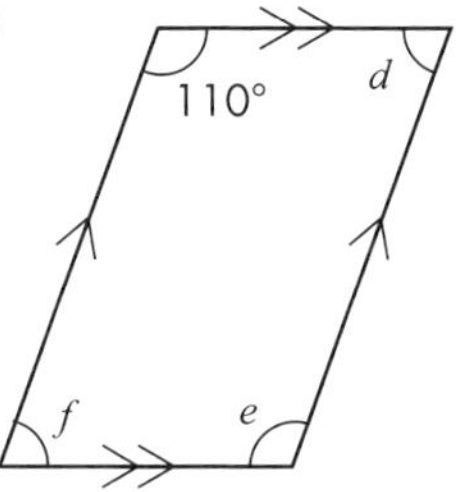

c

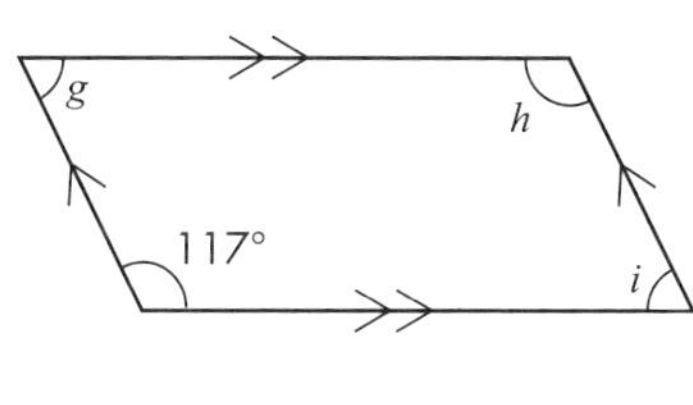

3 For each of these kites, calculate the sizes of the lettered angles.

a

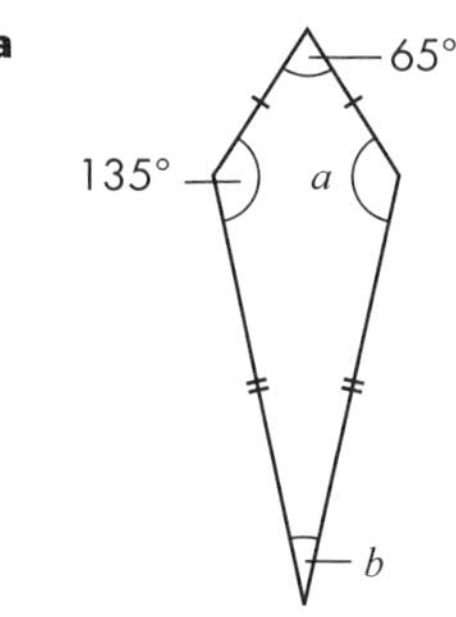

b

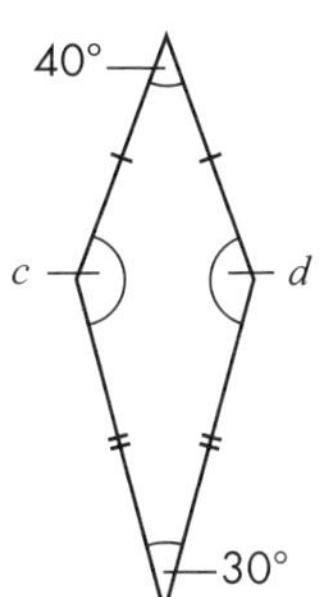

c

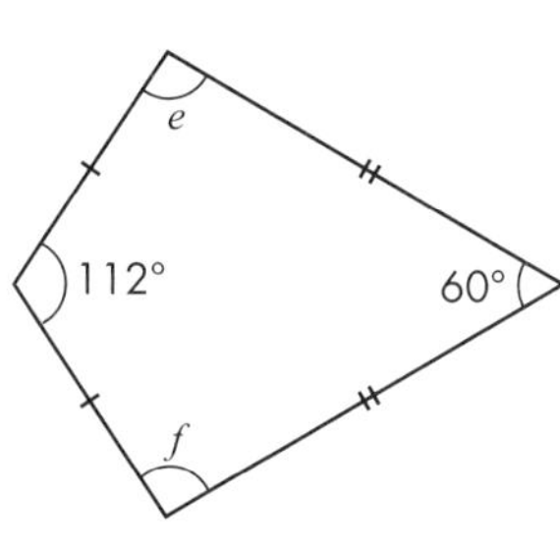

4 For each of these rhombuses, calculate the sizes of the lettered angles.

a

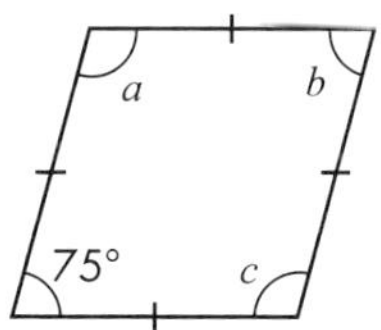

b

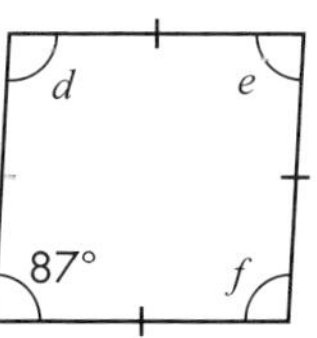

c

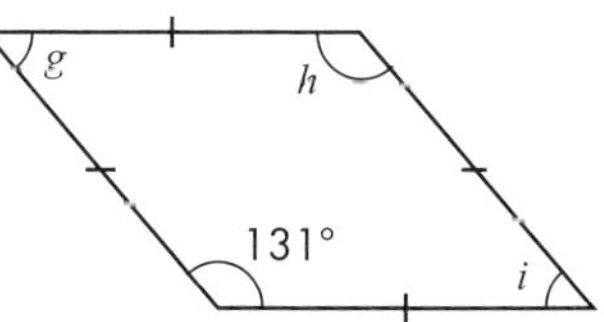

5 For each of these shapes, calculate the sizes of the lettered angles.

a

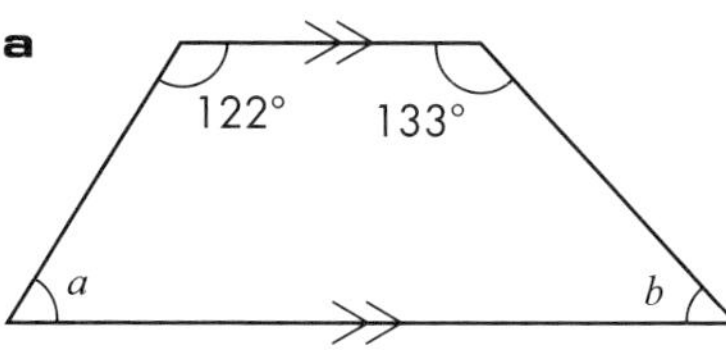

b

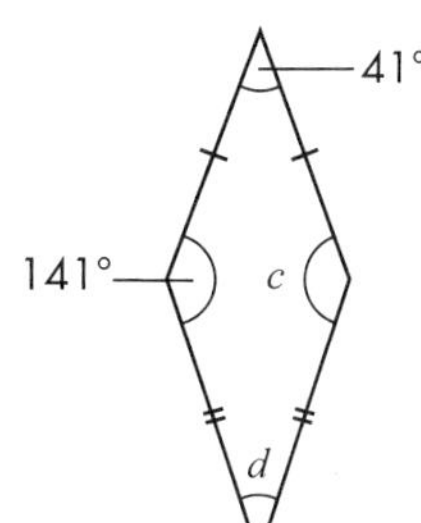

c

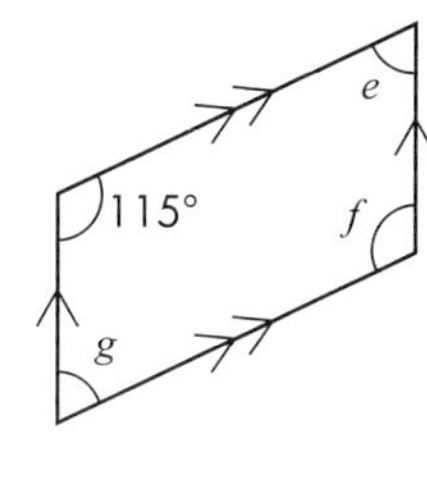

D

6 Dani is making a kite.

She needs angle C to be half the size of angle A.

Work out the size of angles B and D.

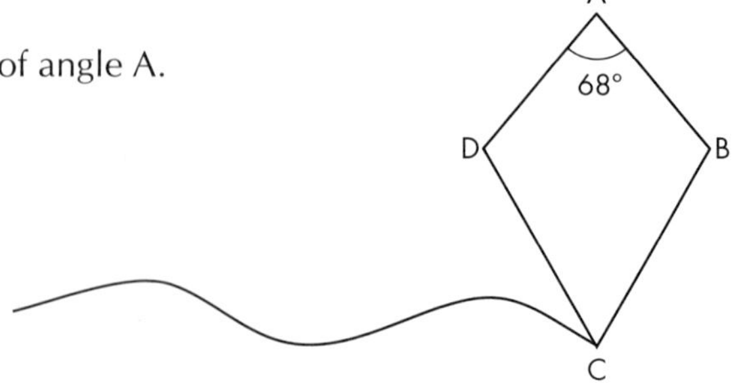

AU PS **7** David says that a parallelogram is a special type of rectangle.

Marie says that he is wrong and that a rectangle is a special type of parallelogram.

Who is correct?

Give a reason for your answer.

AU **8** The diagram shows a quadrilateral ABCD.

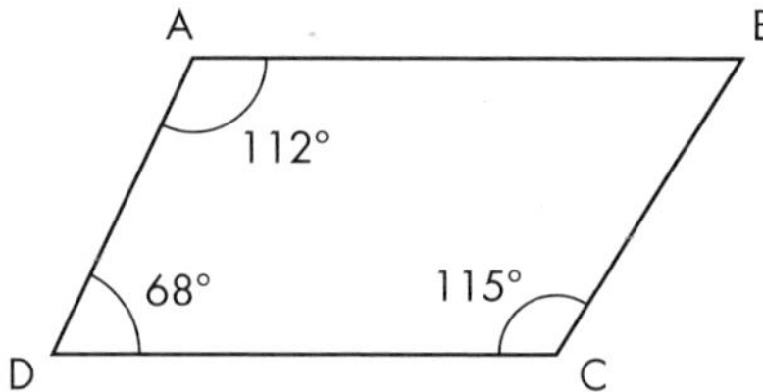

a Calculate the size of angle B.

b What special name is given to the quadrilateral ABCD? Explain your answer.

10.8 Bearings

This section will show you how to:

- use a bearing to specify a direction

Key words
bearing
three-figure bearing

The **bearing** of a point B from a point A is the angle through which you turn *clockwise* as you change direction from *due north* to the direction of B.

For example, in this diagram the bearing of B from A is 060°.

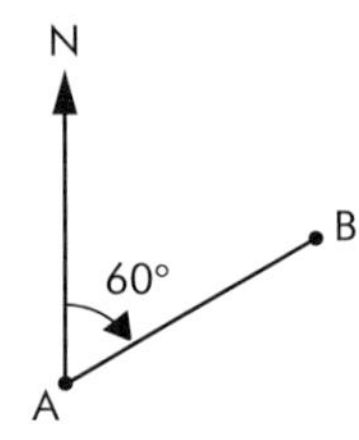

As a bearing can have any value from 0° to 360°, it is customary to give all bearings in three figures. This is known as a **three-figure bearing**. So, in the example on the previous page, the bearing is written as 060°, using three figures. Here are three more examples.

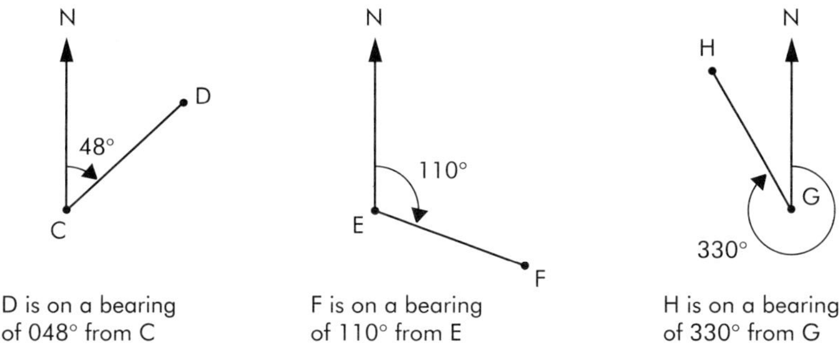

D is on a bearing of 048° from C

F is on a bearing of 110° from E

H is on a bearing of 330° from G

There are eight bearings with which you should be familiar. They are shown in the diagram.

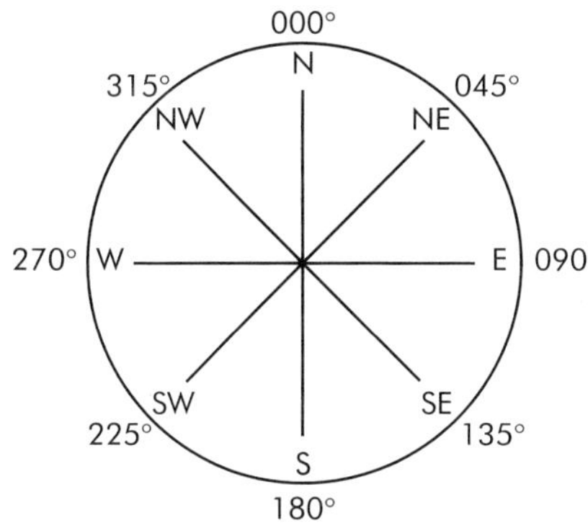

EXAMPLE 9

A, B and C are three towns.

a Write down the bearing of B from A and the bearing of C from A.

The bearing of B from A is 070°.

The bearing of C from A is $360° - 115° = 245°$.

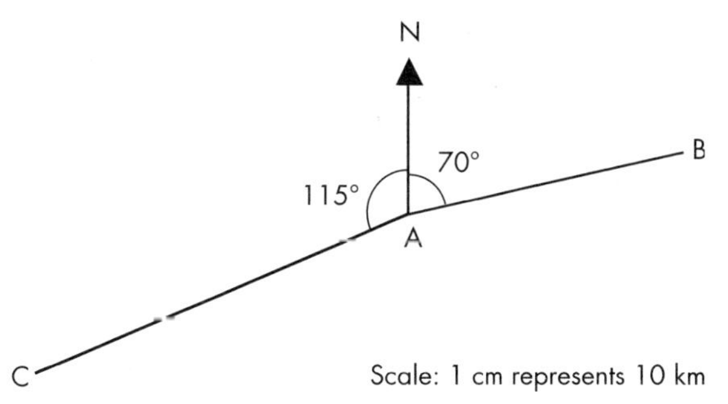

b Use the scale to work out the actual distances between **i** A and B **ii** A and C.

i On the diagram AB is 3 cm, so the actual distance between A and B is 30 km.

ii On the diagram AC is 4 cm, so the actual distance between A and C is 40 km.

EXERCISE 10H

D

1 Look at this map. By measuring angles, find the following bearings.

N
Greystones
Abbey
Ecclesall
Dore
Millhouses
Totley

a Totley from Dore
b Dore from Ecclesall
c Millhouses from Dore
d Greystones from Abbey
e Millhouses from Greystones
f Totley from Millhouses

2 Draw sketches to illustrate the following situations.

a Castleton is on a bearing of 170° from Hope.

b Bude is on a bearing of 310° from Wadebridge.

3 A is due north from C. B is due east from A. B is on a bearing of 045° from C. Sketch the layout of the three points, A, B and C.

4 Captain Bird decided to sail his ship around the four sides of a square kilometre.

a Assuming he started sailing due north, write down the further three bearings he would use in order to complete the square in a clockwise direction.

b Assuming he started sailing on a bearing of 090°, write down the further three bearings he would use in order to complete the square in an anticlockwise direction.

5 The map shows a boat journey around an island, starting and finishing at S. On the map, 1 centimetre represents 10 kilometres. Measure the distance and bearing of each leg of the journey. Copy and complete the table below.

Leg	Actual distance	Bearing
1		
2		
3		
4		
5		

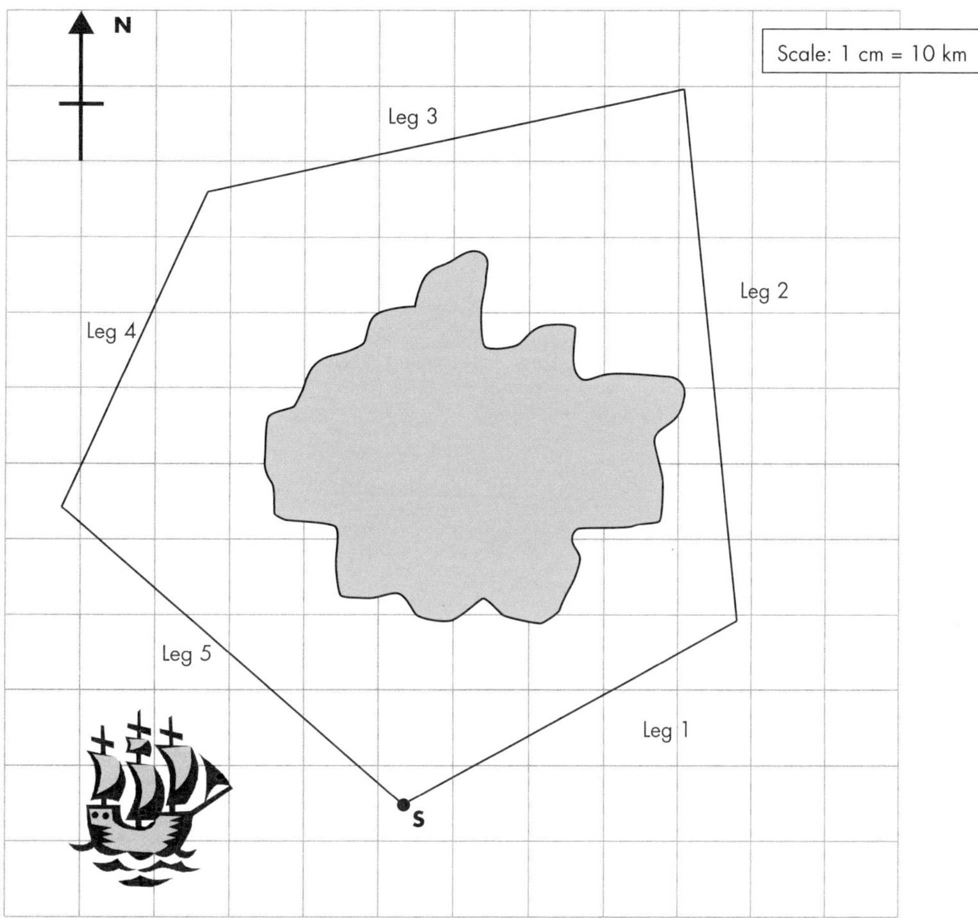

6 The diagram shows a port P and two harbours X and Y on the coast.

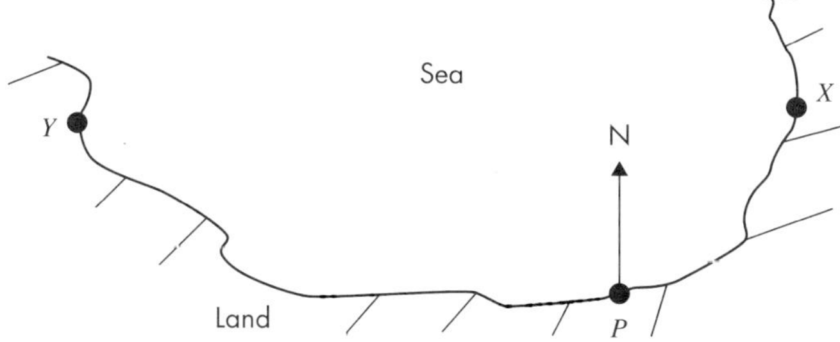

a A fishing boat sails to X from P.

What is the three-figure bearing of X from P?

b A yacht sails to Y from P.

What is the three-figure bearing of Y from P?

D

7 Draw diagrams to solve the following problems.

a The three-figure bearing of A from B is 070°. Work out the three-figure bearing of B from A.

b The three-figure bearing of P from Q is 145°. Work out the three-figure bearing of Q from P.

c The three-figure bearing of X from Y is 324°. Work out the three-figure bearing of Y from X.

8 The diagram shows the position of Kim's house H and the college C.

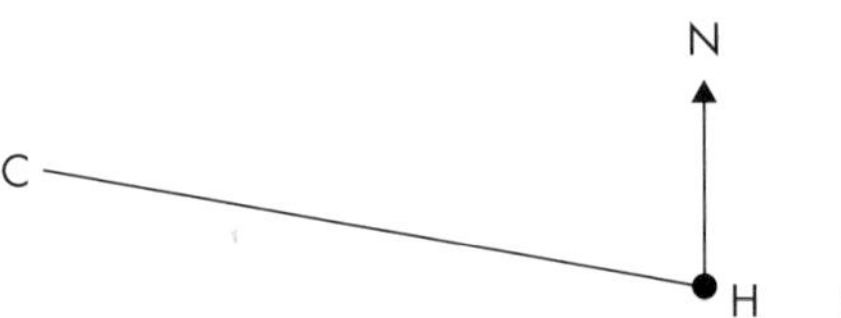

Scale: 1 cm represents 200 m

a Use the diagram to work out the actual distance from Kim's house to the college.

b Measure and write down the three-figure bearing of the college from Kim's house.

c The supermarket S is 600 m from Kim's house on a bearing of 150°.

Mark the position of S on a copy of the diagram.

FM 9 Trevor is flying a plane on a bearing of 072°.

He is instructed by a control tower to turn and fly due south towards an airport.

Through what angle does he need to turn?

C

PS 10 Apple Bay (A), Broadside (B) and Caverly (C) are three villages in a bay.

The villages lie on the vertices of a square.

The bearing of B from A is 030°.

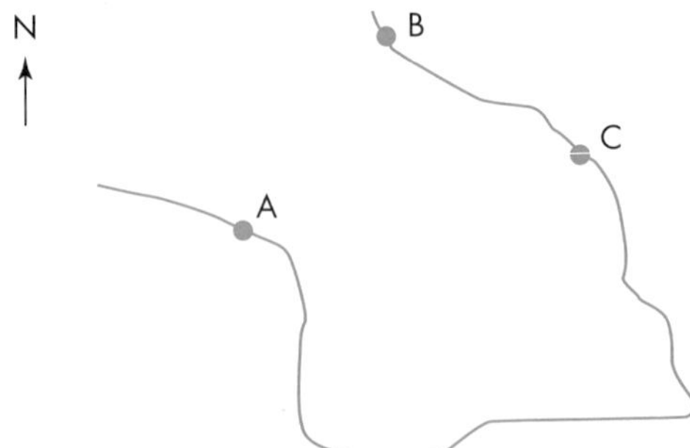

Work out the bearing of Apple Bay from Caverly.

GRADE BOOSTER

- **F** You can measure and draw angles
- **F** You know that the sum of the angles on a line is 180°
- **F** You know that the sum of the angles at a point is 360°
- **E** You know that the sum of the angles in a triangle is 180°
- **E** You know that the sum of the angles in a quadrilateral is 360°
- **E** You can find the exterior angle of a triangle and quadrilateral
- **D** You can find angles in parallel lines
- **D** You know all the properties of special quadrilaterals
- **D** You can find interior and exterior angles in regular polygons
- **D** You can use three-figure bearings
- **C** You can use interior angles and exterior angles to find the number of sides in a regular polygon

What you should know now

- How to measure and draw angles
- How to find angles on a line or at a point
- How to find angles in triangles, quadrilaterals and polygons
- How to find interior and exterior angles in polygons
- How to use bearings

1 Which of the marked angles is an acute angle? (1)

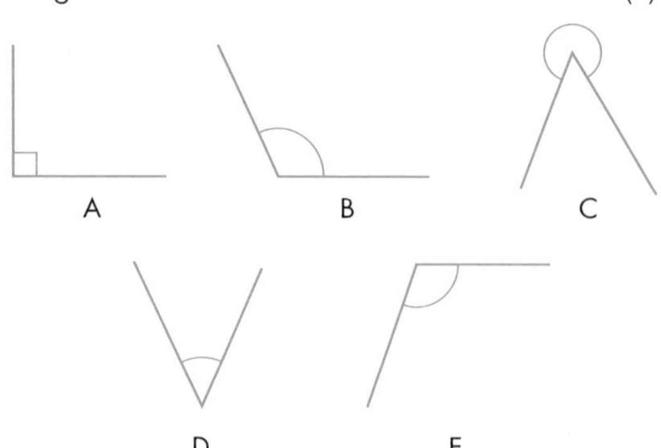

(Total 1 mark)

Edexcel, March 2009, Paper 7 Foundation, Unit 2 Stage 1, Question 5

2 The diagram shows three angles on a straight line AB.

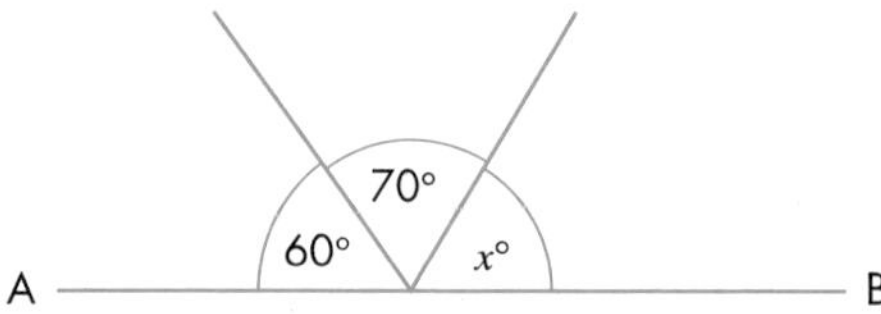

Work out the value of x.

3

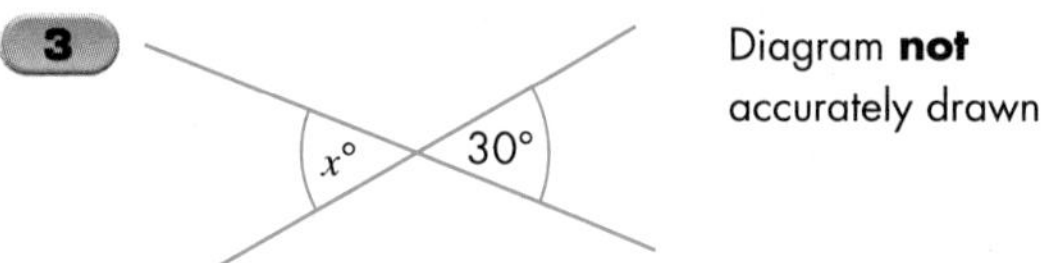

Diagram **not** accurately drawn

a **i** Write down the value of x. (1)

ii Give a reason for your answer. (1)

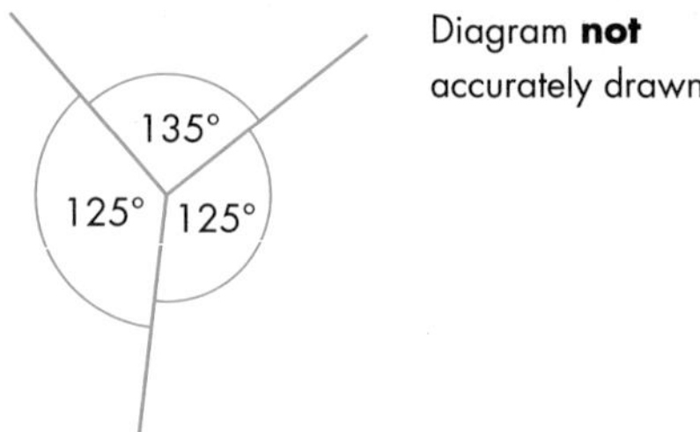

Diagram **not** accurately drawn

This diagram is wrong.

b Explain why. (1)

(Total 3 marks)

Edexcel, June 2008, Paper 2 Foundation, Question 14

4

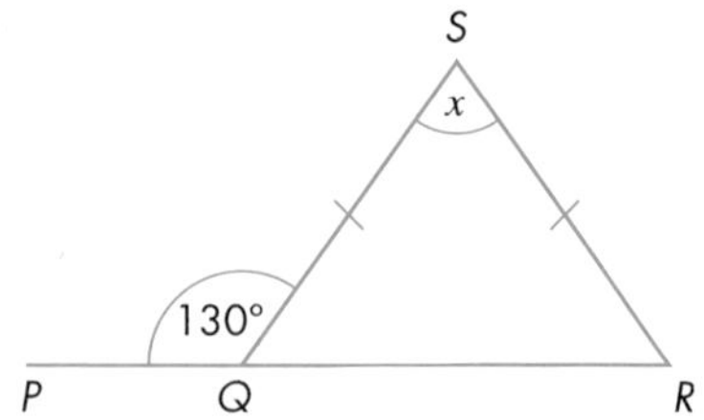

Diagram **not** accurately drawn

PQR is a straight line.

SQ = *SR*.

What is the size of the angle marked x? (1)

50°	130°	70°	80°	60°
A	**B**	**C**	**D**	**E**

(Total 1 mark)

Edexcel, March 2009, Paper 7 Foundation, Unit 2 Stage 1, Question 20

5

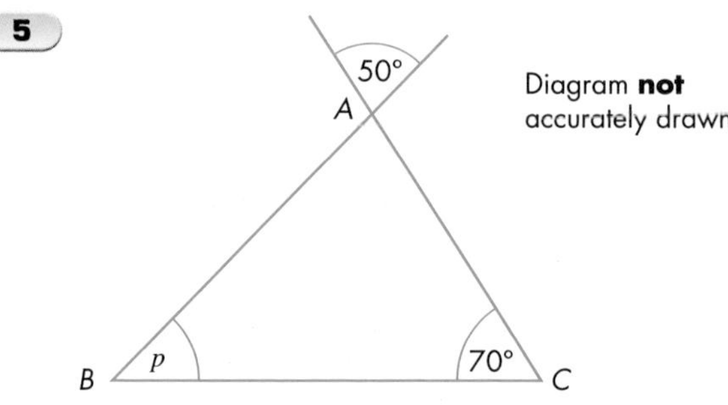

ABC is a triangle.

Work out the size of the angle marked p. (2)

(Total 2 marks)

Edexcel, June 2008, Paper 10 Foundation, Unit 3 Test, Question 7

6

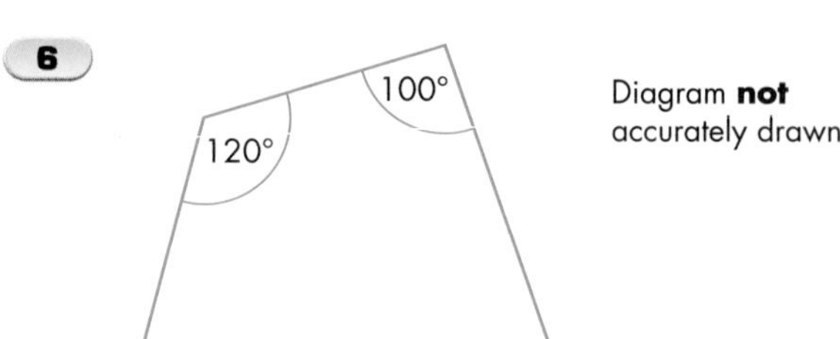

Work out the value of y. (2)

(Total 2 marks)

Edexcel, May 2008, Paper 12 Foundation, Question 9

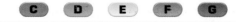

7 The diagram shows a kite.

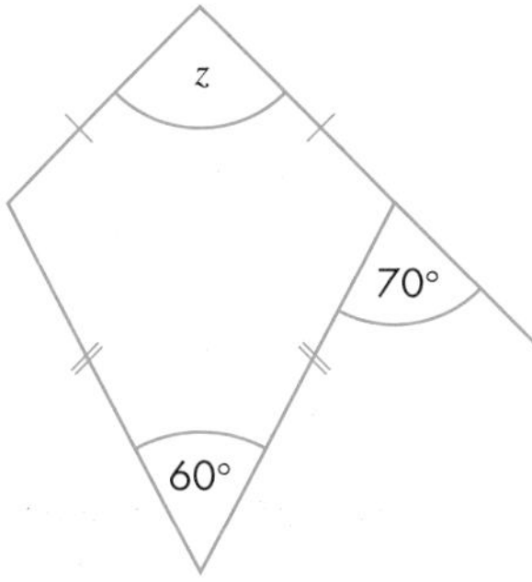

Calculate the size of the angle marked z.

8 The lines AB and CD are parallel.
Angle BAD = 35°. Angle BCD = 40°.

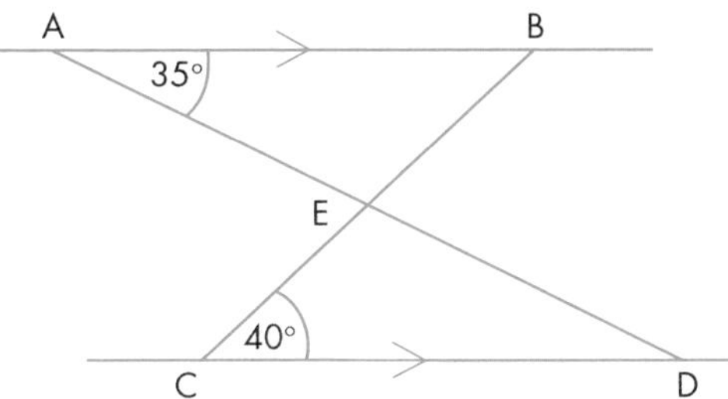

Show clearly why angle AEC is 75°.

9

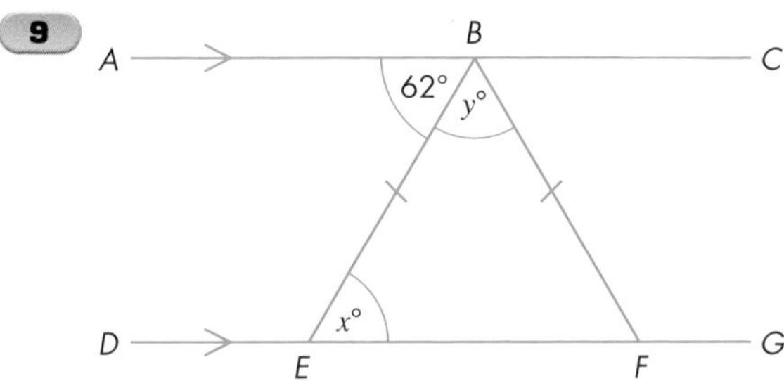

Diagram **not** accurately drawn

ABC and *DEFG* are straight lines.

AC is parallel to *DG*.

BE = *BF*.

Angle *ABE* = 62°.

a **i** Find the value of x. (1)

ii Give a reason for your answer. (1)

b Work out the value of y. (2)

(Total 4 marks)

Edexcel, March 2007, Paper 10 Foundation, Unit 3 Test, Question 7

10 The diagram shows the position of two airports, *A* and *B*.

A plane flies from airport *A* to airport *B*.

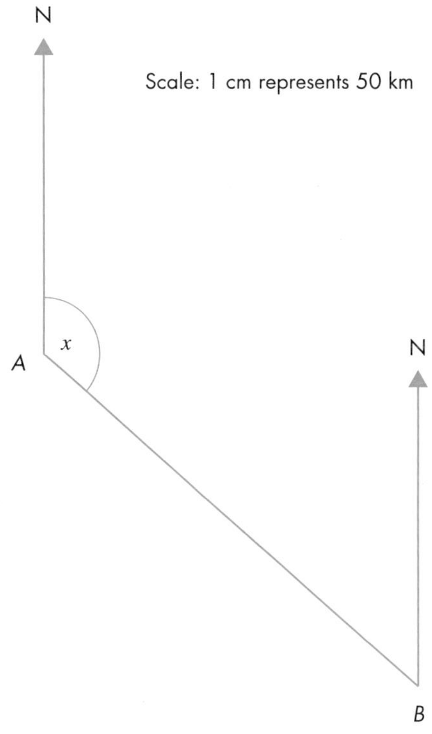

a Measure the size of the angle marked x. (1)

b Work out the real distance between airport *A* and airport *B*.

Use the scale 1 cm represents 50 km. (2)

Airport C is 350 km on a bearing of 060° from airport B.

c On a diagram, mark airport *C* with a cross (X).

Label it C. (2)

(Total 5 marks)

Edexcel, May 2008, Paper 1 Foundation, Question 17

Worked Examination Questions

1 ABC is a **triangle**.
D is a **point** on AB such that BC = BD.

a Work out the value of x.

b Work out the value of y.

c Is it true that AD = DC?
Give a reason for your answer.

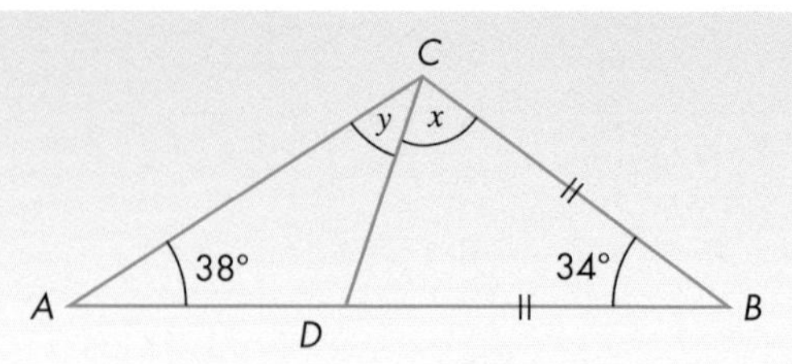

a Triangle BCD is isosceles, so angle BDC is also x.

Angles in a triangle = 180°,

so $2x + 34° = 180°$

$2x = 146°$

So $x = 73°$

You get 1 mark for setting up an equation.

180 – 34 = 146
You get 1 mark for first step of solving the equation.

You get 1 mark for the correct answer.

3 marks

b Angle ADC = 180° – 73° = 107° (*angles on a line*)

$y + 38° + 107° = 180°$ (*angles in a triangle*)

$y + 145° = 180°$

So $y = 35°$

You get 1 method mark for finding 107°.

You get 1 mark for finding $y = 35°$.

2 marks

c No, since triangle ACD is not an isosceles triangle.

You get 1 mark for stating that triangle ACD is not an isosceles triangle.

1 mark

Total: 6 marks

Worked Examination Questions

2 The lines AB and CD are **parallel**.

a Write down the value of a.
Give a reason for your answer.

b Write down the value of b.
Give a reason for your answer.

c Work out the value of c.
Give a reason for your answer.

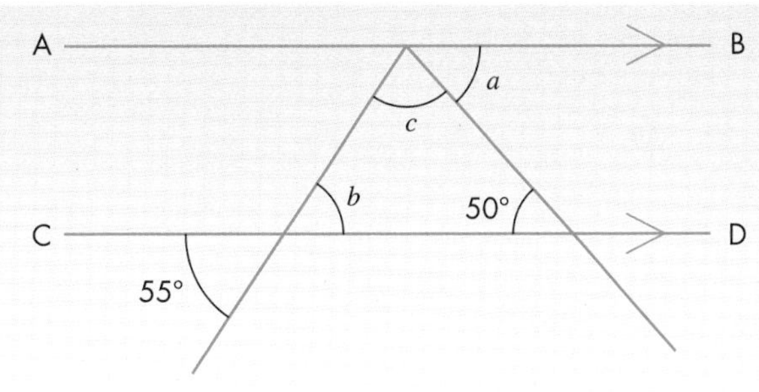

a $a = 50°$

It is an *alternate angle* between the parallel lines.

2 marks

b $b = 55°$

It is an *opposite angle* in two intersecting lines.

2 marks

c $c = 75°$

The three angles in the triangle add up to 180°.

2 marks

Total: 6 marks

Never measure the angles with a protractor as the diagrams in examinations are not drawn accurately.

You get 1 accuracy mark for the correct answer.

You get 1 mark for a correct reason.
Stating "It is a Z angle" is not acceptable in examinations. You must identify it as an alternate angle to get a mark for quality of written communication.

You get 1 accuracy mark for the correct answer.

You get 1 mark for a correct reason.

You get 1 accuracy mark for the correct answer.

You get 1 mark for a correct reason.

10

Functional Maths
Origami

Product designers will make a model of their product before the final product is manufactured. They use these models to trial products and to ensure that their design is suitable and fully-functioning.

In this task you are going to make a model of a chip holder, following instructions and using your knowledge of angles and properties of shapes to calculate unknown angles.

Getting started

- What is an angle?
- If one of the angles in a triangle is 36°, what could the other two angles be?
- Name some quadrilaterals that have at least one pair of parallel sides.
  - How many can you name?
  - What other properties do they have?

Your task

Alan owns a fish and chip shop. He uses plastic containers for his chips, but he wants to switch to paper holders. He has seen a design for a chip holder on the internet and thinks that it will work for him.

1 Follow the instructions to make the chip holder. Write a summary, giving the name of the shape, the effect of changing the size of the holder on the angles, and a description of how effective the holder would be.

2 Alan wants the chip holders to sit in a rack. To do this, he needs to work out the angles of his chip holder. However, he does not have a protractor.

Use your knowledge of angles and properties of shape to calculate the angles for Alan. Then, design a rack to fit the chip holders. You must include the rack's dimensions, angles and at least two different elevations (views) of the final product design.

Instructions for Alan's design

1 Fold a square piece of paper in half along its diagonal.

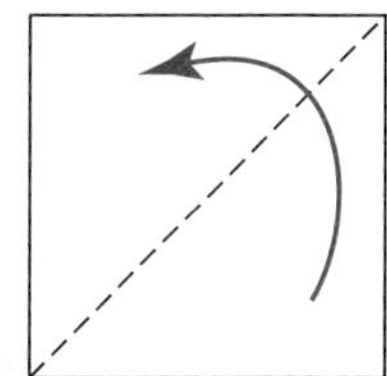

to

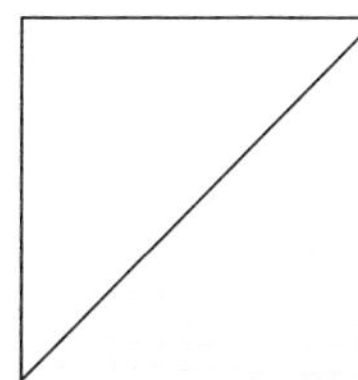

2 Turn the paper so that the longest side is along the bottom. Fold corner A to meet the opposite side, so that the top and bottom edges are parallel.

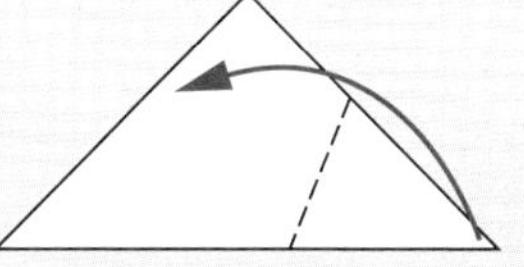

to

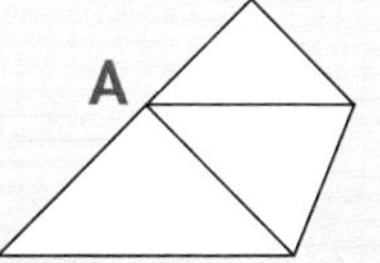

3 Fold corner B to meet point C.

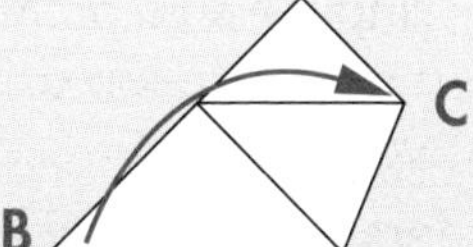

to

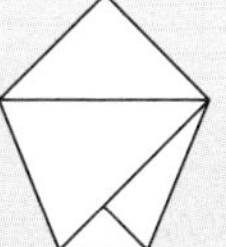

4 Fold the top layer down so that it lies over the folded sides.

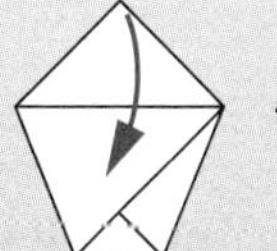

to

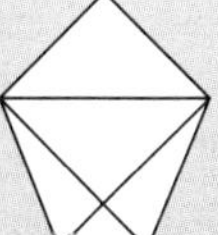

5 Turn over and repeat step 4 for this side.

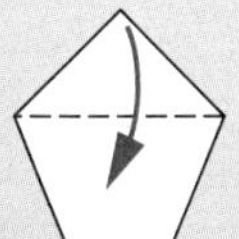

to

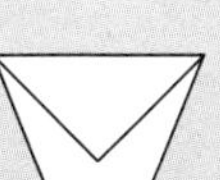

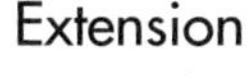

Extension

- Find a way to check the angles that you have worked out.
- In step 2 of the instructions, you need to fold point B. There is a way to deduce the position to which you need to fold point B, without relying on estimating by eye. What is this and why does it work?

Why this chapter matters

Chance is a part of everyday life. Judgements are frequently made based on probability. For example,

- **there is an 80% chance that United will win the game tomorrow**
- **there is a 40% chance of rain tomorrow**
- **she has a 50–50 chance of having a baby girl**
- **there is a 10% chance of the bus being on time tonight.**

Certain probabilities are given to certain events, although two people might give different probabilities to those same events because of their different views. For example, one person might not agree that there is an 80% chance of United winning the game. They might well say that there is only a 70% chance of United winning tomorrow. A lot depends on what that person believes or has experienced.

Chance may be taken into account for lots of different events, from the lottery to tomorrow's weather. Probability is a branch of mathematics that describes the chance of outcomes.

Probability originated from the study of games of chance, such as tossing a dice or spinning a roulette wheel.

Mathematicians in the 16th and 17th centuries started to think about the mathematics of chance in games. Probability theory, as a branch of mathematics, developed in the 17th century when French gamblers asked mathematicians Blaise Pascal and Pierre de Fermat for help in their gambling.

Now, in the 21st century, probability theory is used to control the flow of traffic through road systems (above) or the running of telephone exchanges (right), and to look at patterns of the spread of infections.

There are many other everyday applications and as you work through this chapter you will see how frequently the language of probability is used.

Probability: Probability of events

The grades given in this chapter are target grades.

1 Probability scale

2 Calculating probabilities

3 Probability that an outcome of an event will not happen

This chapter will show you ...

G to E how to use the the language of probability

F to E how to work out the probability of outcomes of events

Visual overview

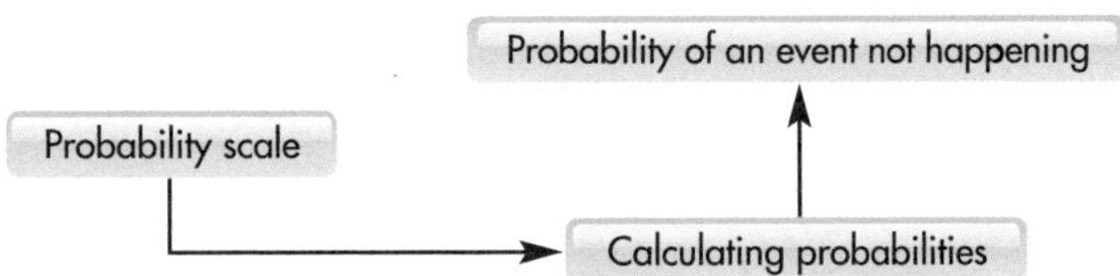

What you should already know

- How to add, subtract and cancel fractions **(KS3 level 4, GCSE grade F)**
- That outcomes of events cannot always be predicted and that the laws of chance apply to everyday events **(KS3 level 4, GCSE grade F)**
- How to list all the outcomes of an event in a systematic manner **(KS3 level 5, GCSE grade E)**

Quick check

1 Simplify the following fractions.

a $\frac{6}{8}$ **b** $\frac{6}{36}$ **c** $\frac{3}{12}$ **d** $\frac{8}{10}$ **e** $\frac{6}{9}$ **f** $\frac{5}{20}$

2 Calculate the following.

a $\frac{1}{8}+\frac{3}{8}$ **b** $\frac{5}{12}+\frac{3}{12}$ **c** $\frac{5}{36}+\frac{3}{36}$ **d** $\frac{2}{9}+\frac{1}{6}$ **e** $\frac{3}{5}+\frac{3}{20}$

3 Frank likes to wear brightly coloured hats and socks.

He has two hats, one is green and the other is yellow.
He has three pairs of socks, which are red, purple and pink.

Write down all the six possible combinations of hats and socks Frank could wear.

For example, he could wear a green hat and red socks.

11.1 Probability scale

This section will show you how to:

- use the probability scale and the basic language of probability

Key words
certain
chance
event
impossible
likely
outcome
probability
probability scale
unlikely

Almost daily, you hear somebody talking about the probability of whether this or that will happen. They usually use words such as '**chance**', 'likelihood' or 'risk' rather than 'probability'. For example:

"What is the likelihood of rain tomorrow?"
"What chance does she have of winning the 100 metre sprint"
"Is there a risk that his company will go bankrupt?"

You can give a value to the chance of any of these **outcomes** or **events** happening – and millions of others, as well. This value is called the **probability**.

It is true that some things are certain to happen and that some things cannot happen; that is, the chance of something happening can be anywhere between **impossible** and **certain**. This situation is represented on a sliding scale called the **probability scale**, as shown below.

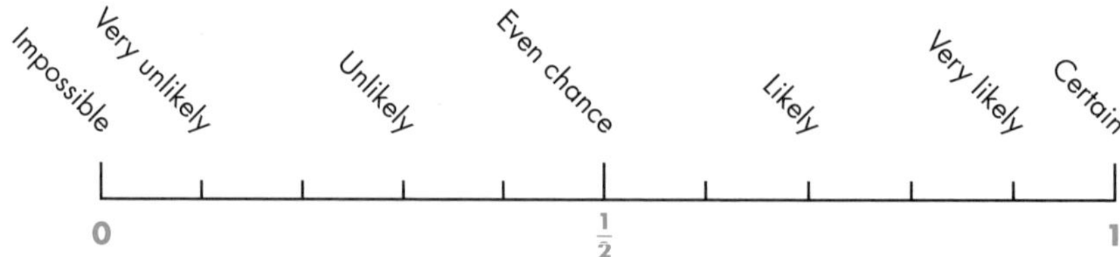

Note: All probabilities lie somewhere in the range of **0** to **1**.

An outcome or an event that cannot happen (is impossible) has a probability of 0.
For example, the probability that pigs will fly is 0.

An outcome or an event that is certain to happen has a probability of 1.
For example, the probability that the sun will rise tomorrow is 1.

EXAMPLE 1

Put arrows on the probability scale to show the probability of each of the outcomes of these events.

- **a** You will get a head when throwing a coin.
- **b** You will get a six when throwing a dice.
- **c** You will have maths homework this week.

a This outcome is an even chance. (Commonly described as a fifty-fifty chance.)

b This outcome is fairly **unlikely**.

c This outcome is **likely**.

The arrows show the approximate probabilities on the probability scale.

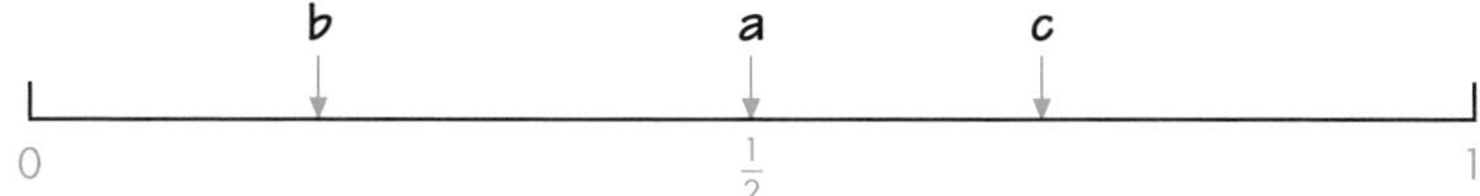

EXERCISE 11A

G

1 State whether each of the following events is impossible, very unlikely, unlikely, even chance, likely, very likely or certain.

- **a** Picking out a Heart from a well-shuffled pack of cards.
- **b** Christmas Day being on the 25th December.
- **c** Someone in your class is left-handed.
- **d** You will live to be 100.
- **e** A score of seven is obtained when throwing a dice.
- **f** You will watch some TV this evening.
- **g** A new-born baby will be a girl.

2 Draw a probability scale and put an arrow to show the approximate probability of each of the following events happening.

- **a** The next car you see will have been made in Europe.
- **b** A person in your class will have been born in the 20th century.
- **c** It will rain tomorrow.
- **d** In the next Olympic Games, someone will run the 1500 m race in 3 minutes.
- **e** During this week, you will have chips with a meal.

G

3 Draw a probability scale and mark an arrow to show the approximate probability of each event.

a The next person to come into the room will be male.

b The person sitting next to you in mathematics is over 16 years old.

c Someone in the class will have mobile phone.

F

4 Give two events of your own for which you think the probability of an outcome is as follows

a impossible **b** very unlikely **c** unlikely

d evens **e** likely **f** very likely

g certain.

5 In August, Janine and her family are going on holiday to Corsica, a French island off the south coast of France. She wonders about the chance of various events happening.

For each of the following, state whether the answer is likely to be:

impossible, very unlikely, unlikely, even chance, likely, very likely, certain.

a It being sunny most days.

b The aeroplane taking off smoothly.

c Her ears popping during the landing (they usually do).

d Hearing most people speak English on the plane.

e Hearing someone speak French in the resort they are going to.

f It raining in France when they arrive.

g Being able to wear her bikini while on holiday.

h The sea being warm when she swims.

i Her becoming the President of France.

j Her meeting her future husband while there.

k Her getting sunburnt if she doesn't put sun cream on but lies in the sun all day.

l Seeing the cost of meals in pounds in restaurants.

AU 6 "I have bought five lottery tickets this week, so I have a good chance of winning."

Explain what is wrong with this statement.

11.2 Calculating probabilities

This section will show you how to:

- calculate the probability of outcomes of events

Key words
equally likely
event
outcome
probability fraction
random

In question 2 of Exercise 11A, you may have had difficulty in knowing exactly where to put some of the arrows on the probability scale. It would have been easier for you if each result of the **event** could have been given a value, from 0 to 1, to represent the probability for that result.

For some events, this can be done by first finding all the possible results, or **outcomes**, for a particular event. For example, when you throw a coin there are two **equally likely** outcomes: heads or tails. If you want to calculate the probability of getting a head, there is only one outcome that is possible. So, you can say that there is a 1 in 2, or 1 out of 2, chance of getting a head. This is usually given as a **probability fraction**, namely $\frac{1}{2}$. So, you would write the event as:

$$P(\text{head}) = \frac{1}{2}$$

Probabilities can also be written as decimals or percentages, so that:

$$P(\text{head}) = \frac{1}{2} \text{ or } 0.5 \text{ or } 50\%$$

It is more usual to give probabilities as fractions in GCSE examinations but you will frequently come across probabilities given as percentages, for example, in the weather forecasts on TV.

The probability of an outcome is defined as:

$$P(\text{event}) = \frac{\text{number of ways the outcome can happen}}{\text{total number of possible outcomes}}$$

This definition always leads to a fraction, which should be cancelled to its simplest form.

Another probability term you will meet is at **random**. This means that the outcome cannot be predicted or affected by anyone.

EXAMPLE 2

A bag contains five red balls and three blue balls. A ball is taken out at random.

What is the probability that it is:

a red **b** blue **c** green?

Use the formula

$$P(\text{event}) = \frac{\text{number of ways the event can happen}}{\text{total number of possible outcomes}}$$

to work out these probabilities.

a There are five red balls out of a total of eight, so $P(\text{red}) = \frac{5}{8}$

b There are three blue balls out of a total of eight, so $P(\text{blue}) = \frac{3}{8}$

c There are no green balls, so this event is impossible: $P(\text{green}) = 0$

EXAMPLE 3

The spinner shown here is spun and the score on the side on which it lands is recorded.

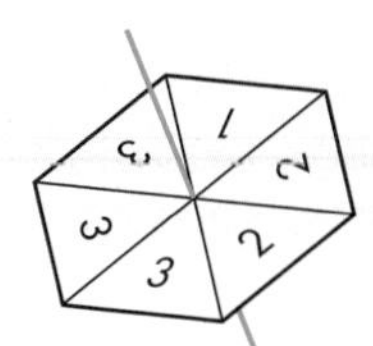

What is the probability that the score is:

a 2
b odd
c less than 5?

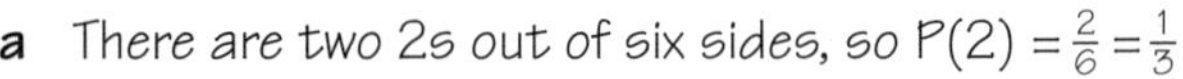

a There are two 2s out of six sides, so $P(2) = \frac{2}{6} = \frac{1}{3}$

b There are four odd numbers, so $P(\text{odd}) = \frac{4}{6} = \frac{2}{3}$

c All of the numbers are less than 5, so this is a certain event.

$P(\text{less than } 5) = 1$

EXAMPLE 4

Bernice is always early, just on time or late for work.

The probability that she is early is 0.1, the probability she is just on time is 0.5.

What is the probability that she is late?

As all the possibilities are covered – that is 'early', 'on time' and 'late' – the total probability is 1. So,

$P(\text{early}) + P(\text{on time}) = 0.1 + 0.5 = 0.6$

So, the probability of Bernice being late is $1 - 0.6 = 0.4$.

EXERCISE 11B

1 What is the probability of each of the following events?

- **a** Throwing a 2 with a fair, six-sided dice.
- **b** Throwing a 6 with a fair, six-sided dice.
- **c** Tossing a fair coin and getting a tail.
- **d** Drawing a Queen from a pack of cards.
- **e** Drawing a Heart from a pack of cards.
- **f** Drawing a black card from a pack of cards.
- **g** Throwing a 2 or a 6 with a fair, six-sided dice.
- **h** Drawing a black Queen from a pack of cards.
- **i** Drawing an Ace from a pack of cards.
- **j** Throwing a 7 with a fair, six-sided dice.

HINTS AND TIPS

If an event is impossible, just write the probability as 0, not as a fraction such as $\frac{0}{6}$. If it is certain, write the probability as 1, not as a fraction such as $\frac{6}{6}$.

HINTS AND TIPS

Remember to cancel the fractions if possible.

2 What is the probability of each of the following events?

- **a** Throwing an even number with a fair, six-sided dice.
- **b** Throwing a prime number with a fair, six-sided dice.
- **c** Getting a Heart or a Club from a pack of cards.
- **d** Drawing the King of Hearts from a pack of cards.
- **e** Drawing a picture card or an Ace from a pack of cards.
- **f** Drawing the seven of Diamonds from a pack of cards.

3 A bag contains only blue balls. If I take one out at random, what is the probability of each of these outcomes?

- **a** I get a black ball.
- **b** I get a blue ball.

4 Number cards with the numbers 1 to 10 inclusive are placed in a hat. Bob takes a number card out of the bag without looking. What is the probability that he draws:

- **a** the number 7
- **b** an even number
- **c** a number greater than 6
- **d** a number less than 3
- **e** a number between 3 and 8?

5 A bag contains one blue ball, one pink ball and one black ball. Craig takes a ball from the bag without looking. What is the probability that he takes out:

- **a** the blue ball
- **b** the pink ball
- **c** a ball that is not black?

HINTS AND TIPS

A ball that is not black must be pink or blue.

F

6 A pencil case contains six red pens and five blue pens. Geoff takes out a pen without looking at what it is. What is the probability that he takes out:

a a red pen **b** a blue pen **c** a pen that is not blue?

7 A bag contains 50 balls. 10 are green, 15 are red and the rest are white. Gemma takes a ball from the bag at random. What is the probability that she takes:

a a green ball **b** a white ball

c a ball that is not white **d** a ball that is green or white?

8 A box contains seven bags of cheese and onion crisps, two bags of beef crisps and six bags of plain crisps. Iklil takes out a bag of crisps at random. What is the probability that he gets:

a a bag of cheese and onion crisps **b** a bag of beef crisps

c a bag of crisps that are not cheese and onion **d** a bag of prawn cracker crisps

e a bag of crisps that is either plain or beef?

9 In a Christmas raffle, 2500 tickets are sold. One family has 50 tickets. What is the probability that that family wins the first prize?

E

10 Ashley, Bianca, Charles, Debbie and Eliza are in the same class. Their teacher wants two students to do a special job.

a Write down all the possible combinations of two people, for example, Ashley and Bianca, Ashley and Charles. (There are 10 combinations altogether).

b How many pairs give two boys?

c What is the probability of choosing two boys?

d How many pairs give a boy and a girl?

e What is the probability of choosing a boy and a girl?

f What is the probability of choosing two girls?

HINTS AND TIPS

Try to be systematic when writing out all the pairs.

11 In a sale at the supermarket, there is a box of 10 unlabelled tins. On the side it says: 4 tins of Creamed Rice and 6 tins of Chicken Soup. Mitesh buys this box. When he gets home he wants to have a lunch of chicken soup followed by creamed rice.

a What is the smallest number of tins he could open to get his lunch?

b What is the largest number of tins he could open to get his lunch?

c The first tin he opens is soup. What is the chance that the second tin he opens is:

i soup

ii rice?

E

12 What is the probability of each of the following events?

a Drawing a Jack from a pack of cards.

b Drawing a 10 from a pack of cards.

c Drawing a red card from a pack of cards.

d Drawing a 10 or a Jack from a pack of cards.

e Drawing a Jack or a red card from a pack of cards.

f Drawing a red Jack from a pack of cards.

13 A bag contains 25 coloured balls. 12 are red, 7 are blue and the rest are green. Martin takes a ball at random from the bag.

a Find:

i P(he takes a red)

ii P(he takes a blue)

iii P(he takes a green).

b Add together the three probabilities. What do you notice?

c Explain your answer to part **b**.

14 The weather tomorrow will be sunny, cloudy or raining.

If P(sunny) = 40%, P(cloudy) = 25%, what is P(raining)?

15 At morning break, Priya has a choice of coffee, tea or hot chocolate.

If P(she chooses coffee) = 0.3 and P(she chooses hot chocolate) = 0.2, what is P(she chooses tea)?

PS 16 The following information is known about the classes at Bradway School.

Year	Y1		Y2		Y3		Y4		Y5		Y6	
Class	**P**	**Q**	**R**	**S**	**T**	**U**	**W**	**X**	**Y**	**Z**	**K**	**L**
Girls	7	8	8	10	10	10	9	11	8	12	14	15
Boys	9	10	9	10	12	13	11	12	10	8	16	17

A class representative is chosen at random from each class.

Which class has the best chance of choosing a boy as the representative?

AU 17 The teacher chooses, at random, a student to ring the school bell.

Tom says: "It's even chances that the teacher chooses a boy or a girl."

Explain why Tom might not be correct.

11.3 Probability that an outcome of an event will not happen

This section will show you how to:

- calculate the probability of an outcome of an event not happening when you know the probability of the outcome happening

Key word

outcome

In some questions in Exercise 11B, you were asked for the probability of something *not* happening. For example, in question 5 you were asked for the probability of picking a ball that is *not* black. You could answer this because you knew how many balls were in the bag. However, sometimes you do not have this type of information.

The probability of throwing a six on a fair, six-sided dice is $P(6) = \frac{1}{6}$.

There are five **outcomes** that are not sixes: 1, 2, 3, 4, 5.

So, the probability of *not* throwing a six on a dice is:

$$P(\text{not a } 6) = \frac{5}{6}$$

Notice that:

$$P(6) = \frac{1}{6} \quad \text{and} \quad P(\text{not a } 6) = \frac{5}{6}$$

So,

$$P(6) + P(\text{not a } 6) = 1$$

If you know that $P(6) = \frac{1}{6}$, then P(not a 6) is:

$$1 - \frac{1}{6} = \frac{5}{6}$$

So, if you know P(outcome happening), then:

P(outcome not happening) = 1 – P(outcome happening)

EXAMPLE 5

What is the probability of not picking an Ace from a pack of cards?

First, find the probability of picking an Ace:

P(picking an ace from a pack of cards) $= \frac{4}{52} = \frac{1}{13}$

Therefore:

P(not picking an ace from a pack of cards) $= 1 - \frac{1}{13} = \frac{12}{13}$

EXERCISE 11C

E

1 **a** The probability of winning a prize in a raffle is $\frac{1}{20}$. What is the probability of not winning a prize in the raffle?

b The probability that snow will fall during the Christmas holidays is 45%. What is the probability that it will not snow?

c The probability that Paddy wins a game of chess is 0.7 and the probability that he draws the game is 0.1. What is the probability that he loses the game?

2 Mary picks a card from a pack of well-shuffled playing cards.

Find the probability that she picks:

a **i** a picture card **ii** a card that is not a picture

b **i** a Club **ii** not a Club

c **i** an Ace or a King **ii** neither an Ace nor a King.

3 The following letter cards are put into a bag.

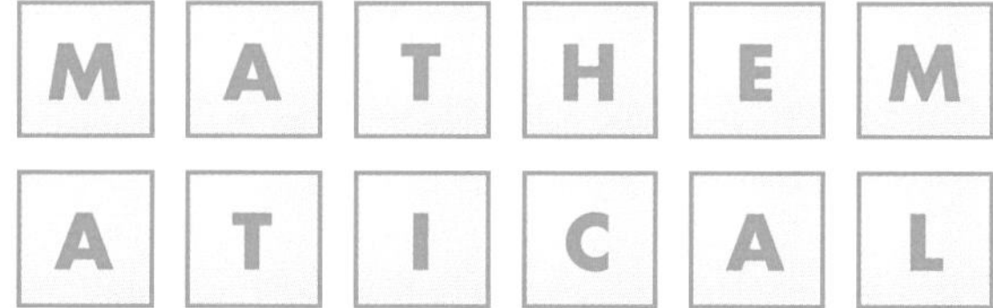

a Steve takes a letter card at random.

i What is the probability he takes a letter A?

ii What is the probability he does not take a letter A?

b Richard picks an M and keeps it. Sue now takes a letter from those remaining.

i What is the probability she takes a letter A?

ii What is the probability she does not take a letter A?

FM **4** The starting point in a board game is:

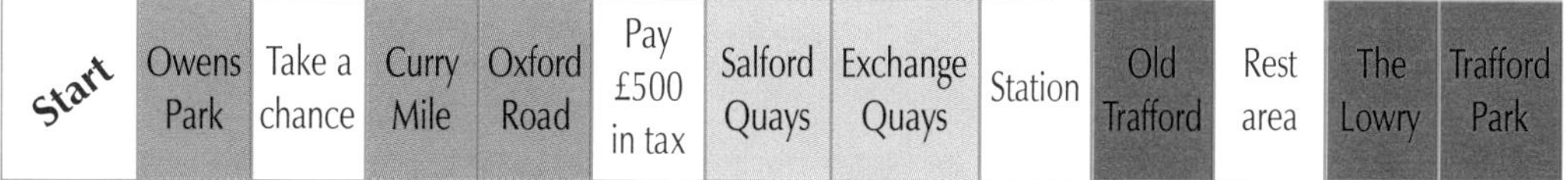

You roll a single dice and move, from the start, the number of places shown by the dice.

What is the probability of *not* landing on:

a a brown square

b the station

c a coloured square?

E

PS 5 Freddie and Taryn are playing a board game. On the next turn Freddie will go to jail if he rolls an even number. Taryn will go to jail if she rolls a 5 or a 6.

Who has the better chance of *not* going to jail on the next turn?

AU 6 Hamzah is told: "The chance of your winning this game is 0.3."

Hamzah says: "So I have a chance of 0.7 of losing."

Explain why Hamzah might be wrong.

GRADE BOOSTER

G You can understand basic terms such as 'certain', 'impossible', 'likely'

F You can understand that the probability scale runs from 0 to 1 and can calculate the probability of events

E You can calculate the probability of an outcome not happening if you know the probability of it happening

What you should know now

- How to use the probability scale and estimate the likelihood of outcomes of events depending on their position on the scale
- How to calculate theoretical probabilities from different situations

EXAMINATION QUESTIONS

1 Some bulbs were planted in October. The ticks in the table shows the months in which each type of bulb flowers.

Type of bulb		Month					
		Jan	Feb	March	April	May	June
	Allium					✓	✓
	Crocus	✓	✓				
	Daffodil		✓	✓	✓		
	Iris	✓	✓				
	Tulip				✓	✓	

a In which months do tulips flower?

b Which type of bulb flowers in March?

c In which month do most types of bulb flower?

d Which type of bulb flowers in the same months as the iris?

Ben puts one of each type of these bulbs in a bag. He takes a bulb from the bag without looking.

e **i** Write down the probability that he will take a crocus bulb.

ii Copy the probability scale and mark with a cross (X) the probability that he will take a bulb which flowers in February.

0 ——————————————— 1

2 Here are some statements.

Draw an arrow from each statement to the word which best describes its likelihood.

One has been done for you. (3)

A head is obtained when a fair coin is thrown once.	Certain
A number less than 7 will be scored when an ordinary six-sided dice is rolled once.	Likely
It will rain every day for a week next July in London.	Even
A red disc is obtained when a disc is taken at random from a bag containing 9 red discs and 2 blue discs.	Unlikely
	Impossible

(Total 3 marks)

Edexcel, June 2008, Paper 5 Foundation, Unit 1 Test, Question 2

3 There are 3 red pens, 4 blue pens and 5 black pens in a box.

Sameena takes a pen, at random, from the box.

Write down the probability that she takes a black pen. (2)

(Total 2 marks)

Edexcel, June 2008, Paper 2 Foundation, Question 18

4 A bag contains some beads which are red or green or blue or yellow.

The table shows the number of beads of each colour.

Colour	Red	Green	Blue	Yellow
Number of beads	3	2	5	2

Samire takes a bead at random from the bag. Write down the probability that she takes a blue bead.

Edexcel, Question 1, Paper 12A Intermediate, March 2005

5 **a** Alice has a spinner that has five equal sections. The numbers 1, 2 and 3 are written on the spinner.

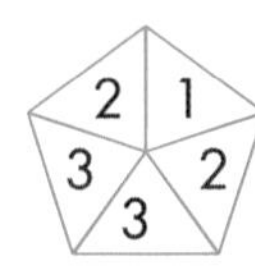

Alice spins the spinner once.
On what number is the spinner least likely to land?

b

There are three numbers.
There are two number 2s.
The chance of getting a 2 is $\frac{2}{3}$.

Alice thinks that the chance of getting a 2 is $\frac{2}{3}$. Explain why Alice is wrong.

6 Fifty people take a maths exam. The table shows the results.

	Pass	Fail
Male	12	16
Female	9	13

a A person is chosen at random from the group.
What is the probability that the person is male?

b A person is chosen at random from the group.
What is the probability that the person passed the test?

7 Doris has a bag in which there are nine counters, all green. Alex has a bag in which there are 15 counters, all red. Jade has a bag in which there are some blue counters.

a What is the probability of picking a red counter from Doris's bag?

b Doris and Alex put all their counters into a box.
What is the probability of choosing a green counter from the box?

c Jade now adds her blue counters to the box.
The probability of choosing a blue counter from the box is now $\frac{1}{3}$.
How many blue counters does Jade put in the box?

8 Marco has a 4-sided spinner.

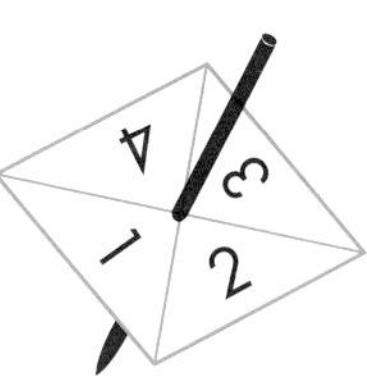

The sides of the spinner are numbered 1, 2, 3 and 4.
The spinner is biased.
The table shows the probability that the spinner will land on each of the numbers 1, 2 and 3.

Number	1	2	3	4
Probability	0.20	0.35	0.20	

Work out the probability that the spinner will land on the number 4. (2)

(Total 2 marks)

Edexcel, June 2008, Paper 5 Foundation, Unit 1 Test, Question 5

9 Here are the ages, in years, of 15 teachers.

35	52	42	27	36
23	31	41	50	34
44	28	45	45	53

a Draw an ordered stem-and-leaf diagram to show this information.
You must include a key. (3)

Key:

One of these teachers is picked at random.

b Work out the probability that this teacher is more than 40 years old. (2)

(Total 5 marks)

Edexcel, May 2008, Paper 1 Foundation, Question 22

Worked Examination Questions

1 Simon has a bag containing blue, green, red, yellow and white marbles.

a Complete the table to show the probability of each colour being chosen at random.

b Which colour of marble is most likely to be chosen at random?

c Calculate the probability that a marble chosen at random is blue or white.

Colour of marble	Probability
Blue	0.3
Green	0.2
Red	0.15
Yellow	
White	0.1

a 0.25

2 marks

The probabilities in the table should add up to 1 so,

$0.3 + 0.2 + 0.15 + 0.1 + \mathbf{?} = 1 \qquad 0.75 + \mathbf{?} = 1$

This calculation gets you 1 method mark. Using the calculation, the missing probability is 0.25. This answer gets you 1 mark.

b Blue marble

1 mark

The blue marble is most likely as it has the largest probability of being chosen. The correct answer is worth 1 mark.

c 0.4

2 marks

The word 'or' means you add probabilities of separate events.
The calculation P(blue or white) = P(blue) + P(white) = 0.3 + 0.1 = 0.4 is worth 1 method mark and the correct answer is worth 1 mark.

Total: 5 marks

FM **2** In a raffle 400 tickets have been sold. There is only one prize.
Mr Raza buys 5 tickets for himself and sells another 40.
Mrs Raza buys 10 tickets for herself and sells another 50.
Mrs Hewes buys 8 tickets for herself and sells just 12 others.

a What is the probability of:

i Mr Raza winning the raffle

ii Mr Raza selling the winning ticket

iii Mr Raza either winning the raffle or selling the winning ticket?

b What is the probability of either Mr or Mrs Raza selling the winning ticket?

c What is the probability of Mrs Hewes not winning the lottery?

a i $\frac{5}{400}$

You get 1 mark for the correct answer.

ii $\frac{40}{400}$

You get 1 mark for the correct answer.

iii $\frac{(5 + 40)}{400} = \frac{45}{400}$ 4 marks

You get 1 method mark for writing out the combined probability. You also get 1 accuracy mark for expressing this as a number out of 400.

b $\frac{(40 + 50)}{400} = \frac{90}{400}$ 2 marks

You get 1 method mark for writing out the combined probability. You also get 1 accuracy mark for expressing this as a number out of 400.

c $1 - \frac{8}{400} = \frac{392}{400}$ 2 marks

You get 1 method mark for subtracting from 1. You also get 1 mark for the final answer.

Total: 8 marks

Worked Examination Questions

PS **3** Here are two fair spinners.

The spinners are spun.

The two numbers are added together.

What is the probability of getting a score of an even number?

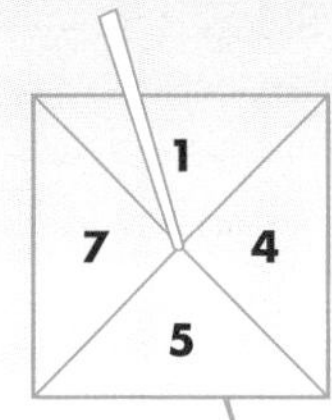

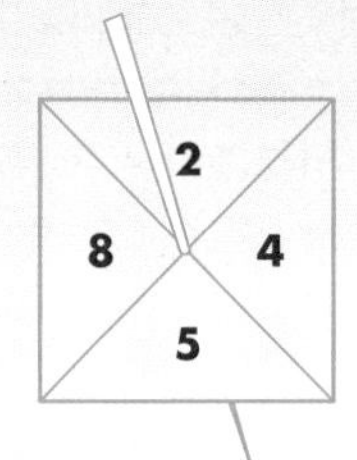

		Spinner A			
		1	**4**	**5**	**7**
Spinner B	**2**	3	6	7	9
	4	5	8	9	11
	5	6	9	10	12
	8	9	12	13	15

You get 1 mark for creating a table or, otherwise, for showing all the possible totals in order to solve the problem.

6

Counting how many even numbers are in the totals will get you 1 mark.

16

Counting how many numbers are in the table altogether will get you 1 mark.

$\frac{6}{16}$

Realising that the probability of getting an even number is $\frac{6}{16}$ will get you 1 mark.

Total: 4 marks

AU **4** Ellie cuts a pack of cards 100 times to see how many times she gets a heart.

She got 25 Hearts and her mother told her, "That's exactly what you would expect to get."

Explain why her mother is not quite correct.

Although the theoretical probability of cutting a Heart is $\frac{1}{4}$ and $\frac{1}{4}$ multiplied by 100 does give an expectation of 25, it would be unusual to cut exactly 25 Hearts. It is more likely that she would get a number close to 25, either just below it or just above it.

You get 1 mark for an explanation that clearly shows this understanding.

Total: 1 mark

11 Functional Maths
Fairground games

Joe had a stall at the local fair and wanted to make a reasonable profit from the game below.

In this game, the player rolls two balls down the sloping board and wins a prize if they land in slots that total more than seven.

Joe wants to know how much he should charge for each go and what the price should be. He would also like to know how much profit he is likely to make.

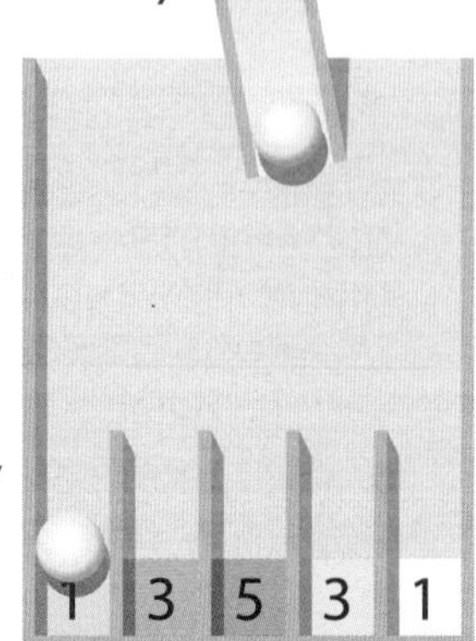

Getting started

Practise calculating probabilities using the spinner and questions below.

- What is the probability of spinning the spinner and getting:
  - a one
  - a five
  - a two
  - a number other than five?

  Give your answer as a fraction and as a decimal.
- If you spun the spinner 20 times, how many times would you expect to get:
  - a five
  - not five
  - an odd number?
- If you spun the spinner 100 times, how many times would you expect to get:
  - a two
  - not a two
  - a prime number?

Now, think about which probabilities you must calculate, in order to help Joe, and to design your own profitable game.

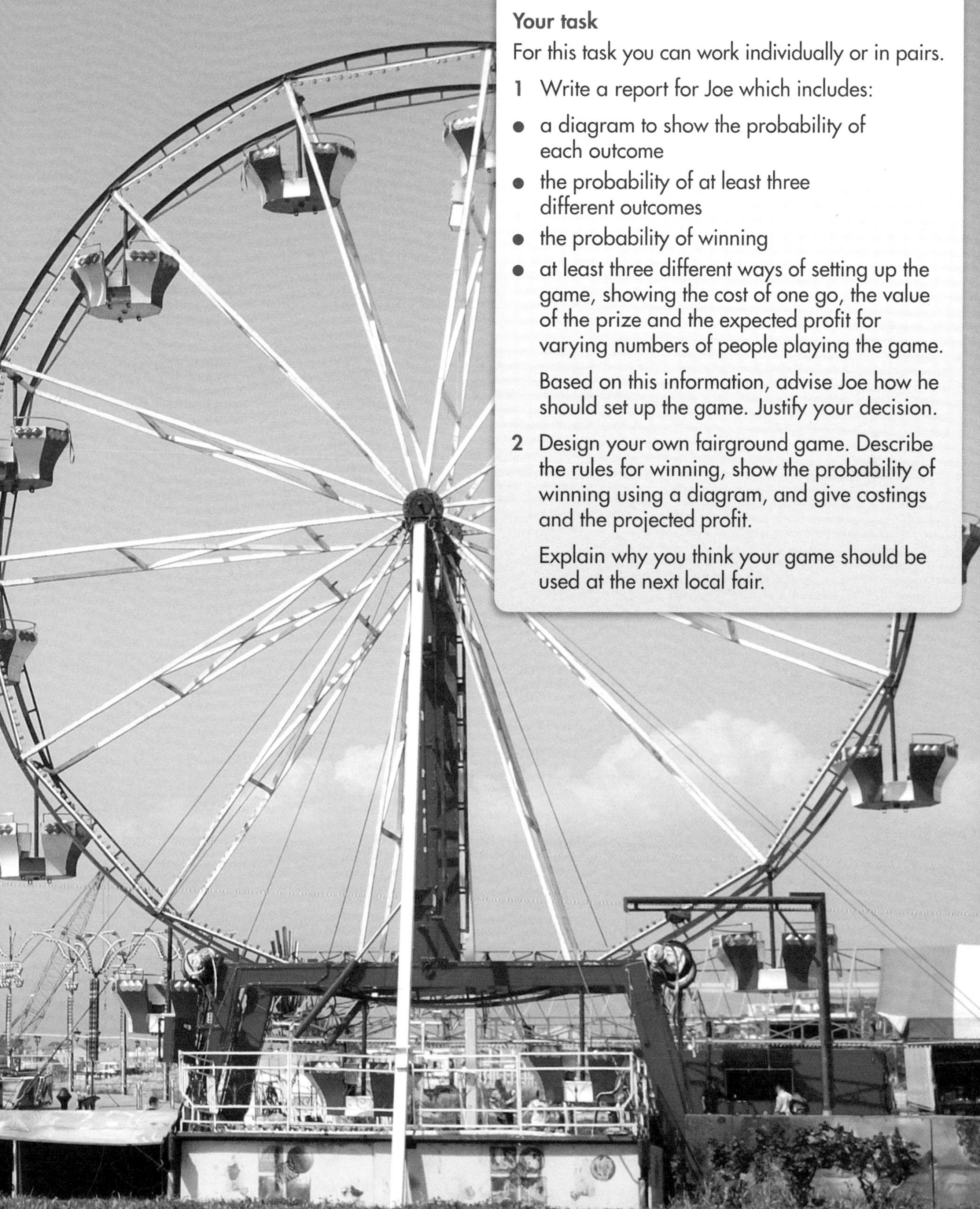

Your task

For this task you can work individually or in pairs.

1 Write a report for Joe which includes:

- a diagram to show the probability of each outcome
- the probability of at least three different outcomes
- the probability of winning
- at least three different ways of setting up the game, showing the cost of one go, the value of the prize and the expected profit for varying numbers of people playing the game.

Based on this information, advise Joe how he should set up the game. Justify your decision.

2 Design your own fairground game. Describe the rules for winning, show the probability of winning using a diagram, and give costings and the projected profit.

Explain why you think your game should be used at the next local fair.

Why this chapter matters

You see transformations every day, whether it is a reflection in a mirror or a miniature version of an object. 2D shapes can also be transformed. The activity below will give you a chance to try a transformation yourself.

How many sides does a strip of paper have?

Take a strip of paper, about 20 cm by 2 cm.

How many sides does it have? Easy! You can see that this has two sides, a top and an underside. If you were to draw a line along one side of the strip, you would have one side with a line 20 cm long on it and one side blank.

Now mark the ends A and B and put a single twist in the strip of paper and tape (or glue) the two ends together as shown.

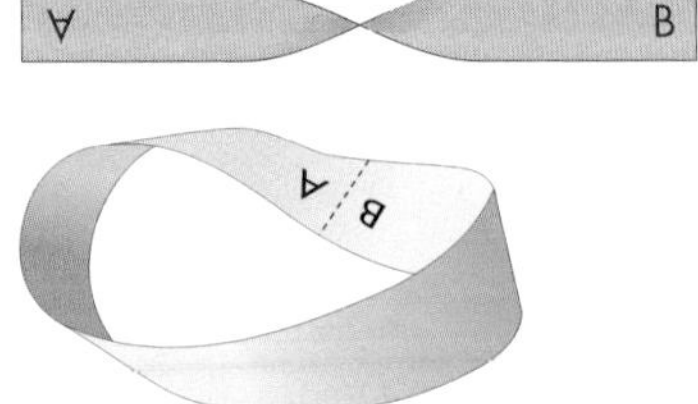

How many sides does this strip of paper have now?

Take a pen and draw a line on the paper, starting at any point you like. Continue the line along the length of the paper – you will eventually come back to your starting point. Your strip has only one side now. There is no blank side.

You have transformed a two-sided piece of paper into a one-sided piece of paper.

This curious shape is called a Möbius strip. It is named after August Ferdinand Möbius, a 19th century German mathematician and astronomer. Möbius, along with others, caused a revolution in geometry.

Möbius strips have a number of surprising applications that exploit this remarkable property of one-sidedness, including conveyer belts in industry as well as in domestic vacuum cleaners.

The Möbius strip has become the universal symbol of recycling. The symbol was created in 1970 by Gary Anderson, who was a senior at the University of Southern California, as part of a contest sponsored by a paper company.

The Möbius strip is one form of transformation. In this chapter, you will look at some other transformations of shapes.

This conveyor belt is used in salt mining.

Chapter

12 Geometry: Transformations 1

The grades given in this chapter are target grades.

1 Congruent shapes

2 Tessellations

This chapter will show you …

F how to recognise congruent shapes

E how 2D shapes tessellate

Visual overview

What you should already know

- How to find the lines of symmetry of a 2D shape **(KS3 level 3, GCSE grade F)**
- How to find the order of rotational symmetry of a 2D shape **(KS3 level 4, GCSE grade F)**
- How to find the equation of a line **(KS3 level 6, GCSE grade E)**

Quick check

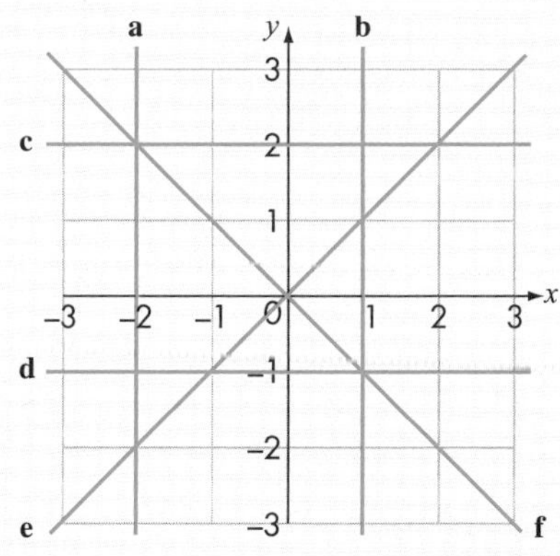

Write down the equations of the lines drawn on the grid.

12.1 Congruent shapes

This section will show you how to:

- recognise congruent shapes

Key word

congruent

Two-dimensional shapes that are exactly the *same* size and shape are said to be **congruent**. For example, although they are in different positions, the triangles below are congruent, because they are all exactly the same size and shape.

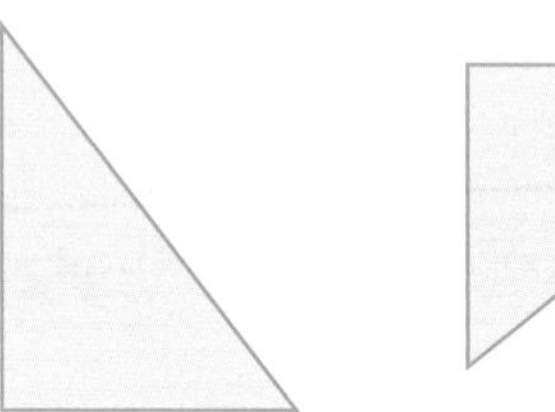
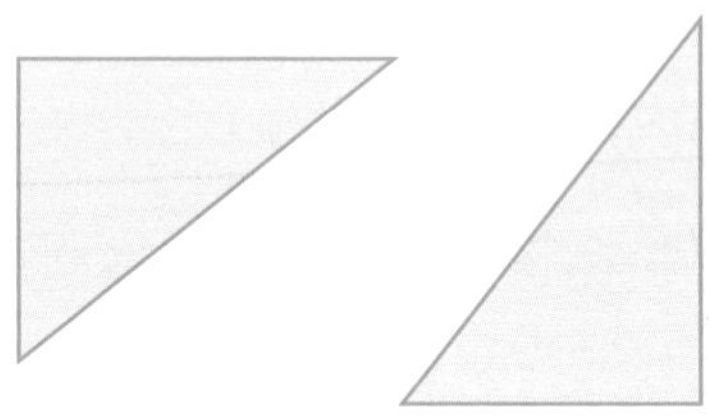
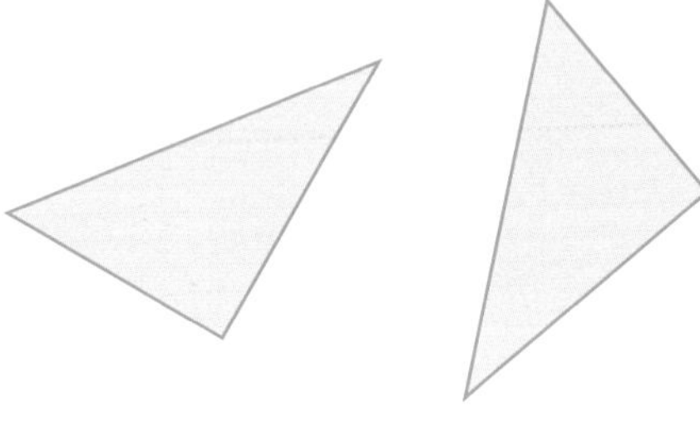

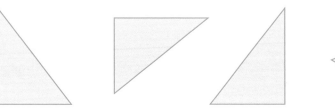

Congruent shapes fit exactly on top of each other. So, one way to see whether shapes are congruent is to trace one of them and check that it covers the other shapes exactly. For some of the shapes, you may have to turn your tracing paper over.

EXAMPLE 1

Which of these shapes is not congruent to the others?

a

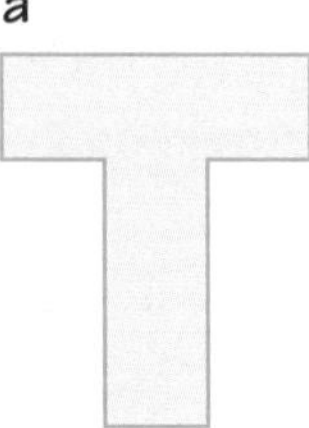

b

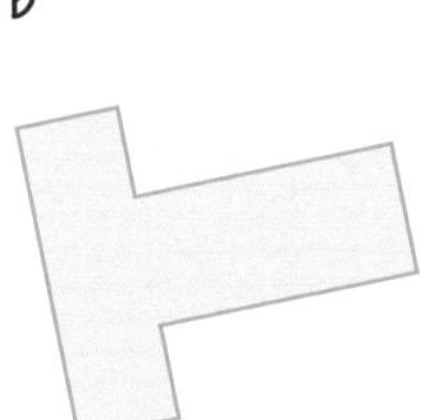

c

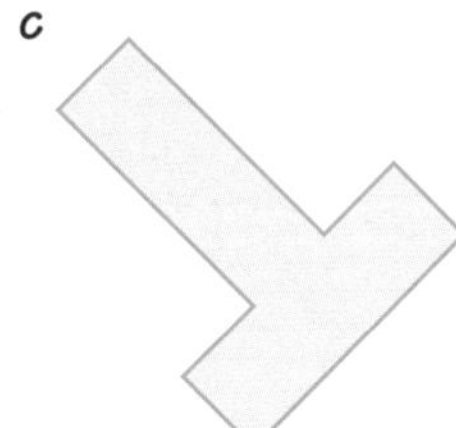

d

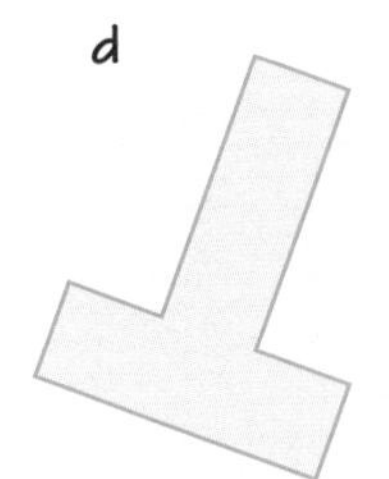

Trace shape **a** and check whether it fits exactly on top of the others.

You should find that shape **b** is not congruent to the others.

FM Functional Maths **AU** (AO2) Assessing Understanding **PS** (AO3) Problem Solving

EXERCISE 12A

1 State whether the shapes in each pair, **a** to **f** are congruent or not.

a

b

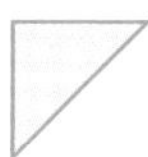

c

d

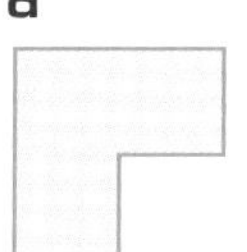

e

f

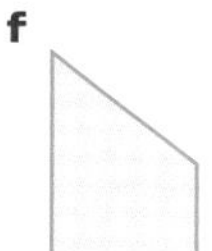

2 Which figure in each group, **a** to **c**, is not congruent to the other two?

a **i**

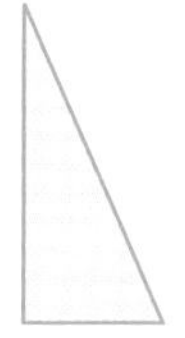

ii

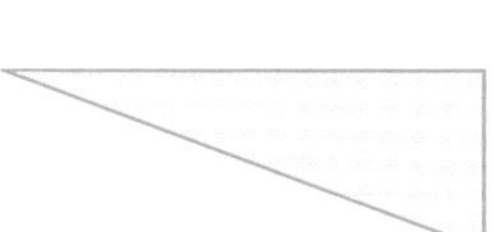

iii

b **i**

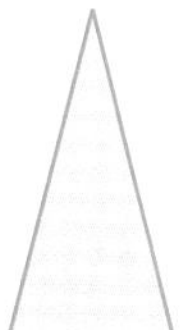

ii

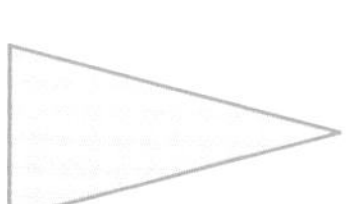

iii

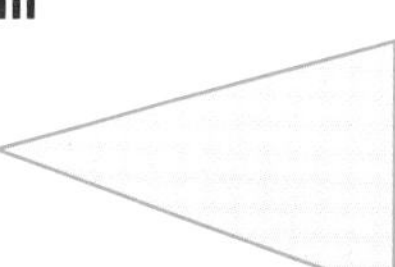

c **i**

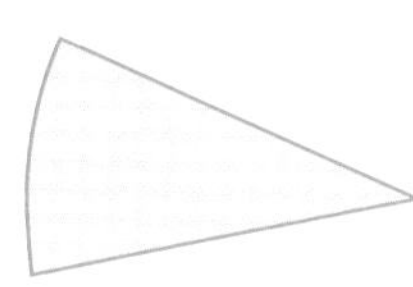

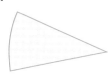

ii

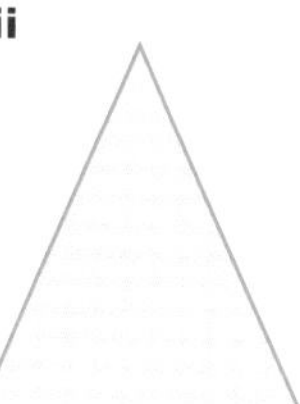

iii

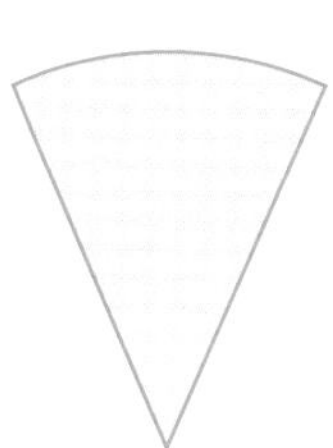

3 For each of the following sets of shapes, write down the numbers of the shapes that are congruent to each other.

a

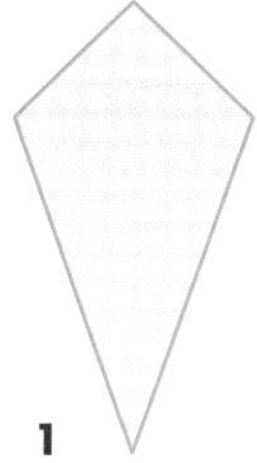

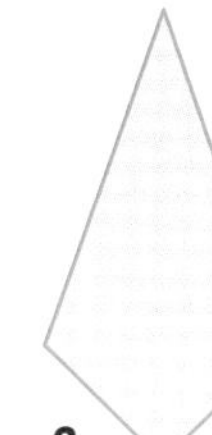

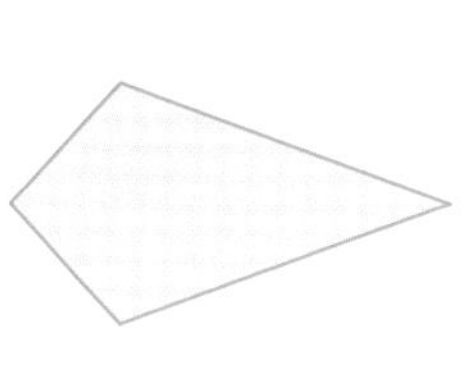

1 2 3 4

G

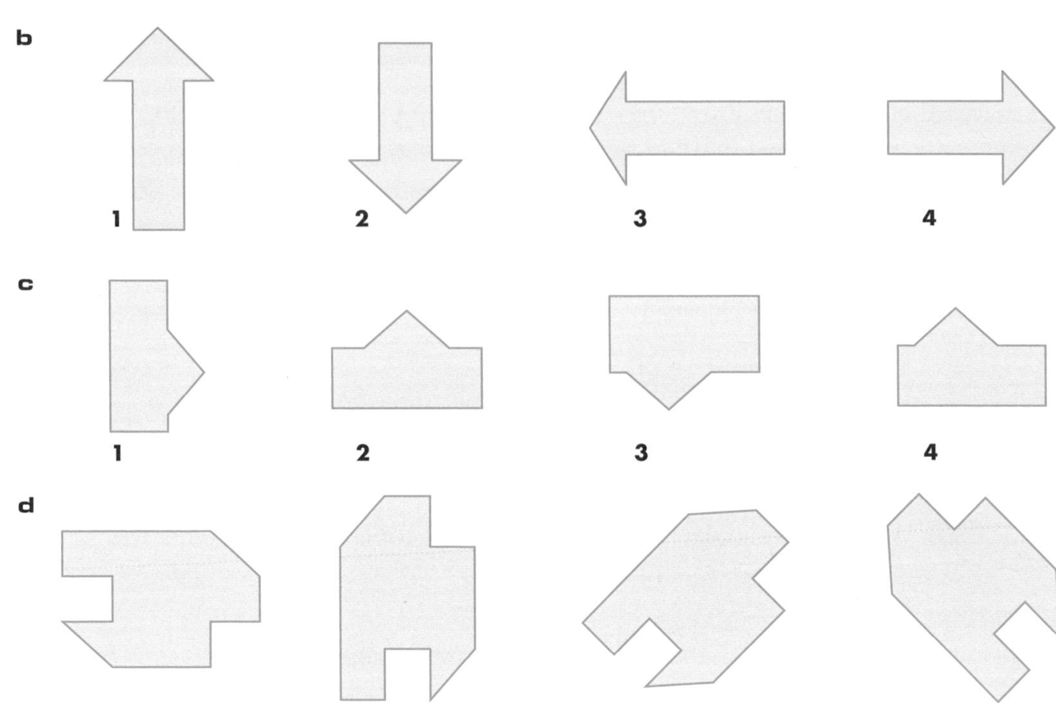

PS 4 There are three letters of the alphabet that are all congruent to each other.

What are they?

AU 5 Billy said, "I have two congruent shapes on my T-shirt, but one is bigger than the other."

What is wrong with Billy's statement?

E

6 Draw a square PQRS. Draw in the diagonals PR and QS. Which triangles are congruent to each other?

7 Draw a rectangle EFGH. Draw in the diagonals EG and FH. Which triangles are congruent to each other?

8 Draw a parallelogram ABCD. Draw in the diagonals AC and BD. Which triangles are congruent to each other?

9 Draw an isosceles triangle ABC where AB = AC. Draw the line from A to the midpoint of BC. Which triangles are congruent to each other?

PS 10 A chessboard is made up of 64 small squares. The area of each square is 1 cm^2.

a How many squares on the board are congruent to a square:

i 2 cm by 2 cm **ii** 3 cm by 3 cm **iii** 4 cm by 4 cm?

b How many different-sized squares are there altogether on the chessboard?

12.2 Tessellations

This section will show you how to:
- tessellate a 2D shape

Key words
tessellate
tessellation

ACTIVITY

Tiling patterns

You need centimetre-squared paper and some card in several different colours.

Make a template for each of the following shapes on the centimetre-squared paper.

Use your template to make about 20 card tiles for each shape, using different colours.

For each shape, put all the tiles together to create a tiling pattern without any gaps.

What do you find?

From this activity you should have found that you could cover as much space as you wanted, using the *same* shape in a repeating pattern. You can say that the shape **tessellates**.

So, a **tessellation** is a regular pattern made with identical plane shapes, which fit together exactly, without overlapping and leaving no gaps.

EXAMPLE 2

Draw tessellations using each of these shapes.

a 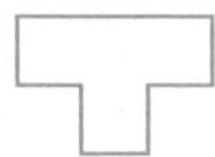

b

c 

These patterns show how each of the shapes tessellates.

a 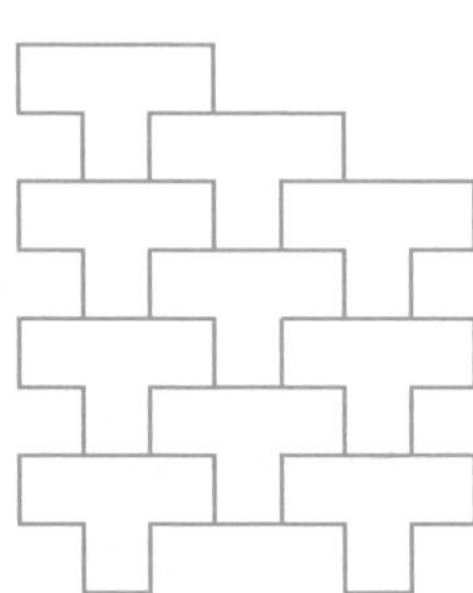

b 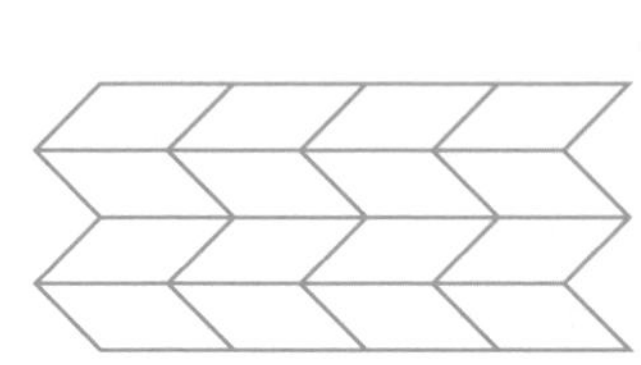

c 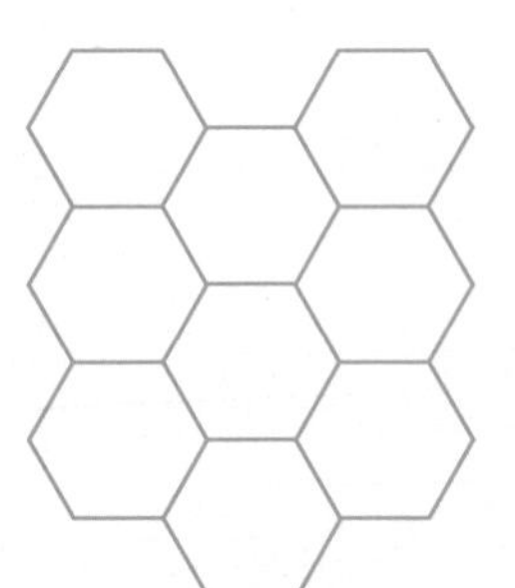

EXERCISE 12B

1 On squared paper, show how each of these shapes tessellates.

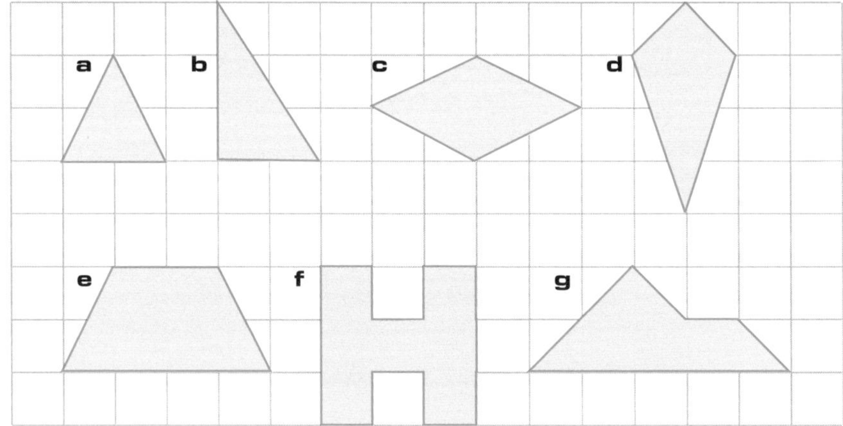

2 Invent some of your own tessellating patterns.

PS 3 'Every quadrilateral will form a tessellation.' Investigate this statement to see whether it is true.

AU 4 Explain why equilateral triangles, squares and regular hexagons tessellate.

PS 5 Tania says, "It's impossible to tessellate a five-sided shape."

Investigate this statement to see if you think it may be true.

D

AU **6** A semi-regular tessellation is made, using two basic shapes.

Explain how this is possible if one of the shapes is a circle.

FM **7** A brick wall is an example of a tessellation.

Bricklayers sometimes like to make interesting patterns in their brickwork.

Using a brick that is three times as long as it is high, sketch a new tessellation that would also give a strong design.

GRADE BOOSTER

G You can recognise congruent shapes

E You know how to tessellate a 2D shape

What you should know now

- How to recognise congruent shapes
- How to tessellate a 2D shape

1 The grid shows six shapes A, B, C, D, E and F.

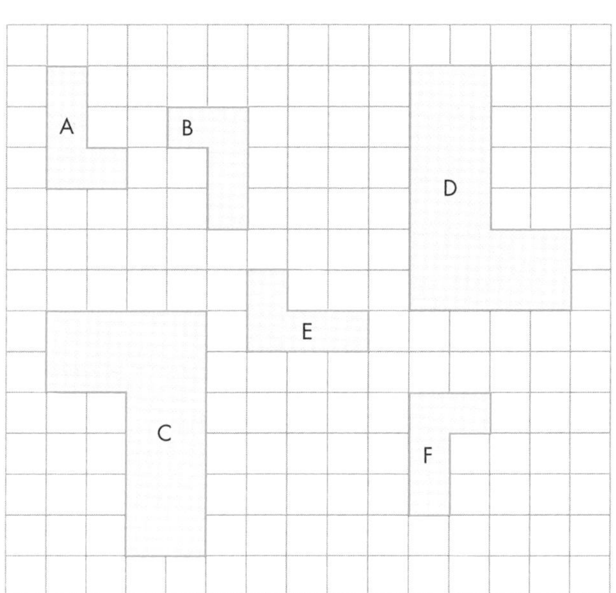

Write down the letters of the shapes that are congruent to shape A.

2 Here are some triangles.

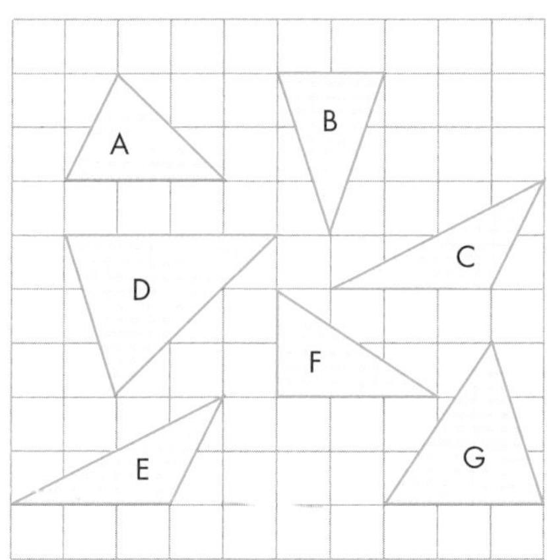

a Write down the letter of the triangle that is

i right-angled

ii isosceles (2)

Two of the triangles are congruent.

b Write down the letters of these two triangles. (1)

(Total 3 marks)

Edexcel, June 2008, Paper 13 Foundation, Question 1

3 Here are 8 shapes.

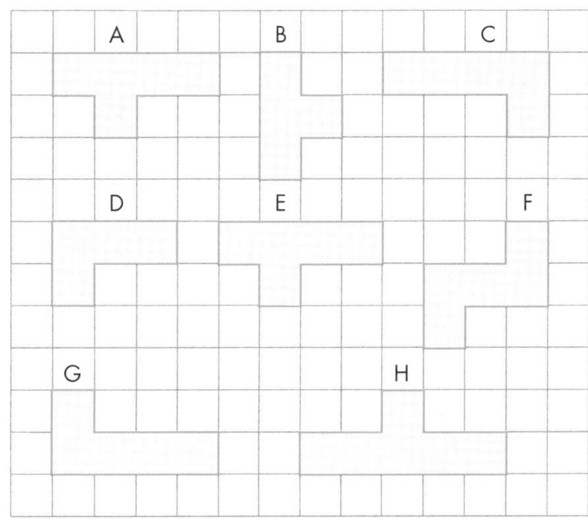

a Write down the letters of two **different** pairs of congruent shapes. (2)

b On the grid, show how the shaded shape will tessellate.

You must draw at least 6 shapes. (2)

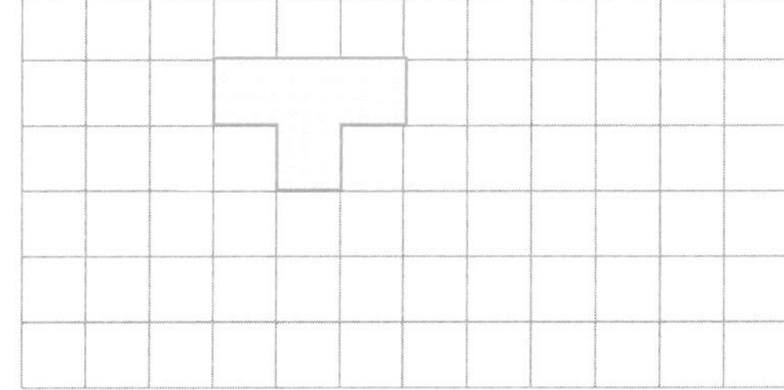

(Total 4 marks)

Edexcel, November 2008, Paper 12 Foundation, Question 5

Worked Examination Questions

1 The grid shows several transformations of the shaded triangle.

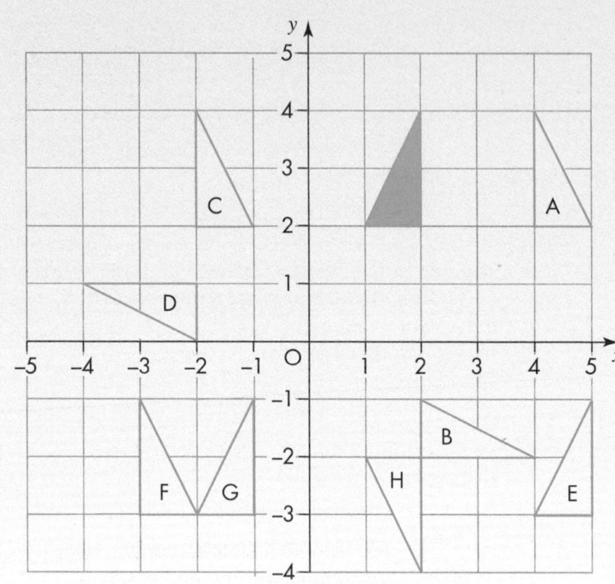

a Write down the letter of the triangle:

i after the shaded triangle is reflected in the line $x = 3$

ii after the shaded triangle is translated by the vector $\begin{pmatrix} 3 \\ -5 \end{pmatrix}$

iii after the shaded triangle is rotated 90° clockwise about O.

b Describe fully the single transformation that takes triangle F onto triangle G.

a i A

ii E

iii B

3 marks

b A reflection in the line $x = -2$.

2 marks

Total: 5 marks

$x = 3$ is the vertical line passing through $x = 3$ on the x-axis.
The correct answer receives 1 independent mark.

Move the triangle 3 squares to the right and 5 squares down.
The correct answer receives 1 independent mark.

Use tracing paper to help you. Trace the shaded triangle, pivot the paper on O holding it in place with your pencil point, and rotate the paper through 90° clockwise.
The correct answer receives 1 independent mark.

The vertical mirror line passes through $x = -2$ on the x-axis.
You get 1 method mark of identifying the reflection and 1 accuracy mark for the mirror line.

Worked Examination Questions

FM **2** Dan went to the local gardening store and saw different paving stones. He looked at some octagonal ones that he thought he would use to pave his back garden.

When he got them back home he found they didn't tessellate.

a Explain why the regular octagon does not tessellate by itself.

b What can he now do to pave his back garden all over?

a Regular octagons placed next to each other will leave gaps but tessellations do not have gaps. The interior angle of the regular octagon is 135° and 360 is not a multiple of 135.

You get 1 mark for stating that there will be gaps or for stating that 360 is not a multiple of 135, or equivalent wording.

1 mark

b The gap in between the regular octagons is in fact a square, so he should buy some square paving stones to fill the gap.

You get 1 mark for indicating that the gap is a square.

Note: Diagrams could have been drawn here; as long as these are interpreted with some words, the marks would be awarded.

1 mark

Total: 2 marks

PS **3** Find the single transformation that is equivalent to a rotation of 90° clockwise about the origin, followed by a reflection in the line $y = x$.

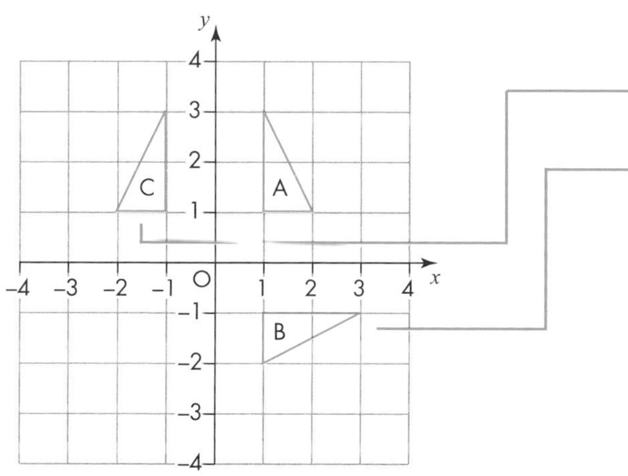

Start with a simple shape on the grid, triangle A.

You get 1 accuracy mark for correctly reflecting triangle B to C.

You get 1 accuracy mark for correctly rotating triangle A to B.

Single transformation is a reflection in the y-axis.

You get 1 method mark for identifying this is a reflection and 1 accuracy mark for identifying the mirror line as the y-axis.

Total: 4 marks

12 Problem Solving
Tessellations

Tessellations are used in art.

M C Escher used repeating patterns and tessellations to create many of his designs, as you can see below. You too can create a design based on tessellations.

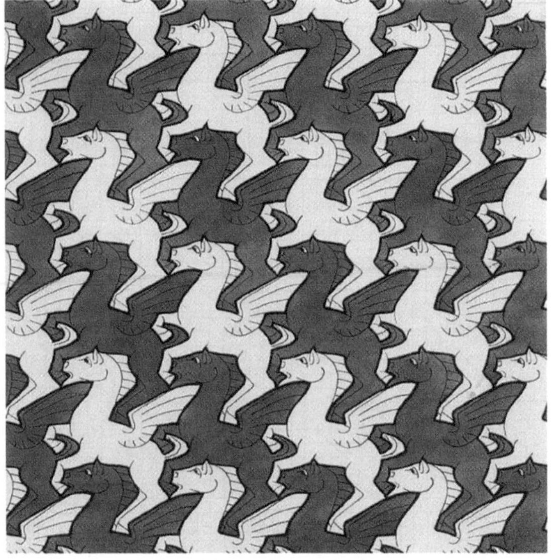

Getting started

Answer the questions below:

- What are the four transformations? Give an example of each transformation.
- What is rotational symmetry?
- Draw a shape that has rotational symmetry of order two.
- What polygon has rotational symmetry of order four?

Your task

In this task you are going to link tessellations with transformations, and then look at creating a design using tessellating shapes.

Fact: Every triangle and quadrilateral tessellates

1. Explain why this is true (using diagrams may be helpful).
2. Using triangles and quadrilaterals, create your own tessellating design. Use the guidance given in the box below to help you.
3. Explain the transformations used in your tessellating design.

Creating a new tessellation

Step 1: Draw any triangle.

Step 2: On any one of the sides, draw a design that has rotational symmetry.

Step 3: Repeat step 2 on the other two sides of the triangle.

By replacing each side of the original triangle with your symmetrical designs, you now have a shape that will tessellate.

Handy hints

- Use a card template to help create a tessellating pattern.
- When you draw your tessellations you need to be fairly accurate or it will not tessellate. Remember, the line you draw will be bigger than the template.
- Esher created many tessellating designs. You can find a number of these on the internet.

Why this chapter matters

For anything, from a bridge to a landscape gardening project, the designer needs to construct plans accurately, to be sure that everything will fit together properly. This will also give the people putting it together a blueprint to work from.

The need for accurate drawings is clear in bridge construction. Bridge engineers are responsible for producing practical bridge designs to meet the requirements of their employers. For example, a bridge intended to carry traffic over a newly constructed railway needs to be strong enough to bear the weight of the traffic and stable enough to counteract the effects of the moving traffic and strong winds. The designers produce a blueprint that has all the measurements, including heights, weights and angles, clearly marked on it. Construction workers then use this blueprint to build the bridge to the exact specifications set by the designers and engineers.

Generally, the construction workers work on both ends of the bridge at the same time, meeting in the middle. The blueprints are therefore essential for making sure that the bridge is safe and that the bridge meets in the middle.

Accurately drawn blueprints were essential in the construction of the Golden Gate Bridge, which crosses the San Francisco Bay. When it was constructed in the 1930s, it was the longest suspension bridge in the world. The bridge engineers had to draw precise blueprints to make sure that they had all the information necessary to build this innovative bridge and that it would be built correctly.

By contrast, a bridge built at a stadium for the Maccabiah Games in Israel was built without proper planning and without accurate blueprints. This led to the bridge collapsing soon after its construction in 1997, killing four athletes and injuring 64 people.

Just like a bridge engineer, you must be accurate in your construction, working with a freshly sharpened pencil and a good pair of compasses, measuring and drawing angles carefully and drawing construction lines as faintly as possible.

Chapter

13 Geometry and measures: Constructions

The grades given in this chapter are target grades.

1 Constructing triangles

This chapter will show you …

D how to construct accurate triangles

Visual overview

Constructing triangles

What you should already know

- The names of common 3-D shapes **(KS3 level 3, GCSE grade G)**
- How to measure lengths of lines **(KS3 level 3, GCSE grade G)**
- How to measure angles with a protractor **(KS3 level 4, GCSE grade F)**

Quick check

1 Measure the following lines.

a

b

c

2 Measure the following angles.

a

b

13.1 Constructing triangles

This section will show you how to:
- construct triangles, using compasses, a protractor and a straight edge

Key words
angle
compasses
construct
side

There are three ways of **constructing** a triangle. Which one you use depends on what information you are given about the triangle.

All three sides known

EXAMPLE 1

Construct a triangle with **sides** that are 5 cm, 4 cm and 6 cm long.

- **Step 1:** Draw the longest side as the base. In this case, the base will be 6 cm, which you draw using a ruler. (The diagrams in this example are drawn at half-size.)

- **Step 2:** Deal with the second longest side, in this case the 5 cm side. Open the **compasses** to a radius of 5 cm (the length of the side), place the point on one end of the 6 cm line and draw a short faint arc, as shown here.

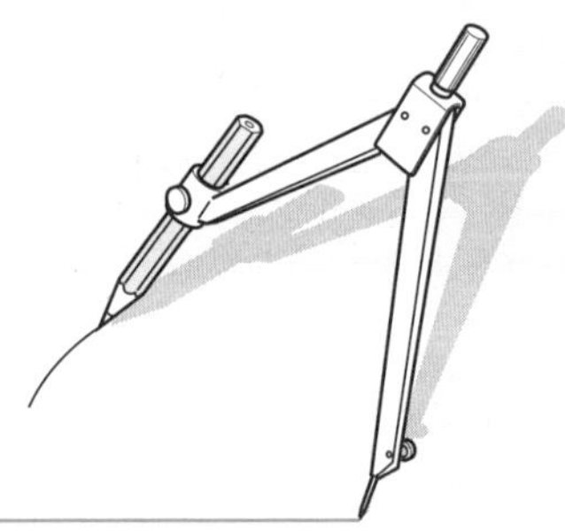

- **Step 3:** Deal with the shortest side, in this case the 4 cm side. Open the compasses to a radius of 4 cm, place the point on the other end of the 6 cm line and draw a second short faint arc to intersect the first arc, as shown here.

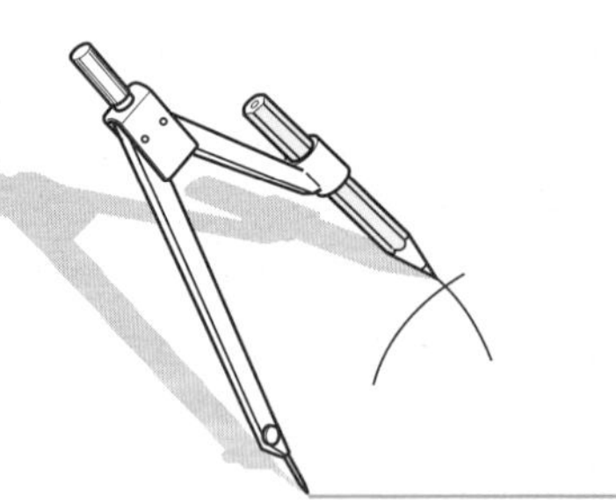

- **Step 4:** Complete the triangle by joining each end of the base line to the point where the two arcs intersect.

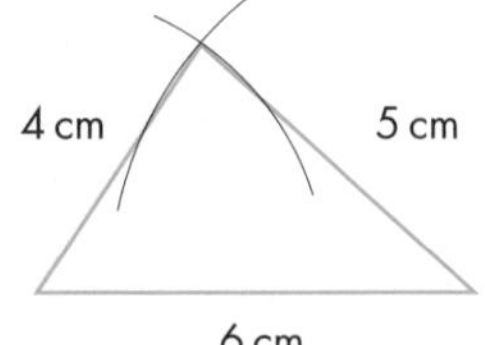

Note: The arcs are construction lines and so are always drawn lightly. They *must* be left in an answer to an examination question to show the examiner how you constructed the triangle.

Two sides and the included angle known

EXAMPLE 2

Draw a triangle ABC, where AB is 6 cm, BC is 5 cm and the included **angle** ABC is 55°. (The diagrams in this example are drawn at half-size.)

- **Step 1:** Draw the longest side, AB, as the base. Label the ends of the base A and B.

- **Step 2:** Place the protractor along AB with its centre on B and make a point on the diagram at the 55° mark.

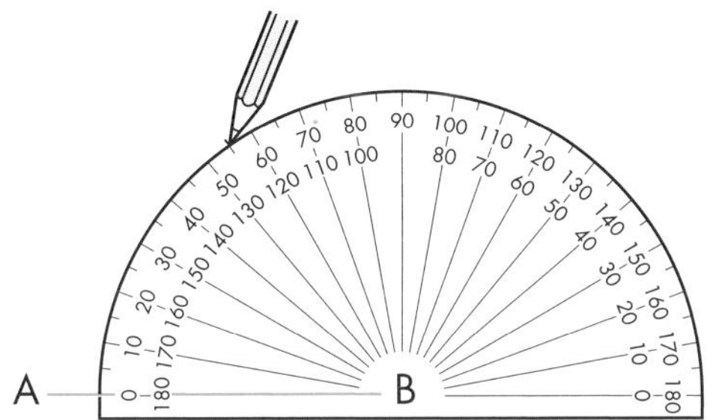

- **Step 3:** Draw a *faint* line from B through the 55° point. From B, using a pair of compasses, measure 5 cm along this line.
- Label the point where the arc cuts the line as C.

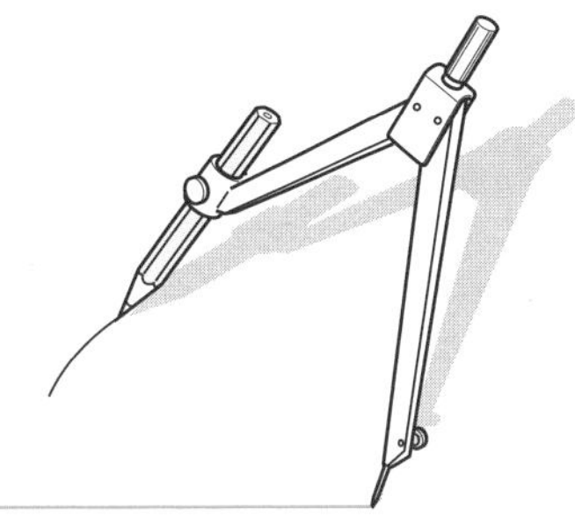

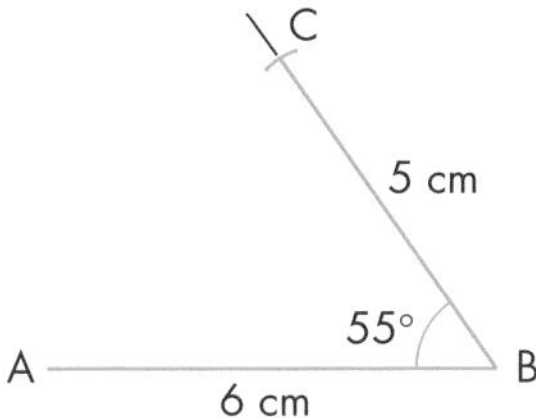

- **Step 4:** Join A and C and make AC and CB into bolder lines.

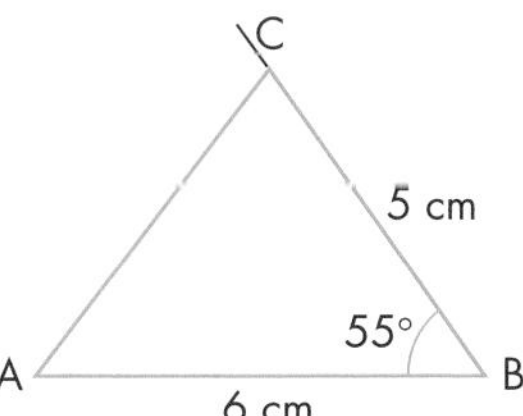

Note: The construction lines are drawn lightly and left in to demonstrate how the triangle has been constructed.

Two angles and a side known

When you know two angles of a triangle, you also know the third.

EXAMPLE 3

Draw a triangle ABC, where AB is 7 cm, angle BAC is 40° and angle ABC is 65°.

- **Step 1:** As before, start by drawing the base, which here has to be 7 cm. Label the ends A and B.

- **Step 2:** Centre the protractor on A and mark the angle of 40°. Draw a faint line from A through this point.

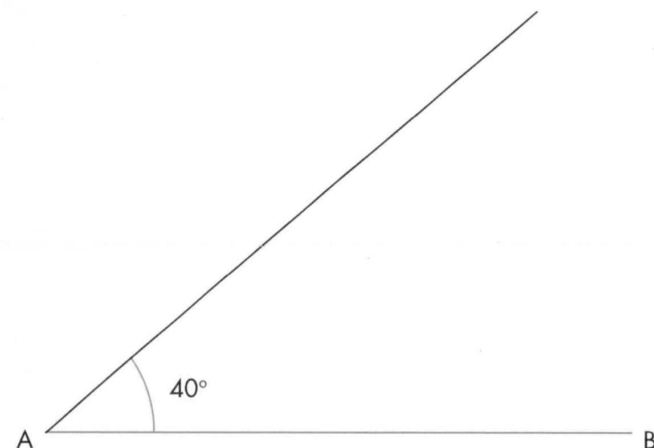

- **Step 3:** Centre the protractor on B and mark the angle of 65°. Draw a faint line from B through this point, to intersect the 40° line drawn from A. Label the point of intersection as C.

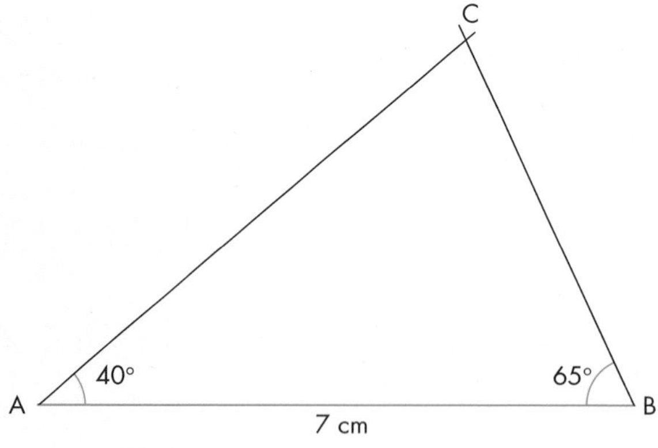

HINTS AND TIPS

You can write angle ABC quickly as ∠ABC, and angle BAC as ∠BAC.

- **Step 4:** Complete the triangle by making AC and BC into bolder lines.

EXERCISE 13A

1 Draw the following triangles accurately and measure the sides and angles not given in the diagram.

> **HINTS AND TIPS**
>
> Always make a sketch if one is not given in the question.

a

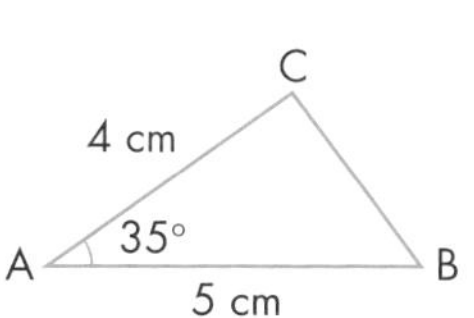

b

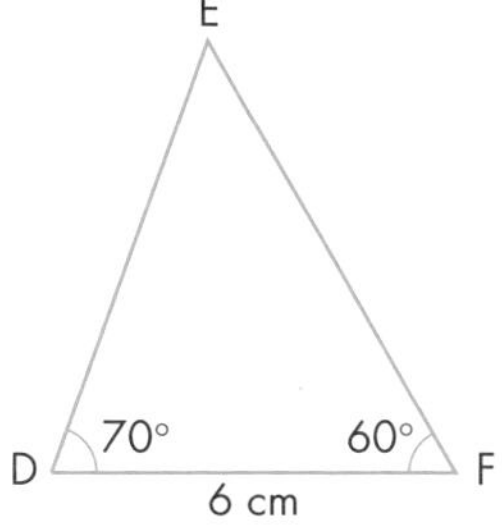

c

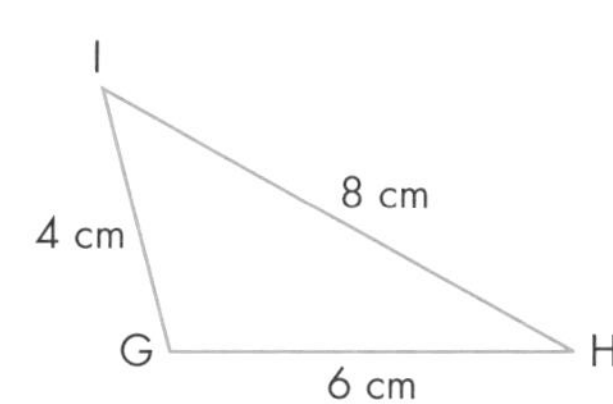

d

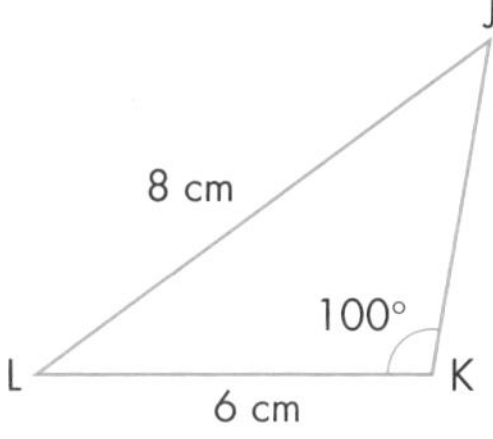

e

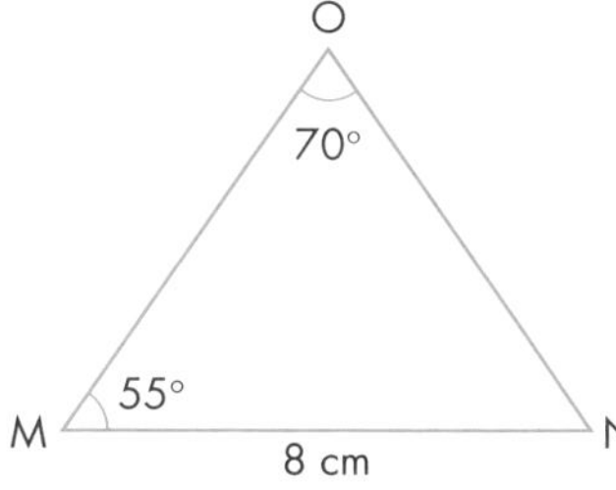

f

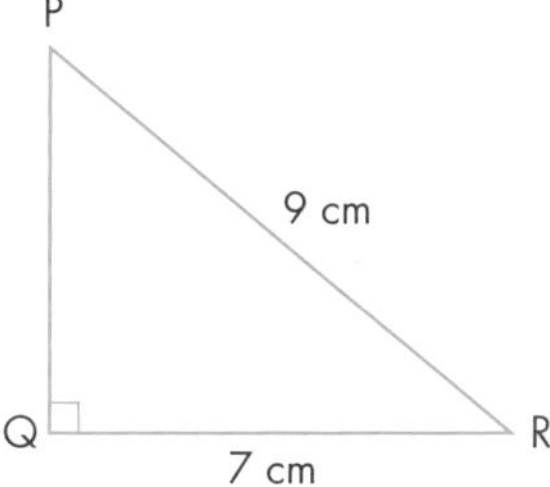

2 **a** Draw a triangle ABC, where AB = 7 cm, BC = 6 cm and AC = 5 cm.

b Measure the sizes of ∠ABC, ∠BCA and ∠CAB.

> **HINTS AND TIPS**
>
> Sketch the triangle first.

3 Draw an isosceles triangle that has two sides of length 7 cm and the included angle of 50°.

a Measure the length of the base of the triangle.

b What is the area of the triangle?

4 A triangle ABC has ∠ABC = 30°, AB = 6 cm and AC = 4 cm. There are two different triangles that can be drawn from this information.

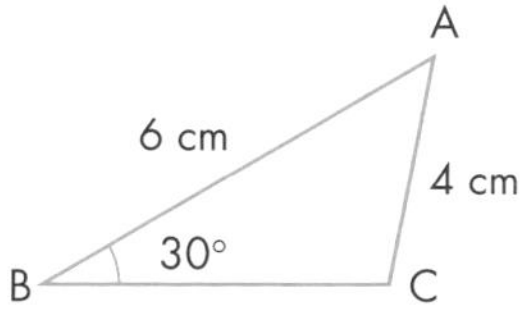

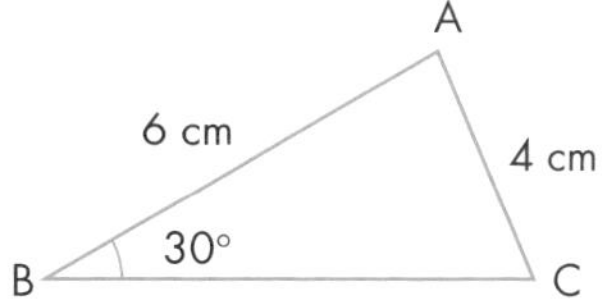

What are the two different lengths that BC can be?

D

5 Construct an equilateral triangle of side length 5 cm.

a Measure the height of the triangle. **b** What is the area of this triangle?

6 Construct a parallelogram with sides of length 5 cm and 8 cm and with an angle of 120° between them.

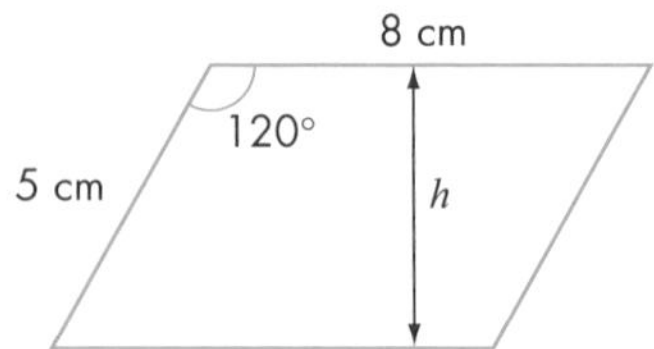

a Measure the height of the parallelogram.

b What is the area of the parallelogram?

7 Groundsmen painting white lines on a sports field may use a knotted rope, like this one.

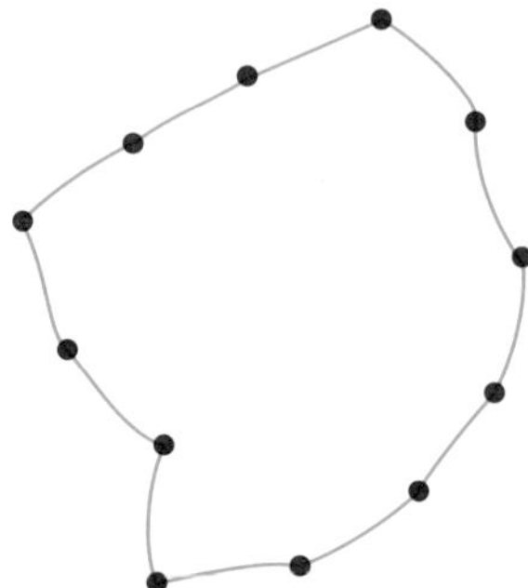

It has 12 equally spaced knots. It can be laid out to give a triangle, like this.

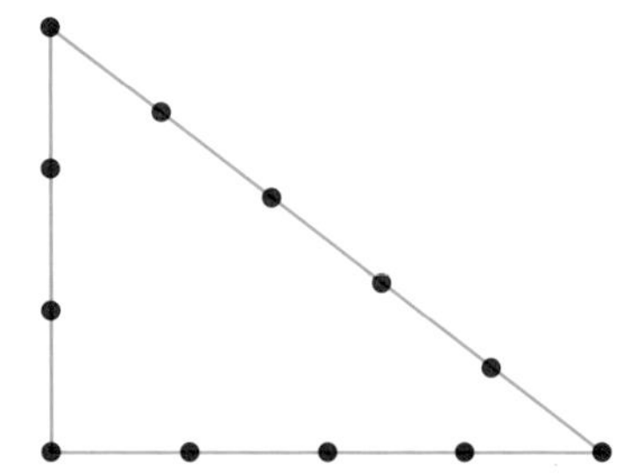

This will always be a right-angled triangle. This helps the groundsmen to draw lines perpendicular to each other.

Here are two more examples of such ropes.

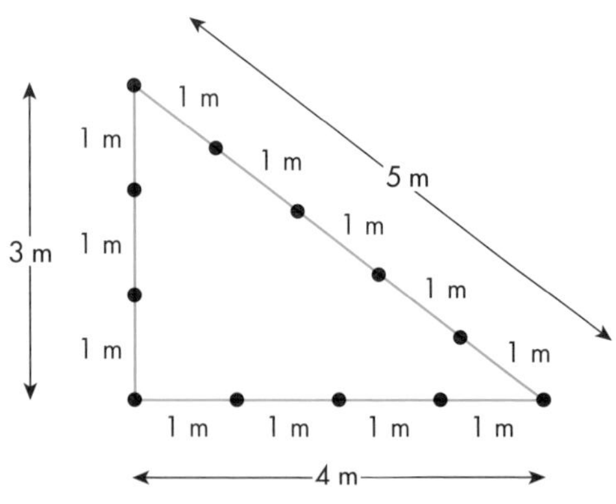

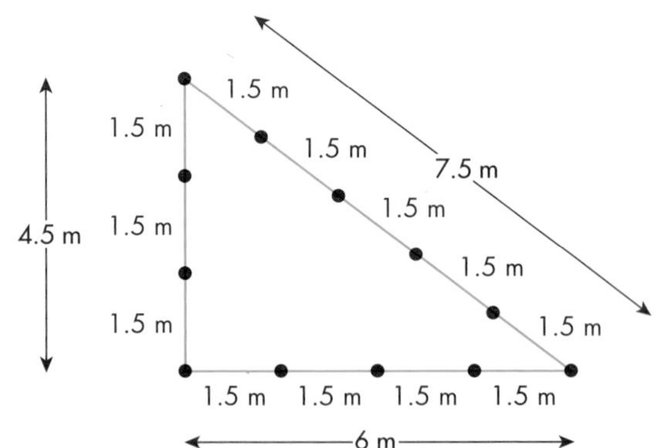

a Show, by constructing each of the above triangles (use a scale of 1 cm ≡ 1 m), that each is a right-angled triangle.

b Choose a different triangle that you think might also be right-angled. Use the same knotted-rope idea to check.

AU 8 Gabriel says that, as long as he knows all three angles of a triangle, he can draw it. Explain why Gabriel is wrong.

PS 9 Construct the triangle with the largest area, which has a total perimeter of 12 cm.

GRADE BOOSTER

D You can construct triangles accurately, using compasses, a protractor and a straight edge

What you should know now

- How to draw scale diagrams and construct accurate diagrams, using mathematical instruments

! Please note: The functional maths activity for 'Constructions' can be found in Book 2.

1 The diagram shows a sketch of triangle ABC.

AB = 5.6 cm.

Angle A = 43°.

Angle B = 108°.

Make an accurate drawing of triangle ABC.

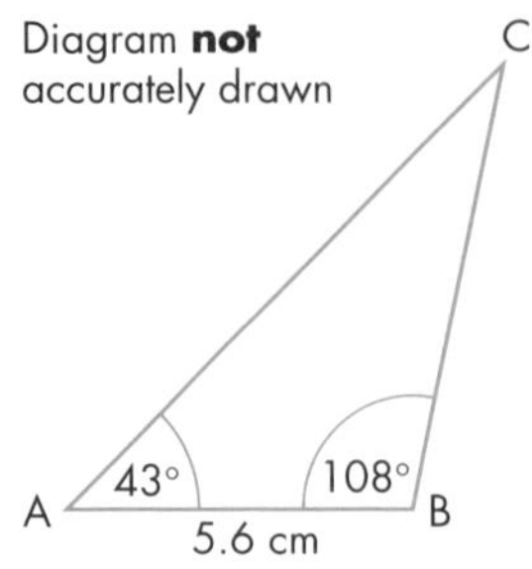

(Total 2 marks)

Edexcel, March 2005, Paper 9A Intermediate, Question 3

2 Use ruler and compasses to construct an equilateral triangle with sides of length 6 centimetres.

You **must** show all your construction lines. (2)

(Total 2 marks)

Edexcel, May 2008, Paper 1 Foundation, Question 27

3 The diagram shows a sketch of triangle *ABC*.

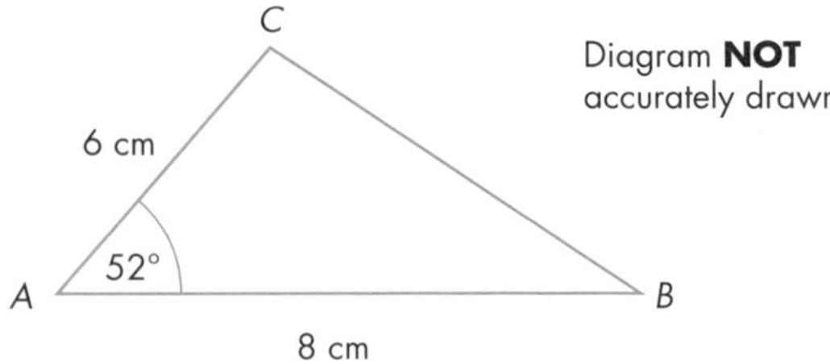

AB = 8 cm.

AC = 6 cm.

Angle *A* = 52°.

Make an accurate drawing of triangle *ABC*. (2)

(Total 2 marks)

Edexcel, November 2008, Paper 13 Foundation, Question 4

Worked Examination Questions

1 Here is a sketch of a triangle. PR = 6.4 cm, QR = 7.7 cm and angle R = 35°.

a Make an accurate drawing of the triangle.

b Measure the size of angle Q on your drawing.

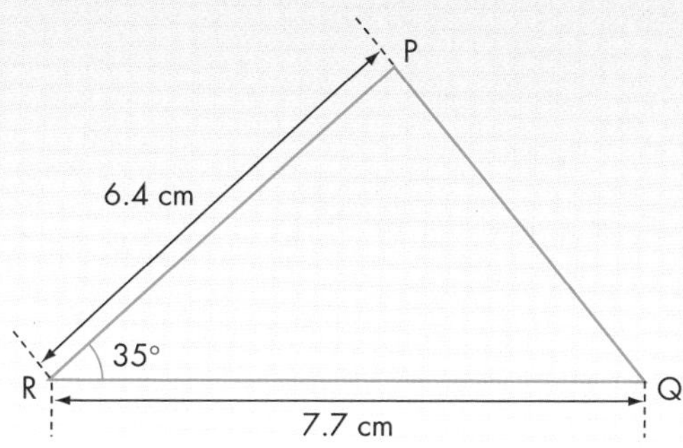

Make an accurate drawing, using these steps.

a **Step 2:** Measure the angle at R as 35°. Draw a faint line at this angle.

You get 1 accuracy mark for creating an angle between 34° and 35°.

Step 3: Using a pair of compasses, draw an arc 6.4 cm long from R. Where this crosses the line from Step 2, make this P.

You get 1 accuracy mark for this arc being within 1 mm.

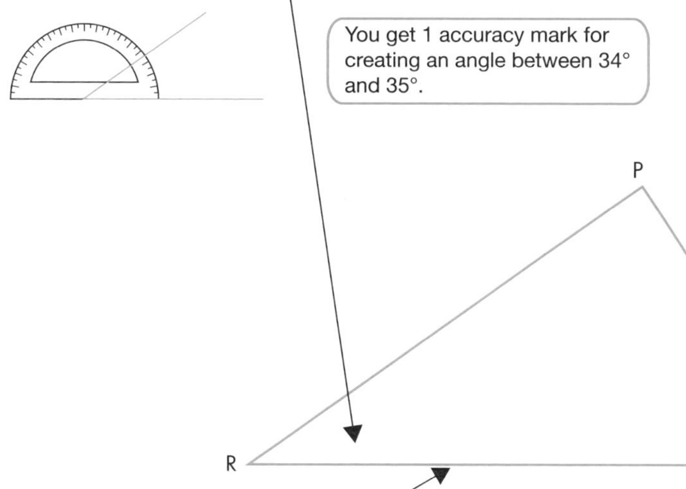

Step 4: Join P to Q.

No marks are awarded for this step.

Step 1: Draw the base as a line 7.7 cm long. You can draw this and measure it with a ruler, although a pair of compasses is more accurate.

No marks are awarded for this step.

2 marks

b 56°

You get 1 accuracy mark for measuring this angle as 56°, ± 2°. If your actual angle was not 56° but you measured it accurately, then you would still get the mark.

1 mark

Total: 3 marks

Why this chapter matters

Most jobs will require you to use some mathematics on a day-to-day basis. Being competent in all your number skills will therefore help you to be more successful in your job.

The mathematics used in the world of work will range from simple mental arithmetic calculations such as addition, subtraction, multiplication and division, to more complex calculations involving calculators, negative numbers and approximation. It will be up to you to select the mathematics that you need to carry out your job. Understanding essential mathematical techniques and how to apply these to a real-life context will make this much easier.

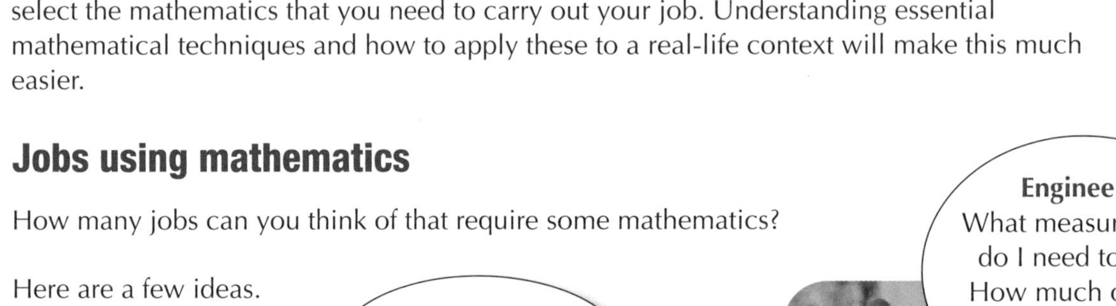

Jobs using mathematics

How many jobs can you think of that require some mathematics?

Here are a few ideas.

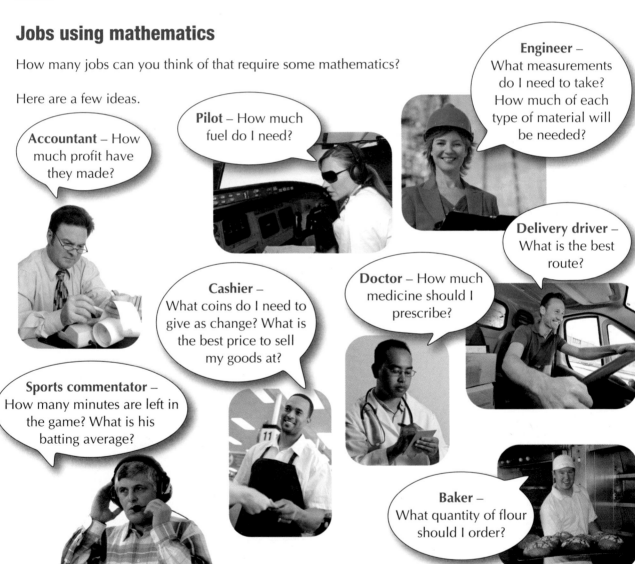

Engineer – What measurements do I need to take? How much of each type of material will be needed?

Pilot – How much fuel do I need?

Accountant – How much profit have they made?

Delivery driver – What is the best route?

Doctor – How much medicine should I prescribe?

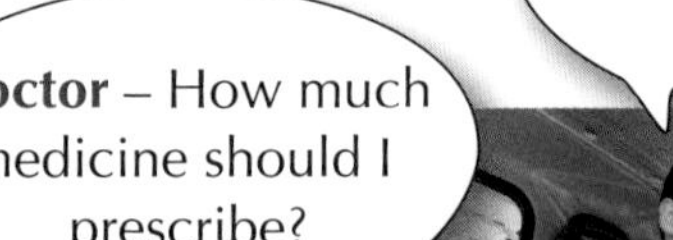

Cashier – What coins do I need to give as change? What is the best price to sell my goods at?

Sports commentator – How many minutes are left in the game? What is his batting average?

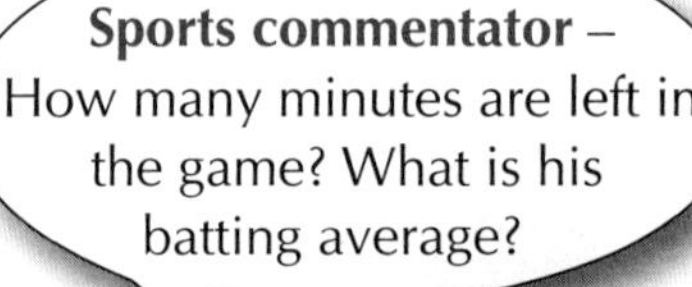

Baker – What quantity of flour should I order?

If you already have a career in mind that you would like to do, think of the questions you will need to ask to carry out your job successfully and the mathematics you will require to find the answers to those questions.

Chapter 14 Number: Further number skills

The grades given in this chapter are target grades.

1 Long multiplication

2 Long division

3 Arithmetic with decimal numbers

4 Arithmetic with fractions

5 Multiplying and dividing with negative numbers

6 Approximation of calculations

This chapter will show you ...

- **F** a reminder of the ways you can multiply a three-digit number by a two-digit number
- **F** a reminder of long division
- **E** how to calculate with decimal numbers
- **E** how to convert between decimals and fractions
- **E** how to multiply and divide negative numbers
- **C** how to use decimal places and significant figures to make approximations
- **C** sensible rounding methods

Visual overview

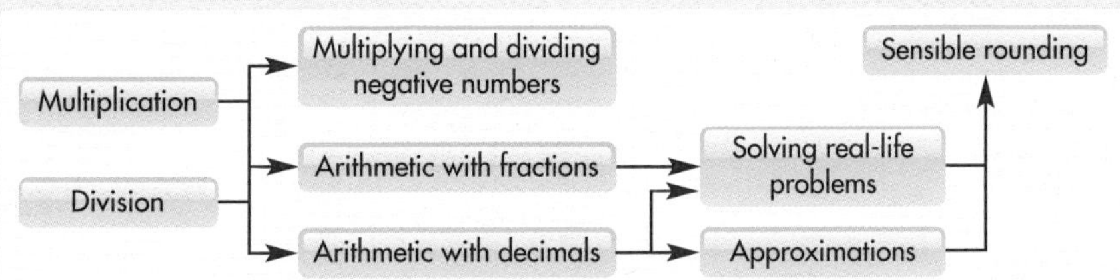

What you should already know

- Multiplication tables up to 10 × 10 **(KS3 level 4, GCSE grade G)**
- How to simplify fractions **(KS3 level 5, GCSE grade G)**

Quick check

1 Write down the first five multiples of 6.

2 Write down the first five multiples of 8.

3 Write down a number that is both a multiple of 3 and a multiple of 5.

4 Write down the smallest number that is a multiple of 4 and a multiple of 5.

5 Write down the smallest number that is a multiple of 4 and a multiple of 6.

6 Simplify the following fractions.

a $\frac{8}{10}$ **b** $\frac{5}{20}$ **c** $\frac{4}{16}$ **d** $\frac{32}{100}$ **e** $\frac{36}{100}$ **f** $\frac{16}{24}$ **g** $\frac{16}{50}$

14.1 Long multiplication

This section will show you how to:

- multiply a three-digit number (e.g. 358) by a two-digit number (e.g. 74) using
  - the grid method (or box method)
  - the column method (or traditional method)
  - the partition method

Key words
column method (or traditional method)
grid method (or box method)
partition method

When you are asked to do long multiplication on the GCSE non-calculator paper, you will be expected to use an appropriate method. The three most common are:

- The **grid method** (or box method), see Example 1 below.
- The **column method** (or traditional method), see Example 2 opposite.
- The **partition method**, see Example 3 opposite.

EXAMPLE 1

Work out 243 × 68 without using a calculator.

Using the grid method split the two numbers into hundreds, tens and units and write them in a grid like the one below. Multiply all the pairs of numbers.

×	200	40	3
60	12 000	2400	180
8	1600	320	24

Add the separate answers to find the total.

```
 12000
  2400
   180
  1600
   320
    24
 -----
 16524
  1 1
```

So, 243 × 68 = 16 524

Note the use of carry marks to help with the calculation. Always try to write carried marks much smaller than the other numbers, so that you don't confuse them with the main calculation.

EXAMPLE 2

Work out 357 × 24 without using a calculator.

There are several different ways to do long multiplication, but the following is perhaps the method that is most commonly used. This is the column method.

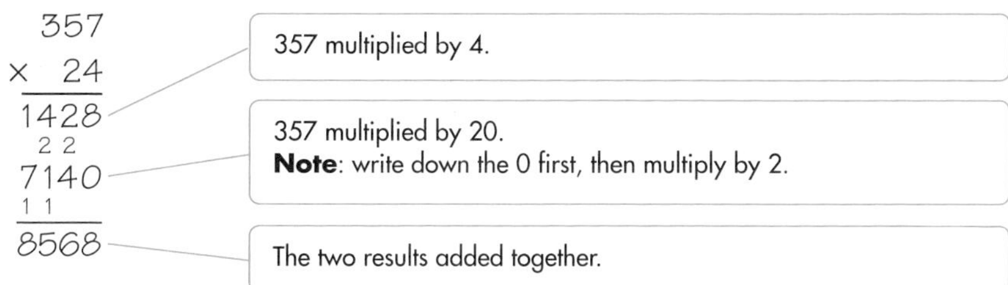
357
× 24
1428
2 2
7140
1 1
8568
357 multiplied by 4.
357 multiplied by 20.
Note: write down the 0 first, then multiply by 2.
The two results added together.

So, 357 × 24 = 8568

EXAMPLE 3

Work out 358 × 74 by the partition method.

Set out a grid as shown.

- Put the larger number along the top and the smaller number down the right-hand side.
- Multiply each possible pair in the grid, putting the numbers into each half as shown.
- Add up the numbers in each diagonal. If a total is larger than nine (in this example there is a total of 14), split the number and put the 1 in the next column on the left ready to be added in that diagonal.
- When you have completed the totalling, the number you are left with is the answer to the multiplication.

3	5	8	
2 / 1	3 / 5	5 / 6	7
1 / 2	2 / 0	3 / 2	4

5
+1 ← 14
2 6 4 9 2

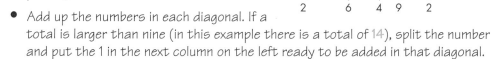

So, 358 × 74 = 26 492

EXERCISE 14A

1 Use your preferred method to calculate the following without using a calculator.

a 357 × 34	**b** 724 × 63	**c** 714 × 42	**d** 898 × 23
e 958 × 54	**f** 676 × 37	**g** 239 × 81	**h** 437 × 29
i 539 × 37	**j** 477 × 55	**k** 371 × 85	**l** 843 × 93
m 507 × 34	**n** 810 × 54	**o** 905 × 73	**p** 1435 × 72
q 2504 × 56	**r** 4037 × 23	**s** 8009 × 65	**t** 2070 × 38

F

F

FM **2** There are 48 cans of soup in a crate. A supermarket had a delivery of 125 crates of soup. The supermarket shelves will hold a total of 2500 cans. How many cans will be in the store room?

FM **3** Greystones Primary School has 12 classes, each of which has 26 students. The school hall will hold 250 students. Can all the students fit in the hall for an assembly?

FM **4** Suhail walks to school each day, there and back. The distance to school is 450 m. Suhail claims he walks over a marathon in distance, to and from school, in a school term of 64 days. A marathon is 42.1 km.

Is he correct?

PS **5** On one page of a newspaper there are seven columns. In each column there are 172 lines, and in each line there are 50 letters. The newspaper has the equivalent of 20 pages of print, excluding photographs and adverts. Are there more than a million letters in the paper?

6 A tank of water was emptied into casks. Each cask held 81 litres. 71 casks were filled and there were 68 litres left over. How much water was there in the tank to start with?

FM **7** Joy was going to do a sponsored walk to raise money for the Macmillan Nurses. She managed to get 18 people to sponsor her, each for 35p per kilometre. She walked a total of 48 km. Did Joy reach her target of £400?

14.2 Long division

This section will show you how to:

- divide, without a calculator, a three-digit or four-digit number by a two-digit number, e.g. 840 ÷ 24

Key words

long division

remainder

There are several different ways of doing **long division**. It is acceptable to use any of them, provided it gives the correct answer and you can show all your working clearly. Two methods are shown in this book. Example 4 shows the *Italian method*, sometimes called the DMSB method (Divide, Multiply, Subtract and Bring down). It is the most commonly used way of doing long division.

Example 5 shows a method of repeated subtraction, which is sometimes called the *chunking method*.

EXAMPLE 4

Work out 840 ÷ 24.

It is a good idea to jot down the appropriate times table before you start the long division. In this case, it will be the 24 times table.

1	2	3	4	5	6	7	8	9
24	48	72	96	120	144	168	192	216

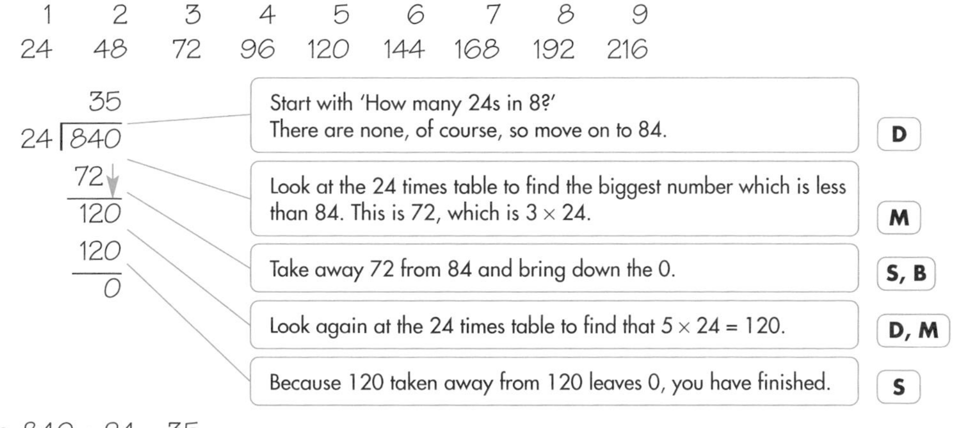

35
24 | 840
72
120
120
0
Start with 'How many 24s in 8?' There are none, of course, so move on to 84.
D
Look at the 24 times table to find the biggest number which is less than 84. This is 72, which is 3 × 24.
M
Take away 72 from 84 and bring down the 0.
S, B
Look again at the 24 times table to find that 5 × 24 = 120.
D, M
Because 120 taken away from 120 leaves 0, you have finished.
S

So, 840 ÷ 24 = 35

You may do a division as a short division, without writing down all the numbers. It will look like this:

$$24 \overline{)\ 8\ 4\ {}^{12}0}\quad = 3\ 5$$

Notice how the **remainder** from 84 is placed in front of the 0 to make it 120.

EXAMPLE 5

Work out 1655 ÷ 35.

Jot down some of the multiples of 35 that may be useful.

1 × 35 = 35   2 × 35 = 70   5 × 35 = 175   10 × 35 = 350   20 × 35 = 700

1655		
− 700	20 × 35	From 1655, subtract a large multiple of 35, such as 20 × 35 = 700.
955		
− 700	20 × 35	From 955, subtract a large multiple of 35, such as 20 × 35 = 700.
255		
− 175	5 × 35	From 255, subtract a multiple of 35, such as 5 × 35 = 175.
80		
− 70	2 × 35	From 80, subtract a multiple of 35, such as 2 × 35 = 70.
10	47	

Once the remainder of 10 has been found, you cannot subtract any more multiples of 35. Add up the multiples to see how many times 35 has been subtracted.

So, 1655 ÷ 35 = 47, remainder 10

Sometimes, as here, you will not need a whole multiplication table, and so you could jot down only those parts of the table that you will need. But, don't forget, you are going to have to work *without* a calculator, so you do need all the help you can get.

EXAMPLE 6

Naseema is organising a coach trip for 640 people. Each coach will carry 46 people. How many coaches are needed?

You need to divide the number of people (640) by the number of people in a coach (46).

```
     13
46 ⟌640
    46
    ---
    180
    138
    ---
     42
```

Start by dividing 64 by 46 … which gives 1 remainder 18.

Now divide 180 by 46 … which gives 3 remainder 42.

We have come to the end of the whole number division to give 13 remainder 42.

This tells Naseema that 14 coaches are needed to take all 640 passengers. (There will be 46 – 42 = 4 spare seats)

EXERCISE 14B

F

1 Solve the following by long division.

a 525 ÷ 21	**b** 480 ÷ 32	**c** 925 ÷ 25	**d** 645 ÷ 15
e 621 ÷ 23	**f** 576 ÷ 12	**g** 1643 ÷ 31	**h** 728 ÷ 14
i 832 ÷ 26	**j** 2394 ÷ 42	**k** 829 ÷ 22	**l** 780 ÷ 31
m 895 ÷ 26	**n** 873 ÷ 16	**o** 875 ÷ 24	**p** 225 ÷ 13
q 759 ÷ 33	**r** 1478 ÷ 24	**s** 756 ÷ 18	**t** 1163 ÷ 43

2 3600 supporters of Barnsley Football Club want to go to an away game by coach. Each coach can hold 53 passengers. How many coaches will they need altogether?

3 How many stamps costing 26p each can I buy for £10?

FM **4** Kirsty is collecting a set of 40 porcelain animals. Each animal costs 45p. Each month she spends up to £5 of her pocket money on these animals. How many months will it take her to buy the full set?

FM **5** Amina wanted to save up to see a concert. The cost of a ticket was £25. She was paid 75p per hour to mind her little sister. For how many hours would Amina have to mind her sister to be able to afford the ticket?

AU **6** The magazine *Teen Dance* costs £2.20 in a newsagent. It costs £21 to get the magazine on a yearly subscription. How much cheaper is the magazine each month if it is bought by subscription?

14.3 Arithmetic with decimal numbers

This section will show you how to:
- identify the information that a decimal number shows
- round a decimal number
- identify decimal places
- add and subtract two decimal numbers
- multiply and divide a decimal number by a whole number less than 10
- multiply a decimal number by a two-digit number
- multiply a decimal number by another decimal number

Key words
decimal fraction
decimal place
decimal point
digit
evaluate
round

The number system is extended by using decimal numbers to represent fractions.

The **decimal point** separates the **decimal fraction** from the whole-number part.

For example, the number 25.374 means:

Tens	Units		Tenths	Hundredths	Thousandths
10	1		$\frac{1}{10}$	$\frac{1}{100}$	$\frac{1}{1000}$
2	**5**	**.**	**3**	**7**	**4**

You already use decimal notation to express amounts of money. For example:

£32.67 means $3 \times$ £10
$2 \times$ £1
$6 \times$ £0.10 (10 pence)
$7 \times$ £0.01 (1 penny)

Decimal places

When a number is written in decimal form, the **digits** to the right of the decimal point are called **decimal places**. For example:

79.4 is written 'with one decimal place'

6.83 is written 'with two decimal places'

0.526 is written 'with three decimal places'.

To **round** a decimal number to a particular number of decimal places, take these steps:

- Count along the decimal places from the decimal point and look at the first digit to be removed.
- When the value of this digit is less than five, just remove the unwanted places.

- When the value of this digit is five or more, add 1 onto the digit in the last decimal place then remove the unwanted places.

Here are some examples.

5.852 rounds to 5.85 to two decimal places

7.156 rounds to 7.16 to two decimal places

0.274 rounds to 0.3 to one decimal place

15.3518 rounds to 15.4 to one decimal place

EXERCISE 14C

F

1 Round each of the following numbers to one decimal place.

a 4.83 **b** 3.79 **c** 2.16 **d** 8.25

e 3.673 **f** 46.935 **g** 23.883 **h** 9.549

i 11.08 **j** 33.509 **k** 7.054 **l** 46.807

m 0.057 **n** 0.109 **o** 0.599 **p** 64.99

q 213.86 **r** 76.07 **s** 455.177 **t** 50.999

HINTS AND TIPS

Just look at the value of the digit in the second decimal place.

2 Round each of the following numbers to two decimal places.

a 5.783 **b** 2.358 **c** 0.977 **d** 33.085 **e** 6.007

f 23.5652 **g** 91.7895 **h** 7.995 **i** 2.3076 **j** 23.9158

k 5.9999 **l** 1.0075 **m** 3.5137 **n** 96.508 **o** 0.009

p 0.065 **q** 7.8091 **r** 569.897 **s** 300.004 **t** 0.0099

3 Round each of the following to the number of decimal places (dp) indicated.

a 4.568 (1 dp) **b** 0.0832 (2 dp) **c** 45.715 93 (3 dp)

d 94.8531 (2 dp) **e** 602.099 (1 dp) **f** 671.7629 (2 dp)

g 7.1124 (1 dp) **h** 6.903 54 (3 dp) **i** 13.7809 (2 dp)

j 0.075 11 (1 dp) **k** 4.001 84 (3 dp) **l** 59.983 (1 dp)

m 11.9854 (2 dp) **n** 899.995 85 (3 dp) **o** 0.0699 (1 dp)

p 0.009 87 (2 dp) **q** 6.0708 (1 dp) **r** 78.392 5 (3 dp)

s 199.9999 (2 dp) **t** 5.0907 (1 dp)

4 Round each of the following to the nearest whole number.

a 8.7 **b** 9.2 **c** 2.7 **d** 6.5 **e** 3.28

f 7.82 **g** 3.19 **h** 7.55 **i** 6.172 **j** 3.961

k 7.388 **l** 1.514 **m** 46.78 **n** 23.19 **o** 96.45

p 32.77 **q** 153.9 **r** 342.5 **s** 704.19 **t** 909.5

F

E

FM 5 Belinda puts the following items in her shopping basket: bread (£1.09), meat (£6.99), cheese (£3.91) and butter (£1.13).

By rounding each price to the nearest pound, work out an estimate of the total cost of the items.

AU 6 Which of the following are correct roundings of the number 3.456?

3 3.0 3.4 3.40 3.45 3.46 3.47 3.5 3.50

PS 7 When a number is rounded to three decimal places the answer is 4.728

Which of these could be the number?

4.71 4.7275 4.7282 4.73

Adding and subtracting with decimals

When you are working with decimals, you must *always* set out your work carefully.

Make sure that the decimal points are in line underneath the first point and each digit is in its correct place or column.

Then you can add or subtract just as you have done before. The decimal point of the answer will be placed directly underneath the other decimal points.

EXAMPLE 7

Work out 4.72 + 13.53.

```
   4.72
+ 13.53
  18.25
   1
```

So, 4.72 + 13.53 = 18.25

Notice how to deal with 7 + 5 = 12, the 1 carrying forward into the next column.

EXAMPLE 8

Work out 7.3 – 1.5.

```
  ⁶7.¹3
–  1.5
   5.8
```

So, 7.3 – 1.5 = 5.8

Notice how to deal with the fact that you cannot take 5 from 3. You have to take one of the units from 7, replace the 7 with a 6 and make the 3 into 13.

Hidden decimal point

Whole numbers are usually written without decimal points. Sometimes you *do* need to show the decimal point in a whole number (see Example 9), in which case it is placed at the right-hand side of the number, followed by a zero.

EXAMPLE 9

Work out 4.2 + 8 + 12.9.

$$\begin{array}{r} 4.2 \\ 8.0 \\ +\,12.9 \\ \hline 25.1 \\ \hline {\scriptstyle 1\,1} \end{array}$$

So, 4.2 + 8 + 12.9 = 25.1

EXERCISE 14D

F

1 Work out each of these.

a 47.3 + 2.5 **b** 16.7 + 4.6 **c** 43.5 + 4.8
d 28.5 + 4.8 **e** 1.26 + 4.73 **f** 2.25 + 5.83
g 83.5 + 6.7 **h** 8.3 + 12.9 **i** 3.65 + 8.5
j 7.38 + 5.7 **k** 7.3 + 5.96 **l** 6.5 + 17.86

HINTS AND TIPS

When the numbers to be added or subtracted do not have the same number of decimal places, put in extra zeros, for example:

$$\begin{array}{r} 3.65 \\ +\,8.50 \\ \hline \end{array} \qquad \begin{array}{r} 8.25 \\ -\,4.50 \\ \hline \end{array}$$

2 Work out each of these.

a 3.8 – 2.4 **b** 4.3 – 2.5 **c** 7.6 – 2.8
d 8.7 – 4.9 **e** 8.25 – 4.5 **f** 19.7 – 13.8
g 9.4 – 5.7 **h** 8.62 – 4.85 **i** 8 – 4.3
j 9 – 7.6 **k** 15 – 3.2 **l** 24 – 8.7

3 **Evaluate** each of the following. (Take care – they are a mixture.)

a 23.8 + 6.9 **b** 8.3 – 1.7 **c** 9 – 5.2 **d** 12.9 + 3.8
e 17.4 – 5.6 **f** 23.4 + 6.8 **g** 35 + 8.3 **h** 9.54 – 2.81
i 34.8 + 3.15 **j** 8.1 – 3.4 **k** 12.5 – 8.7 **l** 198.5 + 12

AU **4** Viki has £4.75 in her purse after buying a watch for £11.99.

a How much did she have in her purse before she bought the watch?

b She then goes home by bus. After paying her bus fare she has £3.35 left in her purse. How much was the bus fare?

FM 5 Pipes are sold in 5.3 m lengths.

Will three pipes be enough to make this arrangement (excluding the corner pieces)? Justify your answer.

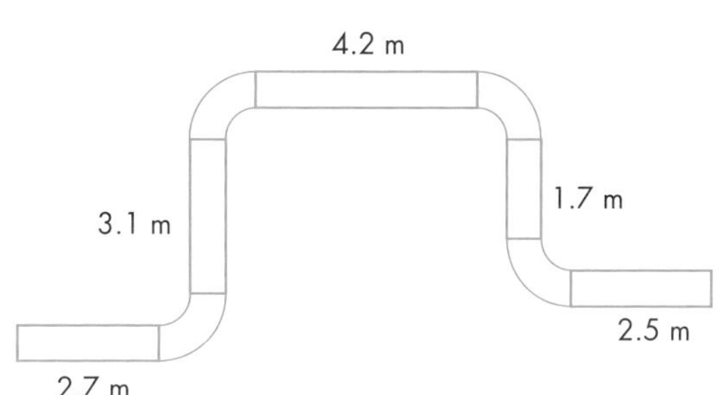

AU 6 Copy and complete the following.

a 12.5 + …… = 17.8 **b** 31.6 + …… = 38.3 **c** 7.16 + …… = 9.21

d …… + 4.2 = 6.1 **e** …… + 10.21 = 15.16 **f** …… + 27.54 = 31.25

AU 7 Copy and complete the following.

a 13.2 – …… = 11.6 **b** 81.4 – …… = 38.7 **c** 7.51 – …… = 3.22

d …… – 3.1 = 9.7 **e** …… – 14.6 = 7.8 **f** …… – 17.32 = 34.65

PS 8 Mark went shopping. He went into three shops and bought one item from each shop.

In total he spent £43.97.

Music Store	
CDs:	£5.98
DVDs:	£7.99

Clothes Store	
Shirt:	£12.50
Jeans	£32.00

Book store	
Magazine:	£2.25
Pen:	£3.98

What did he buy?

Multiplying and dividing decimals by single-digit numbers

You can carry out these operations in exactly the same way as with whole numbers, as long as you remember to put each digit in its correct column.

Again, the decimal point is kept in line underneath or above the first point.

EXAMPLE 10

Work out 4.5 × 3.

```
   4.5
×    3
  13.5
   1
```

So, 4.5 × 3 = 13.5

EXAMPLE 11

Work out 8.25 ÷ 5.

$$5 \overline{)\, 8.{}^{3}2{}^{2}5} \quad = 1.65$$

So, 8.25 ÷ 5 = 1.65

EXAMPLE 12

Work out 5.7 ÷ 2.

$$2 \overline{)\, 5.{}^{1}7{}^{1}0} \quad = 2.85$$

So, 5.7 ÷ 2 = 2.85

HINTS AND TIPS

We add a 0 after the 5.7 in order to continue dividing.
We do not use remainders with decimal places.

EXERCISE 14E

F

1 Evaluate each of these.

a 2.4 × 3 **b** 3.8 × 2 **c** 4.7 × 4 **d** 5.3 × 7
e 6.5 × 5 **f** 3.6 × 8 **g** 2.5 × 4 **h** 9.2 × 6
i 12.3 × 5 **j** 24.4 × 7 **k** 13.6 × 6 **l** 19.3 × 5

2 Evaluate each of these.

a 2.34 × 4 **b** 3.45 × 3 **c** 5.17 × 5 **d** 4.26 × 3
e 0.26 × 7 **f** 0.82 × 4 **g** 0.56 × 5 **h** 0.92 × 6
i 6.03 × 7 **j** 7.02 × 8 **k** 2.55 × 3 **l** 8.16 × 6

3 Evaluate each of these.

a 3.6 ÷ 2 **b** 5.6 ÷ 4 **c** 4.2 ÷ 3 **d** 8.4 ÷ 7
e 4.26 ÷ 2 **f** 3.45 ÷ 5 **g** 8.37 ÷ 3 **h** 9.68 ÷ 8
i 7.56 ÷ 4 **j** 5.43 ÷ 3 **k** 1.32 ÷ 4 **l** 7.6 ÷ 4

4 Evaluate each of these.

a 3.5 ÷ 2 **b** 6.4 ÷ 5 **c** 7.4 ÷ 4 **d** 7.3 ÷ 2

e 8.3 ÷ 5 **f** 5.8 ÷ 4 **g** 7.1 ÷ 5 **h** 9.2 ÷ 8

i 6.7 ÷ 2 **j** 4.9 ÷ 5 **k** 9.2 ÷ 4 **l** 7.3 ÷ 5

> **HINTS AND TIPS**
>
> **Remember** to keep the decimal points in line.

5 Evaluate each of these.

a 7.56 ÷ 4 **b** 4.53 ÷ 3 **c** 1.32 ÷ 5 **d** 8.53 ÷ 2

e 2.448 ÷ 2 **f** 1.274 ÷ 7 **g** 0.837 ÷ 9 **h** 16.336 ÷ 8

i 9.54 ÷ 5 **j** 14 ÷ 5 **k** 17 ÷ 4 **l** 37 ÷ 2

FM **6** Soup is sold in packs of five for £3.25 and packs of eight for £5. Which is the cheaper way of buying soup?

7 **a** Mike took his wife and four children to a theme park. The tickets were £13.25 for each adult and £5.85 for each child. How much did all the tickets cost Mike?

b While in the park, the children wanted ice creams.

Ice creams cost £1.60 each for large ones or £1.20 each for small ones.
Their mother only has £5.

Which size should she buy? Give a reason for your answer.

PS **8** Mary was laying a path through her garden. She bought nine paving stones, each 1.35 m long. She wanted the path to run straight down the garden, which is 10 m long. Has Mary bought too many paving stones? Show all your working.

Long multiplication with decimals

As before, you must put each digit in its correct column and keep the decimal point in line.

EXAMPLE 13

Evaluate 4.27 × 34.

```
     4.27
 ×     34
   ------
    17.08
     1 2
   128.10
      2
   ------
   145.18
   ------
    1
```

So, 4.27 × 34 = 145.18

EXERCISE 14F

E

1 Evaluate each of these.

a 3.72×24 **b** 5.63×53 **c** 1.27×52 **d** 4.54×37

e 67.2×35 **f** 12.4×26 **g** 62.1×18 **h** 81.3×55

i 5.67×82 **j** 0.73×35 **k** 23.8×44 **l** 99.5×19

2 Find the total cost of each of the following purchases.

a 18 ties at £12.45 each

b 25 shirts at £8.95 each

c 13 pairs of tights at £2.30 a pair

HINTS AND TIPS

When the answer is an amount of money in pounds, you must write it with two decimal places. Writing £224.1 will lose you a mark. It should be £224.10.

PS 3 Theo says that he can change multiplications into easier multiplications with the same answer by doubling one number and halving the other number.

For example to work out 8.4×12 he does 16.8×6

Use his method to work out 2.5×14.

D

FM 4 A party of 24 scouts and their leader went into a zoo. The cost of a ticket for each scout was £2.15 and the cost of a ticket for the leader was £2.60. What was the total cost of entering the zoo?

FM AU 5 **a** A market gardener bought 35 trays of seedlings. Each tray cost £3.45. What was the total cost of the trays of seedlings?

b There were 20 seedlings on each tray which are sold at 85p each. How much profit did the market gardener make per tray?

Multiplying two decimal numbers together

Follow these steps to multiply one decimal number by another decimal number.

- First, complete the whole calculation as if the decimal points were not there.
- Then, count the total number of decimal places in the two decimal numbers. This gives the number of decimal places in the answer.

EXAMPLE 14

Evaluate 3.42×2.7.

Ignoring the decimal points gives the following calculation:

$$\begin{array}{r} 342 \\ \times \quad 27 \\ \hline 2394 \\ {\scriptstyle 2\,1} \\ 6840 \\ \hline 9234 \\ \hline {\scriptstyle 1\,1} \end{array}$$

Now, 3.42 has two decimal places (.42) and 2.7 has one decimal place (.7). So, the total number of decimal places in the answer is three.

So, $3.42 \times 2.7 = 9.234$

EXERCISE 14G

1 Evaluate each of these.

a 2.4×0.2 **b** 7.3×0.4 **c** 5.6×0.2 **d** 0.3×0.4

e 0.14×0.2 **f** 0.3×0.3 **g** 0.24×0.8 **h** 5.82×0.52

i 5.8×1.23 **j** 5.6×9.1 **k** 0.875×3.5 **l** 9.12×5.1

FM 2 Jerome is making a jacket. He has £30 to spend and needs 3.4 m of cloth. The cloth is £8.75 per metre.

Can Jerome afford the cloth?

3 For each of the following:

i estimate the answer by first rounding each number to the nearest whole number

ii calculate the exact answer, and then calculate the difference between this and your answers to part **i**.

a 4.8×7.3 **b** 2.4×7.6 **c** 15.3×3.9 **d** 20.1×8.6

e 4.35×2.8 **f** 8.13×3.2 **g** 7.82×5.2 **h** 19.8×7.1

AU PS 4 **a** Use any method to work out 26×22

b Use your answer to part **a** to work out:

i 2.6×2.2

ii 1.3×1.1

iii 2.6×8.8.

E

D

14.4 Arithmetic with fractions

This section will show you how to:

- convert a decimal number to a fraction
- convert a fraction to a decimal
- add and subtract fractions with different denominators
- multiply a mixed number by a fraction
- divide one fraction by another fraction

Key words
decimal
denominator
fraction
mixed number
numerator

Changing a decimal into a fraction

A **decimal** can be changed into a **fraction** by using a place-value table.

For example, $0.32 = \frac{32}{100}$

Units	.	Tenths	Hundredths	Thousandths
0	.	3	2	

EXAMPLE 15

Express 0.32 as a fraction.

$$0.32 = \frac{32}{100}$$

This cancels to $\frac{8}{25}$

So, $0.32 = \frac{8}{25}$

Changing a fraction into a decimal

You can change a fraction into a decimal by dividing the **numerator** by the **denominator**. Example 16 shows how this can be done without a calculator.

EXAMPLE 16

Express $\frac{3}{8}$ as a decimal.

$\frac{3}{8}$ means $3 \div 8$. This is a division calculation:

$$\begin{array}{r} 0.\,3\;7\;5 \\ 8\,\overline{)\,3.{}^{3}0{}^{6}0{}^{4}0} \end{array}$$

So, $\frac{3}{8} = 0.375$

Notice that extra zeros have been put at the end to be able to complete the division.

EXERCISE 14H

1 Change each of these decimals to fractions, cancelling where possible.

a 0.7 **b** 0.4 **c** 0.5 **d** 0.03 **e** 0.06

f 0.13 **g** 0.25 **h** 0.38 **i** 0.55 **j** 0.64

2 Change each of these fractions to decimals. Where necessary, give your answer correct to three decimal places.

a $\frac{1}{2}$ **b** $\frac{3}{4}$ **c** $\frac{3}{5}$ **d** $\frac{9}{10}$ **e** $\frac{1}{3}$

f $\frac{5}{8}$ **g** $\frac{2}{3}$ **h** $\frac{7}{20}$ **i** $\frac{7}{11}$ **j** $\frac{4}{9}$

3 Put each of the following sets of numbers in order, with the smallest first.

a $0.6, 0.3, \frac{1}{2}$ **b** $\frac{2}{5}, 0.8, 0.3$

c $0.35, \frac{1}{4}, 0.15$ **d** $\frac{7}{10}, 0.72, 0.71$

e $0.8, \frac{3}{4}, 0.7$ **f** $0.08, 0.1, \frac{1}{20}$

g $0.55, \frac{1}{2}, 0.4$ **h** $1\frac{1}{4}, 1.2, 1.23$

HINTS AND TIPS

Convert the fractions to decimals first.

FM 4 Two shops are advertising the same T-shirts.

Which shop has the best offer?
Give a reason for your answer.

PS 5 Which is bigger $\frac{7}{8}$ or 0.87?

Show your working.

PS 6 Which is smaller $\frac{2}{3}$ or 0.7?

Show your working.

D

PS 7 Complete this statement with a decimal.

$\frac{1}{4}$ of 50 is the same as ...... × 100

PS 8 Complete this statement with a fraction.

$\frac{2}{3}$ of 60 is the same as ...... of 80

Addition and subtraction of fractions

Fractions can only be added or subtracted after you have converted them to equivalent fractions with the same denominator.

Addition, subtraction, multiplication and division of fractions is a review of topics in Chapter 2.

EXAMPLE 17

i $\frac{2}{3} + \frac{1}{5}$

Note you can change both fractions to equivalent fractions with a denominator of 15.

This then becomes:

$$\frac{2 \times 5}{3 \times 5} + \frac{1 \times 3}{5 \times 3} = \frac{10}{15} + \frac{3}{15} = \frac{13}{15}$$

ii $2\frac{3}{4} - 1\frac{5}{6}$

Split the calculation into $\left(2 + \frac{3}{4}\right) - \left(1 + \frac{5}{6}\right)$.

This then becomes:

$$2 - 1 + \frac{3}{4} - \frac{5}{6}$$

Note you can change both fractions to equivalent fractions with a denominator of 12.

$$= 1 + \frac{9}{12} + \frac{10}{12} = 1 - \frac{1}{12}$$

$$= \frac{11}{12}$$

EXERCISE 14I

1 Evaluate the following.

a $\frac{1}{3} + \frac{1}{5}$ **b** $\frac{1}{3} + \frac{1}{4}$ **c** $\frac{1}{5} + \frac{1}{10}$

d $\frac{2}{3} + \frac{1}{4}$ **e** $\frac{3}{4} + \frac{1}{8}$ **f** $\frac{1}{3} + \frac{1}{6}$

g $\frac{1}{2} - \frac{1}{3}$ **h** $\frac{1}{4} - \frac{1}{5}$ **i** $\frac{1}{5} - \frac{1}{10}$

j $\frac{7}{8} - \frac{3}{4}$ **k** $\frac{5}{6} - \frac{3}{4}$ **l** $\frac{5}{6} - \frac{1}{2}$

m $\frac{5}{12} - \frac{1}{4}$ **n** $\frac{1}{3} + \frac{4}{9}$ **o** $\frac{1}{4} + \frac{3}{8}$

p $\frac{7}{8} - \frac{1}{2}$ **q** $\frac{3}{5} - \frac{8}{15}$ **r** $\frac{11}{12} + \frac{5}{8}$

s $\frac{7}{16} + \frac{3}{10}$ **t** $\frac{4}{9} - \frac{2}{21}$ **u** $\frac{5}{6} - \frac{4}{27}$

2 Evaluate the following.

a $2\frac{1}{7} + 1\frac{3}{14}$ **b** $6\frac{3}{10} + 1\frac{4}{5} + 2\frac{1}{2}$ **c** $3\frac{1}{2} - 1\frac{1}{3}$

d $1\frac{7}{18} + 2\frac{3}{10}$ **e** $3\frac{2}{6} + 1\frac{9}{20}$ **f** $1\frac{1}{8} - \frac{5}{9}$

g $1\frac{3}{16} - \frac{7}{12}$ **h** $\frac{5}{6} + \frac{7}{16} + \frac{5}{8}$ **i** $\frac{7}{10} + \frac{3}{8} + \frac{5}{6}$

j $1\frac{1}{3} + \frac{7}{10} - \frac{4}{15}$ **k** $\frac{5}{14} + 1\frac{3}{7} - \frac{5}{12}$

3 In a class of children, three-quarters are Chinese, one-fifth are Malay and the rest are Indian. What fraction of the class are Indian?

4 **a** In a class election, half of the people voted for Aminah, one-third voted for Janet and the rest voted for Peter. What fraction of the class voted for Peter?

b One of the following is the number of people in the class.

25 28 30 32

How many people are in the class?

C

FM 5 A group of people travelled from Hope to Castletown. One-twentieth of them decided to walk, one-twelfth went by car and all the rest went by bus. What fraction went by bus?

FM 6 A one-litre flask filled with milk is used to fill two glasses, one of capacity half a litre and the other of capacity one-sixth of a litre. What fraction of a litre will remain in the flask?

FM 7 Katie spent three-eighths of her income on rent, and two-fifths of what was left on food. What fraction of her income was left after buying her food?

PS 8 Mick says that $1\frac{1}{3} + 2\frac{1}{4} = 3\frac{2}{7}$.

He is incorrect.

What is the mistake that he has made?

Work out the correct answer.

Multiplication of fractions

Remember:

- To multiply two fractions, multiply the numerators (top numbers) and multiply the denominators (bottom numbers) and cancel if possible.
- When multiplying a **mixed number**, change the mixed number to an improper fraction before you start multiplying.

EXAMPLE 18

Work out

$$1\frac{3}{4} \times \frac{2}{5}$$

Change the mixed number to an improper fraction.

$$1\frac{3}{4} \text{ to } \frac{7}{4}$$

The problem is now:

$$\frac{7}{4} \times \frac{2}{5}$$

So, $\frac{7}{4} \times \frac{2}{5} = \frac{14}{20}$

This cancels to $\frac{7}{10}$.

EXAMPLE 19

A boy had 930 stamps in his collection. $\frac{2}{15}$ of them were British stamps. How many British stamps did he have?

The problem is:

$$\frac{2}{15} \times 930$$

First, calculate $\frac{1}{15}$ of 930.

$$\frac{1}{15} \times 930 = 930 \div 15 = 62$$

So, $\frac{2}{15}$ of $930 = 2 \times 62 = 124$

He had 124 British stamps.

EXERCISE 14J

1 Evaluate the following, leaving each answer in its simplest form.

a $\frac{1}{2} \times \frac{1}{3}$ **b** $\frac{1}{4} \times \frac{2}{5}$ **c** $\frac{3}{4} \times \frac{1}{2}$ **d** $\frac{3}{7} \times \frac{1}{2}$

e $\frac{2}{3} \times \frac{4}{5}$ **f** $\frac{1}{3} \times \frac{3}{5}$ **g** $\frac{1}{3} \times \frac{6}{7}$ **h** $\frac{3}{4} \times \frac{2}{5}$

i $\frac{2}{3} \times \frac{3}{4}$ **j** $\frac{1}{2} \times \frac{4}{5}$

2 Evaluate the following, leaving each answer in its simplest form.

a $\frac{5}{16} \times \frac{3}{10}$ **b** $\frac{9}{10} \times \frac{5}{12}$ **c** $\frac{14}{15} \times \frac{3}{8}$ **d** $\frac{8}{9} \times \frac{6}{15}$

e $\frac{6}{7} \times \frac{21}{30}$ **f** $\frac{9}{14} \times \frac{35}{36}$

3 I walked two-thirds of the way along Pungol Road which is 4.5 km long. How far have I walked?

4 One-quarter of Alan's stamp collection was given to him by his sister. Unfortunately two-thirds of these were torn. What fraction of his collection was given to him by his sister and were not torn?

5 Bilal eats one-quarter of a cake, and then half of what is left. How much cake is left uneaten?

6 A merchant buys 28 crates, each containing three-quarters of a tonne of waste metal. What is the total weight of waste metal?

E

D

D

C

7 Because of illness, on one day two-fifths of a school was absent. If the school had 650 pupils on the register, how many were absent that day?

8 To increase sales, a shop reduced the price of a car stereo radio by $\frac{2}{5}$. If the original price was £85, what was the new price?

9 Two-fifths of a class were boys. If the class contained 30 children, how many were girls?

10 Evaluate the following, giving each answer as a mixed number where possible.

a $1\frac{1}{4} \times \frac{1}{3}$ **b** $1\frac{2}{3} \times 1\frac{1}{4}$ **c** $2\frac{1}{2} \times 2\frac{1}{2}$ **d** $1\frac{3}{4} \times 1\frac{2}{3}$

e $3\frac{1}{4} \times 1\frac{1}{5}$ **f** $1\frac{1}{4} \times 2\frac{2}{3}$ **g** $2\frac{1}{2} \times 5$ **h** $7\frac{1}{2} \times 4$

11 Which is larger, $\frac{3}{4}$ of $2\frac{1}{2}$ or $\frac{2}{5}$ of $6\frac{1}{2}$?

PS 12 After James spent $\frac{2}{5}$ of his pocket money on magazines, and $\frac{1}{4}$ of his pocket money at a football match, he had £1.75 left. How much pocket money did he have in the beginning?

13 Which is the biggest: half of 96, one-third of 141, two-fifths of 120, or three-quarters of 68?

PS 14 At a burger-eating competition, Liam ate 34 burgers in 20 minutes while Ahmed ate 26 burgers in 20 minutes. Assuming they ate the burgers at a steady rate, how long after the start of the competition would they have consumed a total of 21 burgers between them?

PS 15 If £5.20 is two-thirds of three-quarters of a sum of money, what is the sum?

PS 16 Emily lost $\frac{3}{4}$ of her money in the market, but then found $\frac{3}{5}$ of what she had lost. She now had £21 altogether. How much did she start with?

Dividing fractions

Look at the problem $3 \div \frac{3}{4}$. This is like asking, 'How many $\frac{3}{4}$s are there in 3?'

Look at the diagram.

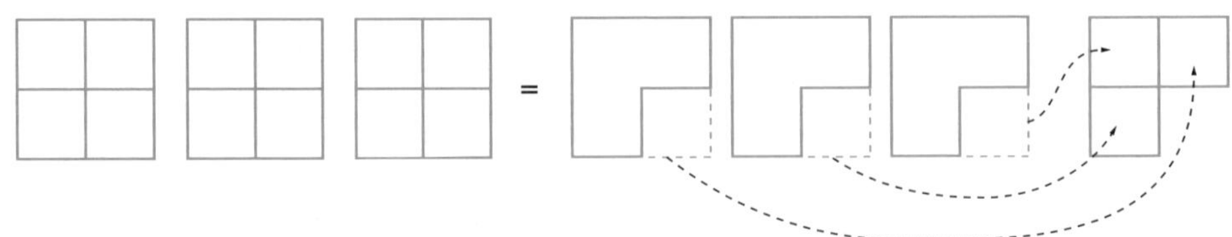

Each of the three whole shapes is divided into quarters. What is the total number of quarters divided by 3?

Can you see that you could fit the four shapes on the right-hand side of the = sign into the three shapes on the left-hand side?

i.e. $3 \div \frac{3}{4} = 4$

or $3 \div \frac{3}{4} = 3 \times \frac{4}{3} = \frac{3 \times 4}{3} = \frac{12}{3} = 4$

So, to divide by a fraction, you turn the fraction upside down (finding its reciprocal), and then multiply.

EXERCISE 14K

1 Evaluate the following, giving your answer as a mixed number where possible.

a $\frac{1}{4} \div \frac{1}{3}$ **b** $\frac{2}{5} \div \frac{2}{7}$ **c** $\frac{4}{5} \div \frac{3}{4}$ **d** $\frac{3}{7} \div \frac{2}{5}$

e $5 \div 1\frac{1}{4}$ **f** $6 \div 1\frac{1}{2}$ **g** $7\frac{1}{2} \div 1\frac{1}{2}$ **h** $3 \div 1\frac{3}{4}$

i $1\frac{5}{12} \div 3\frac{3}{16}$ **j** $3\frac{3}{5} \div 2\frac{1}{4}$

2 A grain merchant has only thirteen and a half tonnes in stock. He has several customers who are all ordering three-quarters of a tonne. How many customers can he supply?

3 For a party, Zahar made twelve and a half litres of lemonade. His glasses could each hold $\frac{5}{16}$ of a litre. How many of the glasses could he fill from the twelve and a half litres of lemonade?

4 How many strips of ribbon, each three and a half centimetres long, can I cut from a roll of ribbon that is fifty-two and a half centimetres long?

5 Joe's stride is three-quarters of a metre long. How many strides does he take to walk the length of a bus twelve metres long?

6 Evaluate the following, giving your answers as a mixed number where possible.

a $2\frac{2}{9} \times 2\frac{1}{10} \times \frac{16}{35}$ **b** $3\frac{1}{5} \times 2\frac{1}{2} \times 4\frac{3}{4}$

c $1\frac{1}{4} \times 1\frac{2}{7} \times 1\frac{1}{6}$ **d** $\frac{18}{25} \times \frac{15}{16} \div 2\frac{2}{5}$

e $\left(\frac{2}{5} \times \frac{2}{5}\right) \times \left(\frac{5}{6} \times \frac{5}{6}\right) \times \left(\frac{3}{4} \times \frac{3}{4}\right)$ **f** $\left(\frac{4}{5} \times \frac{4}{5}\right) \div \left(1\frac{1}{4} \times 1\frac{1}{4}\right)$

C

B

14.5 Multiplying and dividing with negative numbers

This section will show you how to:
- multiply and divide with negative numbers

Key word
negative

The rules for multiplying and dividing with **negative** numbers are very easy.

- When the signs of the numbers are the *same*, the answer is *positive*.
- When the signs of the numbers are *different*, the answer is *negative*.

Here are some examples.

$2 \times 4 = 8 \qquad 12 \div -3 = -4 \qquad -2 \times -3 = 6 \qquad -12 \div -3 = 4$

A common error is to confuse, for example, -3^2 and $(-3)^2$.

$-3^2 = -3 \times 3 = -9$

but,

$(-3)^2 = -3 \times -3 = +9.$

So, this means that if a variable is introduced, for example, $a = -5$, the calculation would be as follows:

$a^2 = -5 \times -5 = +25$

EXAMPLE 20

$a = -2$ and $b = -6$

Work out

a a^2 **b** $a^2 + b^2$ **c** $b^2 - a^2$ **d** $(a - b)^2$

a $a^2 = -2 \times -2 = +4$

b $a^2 + b^2 = +4 + -6 \times -6 = 4 + 36 = 40$

c $b^2 - a^2 = 36 - 4 = 32$

d $(a - b) = (-2 - -6)^2 = (-2 + 6)^2 = (4)^2 = 16$

EXERCISE 14L

E

1 Write down the answers to the following.

a -3×5	**b** -2×7	**c** -4×6	**d** -2×-3	**e** -7×-2
f $-12 \div -6$	**g** $-16 \div 8$	**h** $24 \div -3$	**i** $16 \div -4$	**j** $-6 \div -2$
k 4×-6	**l** 5×-2	**m** 6×-3	**n** -2×-8	**o** -9×-4
p $24 \div -6$	**q** $12 \div -1$	**r** $-36 \div 9$	**s** $-14 \div -2$	**t** $100 \div 4$
u -2×-9	**v** $32 \div -4$	**w** 5×-9	**x** $-21 \div -7$	**y** -5×8

E

2 Write down the answers to the following.

a $-3 + -6$ **b** -2×-8 **c** $2 + -5$ **d** 8×-4 **e** $-36 \div -2$

f -3×-6 **g** $-3 - -9$ **h** $48 \div -12$ **i** -5×-4 **j** $7 - -9$

k $-40 \div -5$ **l** $-40 + -8$ **m** $4 - -9$ **n** $5 - 18$ **o** $72 \div -9$

p $-7 - -7$ **q** $8 - -8$ **r** 6×-7 **s** $-6 \div -1$ **t** $-5 \div -5$

u $-9 - 5$ **v** $4 - -2$ **w** $4 \div -1$ **x** $-7 \div -1$ **y** -4×0

3 What number do you multiply by –3 to get the following?

a 6 **b** –90 **c** –45 **d** 81 **e** 21

4 What number do you divide –36 by to get the following?

a –9 **b** 4 **c** 12 **d** –6 **e** 9

5 Evaluate the following.

a $-6 + (4 - 7)$ **b** $-3 - (-9 - -3)$ **c** $8 + (2 - 9)$

6 Evaluate the following.

a $4 \times (-8 \div -2)$ **b** $-8 -(3 \times -2)$ **c** $-1 \times (8 - -4)$

7 What do you get if you divide –48 by the following?

a –2 **b** –8 **c** 12 **d** 24

AU 8 Write down six different multiplications that give the answer –12.

AU 9 Write down six different divisions that give the answer –4.

10 Find the answers to the following.

a -3×-7 **b** $3 + -7$ **c** $-4 \div -2$ **d** $-7 - 9$ **e** $-12 \div -6$

f $-12 - -7$ **g** 5×-7 **h** $-8 + -9$ **i** $-4 + -8$ **j** $-3 + 9$

k -5×-9 **l** $-16 \div 8$ **m** $-8 - -8$ **n** $6 \div -6$ **o** $-4 + -3$

p -9×4 **q** $-36 \div -4$ **r** -4×-8 **s** $-1 - -1$ **t** $2 - 67$

AU 11 **a** Work out 6×-2

b The average temperature drops by 2 °C every day for six days. How much has the temperature dropped altogether?

c The temperature drops by 6 °C for each of the next three days. Write down the calculation to work out the total drop in temperature over these three days.

PS 12 Put these calculations in order from lowest to highest.

-5×4 $-20 \div 2$ $-16 \div -4$ 3×-6

13 $x = -2$, $y = -3$ and $z = -4$. Work out

a x^2 **b** $y^2 + z^2$ **c** $z^2 - x^2$ **d** $(x - y)^2$

D

14.6 Approximation of calculations

This section will show you how to:

- identify significant figures
- round to one significant figure
- approximate the result before multiplying two numbers together
- approximate the result before dividing two numbers
- round a calculation, at the end of a problem, to give what is considered to be a sensible answer

Key words
approximate
round
significant figure

Rounding to significant figures

You will often use **significant figures** when you want to **approximate** a number with quite a few digits in it.

The following table illustrates some numbers written to one, two and three significant figures (sf).

One sf	8	50	200	90 000	0.000 07	0.003	0.4
Two sf	67	4.8	0.76	45 000	730	0.006 7	0.40
Three sf	312	65.9	40.3	0.0761	7.05	0.003 01	0.400

In the GCSE examination you usually only have to **round** numbers to one significant figure.

The steps taken to round a number to one significant figure are very similar to those used for decimal places.

- From the left, find the second digit. If the original number is less than one, start counting from the first non-zero digit.
- When the value of the second digit is less than five, leave the first digit as it is.
- When the value of the second digit is equal to or greater than five, add 1 to the first digit.
- Put in enough zeros at the end to keep the number the right size.

For example, the following tables show some numbers rounded to one significant figure.

Number	Rounded to 1 sf
78	80
32	30
0.69	0.7
1.89	2
998	1000
0.432	0.4

Number	Rounded to 1 sf
45 281	50 000
568	600
8054	8000
7.837	8
99.8	100
0.078	0.08

EXERCISE 14M

1 Round each of the following numbers to 1 significant figure.

a 46 313 **b** 57 123 **c** 30 569 **d** 94 558 **e** 85 299

f 54.26 **g** 85.18 **h** 27.09 **i** 96.432 **j** 167.77

k 0.5388 **l** 0.2823 **m** 0.005 84 **n** 0.047 85 **o** 0.000 876

p 9.9 **q** 89.5 **r** 90.78 **s** 199 **t** 999.99

2 Round each of the following numbers to 1 significant figure.

a 56 147 **b** 26 813 **c** 79 611 **d** 30 578 **e** 14 009

f 5876 **g** 1065 **h** 847 **i** 109 **j** 638.7

k 1.689 **l** 4.0854 **m** 2.658 **n** 8.0089 **o** 41.564

p 0.8006 **q** 0.458 **r** 0.0658 **s** 0.9996 **t** 0.009 82

3 Write down the smallest and the greatest numbers of sweets that can be found in each of these jars.

a

b

c

4 Write down the smallest and the greatest numbers of people that might live in these towns.

Elsecar	population 800 (to one significant figure)
Hoyland	population 1000 (to one significant figure)
Barnsley	population 200 000 (to one significant figure)

PS 5 A joiner estimates that he has 20 pieces of skirting board in stock. He is correct to one significant figure. He uses three pieces and now has 10 left to one significant figure.

How many could he have had to start with?
Work out all possible answers.

AU 6 There are 500 fish in a pond to one significant figure.

What is the least possible number of fish that could be taken from the pond so that there are 400 fish in the pond to one significant figure?

Approximation of calculations

How would you approximate the value of a calculation? What would you actually do when you try to approximate an answer to a problem?

For example, what is the approximate answer to 35.1×6.58?

To find the approximate answer, you simply round each number to 1 significant figure, then complete the calculation. So in this case, the approximation is:

$35.1 \times 6.58 \approx 40 \times 7 = 280$

Note the symbol $\approx$ which means 'approximately equal to'.

For the division $89.1 \div 2.98$, the approximate answer is $90 \div 3 = 30$.

Sometimes when dividing it can be sensible to round to 2 sf instead of 1 sf. For example,

$24.3 \div 3.87$ using $24 \div 4$ gives an approximate answer of 6

whereas

$24.3 \div 3.87$ using $20 \div 4$ gives an approximate answer of 5.

Both of these answers would be acceptable in the GCSE examination as they are both sensible answers, but generally rounding to one significant figure is easier.

A quick approximation is always a great help in any calculation since it often stops you giving a silly answer.

EXERCISE 14N

D

1 Find approximate answers to the following.

a 5435×7.31 **b** 5280×3.211 **c** $63.24 \times 3.514 \times 4.2$

d 3508×2.79 **e** $72.1 \times 3.225 \times 5.23$ **f** $470 \times 7.85 \times 0.99$

g $354 \div 79.8$ **h** $36.8 \div 1.876$ **i** $5974 \div 5.29$

Check your answers on a calculator to see how close you were.

2 Find the approximate monthly pay of the following people whose annual salaries are given.

a Paul £35 200 **b** Michael £25 600

c Jennifer £18 125 **d** Ross £8420

3 Find the approximate annual pay of the following people who earn:

a Kevin £270 a week **b** Malcolm £1528 a month **c** David £347 a week

AU **4** A farmer bought 2713 kg of seed at a cost of £7.34 per kg. Find the approximate total cost of this seed.

5 By rounding, find an approximate answer to each of the following.

a $\frac{573 + 783}{107}$ **b** $\frac{783 - 572}{24}$ **c** $\frac{352 + 657}{999}$ **d** $\frac{1123 - 689}{354}$

e $\frac{589 + 773}{658 - 351}$ **f** $\frac{793 - 569}{998 - 667}$ **g** $\frac{354 + 656}{997 - 656}$ **h** $\frac{1124 - 661}{355 + 570}$

i $\frac{28.3 \times 19.5}{97.4}$ **j** $\frac{78.3 \times 22.6}{3.69}$ **k** $\frac{3.52 \times 7.95}{15.9}$ **l** $\frac{11.78 \times 77.8}{39.4}$

6 Find the approximate answer to each of the following.

a $208 \div 0.378$ **b** $96 \div 0.48$ **c** $53.9 \div 0.58$

d $14.74 \div 0.285$ **e** $28.7 \div 0.621$ **f** $406.9 \div 0.783$

Check your answers on a calculator to see how close you were.

7 A litre of paint will cover an area of about 8.7 m^2. Approximately how many litre cans will I need to buy to paint a room with a total surface area of 73 m^2?

8 By rounding, find the approximate answer to each of the following.

a $\frac{84.7 + 12.6}{0.483}$ **b** $\frac{32.8 \times 71.4}{0.812}$ **c** $\frac{34.9 - 27.9}{0.691}$ **d** $\frac{12.7 \times 38.9}{0.42}$

FM 9 It took me 6 hours and 40 minutes to drive from Sheffield to Bude, a distance of 295 miles. My car uses petrol at the rate of about 32 miles per gallon. The petrol cost £3.51 per gallon.

a Approximately how many miles did I travel each hour?

b Approximately how many gallons of petrol did I use in going from Sheffield to Bude?

c What was the approximate cost of all the petrol for my journey to Bude and back again?

PS 10 Kirsty arranges for magazines to be put into envelopes. She sorts out 178 magazines between 10.00 am and 1.00 pm. Approximately how many magazines will she be able to sort in a week in which she works for 17 hours?

11 An athlete runs 3.75 km every day. Approximately how far does he run in:

a a week **b** a month **c** a year?

AU 12 1 kg = 1000 g

A box full of magazines weighs 8 kg. One magazine weighs about 15 g. Approximately how many magazines are there in the box?

13 An apple weighs about 280 grams.

a What is the approximate weight of a bag containing a dozen apples?

b Approximately how many apples will there be in a sack weighing 50 kg?

Sensible rounding

Sensible rounding is simply writing or saying answers to questions which have a real-life context so that the answer makes sense and is the sort of thing someone might say in a normal conversation.

For example:

The distance from Rotherham to Sheffield is 9 miles is a sensible statement.

The distance from Rotherham to Sheffield is 8.7864 miles is not sensible.

6 tins of paint is sensible.

5.91 tins of paint is not sensible.

As a general rule if it sounds sensible it will be acceptable.

EXERCISE 14P

D

1 Round each of the following to give sensible answers.

- **a** I am 1.7359 m tall.
- **b** It took me 5 minutes 44.83 seconds to mend the television.
- **c** My kitten weighs 237.97 g.
- **d** The correct temperature at which to drink Earl Grey tea is 82.739 °C.
- **e** The distance from Wath to Sheffield is 15.528 miles.
- **f** The area of the floor is 13.673 m^2.

FM 2 Rewrite the following article using sensible amounts.

It was a hot day, the temperature was 81.699 °F and still rising. I had now walked 5.3289 km in just over 113.98 minutes. But I didn't care since I knew that the 43 275 people watching the race were cheering me on. I won by clipping 6.2 seconds off the record time. This was the 67th time the race had taken place since records first began in 1788. Well, next year I will only have 15 practice walks beforehand as I strive to beat the record by at least another 4.9 seconds.

C

AU 3 A lorry load of scrap metal weighs 39.715 tonnes.
It is worth £20.35 per tonne.

Approximately how much is the load worth?

PS 4 The accurate temperature is 18.2 °C.
David rounds the temperature to the nearest 5 °C.
David says the temperature is about 20 °C.

How much would the temperature rise for David to say that the temperature is about 25 °C?

GRADE BOOSTER

- **F** You can multiply a three-digit number by a two-digit number without using a calculator
- **F** You can divide a three- or four-digit number by a two-digit number
- **F** You can solve real problems involving multiplication and division
- **F** You can round decimal numbers to a specific number of places
- **E** You can evaluate calculations involving decimal numbers
- **E** You can convert decimals to fractions
- **E** You can convert fractions to decimals
- **E** You can order a list containing decimals and fractions
- **E** You can round numbers to one significant figure
- **E** You can multiply and divide negative numbers
- **D** You can estimate the approximate value of a calculation before calculating
- **C** You can give sensible answers to real-life questions

What you should know now

- How to do long multiplication
- How to do long division
- How to perform calculations with decimal numbers
- How to round to a specific number of decimal places
- How to round to a specific number of significant figures
- How to multiply and divide with negative numbers
- How to convert between decimals and fractions
- How to make estimates by suitable rounding

1

Gift shop	
Price list	
Key ring	£3.20
Hat	£3.99
Pencil case	£2.70
Ruler	45p
Pen	60p
Pencil	

Keith buys 3 pens.

a Work out the total cost. (2)

Simon buys a pencil case, a ruler and a pen.

He pays with a £5 note.

b Work out how much change he should get. (3)

The gift shop also sells pencils.

The price of a pencil is $\frac{2}{3}$ of the price of a pen.

c Work out the price of a pencil. (2)

(Total 7 marks)

Edexcel, November 2008, Paper 1 Foundation, Question 9

2 450 people go on a Football trip. Each coach will seat 54 people.

a How many coaches are needed?

b How many seats will be empty?

3 Josh buys 40 litres of milk.

The total cost is £33.20

Work out the cost of 1 litre of the milk. (3)

(Total 3 marks)

Edexcel, June 2008, Paper 13 Foundation, Question 4

PS **4** A coach costs £345 to hire.

There are 55 seats on the coach.

Work out a sensible seat price assuming at least 45 seats are sold.

5 **a** Write $\frac{5}{8}$ as a decimal.

b Write 0.6 as a fraction. Give your answer in its lowest terms.

6 Change the following fractions to decimals.

a $\frac{1}{5}$

b $\frac{5}{13}$

7 Estimate the value of these expressions.

a 15.7×29.2

b 143.1×17.8

8 **a** Work out $13 \times 17 - 11 \times 17$

b Find an approximate value of $\frac{51 \times 250}{82}$

You must show **all** your working.

9 **a** Work out $\frac{1}{3} + \frac{1}{12}$ (2)

b Work out $\frac{3}{4} \times \frac{1}{5}$ (1)

(Total 3 marks)

Edexcel, May 2008, Paper 1 Foundation, Question 20

10 Work out 25.6×1.6

You **must** show **all** your working. (3)

(Total 3 marks)

Edexcel, June 2007, Paper 10 Foundation, Unit 3 Test, Question 8

11 Three pupils use calculators to work out

$$\frac{32.7 + 14.3}{1.28 - 0.49}$$

Arnie gets 43.4, Bert gets 36.2 and Chuck gets 59.5. Use approximations to show which one of them is correct.

12 Work out $3\frac{2}{5} - 1\frac{2}{3}$

13 Use approximations to estimate the value of

$$\sqrt{\frac{323\,407}{0.48}}$$

14 Work out an estimate for

$$\frac{29.8 \times 4.1}{0.21}$$ (3)

(Total 3 marks)

Edexcel, November 2007, Paper 10 Foundation, Unit 3 Test, Question 10

Worked Examination Questions

FM **1** In a **survey** the number of visitors to a website was recorded daily. Altogether it was visited 30 million times. Each day it was visited 600 000 times.

Based on this information, for how many days did the survey last?

$$30\,000\,000 \div 600\,000 = \frac{30\,0\not{0}\not{0}\,\not{0}\not{0}\not{0}}{6\not{0}\not{0}\,\not{0}\not{0}\not{0}}$$

$$= \frac{300}{6} = 50 \text{ days}$$

You need to find the number of days using the following calculation:
total times visited ÷ number of visits a day
You get 1 method mark for showing the calculation.

You can cancel the five zeros on the top and the bottom.

Check that your final answer is sensible. You get 1 accuracy mark for the correct answer of 50 days.

Total: 2 marks

PS AU **2** A theatre has 400 tickets available each night for a show on Friday and Saturday nights. The manager of the theatre wants to raise £1500 from ticket sales each night. She expects to sell all the tickets for Saturday but only three-quarters of the tickets for Friday. If the ticket price is the same for both days, how much should she charge?

$\frac{3}{4} \times 400 = 300$

$300 + 400 = 700$ tickets altogether sold

£3000 needed

$3000 \div 700 = 4.28$

Round up to ensure a profit, so for example £4.50 or £5 per ticket

You get 1 method mark for working out that she sells $\frac{3}{4}$ of 400 tickets on Friday.

You get 1 mark for realising that £3000 (£1500 × 2) needs to raised.

For attempting to work out the exact amount per seat (3000 ÷ 700 or £4.28) you get 1 method mark.

Give a sensible price (e.g. to nearest 10p or with a built-in profit margin) to get 1 mark for accuracy.

Total: 4 marks

14 Functional Maths
In the gym

Healthy eating and regular exercise form part of a balanced lifestyle. Health clubs and gyms are a popular way to keep fit. Exercise machines in gyms often show how many calories are burnt off during a workout.

This activity investigates the relationship between calorie intake and burn-off rates.

Remember

We must not burn off all the calories that we consume: the body needs calories as energy.

Getting started

Before you begin your main task, use the information given on these pages to work out:

- the weight category for each person
- the calories they take in at breakfast and the amount they burn off through exercise.

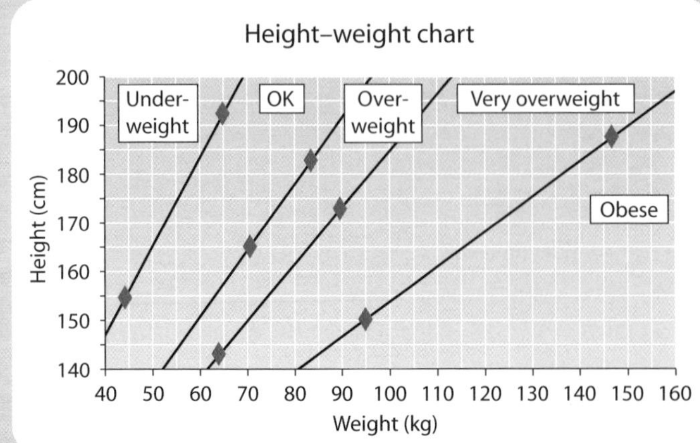

Facts

Calories per 100 g

Apple	46
Bacon	440
Banana	76
Bread	246
Butter	740
Cornflakes	370
Eggs	148
Porridge	368
Sausages	186
Yoghurt	62
Apple juice (100 ml)	41
Orange juice (100 ml)	36
Semi-skimmed milk (100 ml)	48
Skimmed milk (100 ml)	34
Sugar (1 teaspoonful)	20
Tea or coffee (black)	0

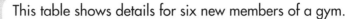

This table shows details for six new members of a gym.

Person	Height & weight	Breakfast	Exercise machine
Lynn	height 162 cm weight 60 kg	150 g toast 20 g butter 250 ml orange juice	cross-trainer (9 cal/min) for 15 minutes
Jess	height 168 cm weight 82 kg	50 g bacon 200 g sausages 50 g eggs tea with 40 ml semi-skimmed milk	walking (6 cal/min) for 30 minutes
Andy	height 180 cm weight 75 kg	100 g apple 100 g banana 150 g yoghurt 200 ml apple juice	running (8 cal/min) for 20 minutes
Dave	height 192 cm weight 95 kg	50 g bacon 150 g bread 20 g butter 200 ml semi-skimmed milk	exercise bike (7 cal/min) for 20 minutes
Pete	height 175 cm weight 60 kg	30 g cornflakes 300 ml semi-skimmed milk tea with 40 ml skimmed milk 2 teaspoons sugar	step machine (9 cal/min) for 15 minutes
Ola	height 165 cm weight 45 kg	75 g porridge 300 ml semi-skimmed milk black coffee 1 teaspoon sugar	rowing (8 cal/min for 20 minutes

Your task

Design a healthy breakfast and then look at how long it would take to burn off the calories contained in this breakfast on each exercise machine.

Use these points to help you in your task:

- Think about the different breakfast options that you find in a canteen, at home or in a cafe.
- Try to calculate the calories in some of these breakfast options.
- Now think about how much exercise you should do to burn off these calories.

Why this chapter matters

This chapter extends the idea of statistical representation (which you first met in Chapter 5) by introducing pie charts and scatter diagrams.

Pie charts

The pie chart first appeared in 1801 in a publication called *The Statistical Breviary* by William Playfair. Do you remember him? He was one of the first to use bar charts, too.

William Playfair used graphical representations of quantitative data, such as bar charts and pie charts, because he believed that "making an appeal to the eye when trying to show data is the best and easiest method of giving any message that might be wanted to show through such diagrams".

He used circles in interesting ways to represent quantitative relationships, varying their sizes and subdividing them into slices to create circular charts – or 'pie' charts.

The term 'pie chart' was not actually used until years later and it is not the only food metaphor that has been used to describe it. The French referred to it as a camembert – a soft, round cheese!

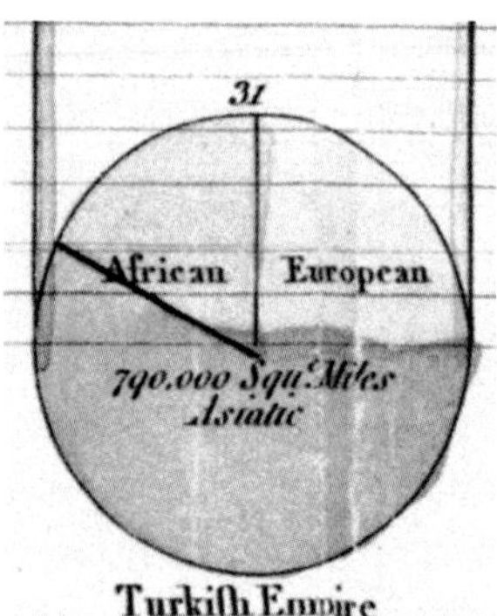

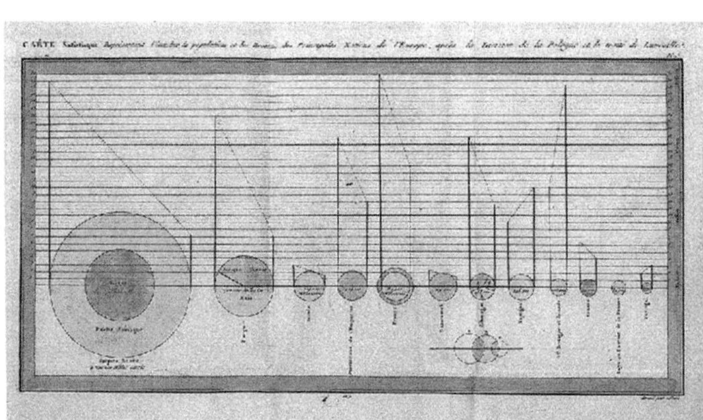

He was said to have been amused at the image of the pie chart (above left), showing a smile, and also liked the use of the pie chart above right, which is now seldom used as it's not clear which area is which.

Scatter diagrams

The word 'scatter' comes to us from Scandinavian influences in the 12th century, but we didn't see any scatter diagrams until one appeared in 1924 in a document from a university in what is now Pakistan. Apparently, they were one of the first establishments to use the technique of plotting points from two sources to see if any connections could be seen between the two.

Scatter diagrams were not used very much until the great energy debate in the late 1960s, when prices and sales of both gas and electricity were being studied. At that time, there was pressure for people to use more electricity, as it was thought to be an infinite power source, whereas gas would seemingly run out one day soon.

We now use both pie charts and scatter diagrams every day, for example, in business presentations, social studies and polls.

Chapter 15

Statistics: Pie charts, scatter diagrams and surveys

The grades given in this chapter are target grades.

This chapter will show you ...

- **E** how to draw and interpret pie charts
- **D** how to design a survey sheet and questionnaire
- **C** how to draw scatter diagrams and lines of best fit
- **C** how to interpret scatter diagrams and the different types of correlation
- **C** how to use the data-handling cycle
- **C** some of the common features of social statistics
- **C** how to describe the data-handling cycle

Visual overview

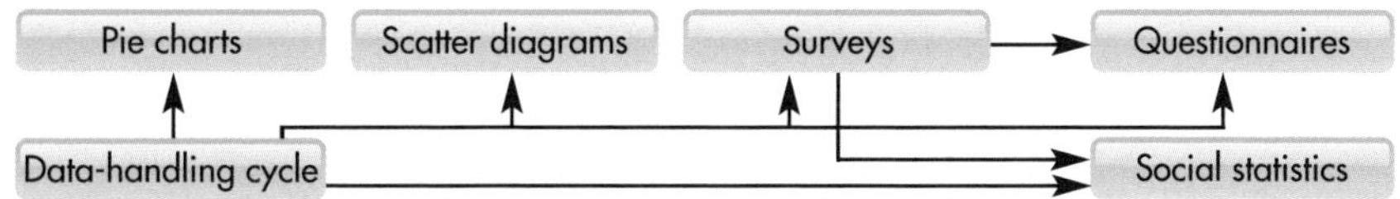

What you should already know

- How to draw and interpret pictograms, bar charts and line graphs **(KS3 level 3, GCSE grade G)**
- How to draw and measure angles **(KS3 level 4, GCSE grade F)**
- How to plot coordinates **(KS3 level 4, GCSE grade G)**

Quick check

1 The bar chart shows how many boys and girls are in five Year 7 forms.

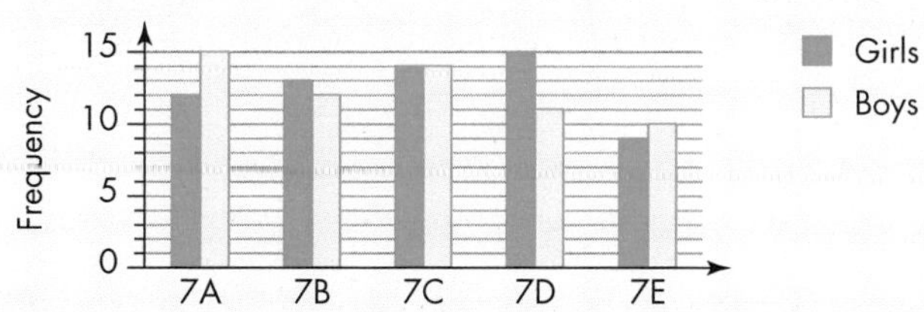

a How many pupils are in 7A?

b How many boys altogether are in the five forms?

2 Draw an angle of 72°.

3 Three points, A, B and C, are shown on the coordinate grid. What are the coordinates of A, B and C?

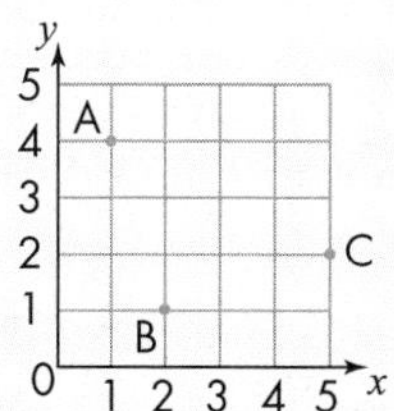

15.1 Pie charts

This section will show you how to:
- draw pie charts

Key words
angle
pie chart
sector

Pictograms, bar charts and line graphs (see Chapter 5) are easy to draw but they can be difficult to interpret when there is a big difference between the frequencies or there are only a few categories. In these cases, it is often more convenient to illustrate the data on a **pie chart**.

In a pie chart, the whole of the data is represented by a circle (the 'pie') and each category of it is represented by a **sector** of the circle (a 'slice of the pie'). The **angle** of each sector is proportional to the frequency of the category it represents.

So, a pie chart cannot show individual frequencies, like a bar chart can, for example. It can only show proportions.

Sometimes the pie chart will be marked off in equal sections rather than angles. In these cases, the numbers are always easy to work with.

EXAMPLE 1

20 people were surveyed about their preferred drink. Their replies are shown in the table.

Drink	Tea	Coffee	Milk	Pop
Frequency	6	7	4	3

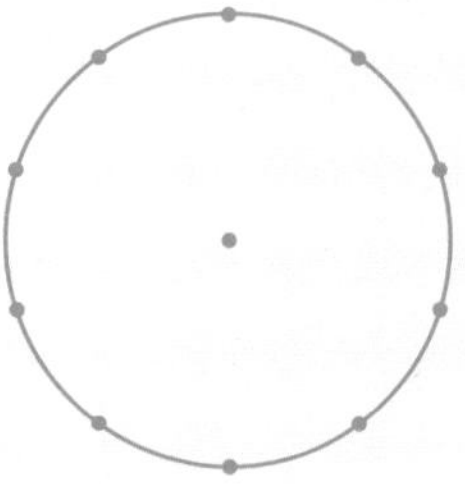

Show the results on the pie chart given.

You can see that the pie chart has 10 equally-spaced divisions.

As there are 20 people, each division is worth two people. So the sector for tea will have three of these divisions. In the same way, coffee will have $3\frac{1}{2}$ divisions, milk will have 2 divisions and pop will have $1\frac{1}{2}$ divisions.

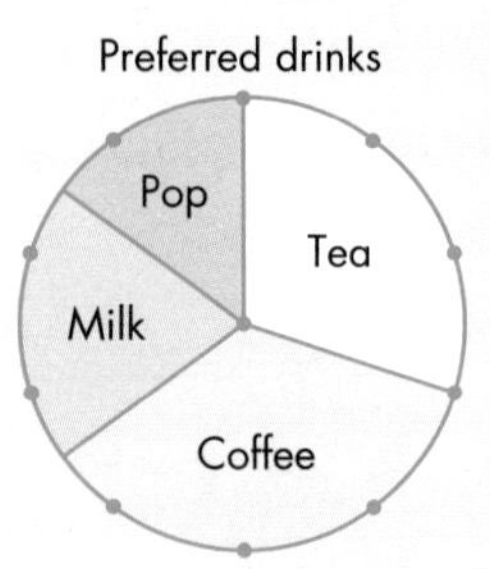

The finished pie chart will look like the one in the diagram.

Note:

- You should always label the sectors of the chart (use shading and a separate key if there is not enough space to write on the chart).
- Give your chart a title.

EXAMPLE 2

In a survey on holidays, 120 people were asked to state which type of transport they used on their last holiday. This table shows the results of the survey. Draw a pie chart to illustrate the data.

Type of transport	Train	Coach	Car	Ship	Plane
Frequency	24	12	59	11	14

You need to find the angle for the fraction of 360° that represents each type of transport. This is usually done in a table, as shown below.

Type of transport	Frequency	Calculation	Angle
Train	24	$\frac{24}{120} \times 360° = 72°$	72°
Coach	12	$\frac{12}{120} \times 360° = 36°$	36°
Car	59	$\frac{59}{120} \times 360° = 177°$	177°
Ship	11	$\frac{11}{120} \times 360° = 33°$	33°
Plane	14	$\frac{14}{120} \times 360° = 42°$	42°
Totals	120		360°

Draw the pie chart, using the calculated angle for each sector.

Note:

- Use the frequency total (120 in this case) to calculate each fraction.
- Check that the sum of all the angles is 360°.
- Label each sector.
- The angles or frequencies do not have to be shown on the pie chart.

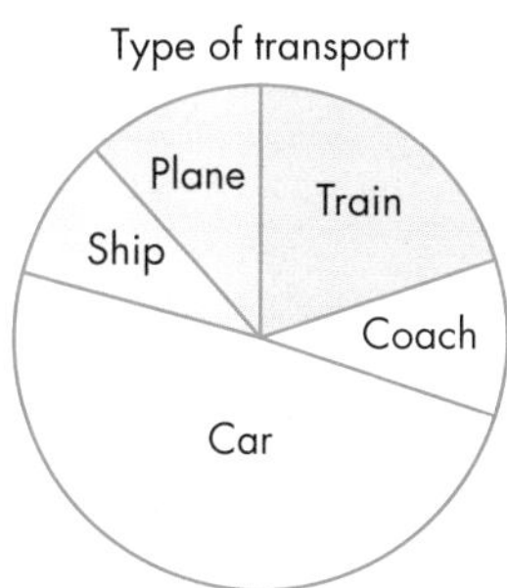

EXERCISE 15A

1 Copy the basic pie chart on the right and draw a pie chart to show each of the following sets of data.

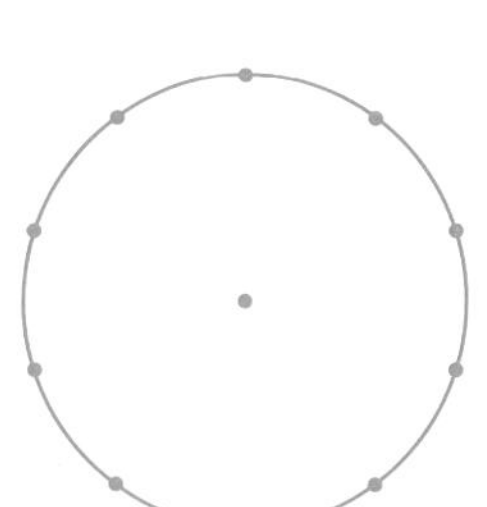

a The favourite pets of 10 children.

Pet	Dog	Cat	Rabbit
Frequency	4	5	1

b The makes of cars of 20 teachers.

Make of car	Ford	Toyota	Vauxhall	Nissan	Peugeot
Frequency	4	5	2	3	6

F

E

c The newspaper read by 40 office workers.

Newspaper	*Sun*	*Mirror*	*Guardian*	*The Times*
Frequency	14	8	6	12

2 Draw a pie chart to represent each of the following sets of data.

HINTS AND TIPS

Remember to complete a table as shown in the examples. Check that all angles add up to 360°.

a The number of children in 40 families.

No. of children	0	1	2	3	4
Frequency	4	10	14	9	3

b The favourite soap-opera of 60 students.

Programme	*Home and Away*	*Neighbours*	*Coronation Street*	*Eastenders*	*Emmerdale*
Frequency	15	18	10	13	4

c How 90 students get to school.

Journey to school	Walk	Car	Bus	Cycle
Frequency	42	13	25	10

3 Mariam asked 24 of her friends which sport they preferred to play. Her data is shown in this frequency table.

Sport	Rugby	Football	Tennis	Squash	Basketball
Frequency	4	11	3	1	5

Illustrate her data on a pie chart.

D

AU 4 Andy wrote down the number of lessons he had per week in each subject on his school timetable.

Mathematics 5 English 5 Science 8 Languages 6
Humanities 6 Arts 4 Games 2

a How many lessons did Andy have on his timetable?

b Draw a pie chart to show the data.

c Draw a bar chart to show the data.

d Which diagram better illustrates the data? Give a reason for your answer.

AU 5 In the run up to an election, 720 people were asked in a poll which political party they would vote for. The results are given in the table.

Conservative	248
Labour	264
Liberal-Democrat	152
Green Party	56

a Draw a pie chart to illustrate the data.

b Why do you think pie charts are used to show this sort of information during elections?

D

6 This pie chart shows the proportions of the different shoe sizes worn by 144 pupils in Year 11 in a London school.

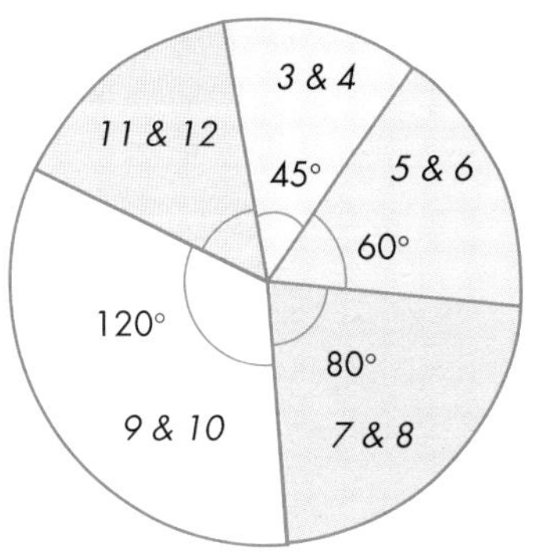

a What is the angle of the sector representing shoe sizes 11 and 12?

b How many pupils had a shoe size of 11 or 12?

c What percentage of pupils wore the modal size?

AU 7 The table below shows the numbers of candidates, at each grade, taking music examinations in Strings and Brass.

	Grades					
	3	**4**	**5**	**6**	**7**	**Total number of candidates**
Strings	300	980	1050	600	70	3000
Brass	250	360	300	120	70	1100

a Draw a pie chart to represent each of the two examinations.

b Compare the pie charts to decide which group of candidates, Strings or Brass, did better overall. Give reasons to justify your answer.

PS 8 In a survey, a rail company asked passengers whether their service had improved.

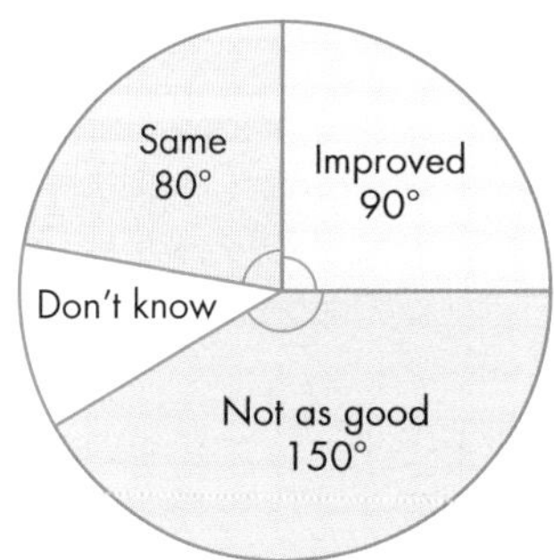

What is the probability that a person picked at random from this survey answered "Don't know"?

AU 9 You have been asked to draw a pie chart representing the different ways in which students come to school one morning.

What data would you collect to do this?

15.2 Scatter diagrams

This section will show you how to:

- draw, interpret and use scatter diagrams

Key words
correlation
line of best fit
negative correlation
no correlation
positive correlation
scatter diagram
variable

A **scatter diagram** (also called a scattergraph or scattergram) is a method of comparing two **variables** by plotting their corresponding values on a graph. These values are usually taken from a table.

In other words, the variables are treated just like a set of (x, y) coordinates. This is shown in the scatter diagram that follows, in which the marks scored in an English test are plotted against the marks scored in a mathematics test.

This graph shows **positive correlation**. This means that pupils who get high marks in mathematics tests also tend to get high marks in English tests.

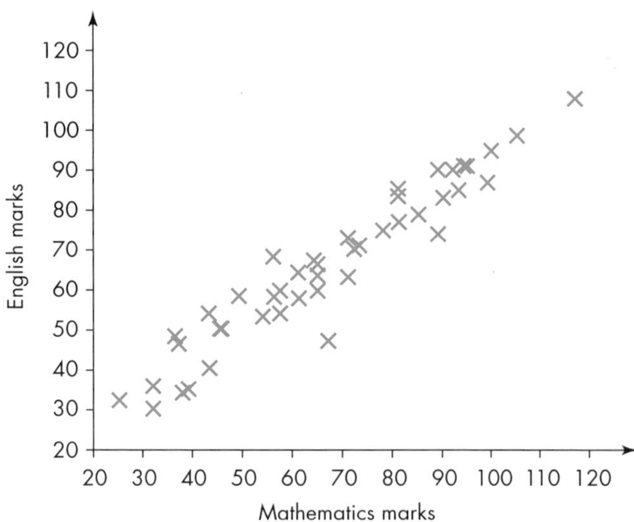

Correlation

There are different types of **correlation**. Here are three statements that may or may not be true.

- The taller people are, the wider their arm span is.
- The older a car is, the lower its value will be.
- The distance you live from your place of work will affect how much you earn.

These relationships could be tested by collecting data and plotting the data on a scatter diagram. For example, the first statement may give a scatter diagram like the first one below.

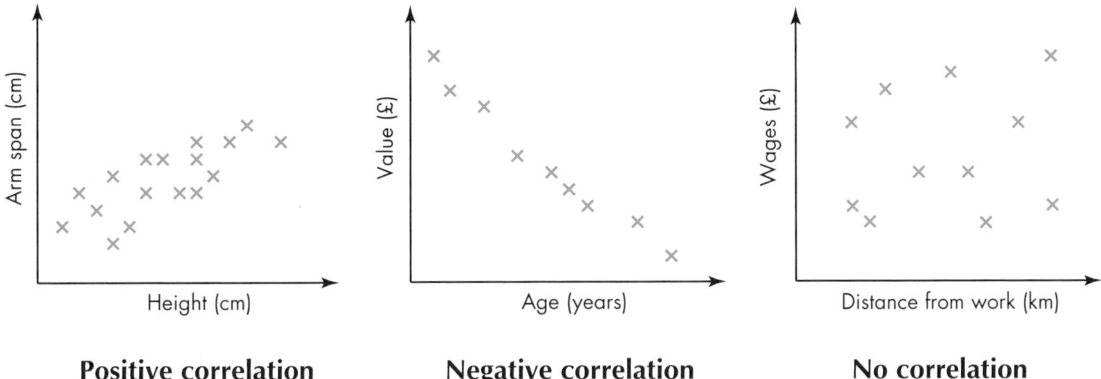

Positive correlation **Negative correlation** **No correlation**

This first diagram has **positive correlation** because, as one quantity increases, so does the other. From such a scatter diagram, you could say that the taller someone is, the wider their arm span.

Testing the second statement may give a scatter diagram like the middle one above. This has **negative correlation** because, as one quantity increases, the other quantity decreases. From such a scatter diagram, you could say that, as a car gets older, its value decreases.

Testing the third statement may give a scatter diagram like the one on the right, above. This scatter diagram has **no correlation**. There is no relationship between the distance a person lives from their work and how much they earn.

EXAMPLE 3

The graphs below show the relationship between the temperature and the amount of ice-cream sold, and that between the age of people and the amount of ice-cream they eat.

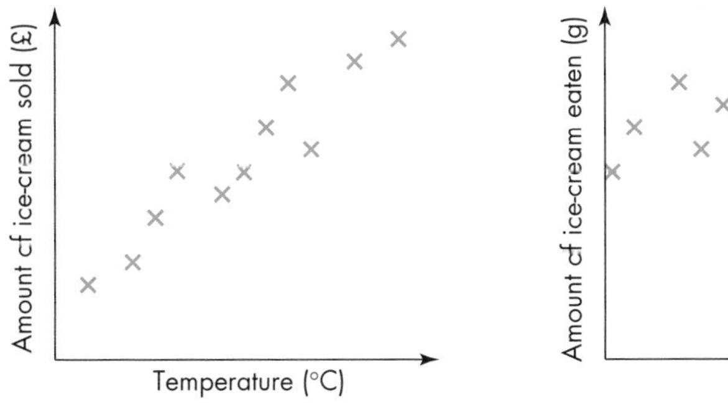

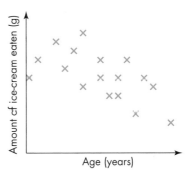

a Comment on the correlation of each graph.

b What does each graph tell you?

The first graph has positive correlation and tells us that, as the temperature increases, the amount of ice-cream sold increases.

The second graph has negative correlation and tells us that, as people get older, they eat less ice-cream.

Line of best fit

A **line of best fit** is a straight line that goes between all the points on a scatter diagram, passing as close as possible to all of them. You should try to have the same number of points on both sides of the line. Because you are drawing this line by eye, examiners make a generous allowance around the correct answer. The line of best fit for the scatter diagram at the start of this section is shown below, left.

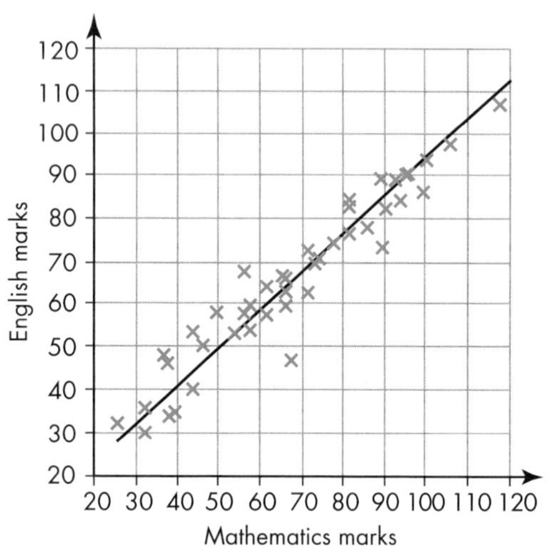

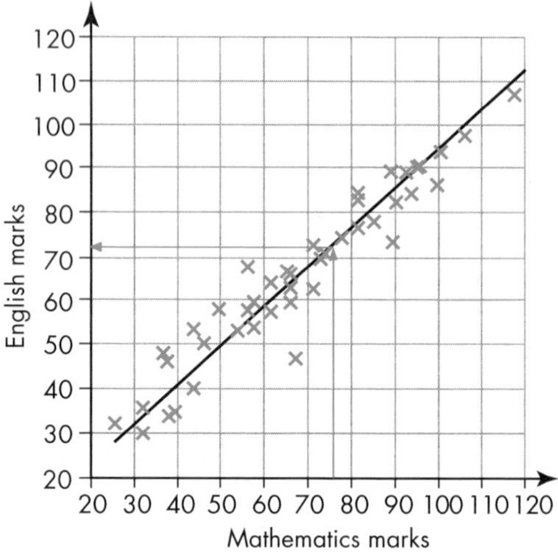

The line of best fit can be used to answer questions such as: "A girl took the mathematics test and scored 75 marks but was ill for the English test. How many marks was she likely to have scored?"

The answer is found by drawing a line up from 75 on the mathematics axis to the line of best fit and then drawing a line across to the English axis as shown in the graph above, right. This gives 73, which is the mark she is likely to have scored in the English test.

EXERCISE 15B

E

1 Describe the correlation of each of these four graphs.

a

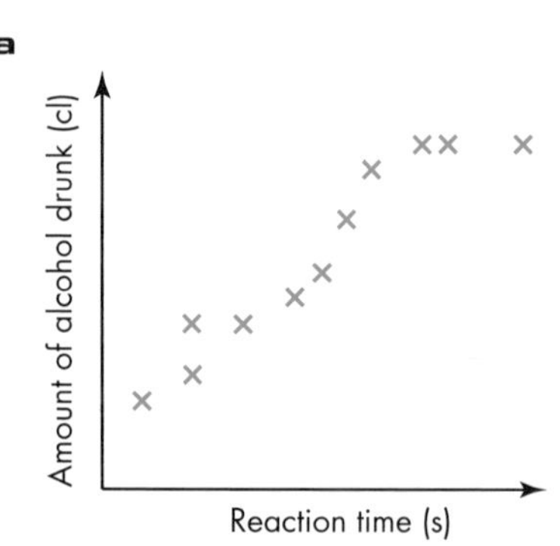

b

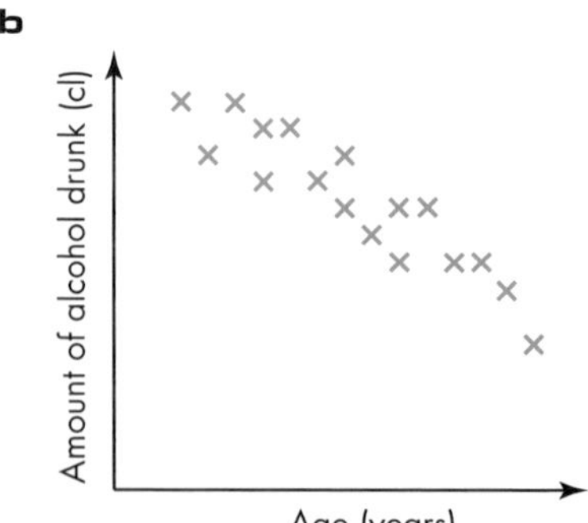

c

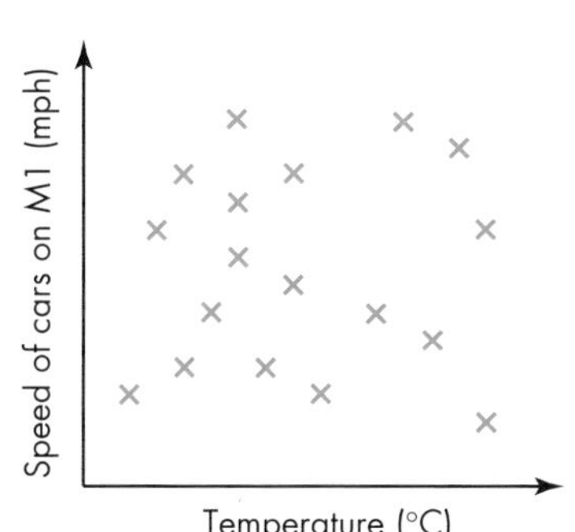

d

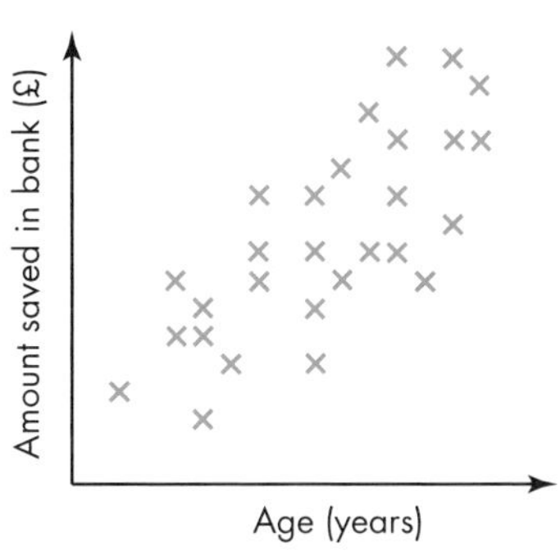

2 Write in words what each graph in question 1 tells you.

3 The table below shows the results of a science experiment in which a ball is rolled along a desk top. The speed of the ball is measured at various points.

Distance from start (cm)	10	20	30	40	50	60	70	80
Speed (cm/s)	18	16	13	10	7	5	3	0

a Plot the data on a scatter diagram.

b Draw the line of best fit.

c If the ball's speed had been measured at 5 cm from the start, what is it likely to have been?

d Estimate how far the ball was from the start when its speed was 12 cm/s.

HINTS AND TIPS

Often in exams axes are given and most, if not all, of the points are plotted.

4 The heights, in centimetres, of 20 mothers and their 15-year-old daughters were measured. These are the results.

Mother	153	162	147	183	174	169	152	164	186	178
Daughter	145	155	142	167	167	151	145	152	163	168
Mother	175	173	158	168	181	173	166	162	180	156
Daughter	172	167	160	154	170	164	156	150	160	152

a Plot these results on a scatter diagram. Take the x-axis for the mothers' heights from 140 to 200. Take the y-axis for the daughters' heights from 140 to 200.

b Is it true that the tall mothers have tall daughters?

C

FM **5** The table below shows the marks for ten students in their mathematics and geography examinations.

Student	Anna	Becky	Cath	Dema	Emma	Fatima	Greta	Hannah	Imogen	Sitara
Maths	57	65	34	87	42	35	59	61	25	35
Geog	45	61	30	78	41	36	35	57	23	34

a Plot the data on a scatter diagram. Take the x-axis for the mathematics scores and the y-axis for the geography scores.

b Draw the line of best fit.

c One of the students was ill when she took the geography examination. Which student was it most likely to be?

d If another student, Kate, was absent for the geography examination but scored 75 in mathematics, what mark would you expect her to have scored in geography?

e If another student, Lina, was absent for the mathematics examination but scored 65 in geography, what mark would you expect her to have scored in mathematics?

FM **6** A form teacher carried out a survey of 20 students from his class and asked them to say how many hours per week they spent playing sport and how many hours per week they spent watching TV. This table shows the results of the survey.

Student	1	2	3	4	5	6	7	8	9	10
Hours playing sport	12	3	5	15	11	0	9	7	6	12
Hours watching TV	18	26	24	16	19	27	12	13	17	14

Student	11	12	13	14	15	16	17	18	19	20
Hours playing sport	12	10	7	6	7	3	1	2	0	12
Hours watching TV	22	16	18	22	12	28	18	20	25	13

a Plot these results on a scatter diagram. Take the x-axis as the number of hours playing sport and the y-axis as the number of hours watching TV.

AU **b** If you knew that another student from the form watched 8 hours of TV a week, would you be able to predict how long they spent playing sport? Explain why.

FM **7** The table shows the times taken and distances travelled by a taxi driver in 10 journeys on one day.

Distance (km)	1.6	8.3	5.2	6.6	4.8	7.2	3.9	5.8	8.8	5.4
Time (minutes)	3	17	11	13	9	15	8	11	16	10

a Draw a scatter diagram of this information, with time on the horizontal axis.

b Draw a line of best fit on your diagram.

c If a taxi journey takes 5 minutes, how far, in kilometres, would you expect the journey to have been?

d How much time would you expect a journey of 4 km to take?

C

PS **8** Omar records the time taken, in hours, and the average speed, in miles per hour (mph), for several different journeys.

Time (h)	0.5	0.8	1.1	1.3	1.6	1.75	2	2.4	2.6
Speed (mph)	42	38	27	30	22	23	21	9	8

Estimate the average speed for a journey of 90 minutes.

AU **9** Describe what you would expect the scatter graph to look like if someone said that it showed negative correlation.

15.3 Surveys

This section will show you how to:

- conduct surveys
- ask good questions in order to collect reliable and valid data

Key words
data-collection sheet
leading question
questionnaire
response
survey

A **survey** is an organised way of asking a lot of people a few, well-constructed questions, or of making a lot of observations in an experiment, in order to reach a conclusion about something. Surveys are used to test out people's opinions or to test a hypothesis.

Simple data-collection sheet

If you need to collect some data to analyse, you will have to design a simple **data-collection sheet**.

Look at this example: "Where do you want to go for the Year 10 trip at the end of term – Blackpool, Alton Towers, The Great Western Show or London?"

You would put this question on the same day to a lot of Year 10 students and enter their answers straight onto a data-collection sheet, as below.

Place	Tally	Frequency
Blackpool	𝍸 𝍸 𝍸 𝍸 III	23
Alton Towers	𝍸 𝍸 𝍸 𝍸 𝍸 𝍸 𝍸 𝍸 𝍸 I	46
The Great Western Show	𝍸 𝍸 IIII	14
London	𝍸 𝍸 𝍸 𝍸 II	22

Notice how plenty of space is left for the tally marks and how the tallies are 'gated' in groups of five to make counting easier when the survey is complete.

This is a good, simple data-collection sheet because:

- only one question is asked ("Where do you want to go?")
- all the possible venues are listed
- the answer from each interviewee can be easily and quickly tallied, and the next interviewee questioned.

Notice too that, since the question listed specific places, they must all appear on the data collection sheet. You would lose marks in an examination if you just asked the open question: "Where do you want to go?"

Data sometimes needs to be collected to obtain **responses** for two different categories. The data-collection sheet is then in the form of a simple two-way table.

EXAMPLE 4

The head of a school wants to find out how long his students spend doing homework in a week. He carries out a survey on 60 students. He uses the two-way table to show the result.

	0–5 hours	0–10 hours	10–20 hours	More than 20 hours
Year 7				

This is not a good table as the categories overlap. A student who does 10 hours' work a week could tick any of two columns. Response categories should not overlap and there should be only one possible place to put a tick.

A better table would be:

	0 up to 5 hours	More than 5 and up to 10 hours	More than 10 and up to 15 hours	More than 15 hours
Year 7	𝍸 II	𝍸		
Year 8	𝍸	𝍸 II		
Year 9	III	𝍸 II	II	
Year 10	III	𝍸	III	I
Year 11	II	IIII	IIII	II

This gives a more accurate picture of the amount of homework done in each year group.

Using your computer

Once the data has been collected for your survey, it can be put into a computer database. This allows the data to be stored and amended or updated at a later date if necessary.

From the database, suitable statistical diagrams can easily be drawn within the software and averages calculated for you. Your results can then be published in, for example, the school magazine.

EXERCISE 15C

FM **1** "People like the supermarket to open on Sundays."

a To see whether this statement is true, design a data-collection sheet that will allow you to capture data while standing outside a supermarket.

b Does it matter on which day you collect data outside the supermarket?

2 The school tuck shop wanted to know which types of chocolate it should order to sell – plain, milk, fruit and nut, wholenut or white chocolate.

a Design a data-collection sheet that you could use to ask the pupils in your school which of these chocolate types are their favourite.

b Invent the first 30 entries on the chart.

> **HINTS AND TIPS**
> Include space for tallies.

3 What type of television programme do people in your age group watch the most? Is it crime, romance, comedy, documentary, sport or something else? Design a data-collection sheet to be used in a survey of your age group.

4 On what do people of your age tend to spend their money? Is it sport, magazines, clubs, cinema, sweets, clothes or something else? Design a data-collection sheet to be used in a survey of your age group.

5 Design two-way tables to show the following. Invent about 40 entries for each one.

a How students in different year groups travel to school in the morning.

b The type of programme that different age groups prefer to watch on TV.

c The favourite sport of boys and girls.

d How much time students in different year groups spend on the computer in the evening.

> **HINTS AND TIPS**
> Make sure all possible responses are covered.

FM **6** Carlos wanted to find out who eats healthy food.
He decided to investigate the hypothesis:

"Boys are less likely to eat healthy food than girls are."

a Design a data-capture form that Carlos could use to help him do this.

b Carlos records information from a sample of 40 boys and 25 girls. He finds that 17 boys and 15 girls eat healthy food. Based on this sample, is the hypothesis correct? Explain your answer.

AU **7** Show how you would find out what kind of tariffs your classmates use on their mobile phones.

AU **8** You have been asked to find out which shops the parents of the students at your school like to use. When creating a data-collection sheet, what two things must you include?

D

C

Questionnaires

When you are putting together a **questionnaire**, you must think very carefully about the sorts of question you are going to ask, to put together a clear, easy-to-use questionnaire.

Here are five rules that you should *always* follow.

- Never ask a **leading question** designed to get a particular response.
- Never ask a personal, irrelevant question.
- Keep each question as simple as possible.
- Include questions that will get a response from whomever is asked.
- Make sure the categories for the responses do not overlap and keep the number of choices to a reasonable number (six at the most).

The following questions are *badly constructed* and should *never* appear in any questionnaire.

✘ *What is your age?* This is personal. Many people will not want to answer. It is always better to give a range of ages such as:

☐ Under 15 ☐ 16–20 ☐ 21–30 ☐ 31–40 ☐ Over 40

✘ *Slaughtering animals for food is cruel to the poor defenceless animals. Don't you agree?* This is a leading question, designed to get a 'yes'. It is better ask an impersonal question such as:

Are you a vegetarian? ☐ Yes ☐ No

✘ *Do you go to discos when abroad?* This can be answered only by those who have been abroad. It is better to ask a starter question, with a follow-up question such as:

Have you been abroad for a holiday? ☐ Yes ☐ No

If 'Yes', did you go to a disco whilst you were away? ☐ Yes ☐ No

✘ *When you first get up in a morning and decide to have some sort of breakfast that might be made by somebody else, do you feel obliged to eat it all or not?* This question is too complicated. It is better to ask a series of shorter questions such as:

What time do you get up for school? ☐ Before 7 ☐ Between 7 and 8 ☐ After 8

Do you have breakfast every day? ☐ Yes ☐ No

If 'No', on how many schooldays do you have breakfast? ☐ 0 ☐ 1 ☐ 2 ☐ 3 ☐ 4 ☐ 5

A questionnaire is usually put together to test a hypothesis or a statement. For example: "People buy cheaper milk from the supermarket as they don't mind not getting it on their doorstep. They'd rather go out to buy it."

A questionnaire designed to test whether this statement is true or not should include these questions:

✓ *Do you have milk delivered to your doorstep?*
✓ *Do you buy cheaper milk from the supermarket?*
✓ *Would you buy your milk only from the supermarket?*

Once the data from these questions has been collected, it can be looked at to see whether or not the majority of people hold views that agree with the statement.

EXERCISE 15D

D

FM **1** These are questions from a questionnaire on healthy eating.

a *Fast food is bad for you. Don't you agree?*

☐ Strongly agree ☐ Agree ☐ Don't know

Give two criticisms of the question.

b *Do you eat fast food?* ☐ Yes ☐ No

If 'Yes', how many times on average per week do you eat fast food?

☐ Once or less ☐ 2 or 3 times ☐ 4 or 5 times ☐ More than 5 times

Give two reasons why this is a good question.

2 This is a question from a survey on pocket money:

How much pocket money do you get each week?

☐ £0–£2 ☐ £0–£5 ☐ £5–£10 ☐ £10 or more

a Give a reason why this is not a good question.

AU **b** Rewrite the question to make it a good question.

C

3 Design a questionnaire to test the following statement.

People under sixteen do not know what is meant by all the jargon used in the business news on TV, but the over-twenties do.

HINTS AND TIPS

Keep questions simple with clear response categories and no overlapping.

4 *The under-twenties feel quite at ease with computers, while the over-forties would rather not bother with them. The twenty-to-forties always try to look good with computers.*

Design a questionnaire to test this statement.

5 Design a questionnaire to test the following hypothesis.

"The older you get, the less sleep you need."

PS **6** Design a questionnaire to test the following hypothesis:

"People with back problems do not sit properly."

AU **7** For a survey, an assistant gave every customer leaving a store a questionnaire that included the following question.

Question: How much do you normally spend in this shop?		
Response:	Less than £15 ☐	More than £25 ☐
	Less than £25 ☐	More than £50 ☐

Explain why the response section of this questionnaire is poor.

15.4 The data-handling cycle

This section will show you how:

- use the data-handling cycle to test a hypothesis

Key words
bias
hypothesis
population
primary data
sample
secondary data

The data-handling cycle

Testing out a **hypothesis** involves a cycle of planning, collecting data, evaluating the significance of the data and then interpreting the results, which may or may not show the hypothesis to be true. This cycle often leads to a refinement of the problem, which starts the cycle all over again.

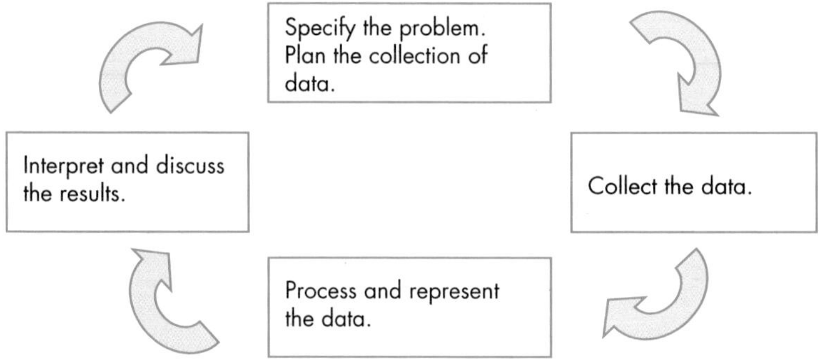

There are four parts to the data-handling cycle.

1 State the hypothesis, which is the idea being tested, outlining the problem and planning what needs to be done.

2 Plan the data collection and collect the data. Record the data collected clearly.

3 Choose the best way to process and represent the data. This will normally mean calculating averages (mean, median, mode) and measures of spread, then representing data in suitable diagrams.

4 Interpret the data and make conclusions.

Then the hypothesis can be refined or changes made to the data collected, for example a different type of data can be collected or the same data can be collected in a different way. In this way, the data-handling cycle helps to improve reliability in the collection and interpretation of data.

EXAMPLE 5

A gardener grows tomatoes, both in a greenhouse and outside.

He wants to investigate the following hypothesis:

"Tomato plants grown inside the greenhouse produce more tomatoes than those grown outside."

Describe the data-handling cycle that may be applied to this problem.

Plan the data collection. Consider 10 tomato plants grown in the greenhouse, and 10 plants grown outside. Count the tomatoes on each plant.

Collect the data. Record the numbers of tomatoes collected from the plants between June and September. Only count those that are 'fit for purpose'.

Choose the best way to process and represent the data. Calculate the mean number collected per plant, as well as the range.

Interpret the data and make conclusions. Look at the statistics. What do they show? Is there a clear conclusion or do you need to alter the hypothesis in any way? Discuss the results, refine the method and continue the cycle.

As you see, in describing the data-handling cycle, you must refer to each of the four parts.

Data Collection

Data that you collect yourself is called **primary data**. You control it, in terms of accuracy and amount.

Data collected by someone else is called **secondary data**. Generally, there is a lot of this type of data available on the internet or in newspapers. This provides a huge volume of data but you have to rely on the sources being reliable, for accuracy.

EXERCISE 15E

Use the data-handling cycle to describe how you would test each of the following hypotheses. In each case state whether you would use primary or secondary data.

1. August is the hottest month of the year.
2. Boys are better than girls at estimating distances.
3. More men go to football matches than women.
4. Tennis is watched by more women than men.
5. The more revision you do, the better your exam results.
6. The older you are, the more likely you are to shop at a department store.

C

15.5 Other uses of statistics

This section will show you how:

- statistics are used in everyday life and what information the government needs about the population

Key words
margin of error
national census
polls
retail price index
social statistics
time series

This section will explain about **social statistics** and introduce some of the more common ones in daily use.

In daily life, many situations occur in which statistical techniques are used to produce data. The results of surveys appear in newspapers every day. There are many on-line **polls** and phone-ins that give people the chance to vote, such as in reality TV shows.

Results for these are usually given as a percentage with a **margin of error**, which is a measure of how accurate the information is.

Some common social statistics in daily use are briefly described below.

General index of retail prices

This is also know as the **retail price index** (RPI) and it measures how much the daily cost of living increases (or decreases). One year is chosen as the base year and given an index number, usually 100. The corresponding costs in subsequent years are compared to this and given a number proportional to the base year, such as 103.

Note: The numbers do not represent actual values but just compare current prices to those in the base year.

Time series

Like the RPI, a **time series** measures changes in a quantity over time. Unlike the RPI, though, the actual values of the quantity are used. A time series might track, for example, how the exchange rate between the pound and the dollar changes over time.

National census

A **national census** is a survey of all people and households in a country. Data about categories such as age, gender, religion and employment status is collected to enable governments to plan where to allocate future resources. In Britain a national census is taken every 10 years. The most recent census was in 2001.

EXERCISE 15F

1 In 2004 the cost of a litre of petrol was 78p. Using 2004 as a base year, the price index of petrol for each of the next five years is shown in this table.

Year	2004	2005	2006	2007	2008	2009
Index	100	103	108	109	112	120
Price	78p					

Work out the price of petrol in each subsequent year.

Give your answers to 1 decimal place.

2 The following is taken from the UK government statistics website.

In mid-2004 the UK was home to 59.8 million people, of which 50.1 million lived in England. The average age was 38.6 years, an increase on 1971 when it was 34.1 years. In mid-2004 approximately one in five people in the UK were aged under 16 and one in six people were aged 65 or over.

Use this extract to answer the following questions about the UK in 2004.

a How many of the population of the UK *did not* live in England?

b By how much had the average age increased since 1971?

c Approximately how many of the population were aged under 16?

d Approximately how many of the population were aged over 65?

3 The graph shows the exchange rate for the dollar against the pound for each month in one year.

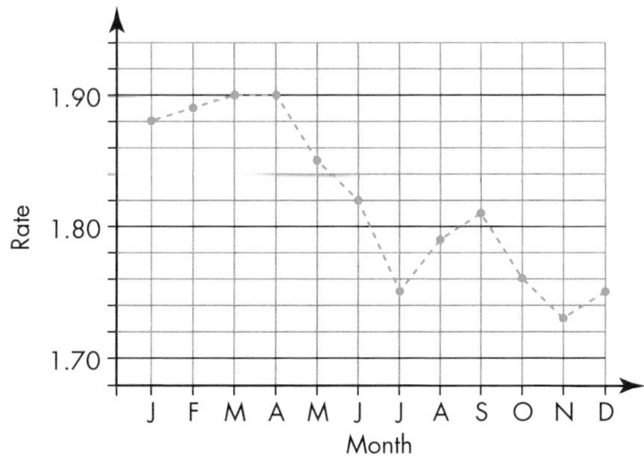

a What was the exchange rate in January?

b Between which two consecutive months did the exchange rate fall most?

c Explain why you could not use the graph to predict the exchange rate in the January following this year.

D

4 The general index of retail prices started in January 1987, when it was given a base number of 100. In January 2006 the index number was 194.1.

If the 'standard weekly shopping basket' cost £38.50 in January 1987, how much would it have cost in January 2006?

C

FM 5 The time series shows car production in Britain from November 2008 to November 2009.

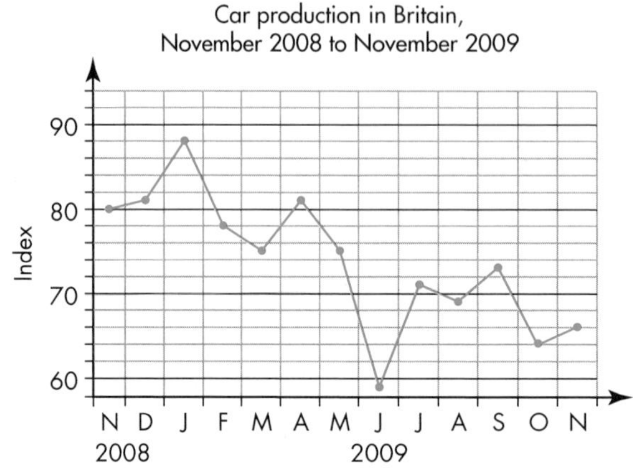

a Why was there a sharp drop in production in June 2009?

b The average production over the first three months shown was 172 000 cars.

i Work out an approximate value for the average production over the last three months shown.

ii The base month for the index is January 2005 when the index was 100. What was the approximate production in January 2005?

AU 6 The retail price index measures how much the daily cost of living increases or decreases. If 2008 is given a base index number of 100, then 2009 is given 98. What does this mean?

PS 7 On one day in 2009 Alex spent £51 in a supermarket.

She knew that the price index over the last three years was:

2007	2008	2009
100	103	102

How much would she have paid in the supermarket for the same goods in 2008?

GRADE BOOSTER

- **F** You can interpret a simple pie chart
- **E** You can draw a pie chart
- **E** You can recognise the different types of correlation
- **D** You can design a data-collection sheet
- **C** You can use a line of best fit and use it to make predictions
- **C** You can interpret a scatter diagram
- **C** You can design and criticise questions for questionnaires
- **C** You can describe the data-handling cycle

What you should know now

- How to read and draw pie charts
- How to plot scatter diagrams, recognise correlation, draw lines of best fit and use them to predict values
- How to design questionnaires and know how to ask suitable questions

EXAMINATION QUESTIONS

1 The table gives information about the medals won by Austria in the 2002 Winter Olympic Games.

Draw an accurate pie chart to show this information.

Medal	Frequency
Gold	3
Silver	4
Bronze	11

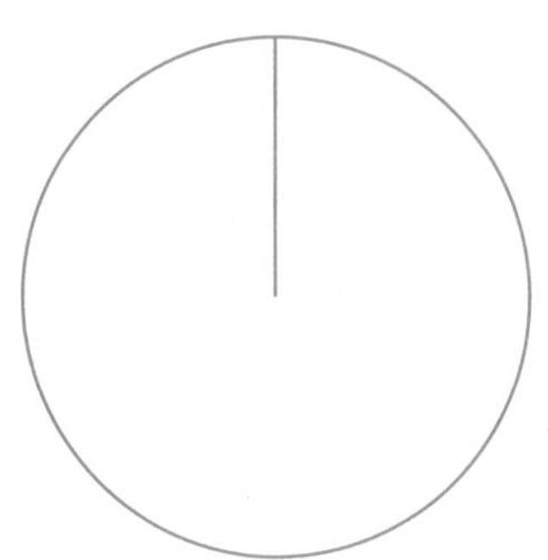

Edexcel, June 2005, Paper 4 Intermediate, Question 5

2 The table below shows how a number of men and women on a cruise ship rated their understanding of the rules of croquet.

	Totally understand	Understand	Understand some	Understand a little	Do not understand at all	Total
Men	160	520	560	320	40	1600
Women	160	240	200	80	40	720

The pie chart for men has been drawn for you.

a Copy and complete the pie chart for women.

b Which group, men or women, do you think had a better overall understanding of the rules of croquet? Give *one* reason to justify your answer.

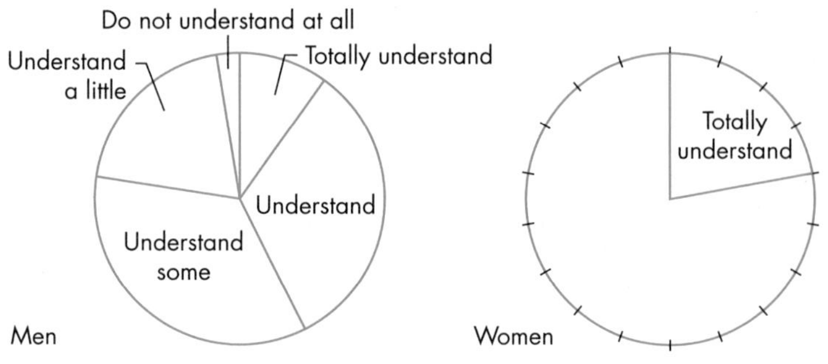

3 Some male students were asked to choose their favourite sport.

The pie chart shows information about the results.

The pie chart is drawn accurately.

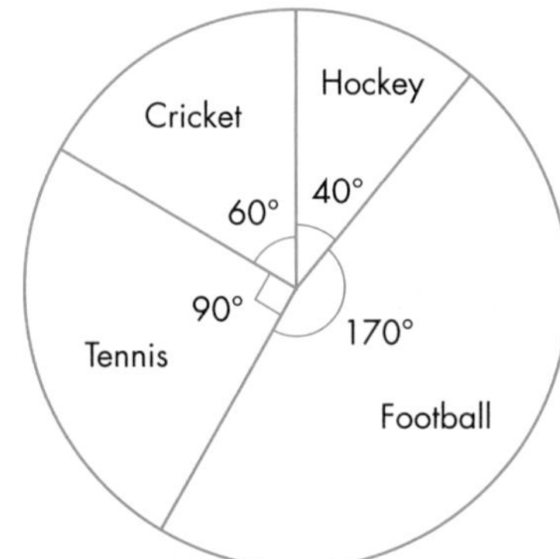

a 12 male students chose hockey.

Work out the number of male students who chose tennis. (3)

b A second pie is to be drawn for some female students.

There are 240 female students.

130 of the female students chose hockey.

Calculate the angle in the second pie chart for 130 female students. (2)

(Total 5 marks)

Edexcel, June 2009, IGCSE, Paper 2 Foundation, Question 13

C D E

4 The pie chart gives information about the mathematics exam grades of some students.

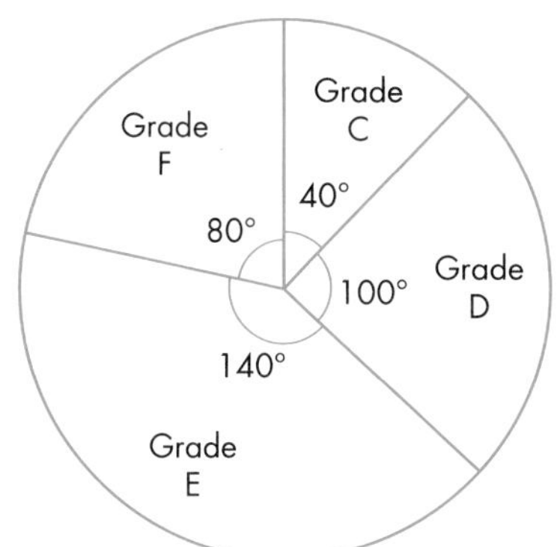

Diagram **not** accurately drawn

a What grade was the mode? (1)

b What fraction of the students got grade D? (1)

8 of the students got grade C.

c **i** How many of the students got grade F?

ii How many of the students took the exam? (3)

This accurate pie chart gives information about the English exam grades for a different set of students.

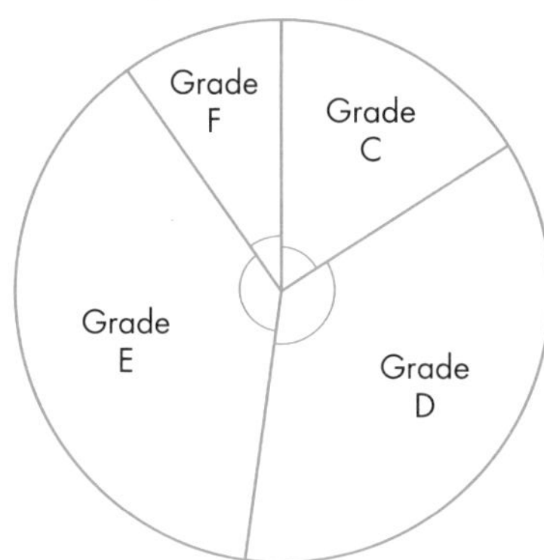

Sean says "More students got a grade D in English than in mathematics."

d Sean could be **wrong**.

Explain why. (1)

(Total 6 marks)

Edexcel, June 2008, Paper 2 Foundation, Question 16

5 The scatter diagrams below show the results of a survey on the average number of hours of sunshine in a week during the summer weeks in Bournemouth.

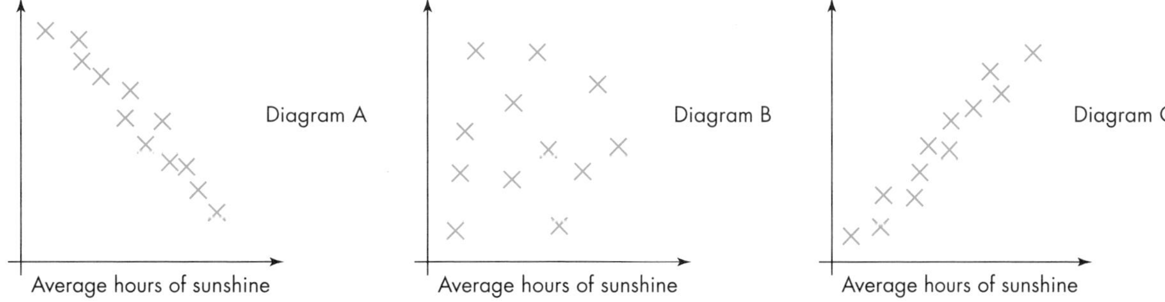

a Which scatter diagram shows the average hours of sunshine plotted against:

i the number of ice creams sold?

ii the number of umbrellas sold?

iii the number of births in the town?

b State which one of the diagrams shows a negative correlation.

6 Joy wants to find out who eats vegetarian food. She decides to investigate this hypothesis:

> Girls are more likely than boys to eat vegetarian food.

a Design a two-way table that Joy might use to help her do this.

b Joy records information from a sample of 40 boys and 30 girls. She finds that 18 boys and 16 girls eat vegetarian food. Based on this sample, is the hypothesis correct? Explain your answer.

7 Naomi wants to find out how often adults go to the cinema.

She uses this question on a questionnaire.

> "How many times do you go to the cinema?"
> Not very often ☐ Sometimes ☐
> A lot ☐

a Write down **two** things wrong with this question. (2)

b Design a better question for her questionnaire to find out how often adults go to the cinema.

You should include some response boxes. (2)

(Total 4 marks)

Edexcel, November 2008, Paper 1 Foundation, Question 24

8 Mr Brown owns a café in the town centre.

He wants to find out what people think of the service in the café.

He uses this question on his questionnaire.

> What do you think of the service in the café?
> Excellent ☐ Very good ☐ Good ☐

a Write down one thing that is wrong with this question. (1)

Mr Brown wants to find out how often people visit the town centre.

b Design a suitable question for his questionnaire to find out how often people visit the town centre.

You must include some response boxes. (2)

(Total 3 marks)

Edexcel, March 2007, Paper 8 Foundation, Unit 2 Test, Question 4

9 James wants to find out how many text messages people send.

He uses this question on a questionnaire.

> "How many text messages do you send?"
> 1 to 10 ☐ 11 to 20 ☐ 21 to 30 ☐
> more than 30 ☐

a Write down **two** things wrong with this question. (2)

James asks 10 students in his class to complete his questionnaire.

b Give **one** reason why this may not be a suitable sample. (1)

(Total 3 marks)

Edexcel, March 2009, Paper 5 Foundation, Unit 1 Test, Question 4

10 The scatter graph shows some information about six new-born baby apes. For each baby ape, it shows the mother's leg length and the baby ape's birth weight.

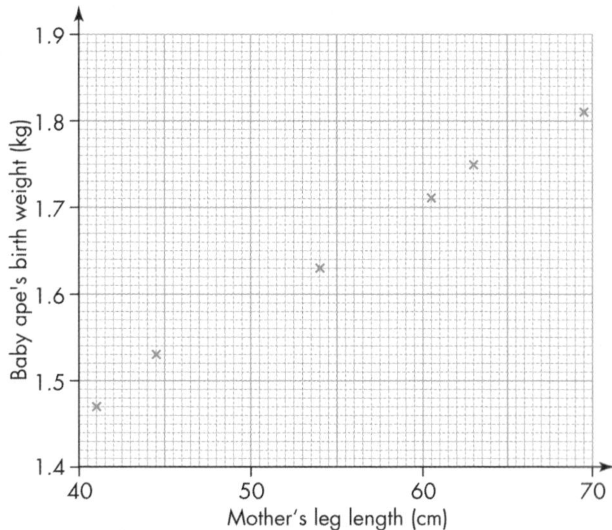

The table shows the mother's leg length and the birth weight of two more baby apes.

Mother's leg length (cm)	50	65
Baby ape's birth weight (kg)	1.6	1.75

a Copy the graph onto graph paper and plot the information from the table.

b Describe the correlation between a mother's leg length and her baby ape's birth weight.

c Draw a line of best fit on your graph.

A mother's leg length is 55 cm.

d Use your line of best fit to estimate the birth weight of her baby ape.

11 The table shows the time taken and distance travelled by a taxi driver for 10 journeys one day.

Time (min)	Distance (km)
3	1.7
17	8.3
11	5.1
13	6.7
9	4.7
15	7.3
8	3.8
11	5.7
16	8.7
10	5.3

a Plot on a grid, a scatter diagram with time, on the horizontal axis, from 0 to 20, and distance, on the vertical axis, from 0 to 10.

b Draw a line of best fit on your diagram.

c A taxi journey takes 4 minutes. How many kilometres is the journey?

d A taxi journey is 10 kilometres. How many minutes will it take?

12 Sanji goes fishing for pike.

The scatter graph shows information about the weights and lengths of some of the pike Sanji caught.

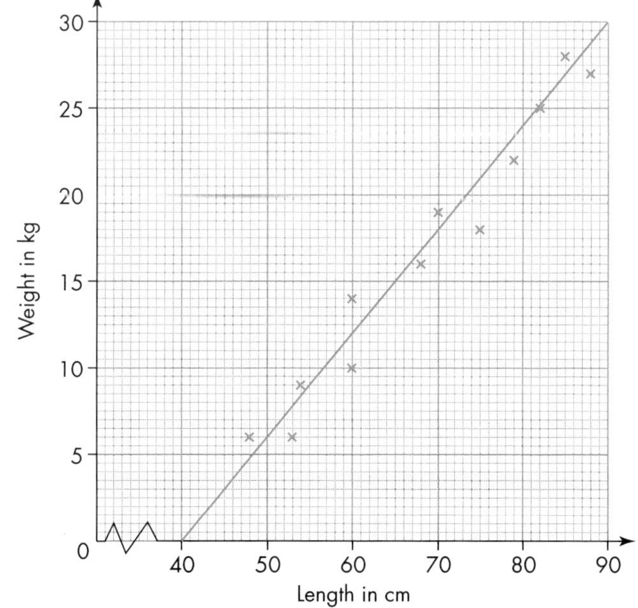

a Describe the relationship between the weight and the length of these pike. (1)

Sanji also caught a pike of weight 24 kg and length 78 cm.

b Show this information in the scatter graph. (1)

A pike has a length of 65 cm.

c Estimate the weight of this pike. (2)

(Total 4 marks)

Edexcel, June 2009, Paper 5 Foundation, Unit 1 Test, Question 4

13 The scatter graph shows information about eight sheep.

It shows the height and the length of each sheep.

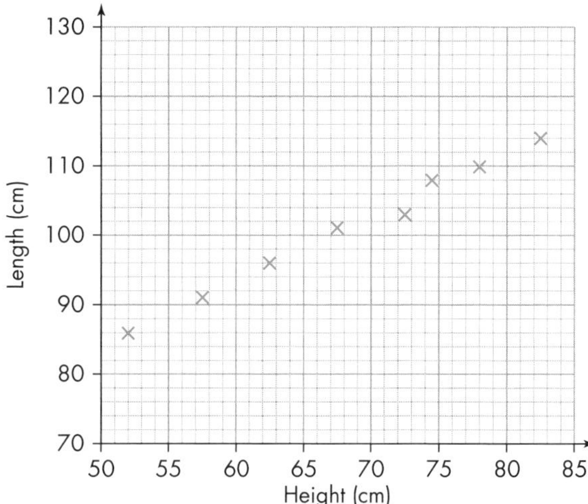

The table gives the height and the length of two more sheep.

Height (cm)	65	80
Length (cm)	100	110

a On the scatter graph, plot the information from the table. (1)

b Describe the relationship between the height and the length of these sheep. (1)

The height of a sheep is 76 cm.

c Estimate the length of this sheep. (2)

(Total 4 marks)

Edexcel, June 2009, Paper 2 Foundation, Question 21

Worked Examination Questions

1 The scatter diagram shows the relationship between the total mileage of a car and its value as a percentage of its original value.

a Which of the four points, A, B, C or D, represents each of the statements below?

Alf: I have a rare car. It has done a lot of miles but it is still worth a lot of money.

Belinda: My car is quite new. It hasn't done many miles.

Charles: My car hasn't done many miles but it is really old and rusty.

b Write a statement to match the fourth point.

c What does the graph tell you about the relationship between the mileage of a car and the percentage of its original value?

d Draw scatter diagrams to show the relationship between:

i the amount of petrol used and the distance driven

ii the value of a car and the age of the driver.

a Alf is represented by point D.
Belinda is represented by point A.
Charles is represented by point B.

3 marks

Read both axes. The horizontal axis is total mileage and the vertical axis is percentage of original value. So, Alf would be to the right of the horizontal axis and to the top of the vertical axis. Use similar reasoning for the other people. You get 1 mark for each correct person linked to a point.

b 'My car has done a lot of miles and isn't worth very much.'

1 mark

The fourth point, C, is a car that has high mileage with low value.
Any statement that says something like this answer given would get the 1 mark available.

c The more mileage a car has done, the less is its value.

1 mark

The graph basically shows weak negative correlation, so as one variable increases, the other decreases.
Using this principle to give the correct answer earns you 1 mark.

d i

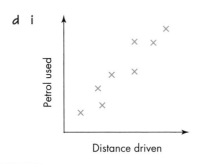

ii

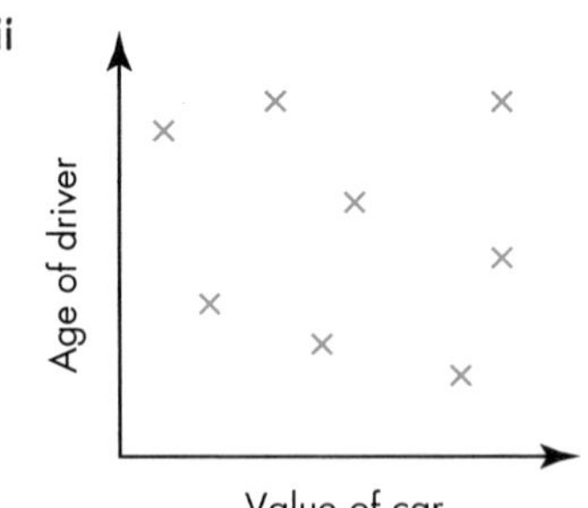

The first diagram shows positive correlation, as the more miles are driven, the more petrol is used.
The second diagram shows no correlation as there is no relationship.
You earn 1 method mark for each correct diagram.

2 marks

Total: 7 marks

Worked Examination Questions

AU **2** The table below shows the number of learners at each grade for two practice driving tests, Theory and Practical.

	Grades					
	Excellent	**Very good**	**Good**	**Pass**	**Fail**	**Total number of learners**
Theory	208	888	1032	696	56	2880
Practical	240	351	291	108	90	1080

a Represent each of the two practice tests in a pie chart.

b By comparing the pie charts, on which test, Theory or Practical, do you think learners did better overall? Give reasons to justify your answer.

a

	Theory		Practical	
Grade	**Frequency**	**Angle**	**Frequency**	**Angle**
Excellent	208	360° × 208 ÷ 2880 = 26°	240	360° × 240 ÷ 1080 = 80°
Very good	888	360° × 888 ÷ 2880 = 111°	351	360° × 351 ÷ 1080 = 117°
Good	1032	360° × 1032 ÷ 2880 = 129°	291	360° × 291 ÷ 1080 = 97°
Pass	696	360° × 696 ÷ 2880 = 87°	108	360° × 108 ÷ 1080 = 36°
Fail	56	360° × 56 ÷ 2880 = 7°	90	360° × 90 ÷ 1080 = 30°

You will earn 1 mark for showing in at least five places the correct process of 360° × frequency ÷ total frequency.

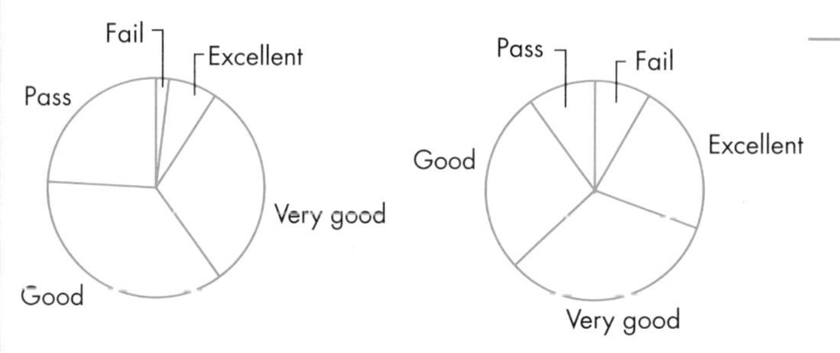

3 accuracy marks are available for the correct angles calculated. You will lose a mark for each incorrect one, to a minimum of 0.
Also available are: 1 method mark for drawing two separate pie charts.
1 method mark for correctly labelling both charts.
1 accuracy mark for each chart being correctly drawn.
Note that each angle can be up to 2° out for these marks to be given.

7 marks

b A greater proportion of learners got "Good" or better on the practical, but learners did better in the theory test as a much smaller proportion failed.

You earn 1 mark for stating one comparison correctly.

So, I would say the learners did better overall in the practical test.

You earn 1 mark for giving either of the tests as better overall but only with a justification as here.

2 marks

Total: 9 marks

Worked Examination Questions

PS **3**
- The older you are, the higher you score on the Speed test.
- The higher the score on the Speed test, the less TV you watch.
- The more TV you watch, the more hours you will sleep.

Suppose the above were all true. Sketch a scatter diagram to illustrate the relationship between age and hours slept.

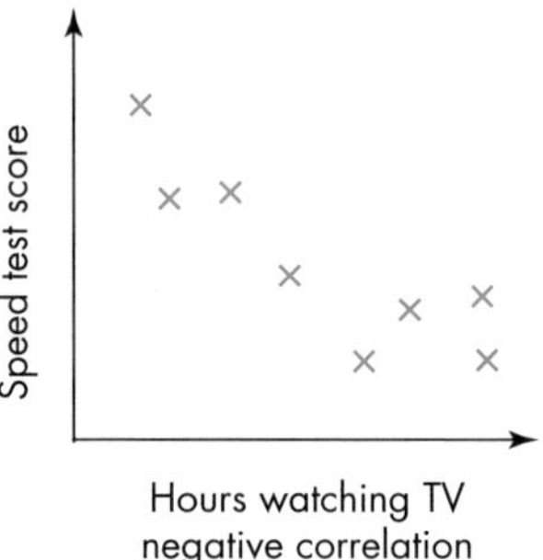

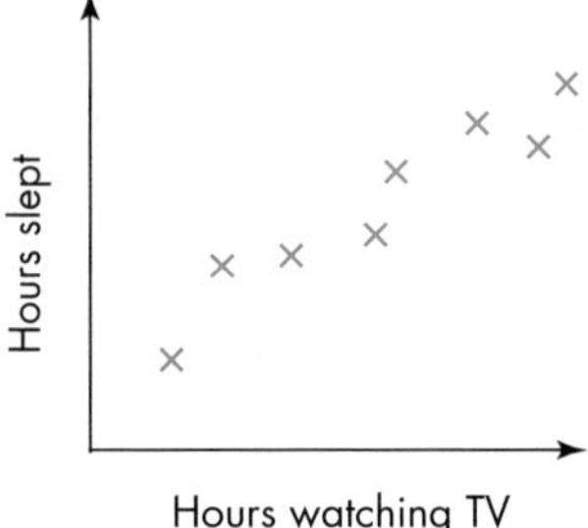

Sketch a scatter diagram for each bullet point given in the question. These three diagrams will help to show the relationships in the question.
You get 1 mark for showing some pictures similar to these.

Older person scores high on Speed test.
High score on Speed test means the person watches TV for a short time.

Imagine the extremes of age starting with the first diagram.
You get 2 independent marks for showing the link for extremes like this.

Watching TV for a short time gives short sleep time.
Hence "High age relates to small hours sleep".
Similarly, "Low age relates to long sleep".

You then get 1 mark for stating for each extreme.

So, the scatter diagram will show negative correlation and can be sketched as:

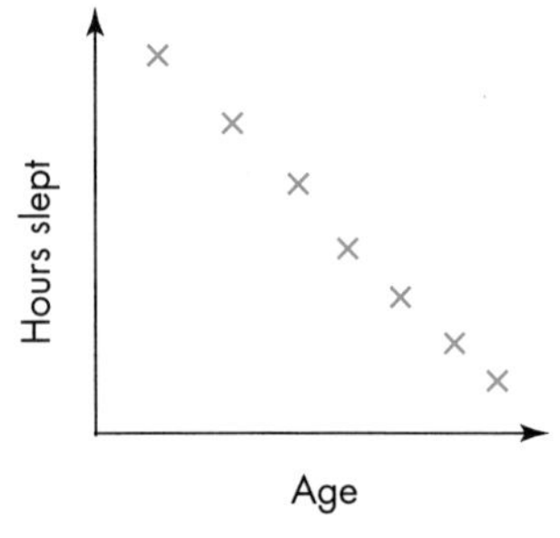

You get 1 mark for stating negative correlation, or suitable words.
You get 1 mark for sketching a diagram showing negative correlation.

Total: 7 marks

Worked Examination Question

AU **4** Godwin is asked to do a survey to investigate the hypothesis:

"As boys get older they lose interest in trains."

Explain the difficulties in trying to create a questionnaire to test out this hypothesis.

4 The difficulties are:

Measuring the level of boys' interest

How can you compare one person's interest with that of another?

Measuring the level of interest will earn the first mark here.
You get 1 mark for independent workings.

If you just ask the question to see what people's thoughts are on this then, again, you have not really tested the hypothesis other than asking peoples thoughts.

You get 1 mark for independent workings will be available for another suitable comment about difficulties as seen.

Total: 2 marks

15 Functional Maths
Riding stables

Riding stables must carefully manage the welfare of their horses, including housing them in stables of the correct size and feeding them the correct amount of food for their weight and intended workload. Mr Owen owns a riding stable and wishes to understand his horses' needs, his customers and how he might go about gathering the right information to help him successfully expand his business.

Getting started

In pairs, consider the following points to help you in planning Mr Owen's expansion of his riding stables.

- There are several different elements to consider when running a riding stable: buying or breeding horses, looking after the horses (including feeding and cleaning), running and training the horses and running classes for different types of students.
- How would you find out if there is correlation between girth, weight and height? How could Mr Owen use this information to predict the body weight of horses of different girths? How useful would this information be to Mr Owen?
- How could you graphically represent the record that Mr Owen kept last summer of the different abilities of his customers? How could he use this information to help in planning his expansion?
- If Mr Owen's customers were asked what they wanted most from the stables, how could you collect, represent and interpret this information?

Your task

Mr Owen needs to work out how much feed to give his horses. He owns six horses, as detailed below. Working in pairs or individually, use the information on the six horses, the Bodyweight calculator and the Feed chart to find the best way to calculate the amount of feed each horse needs.

Feed chart

Body weight of horse (kg)	Weight of feed (kg) at different levels of work	
	Medium	Hard
300	2.4	3.0
350	2.8	3.5
400	3.2	4.0
450	3.6	4.5
500	4.0	5.0
Extra feed per 50 kg	300 g	400 g

Your task

Mr Owen also wants to understand the different abilities and numbers of riders who come to his riding stable, as he wishes to expand his business next year. He keeps a record of his students and the workload they place on the horses during one week of the summer holidays (see Riders table below). How would Mr Owen use this information to make sure he buys the right type of new horses to suit his riders?

Mr Owen is also aware that different riders look for different facilities at the stables. Help him to work out how he can decide which new facilities different riders would value the most.

Help Mr Owen to plan his expansion, using surveys and diagrams.

Bodyweight calculator

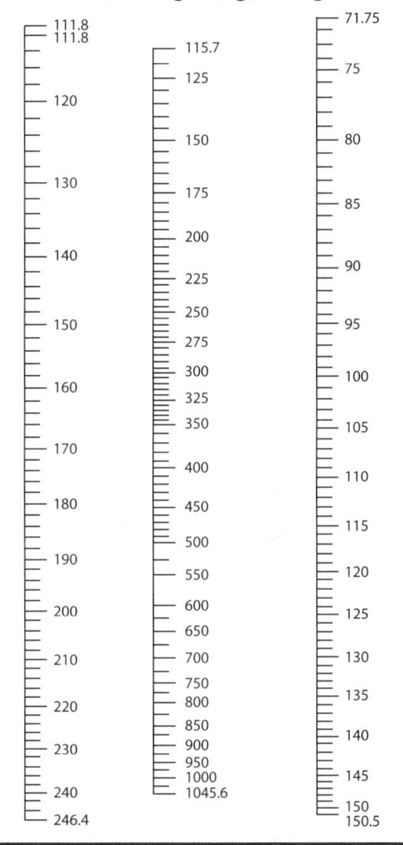

Using the bodyweight calculator

Place a ruler from your measurement for girth line to your measurement for the length line. The point at which the ruler crosses the weight line is the reading for the approximate weight of the horse. Note that the ruler will cross the weight line at an angle.

Riders	Ability/work	Total
Children	Medium	45
Female novice	Medium	20
Female experienced	Hard	61
Male novice	Medium	15
Male experienced	Hard	29

Horse:	Summer	Sally	Skip	Simon	Barney	Teddy
Girth:	220 cm	190 cm	200 cm	180 cm	160 cm	190 cm
Length:	142 cm	95 cm	114 cm	124 cm	110 cm	140 cm
Height: hands and inches	17 h 2 in	14 h 3 in	16 h 0 in	15 h 3 in	15 h 1 in	16 h 2 in
Work:	Medium	Hard	Hard	Medium	Medium	Hard

Answers to Chapter 1

1.1 Adding with grids

Exercise 1A

1

a

1	3	7	11
9	2	8	19
6	5	4	15
16	10	19	45

b

0	6	7	13
8	1	4	13
9	5	3	17
17	12	14	43

c

0	8	7	15
1	6	2	9
9	3	4	16
10	17	13	40

d

2	4	6	12
3	5	7	15
8	9	1	18
13	18	14	45

e

5	9	3	17
6	1	8	15
2	7	4	13
13	17	15	45

f

0	8	3	11
7	2	4	13
1	6	5	12
8	16	12	36

g

9	4	8	21
7	0	5	12
1	6	3	10
17	10	16	43

h

0	8	6	14
7	1	4	12
5	9	2	16
12	18	12	42

i

1	8	7	16
6	2	5	13
0	9	3	12
7	19	15	41

2 **a** Yes (£33) with £1 left over **b** 23

3 **a** 7 and 11
b 31
c 7 + 11 + 12 = 30 (largest two odds and one even gives larger total than largest three evens)

4 Possible answer: 4, 6, 8

5

a

1	7	8	16
0	3	6	9
5	4	2	11
6	14	16	36

b

1	2	3	6
6	5	4	15
7	8	9	24
14	15	16	45

c

9	3	6	18
4	0	5	9
1	2	8	11
14	5	19	38

d

9	1	6	16
2	7	4	13
8	5	0	13
19	13	10	42

e

2	9	6	17
4	1	3	8
5	0	8	13
11	10	17	38

f

1	7	8	16
6	2	4	12
5	9	3	17
12	18	15	45

g

0	2	1	3
9	6	7	22
8	4	5	17
17	12	13	42

h

1	0	3	4
8	7	4	19
9	6	5	20
18	13	12	43

i

1	5	4	10
6	2	3	11
8	7	0	15
15	14	7	36

6 Possible answer: 9, 11, 16

7 Possible answer: 11 and 13

1.2 Multiplication tables check

Exercise 1B

1 **a** 20 **b** 21 **c** 24 **d** 15 **e** 16 **f** 12
g 10 **h** 42 **i** 24 **j** 18 **k** 30 **l** 28
m 18 **n** 56 **o** 25 **p** 45 **q** 27 **r** 30
s 49 **t** 24 **u** 36 **v** 35 **w** 32 **x** 36
y 48 **z** £48 × 2 = £96, yes

2 **a** 5 **b** 4 **c** 6 **d** 6 **e** 5 **f** 4
g 7 **h** 6 **i** 2 **j** 3 **k** 7 **l** 8
m 9 **n** 5 **o** 8 **p** 9 **q** 4 **r** 7
s 7 **t** 9 **u** 5 **v** 4 **w** 5 **x** 7
y 6 **z** £6 per hour, 10 hours

3 **a** 12 **b** 15 **c** 21 **d** 13 **e** 8 **f** 7
g 14 **h** 3 **i** 30 **j** 6 **k** 35 **l** 5
m 16 **n** 7 **o** 16 **p** 15 **q** 27 **r** 6
s 15 **t** 24 **u** 40 **v** 6 **w** 17 **x** 72
y 46 **z** Ahmed is paid more (£33) than Ben (£32)

4 **a** 86 **b** 56 **c** 358 + 6

5 **a** 30 **b** 50 **c** 80 **d** 100 **e** 120 **f** 180
g 240 **h** 400 **i** 700 **j** 900 **k** 1000 **l** 1400
m 2400 **n** 7200 **o** 10 000 **p** 2 **q** 7 **r** 9
s 17 **t** 30 **u** 3 **v** 8 **w** 12 **x** 29
y 50

6 **a** 8 (3 × 8) **b** 900 (900 ÷ 10)

7 For example: 4 × 5 = 20 and 5 × 6 = 30

8 For example: 3 × 4 × 5 = 60 = 10 × 6 and 5 × 6 × 7 = 210 = 35 × 6

1.3 Order of operations and BIDMAS/BODMAS

Exercise 1C

1 **a** 11 **b** 6 **c** 10 **d** 12 **e** 11 **f** 13
g 11 **h** 12 **i** 12 **j** 4 **k** 13 **l** 3

2 **a** 16 **b** 2 **c** 10 **d** 10 **e** 6 **f** 18
g 6 **h** 15 **i** 9 **j** 12 **k** 3 **l** 8

3 **b** 3 + 2 × 4 = 11 **c** 9 ÷ 3 − 2 = 1
d 9 − 4 ÷ 2 = 7 **e** 5 × 2 + 3 = 13
f 5 + 2 × 3 = 11 **g** 10 ÷ 5 − 2 = 0
h 10 − 4 ÷ 2 = 8 **i** 4 × 6 − 7 = 17
j 7 + 4 × 6 = 31 **k** 6 ÷ 3 ÷ 7 = 9
l 7 + 6 ÷ 2 = 10

4 **a** 38 **b** 48 **c** 3 **d** 2 **e** 5 **f** 14
g 10 **h** 2 **i** 5 **j** 19 **k** 15 **l** 2
m 20 **n** 19 **o** 54 **p** 7 **q** 2 **r** 7
s 7 **t** 38 **u** 42 **v** 10 **w** 2 **x** 10
y 10 **z** 24

5 **a** (4 + 1) **b** No brackets needed
c (2 + 1) **d** No brackets needed
e (4 + 4) **f** (16 − 4)
g No brackets needed **h** No brackets needed
i (20 − 10) **j** No brackets needed
k (5 + 5) **l** (4 + 2)
m (15 − 5) **n** (7 − 2)
o (3 + 3) **p** No brackets needed
q No brackets needed **r** (8 − 2)

6 **a** 8 **b** 6 **c** 6 **d** 13 **e** 11 **f** 9
g 12 **h** 8 **i** 15 **j** 16 **k** 1 **l** 7

7 No, correct answer is 5 + 42 = 47

8 **a** $2 + 3 \times 4 = 14$ **b** $8 - 4 \div 4 = 7$ (correct)
c $6 + 3 \times 2 = 12$ (correct) **d** $7 - 1 \times 5 = 2$
e $2 \times 7 + 2 = 16$ (correct) **f** $9 - 3 \times 3 = 0$

9 **a** $2 \times 3 + 5 = 11$ **b** $2 \times (3 + 5) = 16$
c $2 + 3 \times 5 = 17$ **d** $5 - (3 - 2) = 4$
e $5 \times 3 - 2 = 13$ **f** $5 \times 3 \times 2 = 30$

10 $4 + 5 \times 3 = 19$
$(4 + 5) \times 3 = 27$. So $4 + 5 \times 3$ is smaller

11 $(5 - 2) \times 6 = 18$

12 $10 \div (5 - 3) = 4$

13 $10 - 3 \times 1.5$ and $10 - 1.5 - 1.5 - 1.5$

1.4 Place value and ordering numbers

Exercise 1D

1 **a** 40 **b** 5 units **c** 100 **d** 90 **e** 80 **f** 9 units
g 80 **h** 500 **i** 0 **j** 5000 **k** 0 **l** 4 units
m 300 **n** 90 **o** 80 000

2 **a** Forty-three, two hundred
b One hundred and thirty-six; four thousand and ninety-nine
c Two hundred and seventy-one; ten thousand, seven hundred and forty-four

3 **a** Five million, six hundred thousand
b Four million, seventy-five thousand, two hundred
c Three million, seven thousand, nine hundred and fifty
d Two million, seven hundred and eighty-two

4 **a** 8 200 058 **b** 9 406 107 **c** 1 000 502
d 2 076 040

5 **a** 9, 15, 21, 23, 48, 54, 56, 85
b 25, 62, 86, 151, 219, 310, 400, 501
c 97, 357, 368, 740, 888, 2053, 4366

6 **a** 95, 89, 73, 52, 34, 25, 23, 7
b 700, 401, 174, 117, 80, 65, 18, 2
c 6227, 3928, 2034, 762, 480, 395, 89, 59

7 **a** Larger **b** Larger **c** Smaller
d Larger **e** Larger **f** Smaller
g Larger **h** Smaller **i** Smaller

8 **a** Number 4 (£128 250)
b Number 1 (£129 100)
c £850

9 **a** 368, 386, 638, 683, 836, 863
b 368
c 863

10 408, 480, 804, 840

11 33, 35, 38, 53, 55, 58, 83, 85, 88

12 7045 or 7405

13 **a** $973 + 85 = 1058$ or
$975 + 83 = 1058$ or
$985 + 73 = 1058$ or
$983 + 75 = 1058$
b $357 - 98 = 259$

1.5 Rounding

Exercise 1E

1 **a** 20 **b** 60 **c** 80 **d** 50 **e** 100 **f** 20
g 90 **h** 70 **i** 10 **j** 30 **k** 30 **l** 50
m 80 **n** 50 **o** 90 **p** 40 **q** 70 **r** 20
s 100 **t** 110

2 **a** 200 **b** 600 **c** 800 **d** 500 **e** 1000 **f** 100
g 600 **h** 400 **i** 1000 **j** 1100 **k** 300 **l** 500
m 800 **n** 500 **o** 900 **p** 400 **q** 700 **r** 800
s 1000 **t** 1100

3 **a** 1 **b** 2 **c** 1 **d** 1 **e** 3 **f** 2
g 3 **h** 2 **i** 1 **j** 1 **k** 3 **l** 2
m 74 **n** 126 **o** 184

4 **a** 2000 **b** 6000 **c** 8000 **d** 5000 **e** 10 000 **f** 1000
g 6000 **h** 3000 **i** 9000 **j** 2000 **k** 3000 **l** 5000
m 8000 **n** 5000 **o** 9000 **p** 4000 **q** 7000 **r** 8000
s 1000 **t** 2000

5 **a** 230 **b** 570 **c** 720 **d** 520 **e** 910 **f** 230
g 880 **h** 630 **i** 110 **j** 300 **k** 280 **l** 540
m 770 **n** 500 **o** 940 **p** 380 **q** 630 **r** 350
s 1010 **t** 1070

6 **a** True **b** False **c** True **d** True **e** True **f** False

7 Welcome to Swinton population 1400 (to the nearest 100)

8 **a** Man Utd v West Brom
b Blackburn v Fulham
c 40 000, 19 000, 42 000, 26 000, 40 000, 68 000, 35 000, 25 000, 20 000
d 39 600, 19 000, 42 100, 26 100,40 400, 67 800, 34 800, 25 500, 20 200

9 **a** 35 min **b** 55 min **c** 15 min
d 50 min **e** 10 min **f** 15 min
g 45 min **h** 35 min **i** 5 min
j 0 min

10 **a** 375
b 25 (350 to 374 inclusive)

11 A number between 75 and 84 inclusive added to a number between 45 and 54 inclusive with a total not equal to 130, for example $79 + 49 = 128$

1.6 Adding and subtracting numbers with up to four digits

Exercise 1F

1 **a** 713 **b** 151 **c** 6381
d 968 **e** 622 **f** 1315
g 8260 **h** 818 **i** 451
j 852

2 **a** 646 **b** 826 **c** 3818
d 755 **e** 2596 **f** 891
g 350 **h** 2766 **i** 8858
j 841 **k** 6831 **l** 7016
m 1003 **n** 4450 **o** 9944

3 **a** 450 **b** 563 **c** 482
d 414 **e** 285 **f** 486
g 244 **h** 284 **i** 333
j 216 **k** 2892 **l** 4417
m 3767 **n** 4087 **o** 1828

4 **a** 128 **b** 29 **c** 334
d 178 **e** 277 **f** 285
g 335 **h** 399 **i** 4032
j 4765 **k** 3795 **l** 5437

5 **a** 558 miles **b** 254 miles

6 252

7 **a** 6, 7 **b** 4, 7 **c** 4, 8
d 7, 4, 9 **e** 6, 9, 7 **f** 6, 2, 7
g 2, 6, 6 **h** 4, 5, 9 **i** 4, 8, 8
j 4, 4, 9, 8

8 Units digit should be 6 (from 14 – 8)

9 **a** 3, 5 **b** 8, 3 **c** 5, 8
d 8, 5, 4 **e** 6, 7, 5 **f** 1, 2, 1
g 2, 7, 7 **h** 5, 5, 6 **i** 8, 3, 8
j 1, 8, 8, 9

10 For example: 181 – 27 = 154

1.7 Multiplying and dividing by single-digit numbers

Exercise 1G

1 **a** 56 **b** 65 **c** 51
d 38 **e** 108 **f** 115
g 204 **h** 294 **i** 212
j 425 **k** 150 **l** 800
m 960 **n** 1360 **o** 1518

2 **a** 294 **b** 370 **c** 288
d 832 **e** 2163 **f** 2520
g 1644 **h** 3215 **i** 3000
j 2652 **k** 3696 **l** 1880
m 54 387 **n** 21 935 **o** 48 888

3 **a** 219 **b** 317 **c** 315
d 106 **e** 99 **f** 121
g 252 **h** 141 **i** 144
j 86 **k** 63 **l** 2909
m 416 **n** 251 **o** 1284

4 **a** 705 miles **b** £3525

5 **a** 47 miles
b Three numbers with a total of 125. First number must be less than 50. second number less than first number, third number less than second number, for example 48, 42 and 35.

6 **a** 119 **b** 96 **c** 144
d 210 **e** 210

7 **a** 13 **b** 37 weeks **c** 43 m
d 36 **e** 45

8 **a** $152 + 190 = 324$
b $(10 \times 190) + 76 = 1976$
c $(100 \times 38) + 190 = 3990$

Examination questions

1 **a** 3000
b 4681
c five thousand and sixty

2 **a** 9374
b Sixty-two thousand five hundred
c 80
d 2200
e 7000

3 **i** 4 **ii** 2 **iii** 9

4 **a** 17 252
b 5400
c 4000

5 **a** Twenty-nine thousand, seven hundred and sixty-five
b **i** 700 **ii** 9000
c 29 800

6 **a** **i** 54 073 **ii** 54 100
b **i** Twenty-one thousand, eight hundred and nine
ii 22 000

7 **a** **i** 28 000 000 **ii** 2800
b **i** 100 **ii** 10

8 7760 metres

9 **a** £1505
b 19 people

10 **a** 7750
b 7849

11 Murray did the addition first, Harry did the multiplication first.

12 **a** 96
b 32
c **i** 8 **ii** 2
d **i** 1080 **ii** 15

13 **a** **i** 105 **ii** 23
b **i** 72 **ii** 31

14 **a** He used BODMAS and did the power first, then the multiplication before the addition.
b **i** $(2 \times 3)^2 + 6 = 42$ **ii** $2 \times (3^2 + 6) = 30$

15 **a** Adam has calculated 5×2 instead of 5^2, Bekki has added 3 to 5 first, instead of doing the power first.
b 26

Answers to Chapter 2

2.1 Recognise a fraction of a shape

Exercise 2A

1 **a** $\frac{1}{4}$ **b** $\frac{1}{3}$ **c** $\frac{5}{8}$ **d** $\frac{7}{12}$ **e** $\frac{4}{9}$

f $\frac{3}{10}$ **g** $\frac{3}{8}$ **h** $\frac{15}{16}$ **i** $\frac{5}{12}$ **j** $\frac{7}{18}$

k $\frac{4}{8} = \frac{1}{2}$ **l** $\frac{4}{12} = \frac{1}{3}$ **m** $\frac{6}{9} = \frac{2}{3}$

n $\frac{6}{10} = \frac{3}{5}$ **o** $\frac{4}{8} = \frac{1}{2}$ **p** $\frac{5}{64}$

2 Check students' diagrams.

3 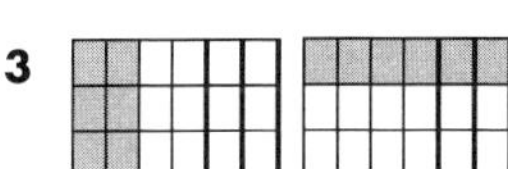

4 **a** g **b** i **c** neither **d** e
e neither **f** m **g** k **h** e

5 Fraction B is not $\frac{1}{3}$. Fraction C does not have a numerator of 4.

6 $\frac{1}{2}$

2.2 Adding and subtracting simple fractions

Exercise 2B

1 **a** $\frac{3}{4}$ **b** $\frac{4}{8} = \frac{1}{2}$ **c** $\frac{3}{5}$ **d** $\frac{8}{10} = \frac{4}{5}$

e $\frac{2}{3}$ **f** $\frac{5}{7}$ **g** $\frac{7}{9}$ **h** $\frac{5}{6}$ **i** $\frac{4}{5}$

j $\frac{7}{8}$ **k** $\frac{5}{10} = \frac{1}{2}$ **l** $\frac{5}{7}$ **m** $\frac{4}{5}$

n $\frac{5}{6}$ **o** $\frac{5}{9}$ **p** $\frac{7}{11}$

2 **a** $\frac{2}{4} = \frac{1}{2}$ **b** $\frac{3}{5}$ **c** $\frac{3}{8}$ **d** $\frac{3}{10}$

e $\frac{1}{3}$ **f** $\frac{4}{6} = \frac{2}{3}$ **g** $\frac{3}{7}$ **h** $\frac{5}{9}$

i $\frac{1}{5}$ **j** $\frac{3}{7}$ **k** $\frac{3}{9} = \frac{1}{3}$ **l** $\frac{6}{10} = \frac{3}{5}$

m $\frac{3}{6} = \frac{1}{2}$ **n** $\frac{2}{8} = \frac{1}{4}$ **o** $\frac{2}{11}$

p $\frac{4}{10} = \frac{2}{5}$

3 + =

$\frac{5}{6}$

4 **a** **b**

c **i** $\frac{3}{4}$ **ii** $\frac{1}{4}$

5 **a** **b**

c **i** $\frac{6}{10} = \frac{3}{5}$ **ii** $\frac{8}{10} = \frac{4}{5}$

iii $\frac{7}{10}$

6 Start with five-fifths, remove four-fifths, put back two-fifths so $\frac{5}{5} - \frac{4}{5} + \frac{2}{5} = \frac{3}{5}$

7 $\frac{1}{6}$ (40 out of the 240 seats)

8 $\frac{1}{8}$ (4 of the 32 students)

9 **a** No, would sell 18 snacks per journey, i.e. 36 for two journeys

b $\frac{18}{30}$ or $\frac{3}{5}$

2.3 Recognise equivalent fractions, using diagrams

Exercise 2C

1 **a** $\frac{4}{24}$ **b** $\frac{8}{24}$ **c** $\frac{3}{24}$ **d** $\frac{16}{24}$ **e** $\frac{20}{24}$

f $\frac{18}{24}$ **g** $\frac{9}{24}$ **h** $\frac{15}{24}$ **i** $\frac{21}{24}$ **j** $\frac{12}{24}$

2 **a** $\frac{11}{24}$ **b** $\frac{9}{24}$ **c** $\frac{7}{24}$ **d** $\frac{19}{24}$ **e** $\frac{23}{24}$

f $\frac{23}{24}$ **g** $\frac{21}{24}$ **h** $\frac{22}{24}$ **i** $\frac{19}{24}$ **j** $\frac{23}{24}$

3 **a** $\frac{5}{20}$ **b** $\frac{4}{20}$ **c** $\frac{15}{20}$ **d** $\frac{16}{20}$ **e** $\frac{2}{20}$

f $\frac{10}{20}$ **g** $\frac{12}{20}$ **h** $\frac{8}{20}$ **i** $\frac{14}{20}$ **j** $\frac{6}{20}$

4 **a** $\frac{9}{20}$ **b** $\frac{14}{20}$ **c** $\frac{11}{20}$ **d** $\frac{19}{20}$ **e** $\frac{19}{20}$

5 **a** $\frac{2}{6}$ **b** $\frac{2}{3}$ **c** none **d** $\frac{6}{8}$

6 **a** $\frac{4}{10}, \frac{8}{20}, \frac{10}{25}$ **b** for example $\frac{12}{30}$

7 **a**

×	2	3	4	5
3	6	9	12	15
4	8	12	16	20

b $\frac{6}{8}, \frac{9}{12}, \frac{12}{16}, \frac{15}{20}$

8 **a** $\frac{1}{12}$

b **i** $\frac{7}{12}$ **ii** $\frac{9}{12}$ (or $\frac{3}{4}$) **iii** $\frac{11}{12}$

iv $\frac{9}{12}$ (or $\frac{3}{4}$) **v** $\frac{7}{12}$

2.4 Equivalent fractions and simplifying fractions by cancelling

Exercise 2D

1 **a** $\frac{8}{20}$ **b** $\frac{3}{12}$ **c** $\frac{15}{40}$ **d** $\frac{12}{15}$ **e** $\frac{15}{18}$

f $\frac{12}{28}$ **g** $\times 2, \frac{6}{20}$ **h** $\times 3, \frac{3}{9}$

i $\times 4, \frac{12}{20}$ **j** $\times 6, \frac{12}{18}$

k $\times 3, \frac{9}{12}$ **l** $\times 5, \frac{25}{40}$

m $\times 2, \frac{14}{20}$ **n** $\times 4, \frac{4}{24}$ **o** $\times 5, \frac{15}{40}$

2 **a** $\frac{1}{2} = \frac{2}{4} = \frac{3}{6} = \frac{4}{8} = \frac{5}{10} = \frac{6}{12}$

b $\frac{1}{3} = \frac{2}{6} = \frac{3}{9} = \frac{4}{12} = \frac{5}{15} = \frac{6}{18}$

c $\frac{3}{4} = \frac{6}{8} = \frac{9}{12} = \frac{12}{16} = \frac{15}{20} = \frac{18}{24}$

d $\frac{2}{5} = \frac{4}{10} = \frac{6}{15} = \frac{8}{20} = \frac{10}{25} = \frac{12}{30}$

e $\frac{3}{7} = \frac{6}{14} = \frac{9}{21} = \frac{12}{28} = \frac{15}{35} = \frac{18}{42}$

3 **a** $\frac{2}{3}$ **b** $\frac{4}{5}$ **c** $\frac{5}{7}$ **d** $\div 6, \frac{2}{3}$

e $25 \div 4, \frac{3}{5}$ **f** $\div 3, \frac{7}{10}$

4 **a** 32 **b** $\frac{1}{2}$ **c** $\frac{5}{6}$

5 **a** $\frac{2}{3}$ **b** $\frac{1}{3}$ **c** $\frac{2}{3}$ **d** $\frac{3}{4}$ **e** $\frac{1}{3}$

f $\frac{1}{2}$ **g** $\frac{7}{8}$ **h** $\frac{4}{5}$ **i** $\frac{1}{2}$ **j** $\frac{1}{4}$

k $\frac{4}{5}$ **l** $\frac{5}{7}$ **m** $\frac{5}{7}$ **n** $\frac{2}{3}$ **o** $\frac{2}{5}$

p $\frac{2}{5}$ **q** $\frac{1}{3}$ **r** $\frac{7}{10}$ **s** $\frac{1}{4}$

t $\frac{3}{2} = 1\frac{1}{2}$ **u** $\frac{2}{3}$ **v** $\frac{2}{3}$ **w** $\frac{3}{4}$

x $\frac{3}{2} = 1\frac{1}{2}$ **y** $\frac{7}{2} = 3\frac{1}{2}$

6 **a** $\frac{1}{2}, \frac{2}{3}, \frac{5}{6}$ **b** $\frac{1}{2}, \frac{5}{8}, \frac{3}{4}$ **c** $\frac{2}{5}, \frac{1}{2}, \frac{7}{10}$

d $\frac{7}{12}, \frac{2}{3}, \frac{3}{4}$ **e** $\frac{1}{6}, \frac{1}{4}, \frac{1}{3}$ **f** $\frac{3}{4}, \frac{4}{5}, \frac{9}{10}$

g $\frac{7}{10}, \frac{4}{5}, \frac{5}{6}$ **h** $\frac{3}{10}, \frac{1}{3}, \frac{2}{5}$

7 **a** $\frac{1}{3} + \frac{1}{4} = \frac{4}{12} + \frac{3}{12} = \frac{7}{12}$
Explanations may involve ruling out other combinations.
b $\frac{1}{2}$ as the smallest denominator is the biggest unit fraction.
Diagrams may be used but must be based on equal sized area.

8 **a** $\frac{1}{5}$ **b** $\frac{1}{77}$

9 **a** $\frac{3}{8}$ **b** $\frac{1}{2}$ **c** $\frac{5}{16}$

2.5 Improper fractions and mixed numbers

Exercise 2E

1 **a** $2\frac{1}{3}$ **b** $2\frac{2}{3}$ **c** $2\frac{1}{4}$ **d** $1\frac{3}{7}$ **e** $2\frac{2}{5}$

f $1\frac{2}{5}$ **g** $2\frac{3}{5}$ **h** $3\frac{3}{4}$ **i** $3\frac{1}{3}$ **j** $2\frac{1}{7}$

k $2\frac{5}{6}$ **l** $3\frac{3}{5}$ **m** $4\frac{3}{4}$ **n** $3\frac{1}{7}$ **o** $1\frac{3}{11}$

p $1\frac{1}{11}$ **q** $5\frac{3}{5}$ **r** $2\frac{5}{7}$ **s** $5\frac{5}{7}$ **t** $8\frac{2}{5}$

u $2\frac{1}{10}$ **v** $2\frac{1}{2}$ **w** $1\frac{2}{3}$ **x** $3\frac{1}{8}$ **y** $2\frac{3}{10}$

z $2\frac{1}{11}$

2 **a** $\frac{10}{3}$ **b** $\frac{35}{6}$ **c** $\frac{9}{5}$ **d** $\frac{37}{7}$ **e** $\frac{41}{10}$

f $\frac{17}{3}$ **g** $\frac{5}{2}$ **h** $\frac{13}{4}$ **i** $\frac{43}{6}$ **j** $\frac{29}{8}$

k $\frac{19}{3}$ **l** $\frac{89}{9}$ **m** $\frac{59}{5}$ **n** $\frac{16}{5}$ **o** $\frac{35}{8}$

p $\frac{28}{9}$ **q** $\frac{26}{5}$ **r** $\frac{11}{4}$ **s** $\frac{30}{7}$ **t** $\frac{49}{6}$

u $\frac{26}{9}$ **v** $\frac{37}{6}$ **w** $\frac{61}{5}$ **x** $\frac{13}{8}$ **y** $\frac{71}{10}$

z $\frac{73}{9}$

3 Students check their own answers.

4 $\frac{27}{4} = 6\frac{3}{4}, \frac{31}{5} = 6\frac{1}{5}, \frac{13}{2} = 6\frac{1}{2}$, so $\frac{27}{4}$ is the biggest since $\frac{1}{5}$ is less than $\frac{1}{2}$ and $\frac{3}{4}$ is greater than $\frac{1}{2}$

5 Any mixed number which is between 7.7272... and 7.9. For example $7\frac{4}{5}$

6 Any mixed number which is between 5.8181... and 5.875. For example $5\frac{16}{19}$
Note that adding the numerators of two fractions and then adding the denominators always gives an answer in between the two fractions.

7 $\frac{9}{4}$ or $\frac{25}{9}$

2.6 Adding and subtracting fractions with the same denominator

Exercise 2F

1 **a** $\frac{5}{7}$ **b** $\frac{7}{9}$ **c** $\frac{4}{5}$ **d** $\frac{6}{7}$ **e** $\frac{7}{11}$

f $\frac{7}{9}$ **g** $\frac{10}{13}$ **h** $\frac{10}{11}$

2 **a** $\frac{3}{7}$ **b** $\frac{1}{9}$ **c** $\frac{4}{11}$ **d** $\frac{7}{13}$ **e** $\frac{4}{7}$

f $\frac{1}{5}$ **g** $\frac{4}{9}$ **h** $\frac{3}{11}$

3 **a** $\frac{6}{8} = \frac{3}{4}$ **b** $\frac{4}{10} = \frac{2}{5}$ **c** $\frac{6}{9} = \frac{2}{3}$

d $\frac{2}{4} = \frac{1}{2}$ **e** $\frac{6}{10} = \frac{3}{5}$ **f** $\frac{6}{12} = \frac{1}{2}$

g $\frac{8}{16} = \frac{1}{2}$ **h** $\frac{10}{16} = \frac{5}{8}$

4 **a** $\frac{4}{8} = \frac{1}{2}$ **b** $\frac{4}{10} = \frac{2}{5}$ **c** $\frac{4}{6} = \frac{2}{3}$

d $\frac{8}{10} = \frac{4}{5}$ **e** $\frac{6}{12} = \frac{1}{2}$ **f** $\frac{2}{8} = \frac{1}{4}$

g $\frac{2}{16} = \frac{1}{8}$ **h** $\frac{6}{18} = \frac{1}{3}$

5 **a** $\frac{12}{10} = \frac{6}{5} = 1\frac{1}{5}$ **b** $\frac{9}{8} = 1\frac{1}{8}$

c $\frac{9}{8} = 1\frac{1}{8}$ **d** $\frac{13}{8} = 1\frac{5}{8}$

e $\frac{11}{8} = 1\frac{3}{8}$ **f** $\frac{7}{6} = 1\frac{1}{6}$

g $\frac{9}{6} = \frac{3}{2} = 1\frac{1}{2}$ **h** $\frac{5}{4} = 1\frac{1}{4}$

6 **a** $\frac{10}{8} = \frac{5}{4} = 1\frac{1}{4}$ **b** $\frac{6}{4} = \frac{3}{2} = 1\frac{1}{2}$

c $\frac{5}{5} = 1$ **d** $\frac{16}{10} = \frac{8}{5} = 1\frac{3}{5}$

e $\frac{10}{8} = \frac{5}{4} = 1\frac{1}{4}$ **f** $\frac{22}{16} = \frac{11}{8} = 1\frac{3}{8}$

g $\frac{16}{12} = \frac{8}{6} = \frac{4}{3} = 1\frac{1}{3}$

h $\frac{18}{16} = \frac{9}{8} = 1\frac{1}{8}$

7 **a** $\frac{5}{8}$ **b** $\frac{5}{10} = \frac{1}{2}$ **c** $\frac{1}{4}$ **d** $\frac{3}{8}$

e $\frac{1}{4}$ **f** $\frac{3}{8}$ **g** $\frac{4}{10} = \frac{2}{5}$ **h** $\frac{5}{16}$

8 $\frac{1}{8}$

9 Ayesha

10 $\frac{4}{6} = \frac{2}{3}$

11 $\frac{3}{8}$

12 $\frac{2}{5}$

13 $\frac{3}{8}$

14 $\frac{4}{11}$

15 $\frac{1}{6}$

16 $\frac{5}{8}$

17 Paul, as $\frac{2}{3}$ is bigger than $\frac{5}{8}$

2.7 Finding a fraction of a quantity

Exercise 2G

1 **a** 18 **b** 10 **c** 18 **d** 28
e 15 **f** 18 **g** 48 **h** 45

2 **a** £1800 **b** 128 g **c** 160 kg
d £116 **e** 65 litres **f** 90 min
g 292 days **h** 21 h **i** 18 h
j 2370 miles

3 **a** $\frac{5}{8}$ of 40 = 25 **b** $\frac{3}{4}$ of 280 = 210

c $\frac{4}{5}$ of 70 = 56 **d** $\frac{5}{6}$ of 72 = 60

e $\frac{3}{5}$ of 95 = 57 **f** $\frac{3}{4}$ of 340 = 255

4 £6080

5 £31 500

6 13 080

7 52 kg

8 **a** 856 **b** 187 675

9 **a** £50 **b** £550

10 **a** 180 g **b** 900 g

11 **a** £120 **b** £240

12 Lion Autos

13 Offer B

2.8 Multiplying and dividing fractions

Exercise 2H

1 **a** $\frac{1}{6}$ **b** $\frac{1}{20}$ **c** $\frac{2}{9}$ **d** $\frac{1}{6}$ **e** $\frac{1}{4}$ **f** $\frac{2}{5}$

g $\frac{1}{2}$ **h** $\frac{1}{2}$ **i** $\frac{3}{14}$ **j** $\frac{35}{48}$ **k** $\frac{8}{15}$ **l** $\frac{21}{32}$

2 **a** 25 sheep **b** $\frac{1}{6}$

3 562 500

4 No one, they all got $\frac{1}{4}$

5 **a** $\frac{3}{4}$ **b** $1\frac{2}{5}$ **c** $1\frac{1}{15}$ **d** $1\frac{1}{14}$ **e** 4 **f** 4

g 5 **h** $1\frac{5}{7}$

6 18

7 **a** $2\frac{2}{15}$ **b** 38 **c** $1\frac{7}{8}$ **d** $\frac{9}{32}$

2.9 One quantity as a fraction of another

Exercise 2I

1 **a** $\frac{1}{3}$ **b** $\frac{1}{5}$ **c** $\frac{2}{5}$ **d** $\frac{5}{24}$ **e** $\frac{2}{5}$

f $\frac{1}{6}$ **g** $\frac{2}{7}$ **h** $\frac{1}{3}$

2 $\frac{3}{5}$

3 $\frac{12}{31}$

4 $\frac{7}{12}$

5 Jon saves $\frac{30}{90} = \frac{1}{3}$

Matt saves $\frac{35}{100}$ which is greater than $\frac{1}{3}$, so Matt saves the greater proportion of his earnings.

6 $\frac{13}{20} = \frac{65}{100}$, $\frac{16}{25} = \frac{64}{100}$, so first mark is better.

Examination questions

1 **a** $\frac{1}{2}$

b 9 squares shaded

2 $\frac{1}{3}, \frac{2}{5}, \frac{1}{2}, \frac{3}{4}$

3 $\frac{3}{10}$

4 **a** £72.96
b 87 miles

5 27 kg

6 **a** $\frac{2}{5}$, because $\frac{2}{5} = 0.4$

b 15

7 $\frac{4}{5}$, because $\frac{4}{5} = 0.8$ and $\frac{3}{4} = 0.75$. One grid has 16 squares shaded from left to right to show 0.8 and the other has 15 squares shaded from top to bottom to show 0.75.

8 120 tissues

9 $\frac{3}{10}$

10 **a** $\frac{1}{2}$

b $\frac{2}{5}$

11 **a** **i** 105 **ii** $\frac{1}{2}$

b $\frac{1}{4}$

c $\frac{3}{5}$

12 £4.80

13 172 g

14 **a** 20

b $\frac{7}{10}$

15 660 g

16 $\frac{2}{3}$

17 Smaller, e.g. if the original packet was 500 g, the new packet size is $\frac{6}{5} \times 500 = 600$.

If this is then reduced by $\frac{1}{5}$, $\frac{4}{5} \times 600 = 480$ g.

Answers to Chapter 3

3.1 Introduction to negative numbers

Activity

1 **a** 860 feet **b** 725 feet **c** 75 feet
d 475 feet **e** 1100 feet **f** 575 feet
g 425 feet **h** 310 feet **i** 700 feet
j 1010 feet

2 480 feet

3 1180 feet

4 Closed gate/Collapsed tunnel and Dead Man's seam/C seam

5 910 feet

6 Dead Man's seam/D seam and B seam/C seam

7 Collapsed tunnel/North gate

Exercise 3A

1 **a** 0 °C **b** 5 °C **c** −2 °C
d −5 °C **e** −1 °C

2 **a** 11 degrees **b** 9 degrees

3 8 degrees

4

	Sump	River Cave	Lost Cave	Echo Cave	Angel Cavern	Bat Cave	Base	Camp 1	Camp 2	Camp 3	Camp 4	Summit
Summit	4030	4010	3930	3910	3795	3665	3420	2300	1070	670	220	0
Camp 4	3810	3790	3710	3690	3575	3445	3200	2080	850	450	0	
Camp 3	3360	3340	3260	3240	3125	2995	2750	1630	400	0		
Camp 2	2960	2940	2860	2840	2725	2595	2350	1230	0			
Camp 1	1730	1710	1630	1610	1495	1365	1120	0				
Base	610	590	510	490	375	245	0					
Bat Cave	365	345	265	245	130	0						
Angel Cavern	235	215	135	115	0							
Echo Cave	120	100	20	0								
Lost Cave	100	80	0									
River Cave	20	0										
Sump	0											

3.2 Everyday use of negative numbers

Exercise 3B

1 −£5

2 −£9

3 Profit

4 −200 m

5 −50 m

6 Above

7 −3 h

8 −5 h

9 After

10 −2 °C

11 −8 °C

12 Above

13 −70 km

14 −200 km

15 North

16 +5 m

17 −5 mph

18 −2

19 **a** You owe the bank £89.72.
b You are paying money out of the account.
c You are paying money into the account.

20 **a** −11 °C **b** 6 degrees

21 **a** −8 **b** 7 **c** 15

22 9 days

23 1.54am

3.3 The number line

Exercise 3C

1 Many different answers to each part

2 Many different answers to each part

3 **a** Is smaller than **b** is bigger than **c** Is smaller than
d Is smaller than **e** is bigger than **f** Is smaller than
g Is smaller than **h** is bigger than **i** is bigger than
j Is smaller than **k** Is smaller than **l** is bigger than

4 **a** Is smaller than **b** Is smaller than **c** Is smaller than
d is bigger than **e** Is smaller than **f** Is smaller than

5 **a** $<$ **b** $>$ **c** $<$ **d** $<$ **e** $<$ **f** $>$
g $<$ **h** $>$ **i** $>$ **j** $>$ **k** $<$ **l** $<$
m $>$ **n** $>$ **o** $<$ **p** $>$

6

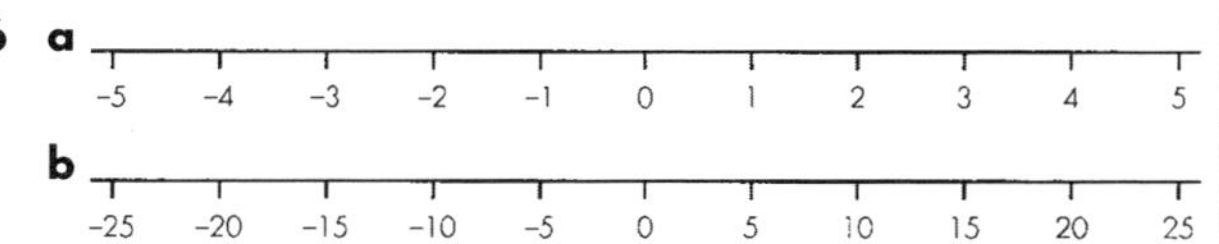

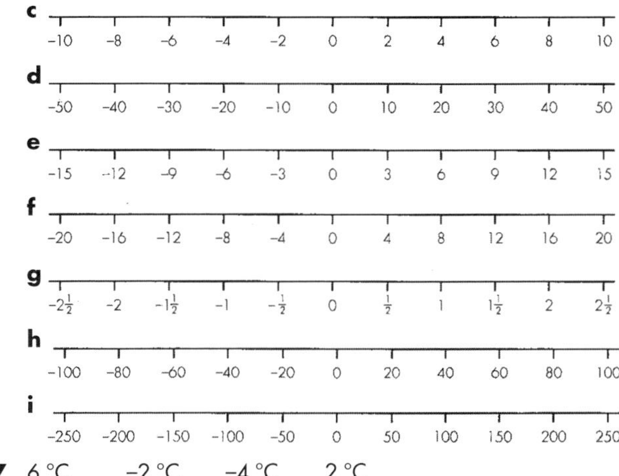

7 6 °C −2 °C −4 °C 2 °C

8

$-4\frac{1}{2}$ $-\frac{1}{4}$ $+3\frac{3}{4}$

−6 −5 −4 −3 −2 −1 0 1 2 3 4 5 6

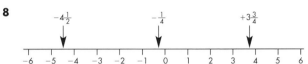

3.4 Arithmetic with negative numbers

Exercise 3D

1 **a** −2° **b** −3° **c** −2° **d** −3° **e** −2° **f** −3°
g 3 **h** 3 **i** −1 **j** −1 **k** 2 **l** −3
m −4 **n** −6 **o** −6 **p** −1 **q** −5 **r** −4
s 4 **t** −1 **u** −5 **v** −4 **w** −5 **x** −5

2 **a** −4 **b** −4 **c** −10 **d** 2 **e** 8 **f** −5
g 2 **h** 5 **i** −7 **j** −12 **k** 13 **l** 25
m −32 **n** −30 **o** −5 **p** −8 **q** −12 **r** 10
s −36 **t** −14 **u** 41 **v** 12 **w** −40 **x** −101

3 **a** 6 **b** −5 **c** 6 **d** −1 **e** −2 **f** −6
g −6 **h** −2 **i** 3 **j** 0 **k** −7 **l** −6
m 8 **n** 1 **o** −9 **p** −9 **q** −5 **r** −80
s −7 **t** −1 **u** −47

4 Students' own check

5 **a** 7 degrees **b** −6 °C

6 **a** 2 − 8
b 2 + 5 − 8 or 2 + 4 − 7 or 8 − 4 − 5 or 8 − 2 − 7 or 5 − 4 − 2
c 2 − 5 − 7 − 8 **d** 2 + 5 − 4 − 7 − 8

7 500 ft

Exercise 3E

1 **a** 6 **b** 7 **c** 8 **d** 6 **e** 8 **f** 10
g 2 **h** −3 **i** 1 **j** 2 **k** −1 **l** −7
m 2 **n** −3 **o** 1 **p** −5 **q** 3 **r** −4
s −3 **t** −8 **u** −10 **v** −9 **w** −4 **x** −9

2 **a** −8 **b** −10 **c** −11 **d** −3 **e** 2 **f** −5
g 1 **h** 4 **i** 7 **j** −8 **k** −5 **l** −11
m 11 **n** 6 **o** 8 **p** 8 **q** −2 **r** −1
s −9 **t** −5 **u** 5 **v** −9 **w** 8 **x** 0

3 **a** 3 °C **b** 0 °C **c** −3 °C **d** −5 °C **e** −11 °C

4 **a** 10 degrees Celsius **b** 7 degrees Celsius
c 9 degrees Celsius

5 −9, −6, −5, −1, 1, 2, 3, 8

6 **a** −3 **b** −4 **c** −2 **d** −7 **e** −14 **f** −6
g −12 **h** −10 **i** 4 **j** −4 **k** 14 **l** 11
m −4 **n** −1 **o** −10 **p** −5 **q** −3 **r** 5
s −4 **t** −8

7 **a** 2 **b** −3 **c** −5 **d** −7 **e** −10 **f** −20

8 **a** 2 **b** 4 **c** −1 **d** −5 **e** −11 **f** 8

9 **a** 13 **b** 2 **c** 5 **d** 4 **e** 11 **f** −2

10 **a** −10 **b** −5 **c** −2 **d** 4 **e** 7 **f** −4

11 Check student's answers

12 Check student's answers

13 **a** −5 **b** 6 **c** 0 **d** 2 **e** 13 **f** 0
g −6 **h** −2 **i** 212 **j** 5 **k** 3 **l** 3
m −67 **n** 7 **o** 25

14 **a** −1, 0, 1, 2, 3 **b** −6, −5, −4, −3, −2
c −3, −2, −1, 0, 1 **d** −8, −7, −6, −5, −4
e −9, −8, −7, −6, −5 **f** 3, 4, 5, 6, 7
g −12, −11, −10, −9, −8 **h** −16, −15, −14, −13, −12
i −2, −1, 0, 1, 2, 3; −4, −3, −2, −1, 0, 1
j −12, −11, −10, −9, −8, −7; −14, −13, −12, −11, −10, −9
k −2, −1, 0, 1, 2, 3; 0, 1, 2, 3, 4, 5
l −8, −7, −6, −5, −4, −3, −2; −5, −4, −3, −2, −1, 0, 1
m −10, −9, −8, −7, −6, −5, −4; −1, 0, 1, 2, 3, 4, 5
n 3, 4, 5, 6, 7, 8, 9; −5, −4, −3, −2, −1, 0, 1

15 **a** −4 **b** 3 **c** 4 **d** −6 **e** 7 **f** 2
g 7 **h** −6 **i** −7 **j** 0 **k** 0 **l** −6
m −7 **n** −9 **o** 4 **p** 0 **q** 5 **r** 0
s 10 **t** −5 **u** 3 **v** −3 **w** −9 **x** 0
y −3 **z** −3

16 **a** +6 + +5 = 11 **b** +6 + −9 = −3
c +6 − −9 = 15 **d** +6 − +5 = 1

17 **a** +5 + +7 − −9 = +21 **b** +5 + −9 − +7 = −11
c +7 + −7, +4 + −4

18 It may not come on as the thermometer inaccuracy might be between 0° and 2° or 2° and 4°

19 −1 and 6

Exercise 3F

1

−1	−9	−2
−5	−4	−3
−6	1	−7

−12

2

1	−4	3
2	0	−2
−3	4	−1

0

3

0	−14	−1
−6	−5	−4
−9	4	−10

−15

4

2	−12	1
−4	−3	−2
−7	6	−8

−9

5

−3	−6	−9
−12	−6	0
−3	−6	−9

−18

6

−2	−18	−1
−6	−7	−8
−13	4	−12

−21

7

−4	−12	−5
−8	−7	−6
−9	−2	−10

−21

8

2	1	−3
−5	0	5
3	−1	−2

0

9

−2	−10	−3
−6	−5	−4
−7	0	−8

−15

10

−8	−1	−3	−14
−8	−9	−7	−2
−11	−6	−4	−5
1	−10	−12	−5

−26

11

−7	5	2	−16
−6	−8	−5	3
−11	−3	0	−2
8	−10	−13	−1

−16

Examination questions

1 **a** Tuesday
b Friday

2 **a** Oslo
b 13 °C

3 **a** Plymouth
b Edinburgh

4 **a** −2

5 **a** 8
b −9
c −7

6 **a** 90 °C
b 540 °C
c Jupiter
d −230 °C

7 **a** −2 °C, −1 °C, 0 °C, 1 °C
b 3 °C

8 **a i** 8 **ii** 9 **iii** −7
b i −7 **ii** 8 **iii** 8

9 **a** 3
b −8

10 **a** 9 °C
b 8 °C
c −1 °C

11 **i** Any number between −196 °C and −210 °C, e.g. −200 °C
ii Any number over −196 °C, e.g. −10 °C or any positive number
iii Any number below −210 °C, e.g. −250 °C

12 **a i** Liquid hydrogen **ii** 70 °C
b i 21 °C **ii** 91 °C

Answers to Chapter 4

4.1 Multiples of whole numbers

Exercise 4A

1 **a** 3, 6, 9, 12, 15 **b** 7, 14, 21, 28, 35 **c** 9, 18, 27, 36, 45 **d** 11, 22, 33, 44, 55 **e** 16, 32, 48, 64, 80

2 **a** 254, 108, 68, 162, 98, 812, 102, 270 **b** 111, 255, 108, 162, 711, 615, 102, 75, 270 **c** 255, 615, 75, 270 **d** 108, 162, 711, 270

3 **a** 72, 132, 216, 312, 168, 144 **b** 161, 91, 168, 294 **c** 72, 102, 132, 78, 216, 312, 168, 144, 294

4 **a** 98 **b** 99 **c** 96 **d** 95 **e** 98 **f** 96

5 **a** 1002 **b** 1008 **c** 1008

6 No, 50 is not a multiple of 6.

7 4 or 5 (as 2, 10 and 20 are not realistic answers)

8 **a** 18 **b** 28 **c** 15

9 66

10 5 numbers: 18, 36, 54, 72, 90

11 **a** 1, 2, 5, 10 **b** 1, 3, 5, 6, 10, 15, 50 **c** 1, 2, 4, 5, 10, 20, 50, 100 **d** For example, 60, 72, 84, 96

4.2 Factors of whole numbers

Exercise 4B

1 **a** 1, 2, 5, 10 **b** 1, 2, 4, 7, 14, 28 **c** 1, 2, 3, 6, 9, 18 **d** 1, 17 **e** 1, 5, 25 **f** 1, 2, 4, 5, 8, 10, 20, 40 **g** 1, 2, 3, 5, 6, 10, 15, 30 **h** 1, 3, 5, 9, 15, 45 **i** 1, 2, 3, 4, 6, 8, 12, 24 **j** 1, 2, 4, 8, 16

2 8 ways (1, 2, 3, 4, 6, 8, 12, 24 per box)

3 **a** 1, 2, 3, 4, 5, 6, 8, 10, 12, 15, 20, 24, 30, 40, 60, 120 **b** 1, 2, 3, 5, 6, 10, 15, 25, 30, 50, 75, 150 **c** 1, 2, 3, 4, 6, 8, 9, 12, 16, 18, 24, 36, 48, 72, 144 **d** 1, 2, 3, 4, 5, 6, 9, 10, 12, 15, 18, 20, 30, 36, 45, 60, 90, 180 **e** 1, 13, 169 **f** 1, 2, 3, 4, 6, 9, 12, 18, 27, 36, 54, 108 **g** 1, 2, 4, 7, 14, 28, 49, 98, 196 **h** 1, 3, 9, 17, 51, 153 **i** 1, 2, 3, 6, 9, 11, 18, 22, 33, 66, 99, 198 **j** 1, 199

4 **a** 55 **b** 67 **c** 29 **d** 39 **e** 65 **f** 80 **g** 80 **h** 70 **i** 81 **j** 50

5 **a** 2 **b** 2 **c** 3 **d** 5 **e** 3 **f** 3 **g** 7 **h** 5 **i** 10 **j** 11

6 4 rows of 3 or 3 rows of 4.

7 It does not have a factor of 3.

8 5

4.3 Prime numbers

Exercise 4C

1 23 and 29

2 97

3 All these numbers are not prime.

4 3, 5, 7

5 Only if all 31 bars are in a single row, as 31 is a prime number and its only factors are 1 and 31.

4.4 Square numbers

Exercise 4D

1 36, 49, 64, 81, 100, 121, 144, 169, 196, 225, 256, 289, 324, 361, 400

2 4, 9, 16, 25, 36, 49

3 **a** 3 **b** 5 **c** 7 **d** Odd numbers

4 **a** 529 **b** 3249 **c** 5929 **d** 15 129 **e** 23 104 **f** 10.24 **g** 90.25 **h** 566.44 **i** 16 **j** 144

5 **a** 169 **b** 196 **c** answer between 169 and 196 (exact answer is 174.24)

6 **a** 50, 65, 82 **b** 98, 128, 162 **c** 51, 66, 83 **d** 48, 63, 80 **e** 149, 164, 181

7 **a** 25, 169, 625, 1681, 3721 **b** Answers in each row are the same

8 £144

9 29

10 36 and 49

Exercise 4E

1 **a** 6, 12, 18, 24, 30 **b** 13, 26, 39, 52, 65 **c** 8, 16, 24, 32, 40 **d** 20, 40, 60, 80, 100 **e** 18, 36, 54, 72, 90

2 1, 4, 9, 16, 25, 36, 49, 64, 81, 100

3 **a** 1, 2, 3, 4, 6, 12 **b** 1, 2, 4, 5, 10, 20
c 1, 3, 9 **d** 1, 2, 4, 8, 16, 32
e 1, 2, 3, 4, 6, 8, 12, 24 **f** 1, 2, 19, 38
g 1, 13 **h** 1, 2, 3, 6, 7, 14, 21, 42
i 1, 3, 5, 9, 15, 45 **j** 1, 2, 3, 4, 6, 9, 12, 18, 36

4 **a** 12, 24, 36 **b** 20, 40, 60 **c** 15, 30, 45
d 18, 36, 54 **e** 35, 70, 105

5 Square numbers

6 13 is a prime number

7 2, 3, 5, 7, 11, 13, 17, 19

8 $1 + 3 + 5 + 7 + 9 = 25$
$1 + 3 + 5 + 7 + 9 + 11 = 36$
$1 + 3 + 5 + 7 + 9 + 11 + 13 = 49$
$1 + 3 + 5 + 7 + 9 + 11 + 13 + 15 = 64$

9

	Square number	Factor of 70
Even number	16	14
Multiple of 7	49	35

10 4761 (69^2)

11 4 packs of sausages, 5 packs of buns

12 24 seconds

13 30 seconds

14 12 minutes; Debbie: 4 and Fred: 3

15 **a** 12 **b** 9 **c** 6 **d** 13 **e** 15 **f** 14
g 16 **h** 10 **i** 18 **j** 17 **k** 8 **l** 21

16 **b** 21, 28, 36, 45, 55

4.5 Square roots

Exercise 4F

1 **a** 2 **b** 5 **c** 7 **d** 1 **e** 9 **f** 10
g 8 **h** 3 **i** 6 **j** 4 **k** 11 **l** 12
m 20 **n** 30 **o** 13

2 **a** +5, −5 **b** +6, −6 **c** +10, −10
d +7, −7 **e** +8, −8 **f** +4, −4
g +3, −3 **h** +9, −9 **i** +1, −1
j +12, −12

3 **a** 81 **b** 40 **c** 100 **d** 14 **e** 36 **f** 15
g 49 **h** 12 **i** 25 **j** 21 **k** 121 **l** 16
m 64 **n** 17 **o** 441

4 **a** 24 **b** 31 **c** 45 **d** 40 **e** 67 **f** 101
g 3.6 **h** 6.5 **i** 13.9 **j** 22.2

5 $\sqrt{50}$, 3^2, $\sqrt{90}$, 4^2

6 4 and 5

7 $\sqrt{324} = 18$

8 15 tiles

4.6 Powers

Exercise 4G

1 **a** 27 **b** 125 **c** 216 **d** 1728 **e** 16 **f** 256
g 625 **h** 32 **i** 2187 **j** 1024

2 **a** 100 **b** 1000 **c** 10 000 **d** 100 000 **e** 1 000 000
f The power is the same as the number of zeros
g **i** 100 000 000 **ii** 10 000 000 000
iii 1 000 000 000 000 000

3 **a** 2^4 **b** 3^5 **c** 7^2 **d** 5^3 **e** 10^7 **f** 6^4
g 4^4 **h** 1^7 **i** 0.5^4 **j** 100^3

4 **a** $3 \times 3 \times 3 \times 3$ **b** $9 \times 9 \times 9$ **c** 6×6
d $10 \times 10 \times 10 \times 10 \times 10$
e $2 \times 2 \times 2 \times 2 \times 2 \times 2 \times 2 \times 2 \times 2 \times 2$
f $8 \times 8 \times 8 \times 8 \times 8 \times 8$ **g** $0.1 \times 0.1 \times 0.1$
h 2.5×2.5 **i** $0.7 \times 0.7 \times 0.7$ **j** 1000×1000

5 **a** 16 **b** 243 **c** 49 **d** 125 **e** 10 000 000
f 1296 **g** 256 **h** 1 **i** 0.0625 **j** 1 000 000

6 **a** 81 **b** 729 **c** 36 **d** 100 000 **e** 1024
f 262 144 **g** 0.001 **h** 6.25 **i** 0.343 **j** 1 000 000

7 10^6

8 10^6

9 4, 8, 16, 32, 64, 128, 256, 512

10 0.001, 0.01, 0.1, 1, 10, 100, 1000, 10 000, 100 000, 1 000 000, 10 000 000, 100 000 000

11 125 m^3

12 **a** answer given **b** 10^2 **c** 2^3 **d** 5^2

13 $6^2 = 36$ or $81 = 9^2$

14 **a** $x = 9$ **b** $y = 8$ **c** $z = 4$
d $w = 4$ **e** $s = 1000$ **f** $r = 1000$

Examination questions

1 **a** 99 **b** 102

2 4, 16, 25, 36, 49, 64, 81

3 **a** 90 **b** 105

4 **a** 4, 6, $4 \times 4 \times 4 \times 4 \times 4$, 1024, 4 **b** 4

5 **a** 6, 12, 18, 24, 30 **b** 1, 2, 3, 4, 6, 12 **c** 25 **d** 23, 29

6 **a** 4, 8 **b** 5, 10 **c** 5, 11 **d** 4, 9 **e** 8

7 **a** 13.69 **b** 53, 59

8 **a** 8 **b** 8

9 **a** 64 **b** 10 **c** 125

10 40

11 5

12 5 packets of pies and 2 packets of breadsticks

Answers to Chapter 5

5.1 Frequency diagrams

Exercise 5A

1 **a**

Goals	0	1	2	3
Frequency	6	8	4	2

b 1 goal **c** 22

2 **a**

Temperature (°C)	14–16	17–19	20–22	23–25	26–28
Frequency	5	10	8	5	2

b 17–19 °C
c Getting warmer in the first half and then getting cooler towards the end.

3 **a** Observation **b** Sampling **c** Observation
d Sampling **e** Observation **f** Experiment

4 **a**

Score	1	2	3	4	5	6
Frequency	5	6	6	6	3	4

b 30 **c** Yes, frequencies are similar.

5 **a**

Height (cm)	151–155	156–160	161–165	166–170	171–175	176–180	181–185	186–190
Frequency	2	5	5	7	5	4	3	1

b 166–170 cm **c** Student's survey results.

6 various answers such as 1–10, 11–20, etc. or 1–20, 21–40, 41–60

7 The ages 20 and 25 are in two different groups.

8 Student's survey results and frequency tables.

5.2 Statistical diagrams

Exercise 5B

1

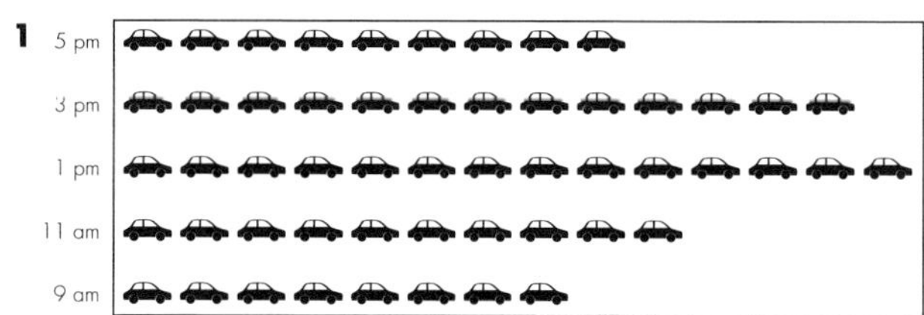

2

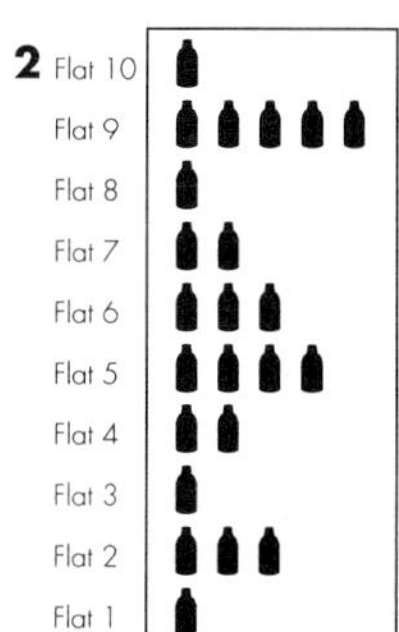

3 **a** May 9 h, Jun 11 h, Jul 12 h, Aug 11 h, Sep 10 h
b July **c** Visual impact, easy to understand.

4 **a** Simon **b** £165
c Difficult to show fractions of a symbol.

5 **a** **i** 12 **ii** 6 **iii** 13
b Check students' pictograms. **c** 63

6 Use a key of 17 students to one symbol.

7 There would be too many symbols to show.

8 **a–c** Student's pictograms.

9 The second teddy bear is much bigger than twice the size of the small teddy.

5.3 Bar charts

Exercise 5C

1 **a** Swimming **b** 74
c For example: limited facilities
d No. It may not include people who are not fit.

2 **a**

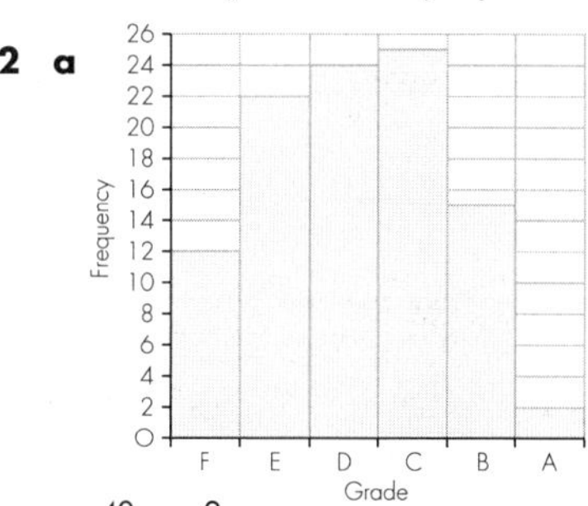

b $\frac{40}{100} = \frac{2}{5}$ **c** Easier to read the exact frequency.

3 **a**

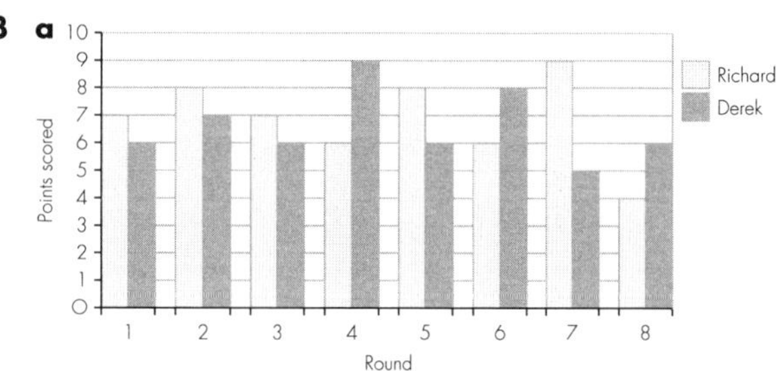

b Richard got more points overall, but Derek was more consistent.

4 **a**

Time (min)	1–10	11–20	21–30	31–40	41–50	51–60
Frequency	4	7	5	5	7	2

b

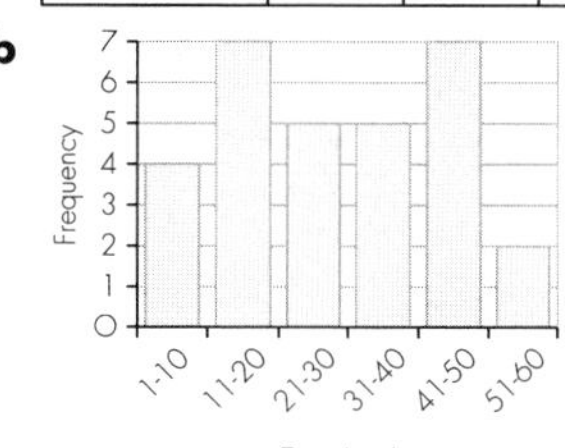

c Some live close to the school. Some live a good distance away and probably travel to school by bus.

5 a

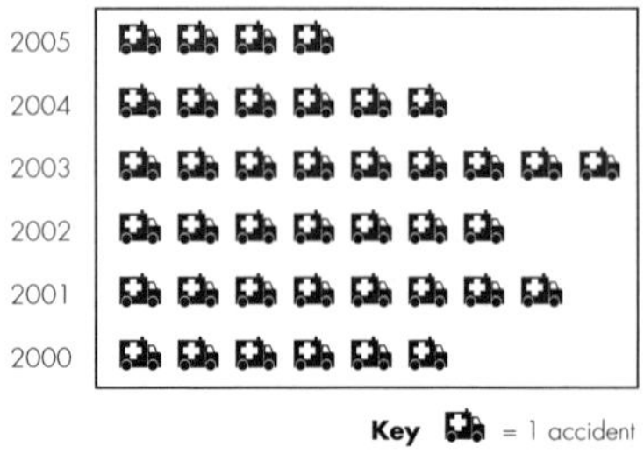

Year	
2005	
2004	
2003	
2002	
2001	
2000	

Key = 1 accident

b

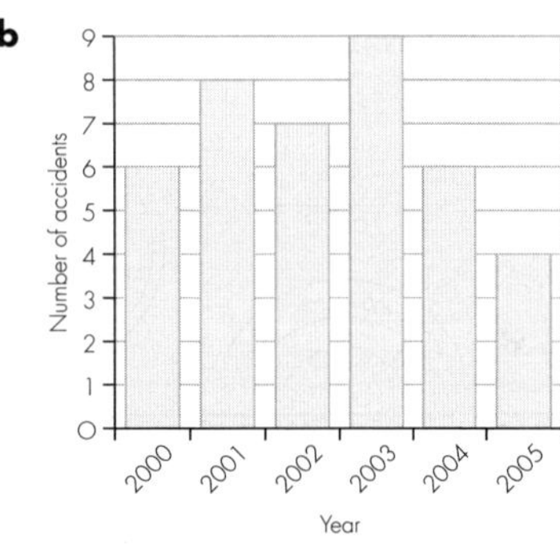

c Use the pictogram because an appropriate symbol makes more impact.

6 Yes. If you double the minimum temperature each time, it is very close to the maximum temperature.

7 The graphs do not start at zero.

8 Student's survey results and pictograms and bar charts.

9 Student's frequency tables and dual bar chart.

10 £44.40

5.4 Line graphs

Exercise 5D

1 a Tuesday, 52p **b** 2p **c** Friday **d** £90

2 a

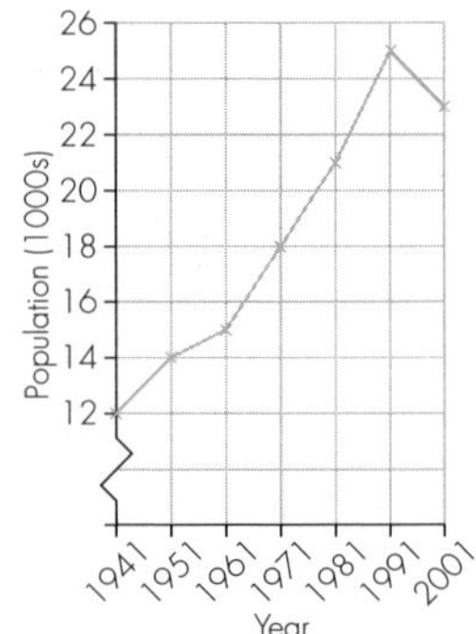

b about 16 500
c 1981 and 1991
d No; do not know the reason why the population started to decrease after 1991

3 a

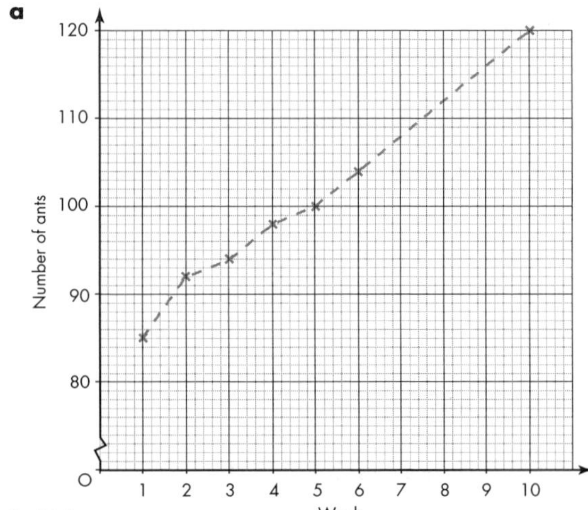

b 112

4 a

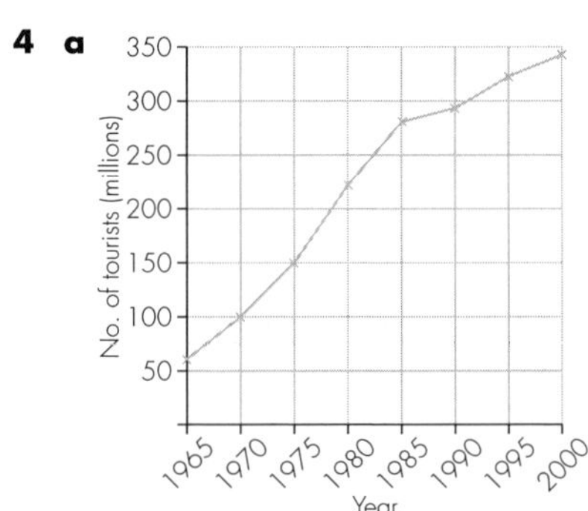

b About 410 million
c 1975 and 1980
d Student's explanation of trend.

5 a

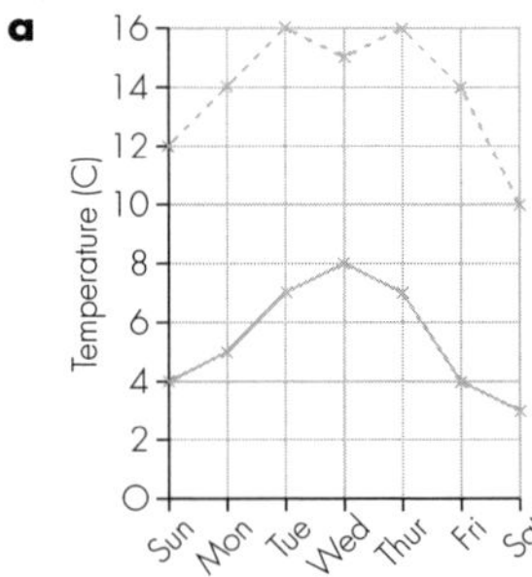

b 7 °C and 10 °C

6 From a graph, about 1040 g

7 All the temperatures were presumably higher than 20 degrees.

5.5 Stem-and-leaf diagrams

Exercise 5E

1 a 17 s **b** 22 s **c** 21 s

2 a 57 **b** 55 **c** 56 **d** 48
e Boys did better, because their marks are higher.

3 **a** 2 8 9
3 4 5 6 8 8 9
4 1 1 3 3 3 8 8
Key 4|3 represents 43 cm
b 48 cm **c** 43 cm **d** 20 cm

4 **a** 0 2 8 9 9 9
1 2 3 7 7 8
2 0 1 2 3
Key 1|2 represents 12 messages
b 23 **c** 9

5 All the data start with a 5 and there are only two digits.

Examination questions

1 Tallies and frequencies: 3, 4, 7, 5, 4, 1

2 **a** 7
b Monday
c Tuesday and Wednesday

3 **a** 20
b 15
c 4 circles on Friday
$2\frac{1}{2}$ circles on Saturday

4 **a** chart filled in appropriately: Blue = 8 squares, Green = 5 squares, Yellow = 3 squares
b Blue

5 **a i** 8 **ii** 10
b Diana = 3 whole boxes
Erikas = 2 whole boxes and one small square

6 **a** Wednesday
Friday
b 320 minutes

7 The buses are all different sizes and there is no key.

8 1

9

2 | 3 5
3 | 1 2 7 8
4 | 0 6 8 9
5 | 6 6

Key: 2 | 5 = £25

10

0 | 7 8 9 9
1 | 0 0 0 2 5 5 6 7 8 9
2 | 1 1 3 8
3 | 0 2

Key: 2 | 5 = 25 years

11 **a** $150 \leq w < 200$
b Tom's tomatoes plotted at these points: (75, 21), (125, 28), (175, 26), (275, 9) and (325, 2) with a line connecting them
c Tom's were generally smaller, but more consistent

Answers to Chapter 6

6.1 Reading scales

Exercise 6A

1	**a i** 4	**ii** 16 g	**iii** 38 mph
	b i 8 kg	**ii** 66 mph	**iii** 60 g
	c i 13 oz	**ii** 85 mph	**iii** 76 kph
	d i 26	**ii** 71.6	**iii** 64

2 **a**

b

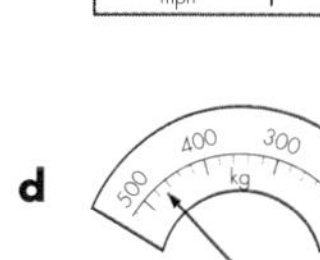

c

d

3 **a** 50 °C **b** 64 °C **c** −10 °C
d 82 °C **e** −16 °C

4 **a** 8 kg **b** 29 mph **c** 230 g
d 12.7 kg

5 360 g **b** weigh out 400 g, then weigh out 300 g

6 **a** 1.2 kg **b** 125 g

7 **a** 125 km/h
b 125 km/h shown on scale; the scale could go up in 10s, 20s or 25s, discuss this with students

8 No, it is pointing to 7.48 m; each division is 0.1 ÷ 5 = 0.02

6.2 Sensible estimates

Exercise 6B

1 Bicycle about 2 m, bus about 10 m, train about 17 m

2 Height about 4 m, length about 18 or 19 m

3 250 g

4 **a** About 4 m **b** About 5 m **c** About 5.5 m

5 About 9.5 g or 10 g

6 About 9 m; the ratio of Joel's height in the photograph to his real height must be the same as the ratio of the height on the statue in the photograph to its actual height

7 About 6 m

6.3 Scale drawings

Exercise 6C

1 **a** Onions: 40 m × 10 m, soft fruit: 50 m × 10 m, apple trees: 20 m × 20 m, lawn: 30 m × 20 m, potatoes: 50 m × 20 m
b Onions: 400 m^2, soft fruit: 500 m^2,
apple trees: 400 m^2,
lawn: 600 m^2,
potatoes: 1000 m^2

2 **a** 33 cm **b** 9 cm

3 **a** 30 cm × 30 cm **b** 40 cm × 10 cm **c** 20 cm × 15 cm
d 30 cm × 20 cm **e** 30 cm × 20 cm **f** 10 cm × 5 cm

4 **a** Student's scale drawing. **b** About 19 m

5 **a** 8.4 km **b** 4.6 km **c** 6.2 km
d 6.4 km **e** 7.6 km **f** 2.4 km

6 **a i** 64 km **ii** 208 km **iii** 116 km **iv** 40 km
b i 84 km **ii** 196 km **iii** 152 km **iv** 128 km

7 **a** 50 km **b** 35 km **c** 45 km

8 **c** 7 cm represents 210 m, so 1 cm represents 30 m or 1 cm represents 3000 cm

Activity: Little and large!
140 miles, 60 miles, 320 miles

6.4 Nets

Exercise 6D

1 **a** **b** **c**

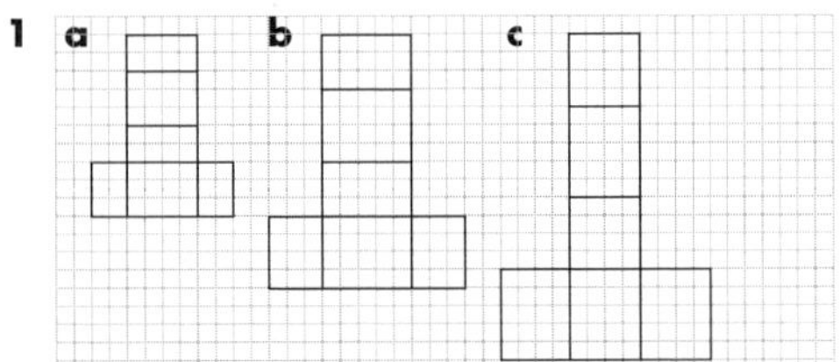

2

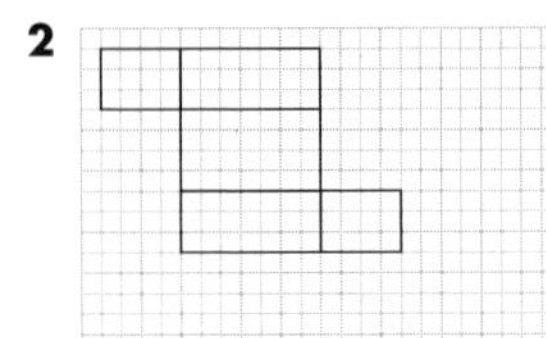

3

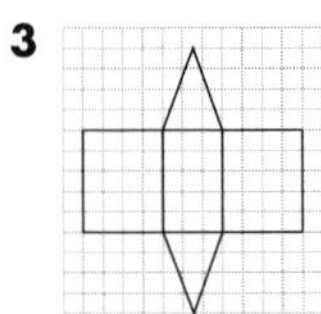

4 **a** **b** **c** **d**

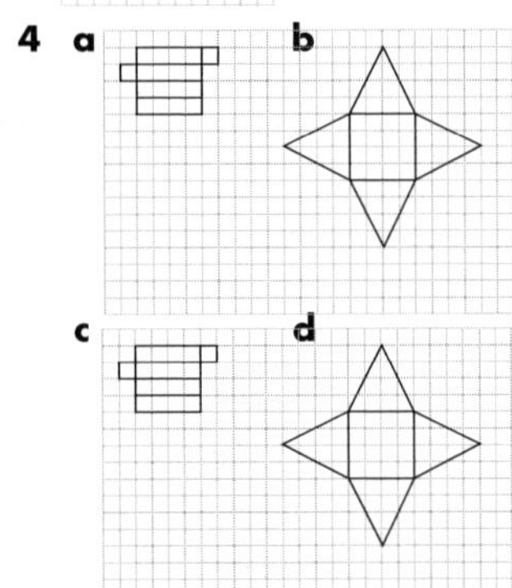

5

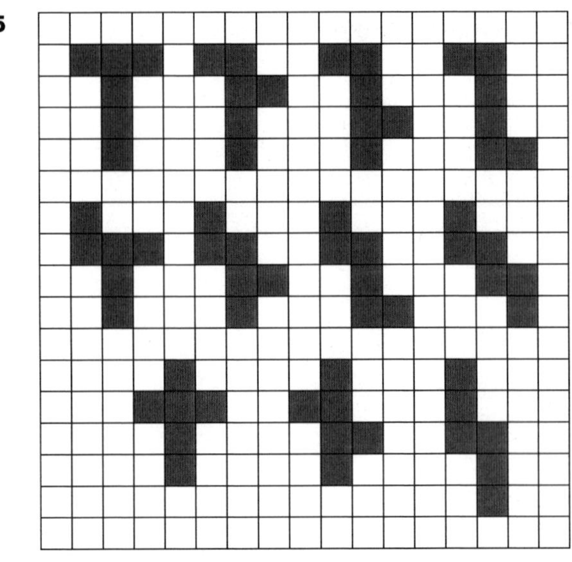

6 **a** and **b**

6.5 Using an isometric grid

Exercise 6E

1 **a** **b** **c**

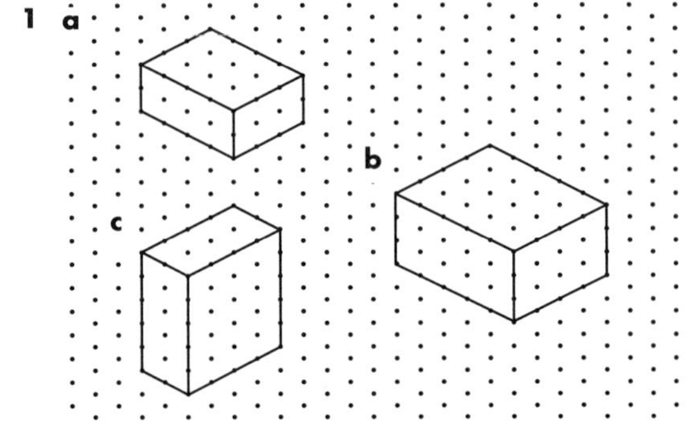

2 **a** **b**

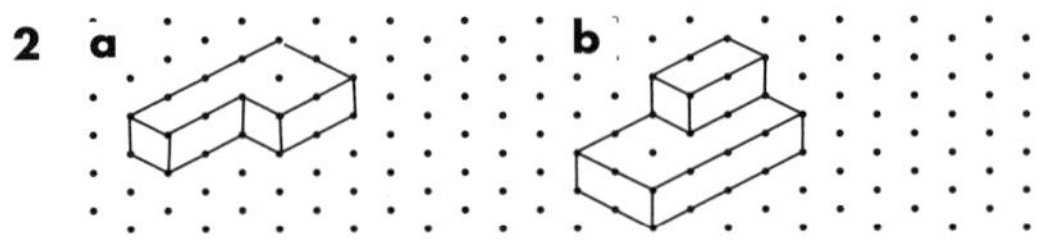

3 **a i** **ii** **iii**
b i **ii** **iii**
c i **ii** **iii**

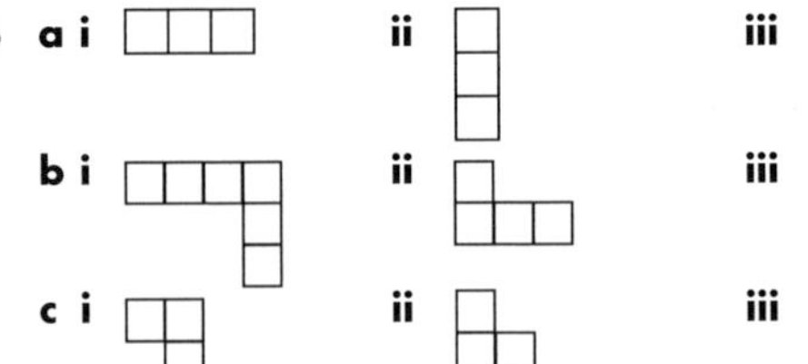

4 Students' own diagram

5

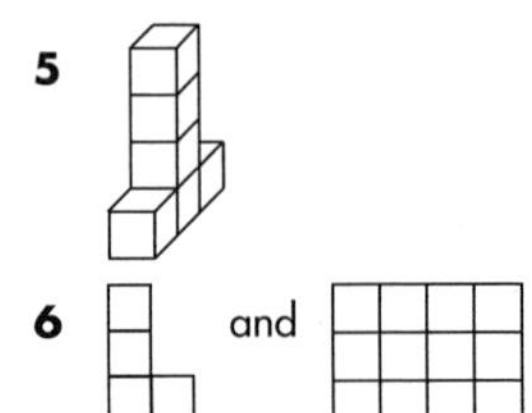

6 and

Examination questions

1 **a** 29
b 120 marked with arrow

2 **a** 17.8 cm
b −2 °C
c 2.8 kg

3

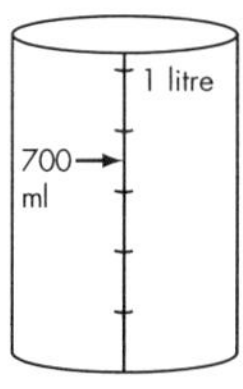

4 **a** 38 °C
b

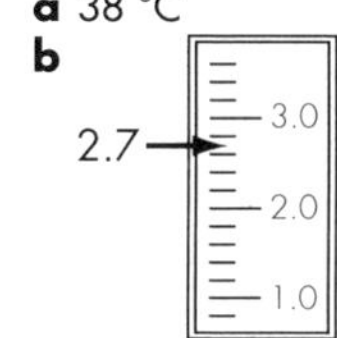

5 A, C, D

6 **a** 8 litres
b £29.10

7 **a** 1.8 m
b about 7 m

8 **a** 250 mm
b 324 m

9

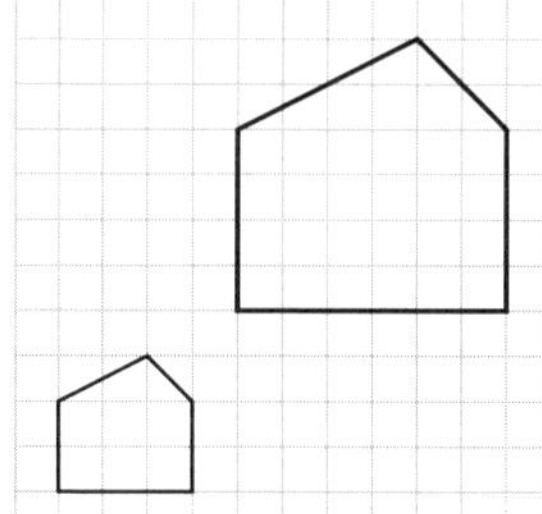

10

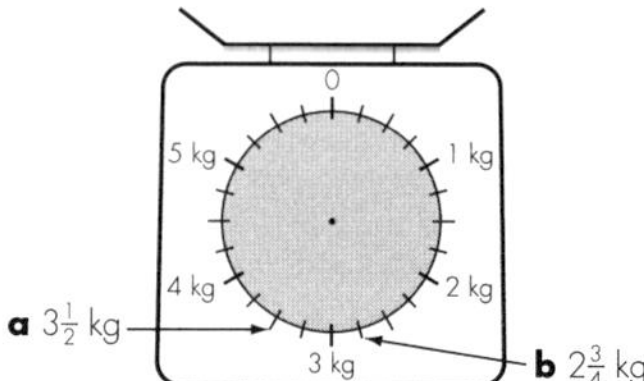

11

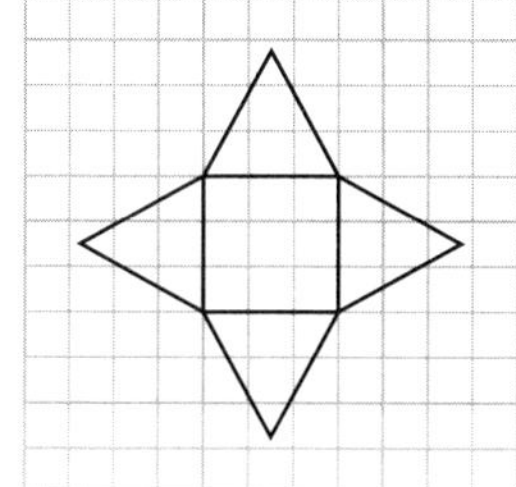

triangles must be equilateral, sides 3 cm

12

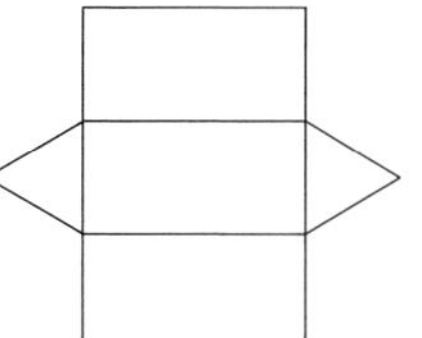

13 The plan view shows a line of 5 squares going horizontally across the page.
The front elevation shows a line of 3 squares going vertically down the page.
The side elevation shows a T-shape made of 5 squares going across and 3 squares going down.

14

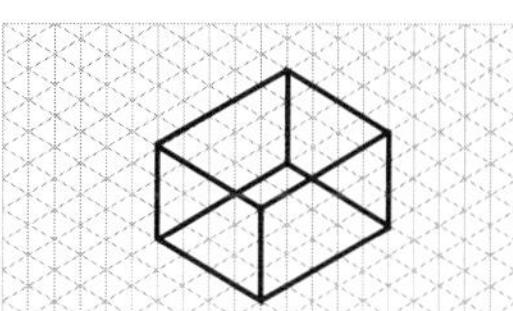

15

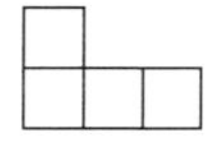

16

Answers to Chapter 7

7.1 Systems of measurement

Exercise 7A

1 **a** metres **b** kilometres **c** millimetres
d kilograms or grams **e** litres
f kilograms **g** tonnes **h** millilitres
i centilitres **j** metres **k** kilograms
l litres **m** grams **n** centilitres
o millimetres

2 Check individual answers.

3 The 5 metre since his height is about 175 cm, the lamp post will be about 525 cm

4 Inches, feet and yards are too small as units; this distance is an approximation and so needs to be a large unit as this is a large distance.

7.2 Metric units

Exercise 7B

1 **a** 1.25 m **b** 8.2 cm **c** 0.55 m
d 2.1 km **e** 2.08 cm **f** 1.24 m
g 4.2 kg **h** 5.75 t **i** 8.5 cl
j 2.58 l **k** 3.4 l **l** 0.6 t
m 0.755 kg **n** 0.8 l **o** 2 l
p 63 cl **q** 8.4 m^3 **r** 35 cm^3
s 1.035 m^3 **t** 0.53 m^3 **u** 34 000 m

2 **a** 3400 mm **b** 135 mm **c** 67 cm
d 7030 m **e** 7.2 mm **f** 25 cm
g 640 m **h** 2400 ml **i** 590 cl
j 84 ml **k** 5200 l **l** 580 g
m 3750 kg **n** 0.000 94 l **o** 2160 cl
p 15 200 g **q** 14 000 l **r** 0.19 ml

3 He should choose the 2000 mm × 15 mm × 20 mm

4 as 1 millilitre = 1 cm^3 and 1 litre = 1000 cm^3

5 1 000 000 000 000

7.3 Imperial units

Exercise 7C

1 **a** 24 in **b** 12 ft **c** 3520 yd
d 80 oz **e** 56 lb **f** 6720 lb
g 40 pt **h** 48 in **i** 36 in
j 30 ft **k** 64 oz **l** 5 ft
m 70 lb **n** 12 yd **o** 224 oz

2 **a** 5 miles **b** 120 pt **c** 5280 ft
d 8 ft **e** 7 st **f** 7 gal
g 2 lb **h** 5 yd **i** 5 tons
j 63 360 in **k** 8 lb **l** 9 gal
m 10 st **n** 3 miles **o** 35 840 oz

3 the 32-ounce bag

4 4 014 489 600

5 1 tonne = 1000 kilograms = 1000 × 2.2 pounds = 2200 pounds; 1 ton = 2240 pounds; 2240 is greater than 2200

7.4 Conversion factors

Exercise 7D

1 **a** 20 cm **b** 13.2 lb **c** 48 km
d 67.5 l **e** 2850 ml **f** 10 gal
g 12 in **h** 50 miles **i** 5 kg
j 3 pints **k** 160 km **l** 123.2 lb
m 180 l **n** 90.9 kg **o** 1100 yd
p 30 cm **q** 6.4 kg **r** 90 cm

2 ton

3 metre

4 **a i** 1000 g **ii** 1 kg
b i 4500 g **ii** 4.5 kg

5 **a** 135 miles **b** 50 mph **c** 2 h 42 min

6 4 hours 10 minutes

7 288

Examination questions

1 metres, kilograms, inches

2 **a** 40 000 m
b 25 miles

3 No. 30 pints ≈ 17.1 litres (30 × 0.570)

4 **a** metres, grams, litres
b 400 cm
c 1.5 kg

5 **a** 4.6 m
b i 2.2 lb **ii** 11 lb

6 **a i** kilometres **ii** litres
b i 50 **ii** 4

7 81.25 mph

8 **a** 1 m
b 2.82 m

9 Centimetres is too small a unit; this distance is an approximation of a large distance and so needs to be a large unit such as kilometres.

10 **a** $7\frac{1}{2}$ miles **b** 2.3 kg of potatoes, 227 grams of butter, 2.25 litres of milk

Answers to Chapter 8

8.1 Lines of symmetry

Exercise 8A

1 **a** 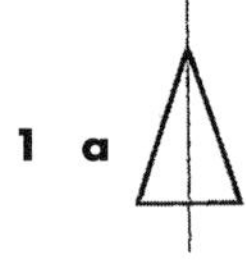 **b** 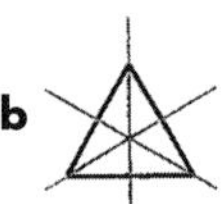

c 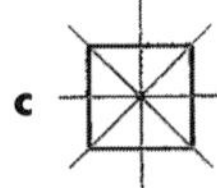 **d** 

e 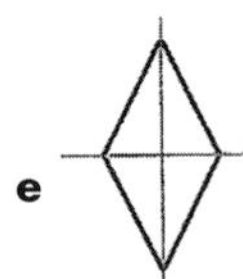 **f** 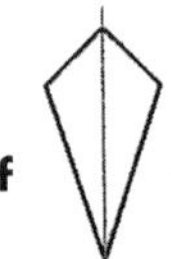 **g** 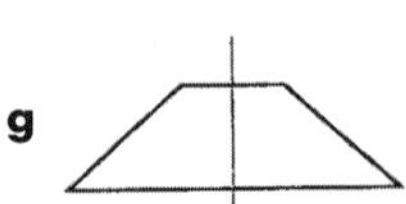

2 Discuss students' answers. There are many possibilities.

3 **a i** 5 **ii** 6 **iii** 8 **b** 10

4 a 

b 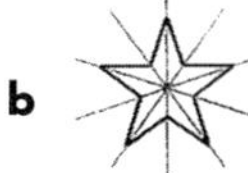

c 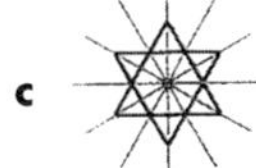

5 a 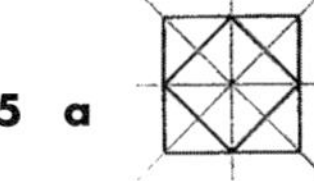

b 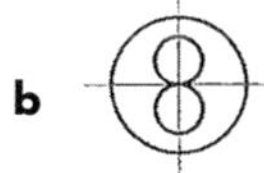

c 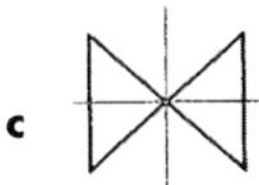

d 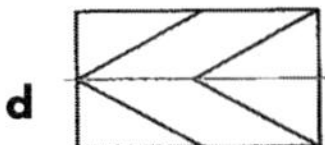

e 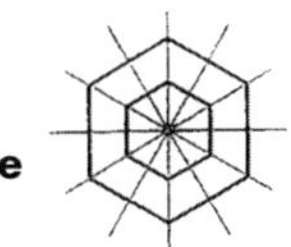

f 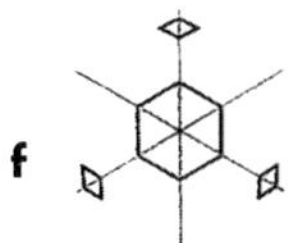

6 2, 1, 1, 2, 0

7 a

b Students' own answers

8 a 1 **b** 5 **c** 1 **d** 6

9 Discuss students' answers. There are many possibilities.

10 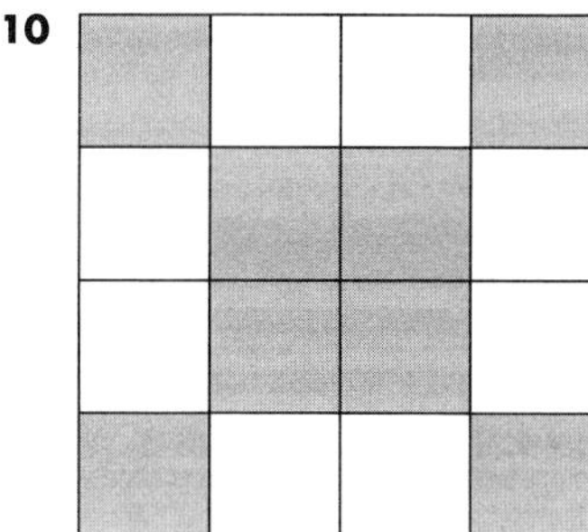

11 No, a triangle may have no lines of symmetry, or one or three, but you cannot draw a triangle with two lines of symmetry.

8.2 Rotational symmetry

Exercise 8B

1 a 4 **b** 2 **c** 2 **d** 3 **e** 6

2 a 4 **b** 5 **c** 6 **d** 4 **e** 6

3 a 2 **b** 2 **c** 2 **d** 2 **e** 2

4 a 4 **b** 3 **c** 8 **d** 2 **e** 4 **f** 2

5 A, B, C, D, E, F, G, J, K, L, M, P, Q, R, T, U, V, W, Y

6 Students' own answers

7 a 6 **b** 9 (the small red circle surrounded by nine 'petals' and 12 (the centre pattern)

8 Students' own answers

9 for example: 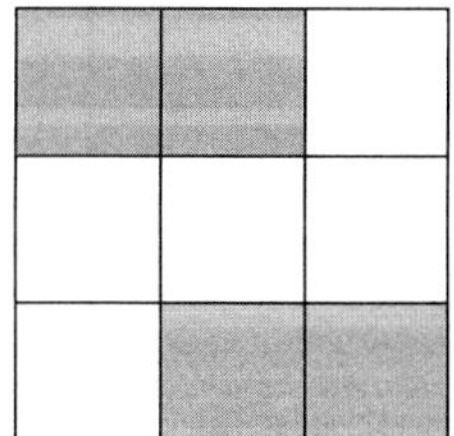

10

		Number of lines of symmetry			
		0	1	2	3
Order of rotational symmetry	1	D	A		
	2	E		B	
	3				C

Examination questions

1 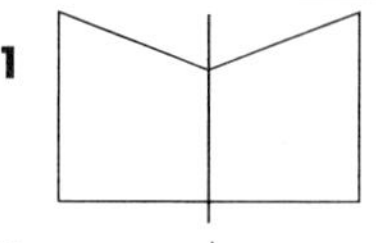

2 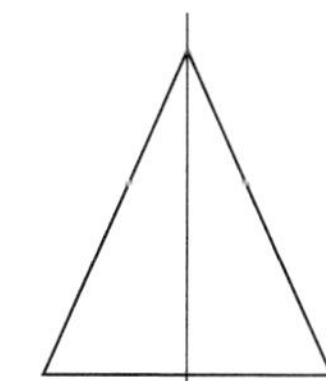

3 i D **ii** B **iii** A

4

5 a 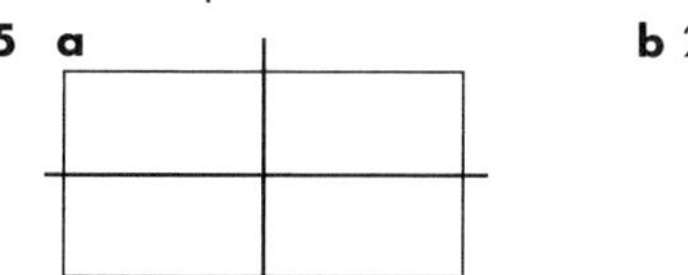

b 2

6 **a** 18
b 11 or 88
c 69

7 **a**

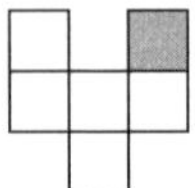

b

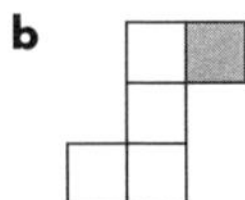

8 **a** 3
b

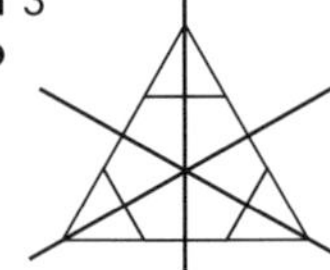

9 **a**

b **i**

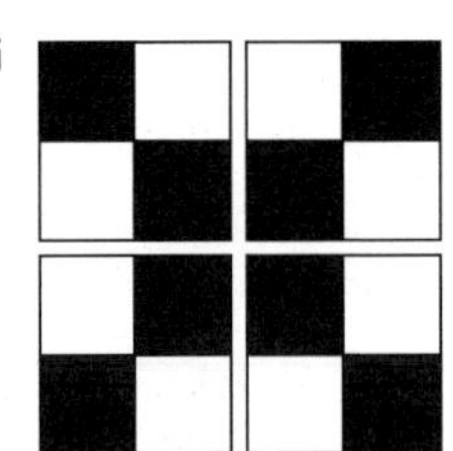

ii 4

Answers to Chapter 9

9.1 Conversion graphs

Exercise 9A

1 **a** **i** $8\frac{1}{4}$ kg **ii** $2\frac{1}{4}$ kg **iii** 9 lb **iv** 22 lb
b 2.2 lb
c Read off the value for12.6 (5.4 kg) and multiply this by 4 (21.6 kg)

2 **a** **i** 10 cm **ii** 23 cm **iii** 2 in **iv** $8\frac{3}{4}$ in
b $2\frac{1}{2}$ cm
c Read off the value for 9 in (23 cm) and multiply this by 2 (46 cm)

3 **a** **i** \$320 **ii** \$100 **iii** £45 **iv** £78
b \$3.20
c It would become less steep.

4 **a** **i** £120 **ii** £82
b **i** 32 **ii** 48

5 **a** **i** £100 **ii** £325
b **i** 500 **ii** 250

6 **a** **i** £70 **ii** £29
b **i** £85 **ii** £38

7 **a** **i** 95 °F **ii** 68 °F **iii** 10 °C **iv** 32 °C
b 32 °F

8 **a** Check student's graph
b 2.15 pm

9 **a** Check student's graph
b £50

10 **a** No trains on Christmas day and Boxing day
b Sudden drop in passenger numbers
c Increased shopping and last-minute present buying

11 No as 100 miles is about 160 km. He has travelled 75 km and has 125 km to go which is a total of 200 miles.

12 **a** Anya: CabCo £8.50, YellaCabs £8.40, so YellaCabs is best
Bettina: CabCo £11.50, YellaCabs £11.60, so CabCo is best
Calista: CabCo £10, YellaCabs £10, so either
b If they shared a cab, the shortest distance is 16 km, which would cost £14.50 with CabCo and £14.80 with Yellacabs.

9.2 Travel graphs

Exercise 9B

1 **a** **i** 2 h **ii** 3 h **iii** 5 h
b **i** 40 km/h **ii** 120 km/h **iii** 40 km/h
c 6.30 am

2 **a** 30 km
b 40 km
c 100 km/h

3 **a** **i** 125 km **ii** 125 km/h
b **i** Between 2 and 3 pm **ii** 25 km/h

4 Jafar started the race quickly and covered the first 1500 metres in 5 minutes. He then took a break for a minute and then finished the race at a slower pace taking 10 minutes overall. Azam started the race at a slower pace, taking 7 minutes to run 1500 metres. He then sped up and finished the race in a total of $9\frac{1}{2}$ minutes, beating Jafar.

5 **a** Araf ran the race at a constant pace, taking 5 minutes to cover the 1000 metres. Sean started slowly, covering the first 500 metres in 4 minutes. He then went faster, covering the last 500 metres in $1\frac{1}{2}$ minutes, giving a total time of $5\frac{1}{2}$ minutes for the race.
b **i** 20 km/h **ii** 12 km/h **iii** 10.9 km/h

6 There are three methods for doing this question.
This table shows the first, which is writing down the distances covered each hour.

Time	9 am	9:30	10:00	10.30	11.00	11.30	12.00	12.30
Walker	0	3	6	9	12	15	18	21
Cyclist	0	0	0	0	7.5	15	22.5	30

The second method is algebra:
Walker takes T hours until overtaken,
so $T = \frac{D}{6}$; Cyclist takes $T - 1.5$ to overtake, so $T - 1.5 = \frac{D}{15}$.
Rearranging gives $15T - 22.5 = 6T$, $9T = 22.5$, $T = 2.5$

The third method is a graph:

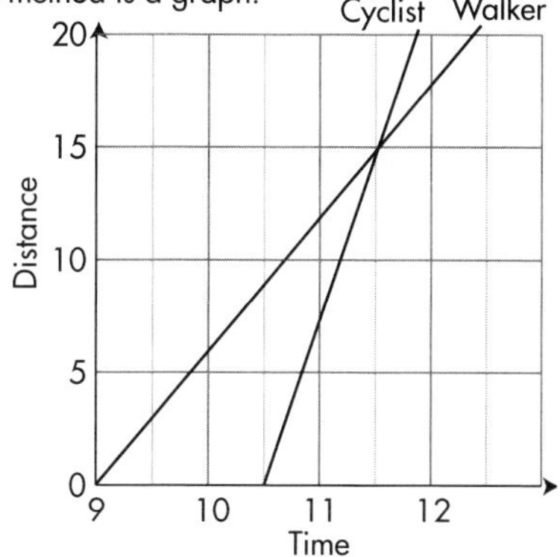

All methods give the same answer of 11:30 when the cyclist overtakes the walker.

7 **a** **i** Because it stopped several times **ii** Ravinder
b Ravinder at 3.55 pm or 3.58 pm, Sue at 4.20 pm, Michael at 4.35 pm
c **i** 24 km/h **ii** 20.6 km/h **iii** 5

9.3 Flow diagrams and graphs

Exercise 9C

1 **a** A(1, 2), B(3, 0), C(0, 1), D(−2, 4), E(−3, 2), F(−2, 0), G(−4, −1), H(−3, −3), I(1, −3), J(4, −2)
b **i** (2, 1) **ii** (−1, −3) **iii** (1, 1)
c $x = -3, x = 2, y = 3, y = -4$
d **i** $x = -\frac{1}{2}$ **ii** $y = -\frac{1}{2}$

2 Values of y: 2, 3, 4, 5, 6

3 Values of y: −2, 0, 2, 4, 6

4 Values of y: 1, 2, 3, 4, 5

5 Values of y: −4, −3, −2, −1, 0

6 **a** Values of y: 0, 4, 8, 12, 16 and 6, 8, 10, 12, 14
b (6, 12)

7 **a** Values of y: −3, −2, −1, 0, 1 and −6, −4, −2, 0, 2
b (3, 0)

8 Points could be (0, −1), (1, 4), (2, 9), (3, 14), (4, 19), (5, 24), etc

9 **a** £20.50
b £25.00, £28.50, £32.00, £35.50, £39.00, £42.50; (£20.00), £23.50, £27.00, £30.50, £34.00, £37.50; (£15.00), £18.50, £22.00, £25.50, £29.00, £32.50; (£10.00), (£13.50), £17.00, £20.50, £24.00, £27.50; (£5.00), (£8.50), (£12.00), £15.50, £19.00, £22.50; (£0.00), (£3.50), (£7.00), (£10.50), (£14.00), £17.50
c Yes, they had 3 cream teas and 4 high teas.

10 **a**

b $y = 3x + 1$
c Graph from (0, 1) to (5, 16)
d Read from 13 on the y-axis across to the graph and down to the x-axis. This should give a value of 4.

11 0, 8, 16, 24, 32, 40, 48

9.4 Linear graphs

Exercise 9D

1 Extreme points are (0, 4), (5, 19)

2 Extreme points are (0, −5), (5, 5)

3 Extreme points are (0, −3), (10, 2)

4 Extreme points are (−3, −4), (3, 14)

5 Extreme points are (−6, 2), (6, 6)

6 **a** Extreme points are (0, −2), (5, 13) and (0, 1), (5, 11)
b (3, 7)

7 **a** Extreme points are (0, −5), (5, 15) and (0, 3), (5, 13)
b (4, 11)

8 **a** Extreme points are (0, −1), (12, 3) and (0, −2), (12, 4)
b (6, 1)

9 **a** Extreme points are (0, 1), (4, 13) and (0, −2), (4, 10)
b Do not cross because they are parallel

10 **a** Values of y: 5, 4, 3, 2, 1, 0. Extreme points are (0, 5), (5, 0)
b Extreme points are (0, 7), (7, 0)

11 **a** Graph from (0, 25) to (8, 265) for Ian and graph from (0, 35) to (8, 255) for John.
b 4 hours

12 **a** Horizontal line through 4, vertical line through 1 and line from origin to (6, 6).
b 4.5 units squared

13 Graph passing through (x, z) (0, 2), (1, 3), (2, 4), (3, 5), (4, 6)

Exercise 9E

1 **a** 39.2 °C **b** Days 4 and 5, steepest line
c Days 8 and 9, steepest line **d** **i** Day 5 **ii** 2 **e** 37 °C

2 **a** 2 **b** $\frac{1}{3}$ **c** −3 **d** 1 **e** −2
f $-\frac{1}{3}$ **g** 5 **h** −5 **i** $\frac{1}{5}$ **j** $-\frac{3}{4}$

3 **a** 1
b −1
They are perpendicular and symmetrical about the axes.

4
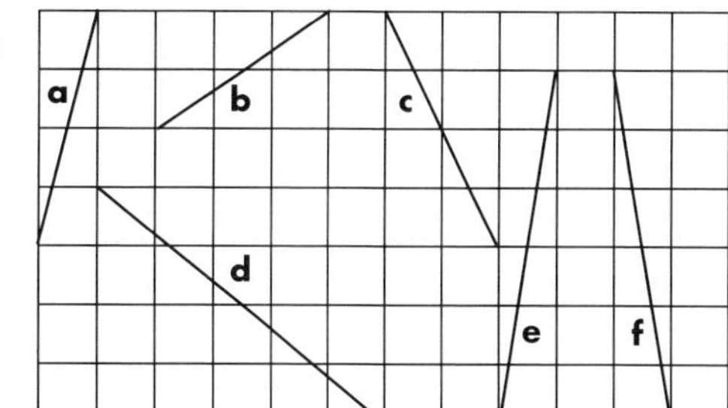

5 Rob has misread the scales. The gradient is actually 2. The line is $y = 2x + 2$. When $x = 10$, $y = 22$.

6 **a** 4 m **b** 1 m **c** **i** 1.375 m **ii** 3.2 m

Examination questions

1 **a** 19 – 21 feet
b 2.3 – 2.4 metres
c i Robert
ii The conversion graph shows that 4 metres is just over 13 feet, which is higher than 12 feet.

2 **a** (4, 3)
b cross drawn at point (–3, 1)
c (1, 0)

3 **a i** (1, 4) **ii** (4, 0)
b i Point P plotted at (3, 2)
ii Point Q plotted at (–4, 3)

4 **a**

x	–1	0	1	2	3
y	–5	–3	–1	1	3

b

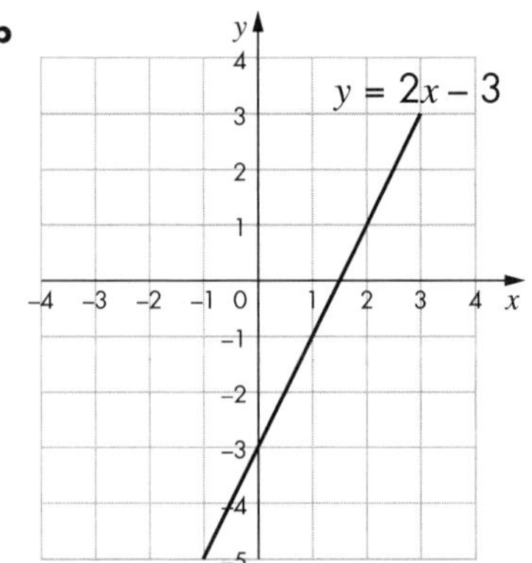

c (0.5, –2)

5 **a**

x	–2	–1	0	1	2
y	–7	–4	–1	2	5

b

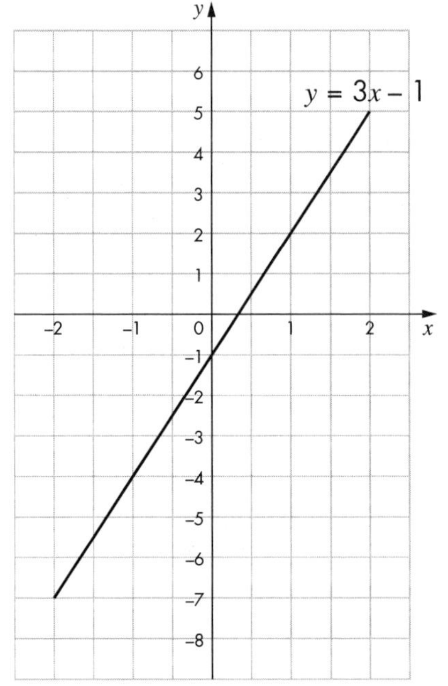

6 **a**

x	–3	–2	–1	0	1	2
y	–8	–5	–2	1	4	7

b

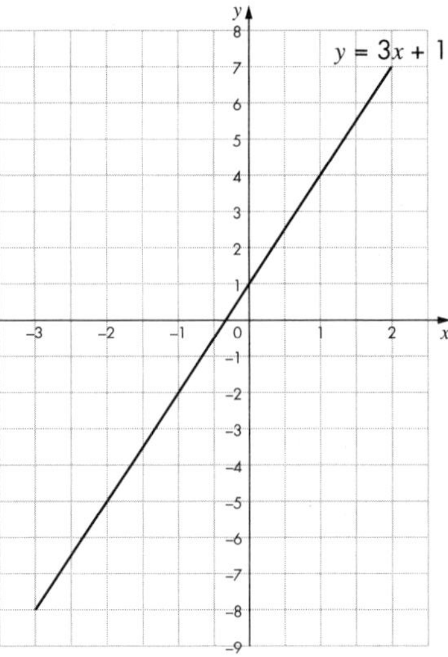

7 **a**

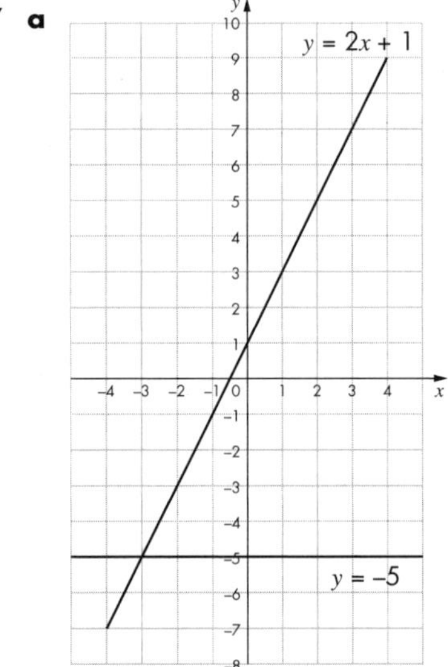

b (–3, –5)

8 **a** 10 km
b i 10:00 **ii** 30 minutes
c 11:20

9 **a** 1:30 pm
b i 10 km **ii** 30 minutes
c 36 km

10

The temperature starts at 0°c and keeps rising.	B
The temperature stays the same for a time and then falls.	D
The temperature rises and then falls quickly.	C
The temperature is always the same.	A
The temperature rises, stays the same for a time and then falls.	F
The temperature rises, stays the same for a time and then rises again.	E

Answers to Chapter 10

10.1 Measuring and drawing angles

Exercise 10A

1 **a** 40° **b** 30° **c** 35° **d** 43° **e** 100° **f** 125° **g** 340° **h** 225°

2 Student's drawings of angles

3 Student's drawing and calculations

4 Yes, the angle is 75°

5 Any angle between 0° and 45°

6 (c) is an obtuse angle, the others are acute

7 **a** 80° **b** 50° **c** 25°

10.2 Angle facts

Exercise 10B

1 **a** 48° **b** 307° **c** 108° **d** 52° **e** 59° **f** 81° **g** 139° **h** 51° **i** 138° **j** 128° **k** 47° **l** 117° **m** 27° **n** 45° **o** 108° **p** 69° **q** 135° **r** 58° **s** 74° **t** 23° **u** 55° **v** 56°

2 **a** 82° **b** 105° **c** 75°

3 45° + 125° = 170° and for a straight line it should be 180°.

4 **a** x = 100° **b** x = 110° **c** x = 30°

5 **a** x = 55° **b** x = 45° **c** x = 12.5°

6 **a** x = 34°, y = 98° **b** x = 70°, y = 120° **c** x = 20°, y = 80°

7 6 × 60° = 360°; imagine six of the triangles meeting at a point

8 x = 35°, y = 75°; $2x$ = 70° (opposite angles), so x = 35° and $x + y$ = 110° (angles on a line), so y = 75°

10.3 Angles in a triangle

Exercise 10C

1 **a** 70° **b** 50° **c** 80° **d** 60° **e** 75° **f** 109° **g** 38° **h** 63°

2 **a** No, total is 190° **b** Yes, total is 180° **c** No, total is 170° **d** Yes, total is 180° **e** Yes, total is 180° **f** No, total is 170°

3 **a** 80° **b** 67° **c** 20° **d** 43° **e** 10° **f** 1°

4 **a** 60° **b** Equilateral triangle **c** Same length

5 **a** 70° each **b** Isosceles triangle **c** Same length

6 x = 50°, y = 80°

7 **a** 109° **b** 130° **c** 135°

8 65°

9 Isosceles triangle; angle DFE = 30° (opposite angles), angle DEF = 75° (angles on a line), angle FDE = 75° (angles in a triangle), so there are two equal angles in the triangle and hence it is an isosceles triangle

10 a = 80° (opposite angles), b = 65° (angles on a line), c = 35° (angles in a triangle)

11 Missing angle = y, $x + y$ = 180° and $a + b + y$ = 180° so $x = a + b$

10.4 Angles in a polygon

Exercise 10D

1 **a** 90° **b** 150° **c** 80° **d** 80° **e** 77° **f** 131° **g** 92° **h** 131°

2 **a** No, total is 350° **b** Yes, total is 360° **c** No, total is 350° **d** No, total is 370° **e** Yes, total is 360° **f** Yes, total is 360°

3 **a** 100° **b** 67° **c** 120° **d** 40° **e** 40° **f** 1°

4 **a** 90° **b** Rectangle **c** Square

5 **a** 120° **b** 170° **c** 125° **d** 136° **e** 149° **f** 126° **g** 212° **h** 114°

6 60° + 60° + 120° + 120° + 120° + 240° = 720°

7 y = 360° − $4x$; $2x + y + 2x$ = 360°, $4x + y$ = 360°, so y = 360° − $4x$

8 **a** $8x$ + 40° = 360° **b** x = 40°, 60°

10.5 Regular polygons

Exercise 10E

1 **a** **i** 45° **ii** 8 **iii** 1080°
b **i** 20° **ii** 18 **iii** 2880°
c **i** 15° **ii** 24 **iii** 3960°
d **i** 36° **ii** 10 **iii** 1440°

2 **a** **i** 172° **ii** 45 **iii** 7740°
b **i** 174° **ii** 60 **iii** 10 440°
c **i** 156° **ii** 15 **iii** 2340°
d **i** 177° **ii** 120 **iii** 21 240°

3 **a** Exterior angle is 7°, which does not divide exactly into 360°
b Exterior angle is 19°, which does not divide exactly into 360°
c Exterior angle is 11°, which does divide exactly into 360°
d Exterior angle is 70°, which does not divide exactly into 360°

4 **a** 7° does not divide exactly into 360°
b 26° does not divide exactly into 360°
c 44° does not divide exactly into 360°
d 13° does not divide exactly into 360°

5 x = 45°, they are the same, true for all regular polygons

6 Three are 135°and two are 67.5°

7 88°; $\frac{1440° - 5 \times 200°}{5}$

8 **a** 36° **b** 10

10.6 Parallel lines

Exercise 10F

1 **a** 40° **b** $b = c = 70°$
c $d = 75°, e = f = 105°$ **d** $g = 50°, h = i = 130°$
e $j = k = l = 70°$ **f** $n = m = 80°$

2 **a** $a = 50°, b = 130°$ **b** $c = d = 65°, e = f = 115°$
c $g = i = 65°, h = 115°$ **d** $j = k = 72°, l = 108°$
e $m = n = o = p = 105°$ **f** $q = r = s = 125°$

3 **a** $a = 95°$ **b** $b = 66°, c = 114°$

4 **a** $x = 30°, y = 120°$ **b** $x = 25°, y = 105°$
c $x = 30°, y = 100°$

5 **a** $x = 50°, y = 110°$ **b** $x = 25°, y = 55°$
c $x = 20°, y = 140°$

6 290°; x is double the angle allied to 35°, so is 2 × 145°

7 $a = 102°$ (angles on a line = 180°), $b = 78°$ (corresponding angle or allied angle), angle BDC = 66° (angles in a triangle = 180°) so $c = 114°$ (angles on a line = 180°), $d = 66°$ (corresponding angle or allied angle)

8 Angle PQD = 64° (alternate angles), so angle DQY = 116° (angles on a line = 180°)

9 Use alternate angles to see b, a and c are all angles on a straight line, and so total 180°

10 Third angle in triangle equals q (alternative angle), angle sum of triangle is 180°.

10.7 Special quadrilaterals

Exercise 10G

1 **a** $a = 110°, b = 55°$ **b** $c = 75°, d = 115°$
c $e = 87°, f = 48°$

2 **a** $a = c = 105°, b = 75°$ **b** $d = f = 70°, e = 110°$
c $g = i = 63°, h = 117°$

3 **a** $a = 135°, b = 25°$ **b** $c = d = 145°$
c $e = f = 94°$

4 **a** $a = c = 105°, b = 75°$ **b** $d = f = 93°, e = 87°$
c $g = i = 49°, h = 131°$

5 **a** $a = 58°, b = 47°$ **b** $c = 141°, d = 37°$
c $e = g = 65°, f = 115°$

6 both 129°

7 Marie, a rectangle must have right angles

8 **a** 65°
b Trapezium, angle A + angle D = 180° and angle B + angle C = 180°

10.8 Bearings

Exercise 10H

1 **a** 110° **b** 250° **c** 091° **d** 270° **e** 130° **f** 180°

2 Student's sketches

3

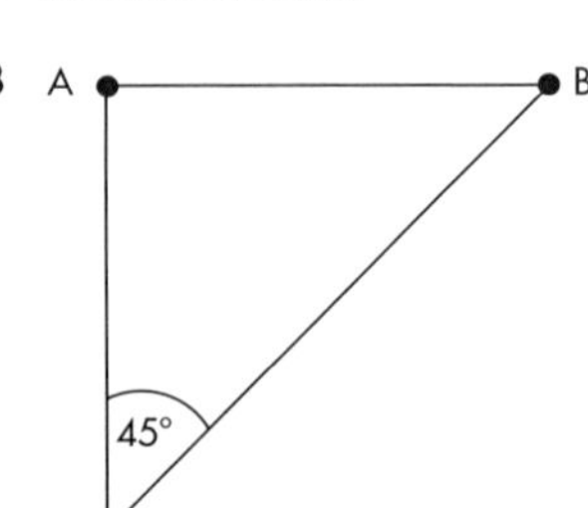

4 **a** 090°, 180°, 270° **b** 000°, 270°, 180°

5

Leg	Actual distance	Bearing
1	50 km	060°
2	70 km	350°
3	65 km	260°
4	46 km	196°
5	60 km	130°

6 **a** 045° **b** 286°

7 **a** 250° **b** 325° **c** 144°

8 **a** 900 m **b** 280° **c** angle NHS = 150° and HS = 3 cm

9 108°

10 255°

Examination questions

1 D

2 **a** 50°

3 **a i** 30° **ii** opposite angles
b 3 angles do not add up to 180°

4 D 80°

5 60°

6 60°

7 70°

8 Angle CDE = 35° (alternate angles), angle CED = 105° (angles in triangle), so angle ACE = 75° (angles on straight line). There are other methods.

9 **a i** 62° **ii** alternate angle
b 56°

10 67°

11 **a** 130° **b** 300 km

Answers to Chapter 11

11.1 Probability scale

Exercise 11A

1 **a** unlikely **b** certain **c** likely
d very unlikely **e** impossible **f** very likely
g evens

2 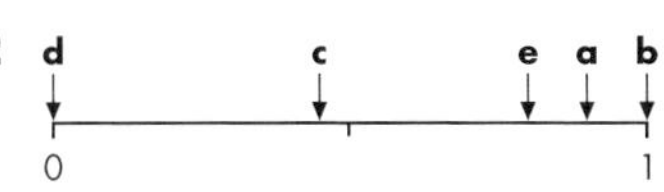

3 b a c
Impossible certain

4 Student to provide own answers.

5 **a** very likely **b** very likely **c** very likely
d very likely **e** certain **f** very unlikely
g certain or very likely **h** very likely
i impossible **j** unlikely **k** certain
l very unlikely

6 As there is so little chance of winning the lottery with one ticket, even though having five tickets increases the chances fivefold, winning is still very unlikely.

11.2 Calculating probabilities

Exercise 11B

1 **a** $\frac{1}{6}$ **b** $\frac{1}{6}$ **c** $\frac{1}{2}$ **d** $\frac{1}{13}$ **e** $\frac{1}{4}$ **f** $\frac{1}{2}$
g $\frac{1}{3}$ **h** $\frac{1}{26}$ **i** $\frac{1}{13}$ **j** 0

2 **a** $\frac{1}{2}$ **b** $\frac{1}{2}$ **c** $\frac{1}{2}$ **d** $\frac{1}{52}$ **e** $\frac{4}{13}$ **f** $\frac{1}{52}$

3 **a** 0 **b** 1

4 **a** $\frac{1}{10}$ **b** $\frac{1}{2}$ **c** $\frac{2}{5}$ **d** $\frac{1}{5}$ **e** $\frac{2}{5}$

5 **a** $\frac{1}{3}$ **b** $\frac{1}{3}$ **c** $\frac{2}{3}$

6 **a** $\frac{6}{11}$ **b** $\frac{5}{11}$ **c** $\frac{6}{11}$

7 **a** $\frac{1}{5}$ **b** $\frac{1}{2}$ **c** $\frac{1}{2}$ **d** $\frac{7}{10}$

8 **a** $\frac{7}{15}$ **b** $\frac{2}{15}$ **c** $\frac{8}{15}$ **d** 0 **e** $\frac{8}{15}$

9 $\frac{1}{50}$

10 **a** AB, AC, AD, AE, BC, BD, BE, CD, CE, DE
b 1 **c** $\frac{1}{10}$ **d** 6 **e** $\frac{3}{5}$ **f** $\frac{3}{10}$

11 **a** 2 **b** 7 **c i** $\frac{5}{9}$ **ii** $\frac{4}{9}$

12 **a** $\frac{1}{13}$ **b** $\frac{1}{13}$ **c** $\frac{1}{2}$ **d** $\frac{2}{13}$ **e** $\frac{7}{13}$ **f** $\frac{1}{26}$

13 **a i** $\frac{12}{25}$ **ii** $\frac{7}{25}$ **iii** $\frac{6}{25}$
b They add up to 1.
c All possible outcomes are mentioned.

14 35%

15 0.5

16 Class U

17 There might not be the same number of boys as girls in the class.

11.3 Probability that an outcome of an event will not happen

Exercise 11C

1 **a** $\frac{19}{20}$ **b** 55% **c** 0.2

2 **a i** $\frac{3}{13}$ **ii** $\frac{10}{13}$ **b i** $\frac{1}{4}$ **ii** $\frac{3}{4}$
c i $\frac{2}{13}$ **ii** $\frac{11}{13}$

3 **a i** $\frac{1}{4}$ **ii** $\frac{3}{4}$ **b i** $\frac{3}{11}$ **ii** $\frac{8}{11}$

4 **a** $\frac{1}{2}$ **b** 1 **c** $\frac{1}{3}$

5 Taryn

6 Because it might be possible for the game to end in a draw.

Examination questions

1 **a** April and May
b Daffodil
c February
d Crocus
e i $\frac{1}{5}$ **ii** 0 × 1

2 A head is obtained when a fair coin is thrown once – Even
A number less than 7 will be scored when an ordinary six-sided dice is rolled once – Certain
A red disc is obtained when a disc is taken at random from a bag containing 9 red discs and 2 blue discs – Likely

3 $\frac{5}{12}$

4 $\frac{5}{12}$

5 **a** 1
b There are five possible outcomes and only two of them are 2.
The probability is $\frac{2}{5}$.

6 **a** $\frac{14}{25}$ **b** $\frac{21}{50}$

7 **a** 0 **b** $\frac{3}{8}$ **c** 12

8 0.25

9 **a**

Stem	Leaf
2	3 7 8
3	1 4 5 6
4	1 2 4 5 5
5	0 2 3

Key: 3 | 7 = 37 years

b $\frac{8}{15}$

Answers to Chapter 12

12.1 Congruent shapes

Exercise 12A

1 **a** yes **b** yes **c** no **d** yes **e** no **f** yes

2 **a** triangle ii **b** triangle iii **c** sector i

3 **a** 1, 3, 4 **b** 2, 4 **c** 1, 4
d 1, 2, 3, 4

4 b, d and p

5 If the shapes are congruent they must be the same size.

6

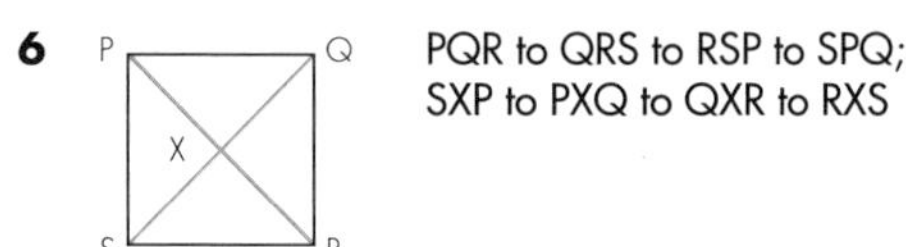

7

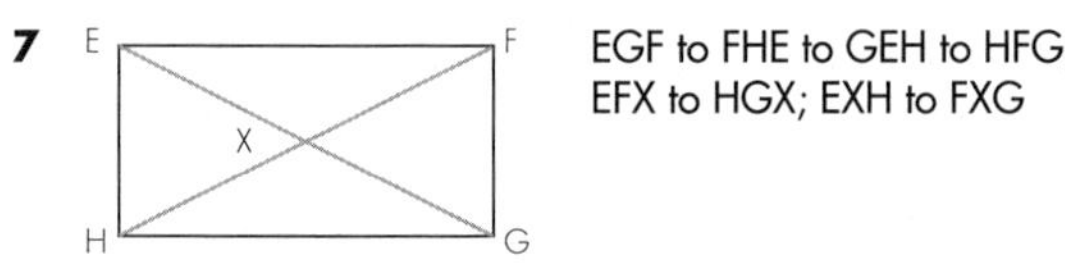

8

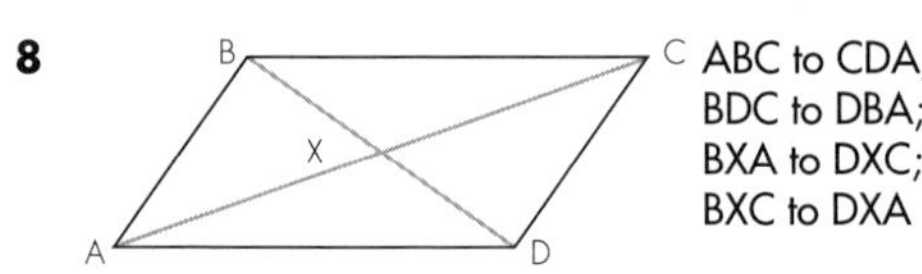

9

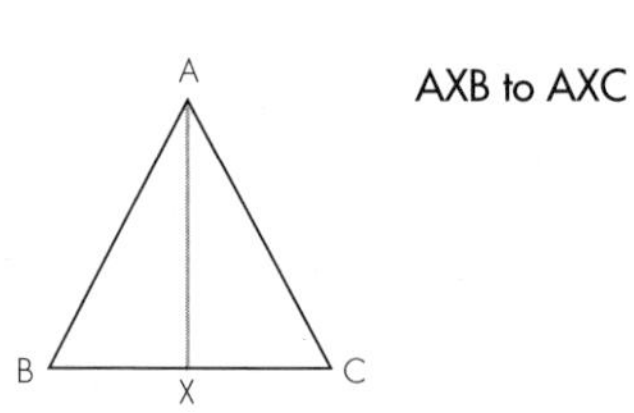

AXB to AXC

10 **a** **i** 49 **ii** 36 **iii** 25
b 204

12.2 Tessellations

Exercise 12B

1 Check students' tessellations.

2 Check students' tessellations.

3 All quadrilaterals tessellate.

4 The interior angle for each shape divides exactly into 360°.

5 It is not true in general but it is impossible to tessellate a regular pentagon.

6 Imagine rows of circles all sitting in perfect columns. The spaces between them all are the same curved four-pointed star shape.

7 There will be many different correct answers, but do not accept bricks arranged in vertical columns as this would not give a strong design.

Examination questions

1 B, E and F

2 **a** **i** F
ii B
b C and E

2 **a** A and E, C and G
b Check student's tessellation

Answers to Chapter 13

13.1 Constructing triangles

Exercise 13A

1 **a** BC = 2.9 cm, ∠B = 53°, ∠C = 92°
b ∠E = 50°, EF = 7.4 cm, ED = 6.8 cm
c ∠G = 105°, ∠H = 29°, ∠I = 46°
d ∠J = 48°, ∠L = 32° JK = 4.3 cm
e ∠N = 55°, ON = OM = 7 cm
f ∠P = 51°, ∠R = 39°, QP = 5.7 cm

2 **b** ∠ABC = 44°, ∠BCA = 79°, ∠CAB = 57°

3 **a** 5.9 cm **b** 18.8 cm^2

4 BC = 2.6 cm, 7.8 cm

5 **a** 4.5 cm **b** 11.25 cm^2

6 **a** 4.3 cm **b** 34.5 cm^2

7 **a** A right-angled triangle constructed with sides 3, 4, 5 and scale marked 1 cm : 1 m.
b A right-angled triangle constructed with sides that add to 12 cm.

8 Even with all three angles, you need to know at least one length.

9 An equilateral triangle of sides 4 cm.

Examination questions

1 Triangle accurately drawn so that line AB = 5.6 cm, line AC = 10.9 cm, line BC = 7.8
Angle A = 43 degrees, angle B = 108 degrees, angle C = 29 degrees

2 Equilateral triangle accurately drawn. Each side is 6 cm long and each angle is 60 degrees

3 Triangle accurately redrawn so that line AB = 8 cm, line AC = 6 cm, line BC = 6.5
Angle A = 52 degrees, angle B = 48 degrees, angle C = 80 degrees

Answers to Chapter 14

14.1 Long multiplication

Exercise 14A

1 **a** 12 138 **b** 45 612 **c** 29 988
d 20 654 **e** 51 732 **f** 25 012
g 19 359 **h** 12 673 **i** 19 943
j 26 235 **k** 31 535 **l** 78 399
m 17 238 **n** 43 740 **o** 66 065
p 103 320 **q** 140 224 **r** 92 851
s 520 585 **t** 78 660

2 3500

3 No, 62 pupils cannot get in

4 Yes, he walks 57.6 km

5 Yes, 1 204 000 letters

6 5819 litres

7 Yes, she raised £302.40

14.2 Long division

Exercise 14B

1 **a** 25 **b** 15 **c** 37
d 43 **e** 27 **f** 48
g 53 **h** 52 **i** 32
j 57 **k** 37 rem 15 **l** 25 rem 5
m 34 rem 11 **n** 54 rem 9 **o** 36 rem 11
p 17 rem 4 **q** 23 **r** 61 rem 14
s 42 **t** 27 rem 2

2 68

3 38

4 4 months

5 34 h

6 £1.75

14.3 Arithmetic with decimal numbers

Exercise 14C

1 **a** 4.8 **b** 3.8 **c** 2.2 **d** 8.3 **e** 3.7 **f** 46.9
g 23.9 **h** 9.5 **i** 11.1 **j** 33.5 **k** 7.1 **l** 46.8
m 0.1 **n** 0.1 **o** 0.6 **p** 65.0 **q** 213.9 **r** 76.1
s 455.2 **t** 51.0

2 **a** 5.78 **b** 2.36 **c** 0.98 **d** 33.09 **e** 6.01 **f** 23.57
g 91.79 **h** 8.00 **i** 2.31 **j** 23.92 **k** 6.00 **l** 1.01
m 3.51 **n** 96.51 **o** 0.01 **p** 0.07 **q** 7.81
r 569.90 **s** 300.00 **t** 0.01

3 **a** 4.6 **b** 0.08 **c** 45.716 **d** 94.85 **e** 602.1
f 671.76 **g** 7.1 **h** 6.904 **i** 13.78 **j** 0.1 **k** 4.002
l 60.0 **m** 11.99 **n** 899.996 **o** 0.1 **p** 0.01
q 6.1 **r** 78.393 **s** 200.00 **t** 5.1

4 **a** 9 **b** 9 **c** 3 **d** 7 **e** 3 **f** 8
g 3 **h** 8 **i** 6 **j** 4 **k** 7 **l** 2
m 47 **n** 23 **o** 96 **p** 33 **q** 154 **r** 343
s 704 **t** 910

5 £1 + £7 + £4 + £1 = £13

6 3, 3.46, 3.5

7 4.7275 or 4.7282

Exercise 14D

1 **a** 49.8 **b** 21.3 **c** 48.3 **d** 33.3 **e** 5.99 **f** 8.08
g 90.2 **h** 21.2 **i** 12.15 **j** 13.08 **k** 13.26 **l** 24.36

2 **a** 1.4 **b** 1.8 **c** 4.8 **d** 3.8 **e** 3.75 **f** 5.9
g 3.7 **h** 3.77 **i** 3.7 **j** 1.4 **k** 11.8 **l** 15.3

3 **a** 30.7 **b** 6.6 **c** 3.8 **d** 16.7 **e** 11.8 **f** 30.2
g 43.3 **h** 6.73 **i** 37.95 **j** 4.7 **k** 3.8 **l** 210.5

4 **a** £16.74 **b** £1.40

5 2.7 m + 2.5 m = 5.2 m (one pipe)
3.1 m + 1.7 m = 4.8 m (second pipe)
4.2 m (third pipe). Three pipes needed.

6 **a** 5.3 **b** 6.7 **c** 2.05 **d** 1.9 **e** 4.95 **f** 3.71

7 **a** 1.6 **b** 42.7 **c** 4.29 **d** 12.8 **e** 22.4 **f** 51.97

8 DVD, shirt and pen

Exercise 14E

1 **a** 7.2 **b** 7.6 **c** 18.8 **d** 37.1 **e** 32.5 **f** 28.8
g 10.0 **h** 55.2 **i** 61.5 **j** 170.8 **k** 81.6 **l** 96.5

2 **a** 9.36 **b** 10.35 **c** 25.85 **d** 12.78 **e** 1.82 **f** 3.28
g 2.80 **h** 5.52 **i** 42.21 **j** 56.16 **k** 7.65 **l** 48.96

3 **a** 1.8 **b** 1.4 **c** 1.4 **d** 1.2 **e** 2.13 **f** 0.69
g 2.79 **h** 1.21 **i** 1.89 **j** 1.81 **k** 0.33 **l** 1.9

4 **a** 1.75 **b** 1.28 **c** 1.85 **d** 3.65 **e** 1.66 **f** 1.45
g 1.42 **h** 1.15 **i** 3.35 **j** 0.98 **k** 2.3 **l** 1.46

5 **a** 1.89 **b** 1.51 **c** 0.264 **d** 4.265 **e** 1.224 **f** 0.182 **g** 0.093 **h** 2.042 **i** 1.908 **j** 2.8 **k** 4.25 **l** 18.5

6 Pack of 8 at £0.625 each

7 **a** £49.90 **b** Small, so that all four children can have an ice cream (cost £4.80)

8 Yes. She only needed 8 paving stones.

Exercise 14F

1 **a** 89.28 **b** 298.39 **c** 66.04 **d** 167.98 **e** 2352.0 **f** 322.4 **g** 1117.8 **h** 4471.5 **i** 464.94 **j** 25.55 **k** 1047.2 **l** 1890.5

2 **a** £224.10 **b** £223.75 **c** £29.90

3 $5 \times 7 = 35$

4 £54.20

5 **a** £120.75 **b** £17 − £3.45 = £13.55

Exercise 14G

1 **a** 0.48 **b** 2.92 **c** 1.12 **d** 0.12 **e** 0.028 **f** 0.09 **g** 0.192 **h** 3.0264 **i** 7.134 **j** 50.96 **k** 3.0625 **l** 46.512

2 Yes, with 25p left over.

3 **a** 35, 35.04, 0.04 **b** 16, 18.24, 2.24 **c** 60, 59.67, 0.33 **d** 180, 172.86, 7.14 **e** 12, 12.18, 0.18 **f** 24, 26.016, 2.016 **g** 40, 40.664, 0.664 **h** 140, 140.58, 0.58

4 **a** 572 **b** **i** 5.72 **ii** 1.43 **iii** 22.88

14.4 Arithmetic with fractions

Exercise 14H

1 **a** $\frac{7}{10}$ **b** $\frac{2}{5}$ **c** $\frac{1}{2}$ **d** $\frac{3}{100}$ **e** $\frac{3}{50}$ **f** $\frac{13}{100}$ **g** $\frac{1}{4}$ **h** $\frac{19}{50}$ **i** $\frac{11}{20}$ **j** $\frac{16}{25}$

2 **a** 0.5 **b** 0.75 **c** 0.6 **d** 0.9 **e** 0.333 **f** 0.625 **g** 0.667 **h** 0.35 **i** 0.636 **j** 0.444

3 **a** 0.3, $\frac{1}{2}$, 0.6 **b** 0.3, $\frac{2}{5}$, 0.8 **c** 0.15, $\frac{1}{4}$, 0.35 **d** $\frac{7}{10}$, 0.71, 0.72 **e** 0.7, $\frac{3}{4}$, 0.8 **f** $\frac{1}{20}$, 0.08, 0.1 **g** 0.4, $\frac{1}{2}$, 0.55 **h** 1.2, 1.23, $1\frac{1}{4}$

4 Just shirts – $\frac{1}{3}$ (0.33) is greater than $\frac{1}{4}$ (0.25)

5 $\frac{7}{8}$ (= 0.875)

6 $\frac{2}{3}$ (= 0.67)

7 0.125

8 $\frac{1}{2}$

Exercise 14I

1 **a** $\frac{8}{15}$ **b** $\frac{7}{12}$ **c** $\frac{3}{10}$ **d** $\frac{11}{12}$ **e** $\frac{7}{8}$ **f** $\frac{1}{2}$ **g** $\frac{1}{6}$ **h** $\frac{1}{20}$ **i** $\frac{1}{10}$ **j** $\frac{1}{8}$ **k** $\frac{1}{12}$ **l** $\frac{1}{3}$ **m** $\frac{1}{6}$ **n** $\frac{7}{9}$ **o** $\frac{5}{8}$ **p** $\frac{3}{8}$ **q** $\frac{1}{15}$ **r** $1\frac{13}{24}$ **s** $\frac{59}{80}$ **t** $\frac{22}{63}$ **u** $\frac{37}{54}$

2 **a** $3\frac{5}{14}$ **b** $10\frac{3}{5}$ **c** $2\frac{1}{6}$ **d** $3\frac{31}{45}$ **e** $4\frac{47}{60}$ **f** $\frac{41}{72}$ **g** $\frac{29}{48}$ **h** $1\frac{43}{48}$ **i** $1\frac{109}{120}$ **j** $1\frac{23}{30}$ **k** $1\frac{31}{84}$

3 $\frac{1}{20}$

4 **a** $\frac{1}{6}$ **b** 30, must be divisible by 2 and 3

5 $\frac{13}{15}$

6 $\frac{1}{3}$

7 $\frac{3}{8}$

8 He has added the numerators and added the denominators instead of using a common denominator. Correct answer is $3\frac{7}{12}$.

Exercise 14J

1 **a** $\frac{1}{6}$ **b** $\frac{1}{10}$ **c** $\frac{3}{8}$ **d** $\frac{3}{14}$ **e** $\frac{8}{15}$ **f** $\frac{1}{5}$ **g** $\frac{2}{7}$ **h** $\frac{3}{10}$ **i** $\frac{1}{2}$ **j** $\frac{2}{5}$

2 **a** $\frac{3}{32}$ **b** $\frac{3}{8}$ **c** $\frac{7}{20}$ **d** $\frac{16}{45}$ **e** $\frac{3}{5}$ **f** $\frac{5}{8}$

3 3 km

4 $\frac{1}{12}$

5 $\frac{3}{8}$

6 21 tonnes

7 260 pupils

8 £51

9 18

10 **a** $\frac{5}{12}$ **b** $2\frac{1}{12}$ **c** $6\frac{1}{4}$ **d** $2\frac{11}{12}$ **e** $3\frac{9}{10}$ **f** $3\frac{1}{3}$ **g** $12\frac{1}{2}$ **h** 30

11 $\frac{2}{5}$ of $6\frac{1}{2} = 2\frac{3}{5}$

12 £5

13 Three-quarters of 68 = 51

14 7 min

15 £10.40

16 £30

Exercise 14K

1 **a** $\frac{3}{4}$ **b** $1\frac{2}{5}$ **c** $1\frac{1}{15}$ **d** $1\frac{1}{14}$ **e** 4 **f** 4
g 5 **h** $1\frac{5}{7}$ **i** $\frac{4}{9}$ **j** $1\frac{3}{5}$

2 18

3 40

4 15

5 16

6 **a** $2\frac{2}{15}$ **b** 38 **c** $1\frac{7}{8}$ **d** $\frac{9}{32}$ **e** $\frac{1}{16}$ **f** $\frac{256}{625}$

14.5 Multiplying and dividing with negative numbers

Exercise 14L

1 **a** −15 **b** −14 **c** −24 **d** 6 **e** 14 **f** 2
g −2 **h** −8 **i** −4 **j** 3 **k** −24 **l** −10
m −18 **n** 16 **o** 36 **p** −4 **q** −12 **r** −4
s 7 **t** 25 **u** 18 **v** −8 **w** −45 **x** 3
y −40

2 **a** −9 **b** 16 **c** −3 **d** −32 **e** 18 **f** 18
g 6 **h** −4 **i** 20 **j** 16 **k** 8 **l** −48
m 13 **n** −13 **o** −8 **p** 0 **q** 16 **r** −42
s 6 **t** 1 **u** −14 **v** 6 **w** −4 **x** 7
y 0

3 **a** −2 **b** 30 **c** 15 **d** −27 **e** −7

4 **a** 4 **b** −9 **c** −3 **d** 6 **e** −4

5 **a** −9 **b** 3 **c** 1

6 **a** 16 **b** −2 **c** −12

7 **a** 24 **b** 6 **c** −4 **d** −2

8 For example: $1 \times (-12)$, -1×12, $2 \times (-6)$, $6 \times (-2)$, $3 \times (-4)$, $4 \times (-3)$

9 For example: $4 \div (-1)$, $8 \div (-2)$, $12 \div (-3)$, $16 \div (-4)$, $20 \div (-5)$, $24 \div (-6)$

10 **a** 21 **b** −4 **c** 2 **d** −16 **e** 2 **f** −5
g −35 **h** −17 **i** −12 **j** 6 **k** 45 **l** −2
m 0 **n** −1 **o** −7 **p** −36 **q** 9 **r** 32
s 0 **t** −65

11 **a** −12 **b** 12 degrees **c** 3×-6

12 -5×4, 3×-6, $-20 \div 2$, $-16 \div -4$

13 **a** 4 **b** 13 **c** 10 **d** 1

14.6 Approximation of calculations

Exercise 14M

1 **a** 50 000 **b** 60 000 **c** 30 000
d 90 000 **e** 90 000 **f** 50
g 90 **h** 30 **i** 100
j 200 **k** 0.5 **l** 0.3
m 0.006 **n** 0.05 **o** 0.0009
p 10 **q** 90 **r** 90
s 200 **t** 1000

2 **a** 60 000 **b** 30 000 **c** 80 000
d 30 000 **e** 10 000 **f** 6000
g 1000 **h** 800 **i** 100
j 600 **k** 2 **l** 4
m 3 **n** 8 **o** 40
p 0.8 **q** 0.5 **r** 0.07
s 1 **t** 0.01

3 **a** 65, 74 **b** 95, 149 **c** 950, 1499

4 Elsecar 750, 849; Hoyland 950, 1499; Barnsley 150 000, 249 999

5 15, 16 or 17

6 1, because there could be 450 then 449

Exercise 14N

1 **a** 35 000 **b** 15 000 **c** 960
d 12 000 **e** 1050 **f** 4000
g 4 **h** 20 **i** 1200

2 **a** £3000 **b** £2000 **c** £1500
d £700

3 **a** £15 000 **b** £18 000 **c** £18 000

4 £21 000

5 **a** 14 **b** 10 **c** 1.1 **d** 1 **e** 5 **f** $\frac{2}{3}$
g 3 or 4 **h** $\frac{1}{2}$ **i** 6 **j** 400 **k** 2 **l** 20

6 **a** 500 **b** 200 **c** 90 **d** 50 **e** 50 **f** 500

7 8

8 **a** 200 **b** 2800 **c** 10 **d** 1000

9 **a** 40 **b** 10 **c** £70

10 1000 or 1200

11 **a** 28 km **b** 120 km **c** 1440 km

12 400

13 **a** 3 kg **b** 200

Exercise 14P

1 **a** 1.7 m **b** 6 min **c** 240 g **d** 80 °C **e** 16 miles **f** 14 m^2

2 82 °F, 5.3 km, 110 min, 43 000 people, 6.2 s, 67th, 1788, 15, 5 s

3 40 × £20 = £800

4 22.5 °C − 18.2 °C = 4.3 °C

Examination questions

1 **a** £1.80
b £1.25
c 40p

2 **a** 9
b 36

3 83p

4 answer in the region of £7.50 to £8

5 **a** 0.625
b $\frac{3}{5}$

6 **a** 0.2
b 0.38

7 **a** 450
b 2800

8 **a** 34
b 150

9 **a** $\frac{5}{12}$
b $\frac{3}{20}$

10 40.96

11 Chuck (30 + 14) ÷ (1.3 −0.5) = 44 ÷ 0.8 = 55

12 $1\frac{11}{15}$

13 800

14 600

Answers to Chapter 15

15.1 Pie charts

Exercise 15A

1 **a**

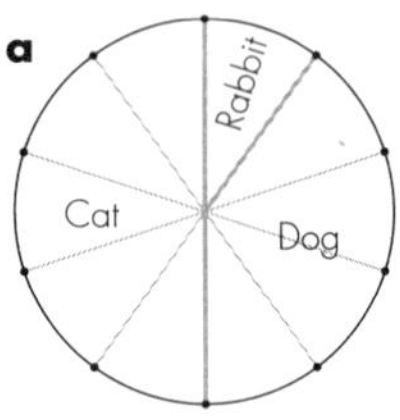

b

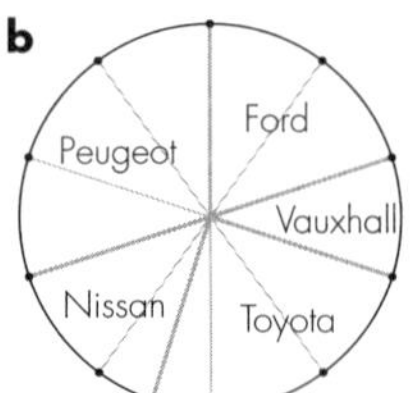

c

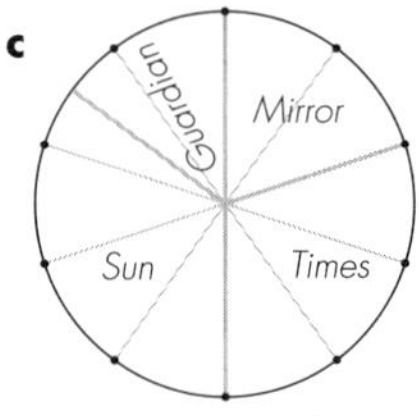

2 Pie charts with following angles:
a 36°, 90°, 126°, 81°, 27°
b 90°, 108°, 60°, 78°, 24°
c 168°, 52°, 100°, 40°

3 Pie charts with these angles: 60°, 165°, 45°, 15°, 75°

4 **a** 36
b Pie charts with these angles: 50°, 50°, 80°, 60°, 60°, 40°, 20°
c Student's bar chart.
d Bar chart, because easier to make comparisons.

5 **a** Pie charts with these angles: 124°, 132°, 76°, 28°
b Split of total data seen at a glance.

6 **a** 55° **b** 22 **c** $33\frac{1}{3}$%

7 **a** Pie charts with these angles:
Strings: 36°, 118°, 126°, 72°, 8°
Brass: 82°, 118°, 98°, 39°, 23°
Overall, the strings candidates did better, as a smaller proportion failed. A higher proportion of Brass candidates scored very good or excellent.

8 $\frac{1}{9}$

9 Choose a class in the school at random and ask each student how they get to school most mornings.

15.2 Scatter diagrams

Exercise 15B

1 **a** Positive correlation **b** Negative correlation
c No correlation **d** Positive correlation

2 **a** A person's reaction time increases as more alcohol is consumed.
b As people get older, they consume less alcohol.
c No relationship between temperature and speed of cars on M1.
d As people get older, they have more money in the bank.

3 **a**, **b** Student's scatter diagram and line of best fit.
c about 20 cm/s
d about 35 cm

4 **a** Student's scatter diagram.
b Yes, usually (good correlation).

5 **a**, **b** Student's scatter diagram and line of best fit.
c Greta **d** about 70 **e** about 72

6 **a** Student's scatter diagram.
b no, because there is no correlation.

7 **a**, **b** Student's scatter diagram and line of best fit.
c about 2.4 km **d** 8 minutes

8 23 mph

9 Points showing a line of best fit sloping down from top left to bottom right.

15.3 Surveys

Exercise 15C

1–5 Student's own answers.

6 **a** The form should look something like this.
Question: Which of the following foods would you normally eat for your main meal of the day?

Name	Sex	Chips	Beef burgers	Vegetables	Pizza	Fish

b Yes, it looks correct as a greater proportion of girls ate healthy food.

7 The data-collection sheet should look something like this.
Question: What kind of tariff do you use on your mobile phone?

Name	Pay as you go		Contract	
	200 or more free texts	Under 200 free texts	200 or more free texts	Under 200 free texts

Various data-collection sheets would achieve the purpose; the student's answer will be accepted, provided he or she has offered some choices that can distinguish one tariff from another.

8 Examples are: shops names, year of student, tally space, frequency.

Exercise 15D

1 **a** Leading question, not enough responses.
b Simple 'yes' and 'no' response, with a follow-up question, responses cover all options and have a reasonable number of choices.

2 **a** Overlapping responses.
b ☐ £0–£2 ☐ over £2 up to £5
☐ over £5 up to £10 ☐ over £10

3–5 Students to provide own questionnaires.

6 A possible solution is:

Do you have a back problem? Yes ☐ No ☐

Tick the diagram that best illustrates how you sit.

☐ ☐ ☐ ☐

7 The groups overlap each other: 'Less than £15' is included in 'Less than £25'.

15.4 The data-handling cycle

Exercise 15E

1 Secondary data. Student to give own description of the data-handling cycle.

2 Primary data. Student to give own description of the data-handling cycle.

3 Primary or secondary. Student to give own description of the data-handling cycle.

4 Primary or secondary. Student to give own description of the data-handling cycle.

5 Primary. Student to give own description of the data-handling cycle.

6 Primary. Student to give own description of the data-handling cycle.

15.5 Other uses of statistics

Exercise 15F

1 Price: 78p, 80.3p, 84.2p, 85p, 87.4p, 93.6p

2 **a** 9.7 million **b** 4.5 years **c** 12 million
d 10 million

3 **a** £1 = $1.88
b Greatest drop was from June to July.
c There is no trend in the data so you cannot tell if it will go up or down.

4 £74.73

5 **a** Holiday month
b i 138–144 thousand **ii** 200–210 thousand

6 The general cost of living in 2009 dropped to 98% of that in 2008.

7 £51.50

Examination questions

1 The angles for the pie chart are: 60° (gold), 80° (silver), 220° (bronze).

2 **a** Understand 6 sections, Understand some 5 sections, Understand a little 2 sections, Do not understand at all 1 section
b Women had a better understanding. Around 85% of the women had some understanding or better, while of the men, the figure was only 75%.

3 **a** 27
b 195°

4 **a** Grade E
b $\frac{100}{360}$ or $\frac{5}{18}$
c **i** 16 **iii** 72
d Reason (e.g. %, not actual numbers; do not know how many students, etc.)

5 **a** **i** Diagram C **ii** Diagram A **iii** Diagram B
b Diagram A

6 **a**

	Eat vegetarian food	Do not eat vegetarian food	Total
Boys			
Girls			
Total			

b Yes, because 53% of the girls said they ate vegetarian compared with only 45% of the boys.

7 **a** 2 of: No time period
Non-exhaustive response boxes
Labels too vague
b Includes time period and proper response boxes

8 **a** There is no way to say the service is poor.
b How often do you visit the town centre?
Everyday ☐ 2, 3 or 4 times a week ☐
once a week ☐ 2, 3 or 4 times a month ☐
once a month ☐ less than once a month ☐

9 **a** no time period and no option for 0
b one of too small a sample, not diverse enough as all from his class, or similar

10 **a** Graph accurately copied and points plotted at (50, 1.6) and (65, 1.75)
b positive, i.e. the longer a mother's leg length, the heavier her baby's birth weight
c Line of best fit drawn with roughly 4 of the crosses on either side
d Around 1.64 kg

11 **a/b** Time on horizontal axis from 0 to 20 and Distance (km) on vertical axis from 0 to 10 with the following points plotted: (3, 1.7) (17, 8.3) (11, 5.1) (13, 6.7) (9, 4.7) (15, 7.3) (8, 3.8) (11, 5.7) (16, 8.7) (10, 5.3) and with line of best fit drawn.
c/d answers depend on student's plotting

12 **a** The longer the pike the more it weighs
b Cross at (78, 24) added to graph
c 15 kg

13 **a** Points (65, 100) and (80, 100) plotted on the graph.
b The greater the height, the longer the sheep.
c Roughly 109 cm

GLOSSARY

above/below A number greater than/less than another number. For example, above average, below freezing (point of water). An object over/under another object.
acute angle An angle with a value between 0° and 90°. (See also **reflex angle**.)
add See **addition**
addition One of the basic operations of arithmetic. The process of combining two or more values to find their total value. Addition is the inverse operation to subtraction.
after/before A number greater than/less than another number. For example 9 is the number after 8, 20 comes before 40. An event occurring later/earlier than another event.
allied angles When two parallel lines are cut by a third line (transversal), two pairs of allied angles are formed between the lines, each pair on one side of the transversal. Each pair of allied angles add up to 180°.
alternate angles When two parallel lines are cut by a third line (transversal), two pairs of alternate angles are formed between the lines. The alternate angles lie one on each side of the transversal. Alternate angles are of equal size.
angle The space (usually measured in degrees [°]) between two intersecting lines or surfaces (planes). The amount of turn needed to move from one line or plane to the other.
angles around a point The angles around a point add up to 360°.
angles on a straight line The angles on a straight line add up to 180°.
approximate An inexact value that is accurate enough for the current situation.
approximation See **approximate**
average speed A single value that represents the speed achieved during a journey. The speed of the whole journey if it had been completed at constant speed.
axes See **axis**
axis A fixed line used for reference, along or from which distances or angles are measured. A pair of coordinate axes is shown.
bar chart A diagram where quantities are represented by rectangles of the same width but different, appropriate heights.
below See **above/below**
bias A die, coin, or spinner has a bias if it is more likely to land on one number/side than another.
brackets The symbols '(' and ')' which are used to separate part of an expression. This may be for clarity or to indicate a part to be worked out individually. When a number and/or value is placed immediately before an expression or value inside a pair of brackets, the two are to be multiplied together. For example, $6a(5b + c) = 30ab + 6ac$
cancel A fraction can be simplified to an equivalent fraction by dividing the numerator and denominator by a common factor. This is called cancelling.
capacity The volume of a liquid or gas.
centilitre (cl) A metric unit of volume or capacity. One hundredth of a litre. 100 cl = 1 litre.
centimetre (cm) A metric unit of length. One hundredth of a metre. 100 cm = 1 m.
certain Definite. An event is definitely going to occur. In this case the probability that the event will occur = 1.
chance The likelihood, or probability, of an event occurring.
class A collection of values grouped under one category or range.
class interval The size or spread of the measurement defining a class. For example, heights could be grouped in 1 cm or 10 cm class intervals.
column A vertical list of numbers or values. The vertical parts of a table. A way of arranging numbers to be added or subtracted.
column method (or traditional method) A method of calculating a 'long multiplication' by multiplying the number by the value of each digit of the multiplier and displaying the results in columns before adding them together to find the result.
compasses Also called a pair of compasses, an instrument used for drawing circles and measuring distances.
congruent Exactly alike in shape and size.
construct To draw angles, lines, or shapes accurately, according to given requirements.
conversion factor A number that is used to convert a measurement in one unit to another unit. For example, $\times \frac{5}{8}$ converts kilometres to miles.
conversion graph A graph that can be used to convert from one unit to another. It is drawn by joining two or more points where the equivalence is known. Sometimes, but not always, it will pass through the origin.
correlation One measurement is affected by or affects another. For example, weight and height may correlate, but weight and hair colour do not.
corresponding angles When two parallel lines are cut by a third line (transversal), four pairs of corresponding angles are formed: a and e, b and f, c and g, and d and h. Corresponding angles are equal.
credit When you credit money to an account, you pay money in. If you pay 'using credit' (or with a credit card), you are paying with money you don't have. (Your bank is crediting you – lending you – money to use. This usually has to be paid back with additional interest.) If your account is 'in credit', you have money in the account. You have a positive amount of money. (See also **debit**.)
cube The result of raising a number to the power of three. For example, two cubed is written: 2^3, which is $2 \times 2 \times 2 = 8$.
data collection sheet A form or table which is used for recording data collected during a survey.
debit When money is taken from an account, the account is debited with that amount. A debit card can only be used to spend money that you have – the cost is debited from the account. If your account is 'in debit', you have spent more than you had. You now have a negative amount – you owe money to the bank.
decagon A polygon with ten straight sides. The internal angles add up to 1440°.
decimal Any number using base 10 for the number system. It usually refers to a number written with one or more decimal places.

decimal fraction Usually refers to the part of a decimal number after (to the right of) the decimal point, that is, the part less than 1.
decimal place Every digit in a number has a place value (hundreds, tens, ones, etc.). The places after (to the right of) the decimal point have place values of tenths, hundredths, etc. These place values are known as decimal places.
decimal point The dot used to separate the integer (whole number) place values from the fraction place values (tenths, etc.)
denominator The number below the line in a fraction. It tells you the denomination, name, or family of the fraction. For example, a denominator of 3 tells you are thinking about thirds; a single unit has been divided into three parts. (See also **denominator** and **numerator**.)
difference The result of the subtraction of two numbers; the amount by which one number is greater than another number.
digit A number symbol. Our (decimal or denary) number system uses the digits 0, 1, 2, 3, 4, 5 , 6, 7, 8, and 9.
discrete data Data that is counted, rather than measured, such as favourite colour or a measurement that occurs in integer values, such as a number of days.
distance–time graph A graph showing the variation of the distance of an object from a given point during an interval of time. Usually, time is measured along the horizontal axis, and distance is measured up the vertical axis.
division One of the basic operations of arithmetic. Division shows the result of sharing. For example, share 12 books among 3 people, they get 4 books each ($12 \div 3 = 4$). It is also used to calculate associated factors. For example, How many threes in twelve? ($12 \div 3 = 4$) There are four threes in twelve (or $4 \times 3 = 12$). It is the inverse operation to multiplication. It can be written using $A \div B$, A/B or $\frac{A}{B}$.
divisions The marks or partitions on a scale that break it into sections.
dual bar chart This shows two bar charts on one set of axes. It might show the heights of boys and the heights of girls, rather than the heights of all children.
elevation An elevation is the view of a 3D shape when it is seen from the front or from another side.
equally likely Two events are described as equally likely if the probabilities of the occurrence of each of the events are equal. For example, when a die is thrown, the outcomes 6 and 2 are equally likely. (They both have a probability of $\frac{1}{6}$.)
equation of a line An equation, usually containing two variables (such as x and y), and from which you can plot a line on graph paper. For example, $y = 2x + 3$.
equivalent The same, equal in value. For example, equivalent fractions, equivalent expressions.
equivalent fractions Equivalent fractions are fractions which can be cancelled down to the same value, such as $\frac{10}{20} = \frac{5}{10} = \frac{1}{2}$.
estimate (As a verb) to state or calculate a value close to the actual value by using experience to judge a distance, weight, etc. or by rounding numbers to make the calculation easier; (as a noun) the value of an estimate.
evaluate Work out the value of something.
event Something that happens. An event could be the toss of a coin, the throw of a die, or a football match.
experimental A result from an experiment.
exterior angle The exterior angles of a polygon are outside the shape. They are formed when a side is produced (extended). An exterior angle and its adjacent interior angle add up to 180°.
factor A whole number that divides exactly into a given number.
factor pair A pair of whole numbers whose product make a given number.
flow diagram A diagram showing the progression of a calculation or a logical path.
foot (ft) An imperial measurement of length, about 15 cm long. 12 inches = 1 foot, 3 feet = 1 yard.
fraction A fraction means 'part of something'. To give a fraction a name, such as 'fifths', we divide the whole amount into equal parts (in this case, five equal parts). A 'proper' fraction represents an amount less than one (the numerator is smaller than the denominator). Any two numbers or expressions can be written as a fraction, i.e. they are written as a numerator and denominator. (See also **numerator** and **denominator**.)
frequency How often something occurs.
frequency table A table showing values (or classes of values) of a variable alongside the number of times each one has occurred.
front elevation The view of a 3D object when seen from the front.
function A function of x is any algebraic expression in which x is the only variable. This is often represented by the function notation $f(x)$ or function of x.
gallon (gal) An imperial measurement of volume and capacity. An average-sized bucket holds about 2 gallons. 8 pints = 1 gallon
gradient How steep a hill or the line of a graph is. The steeper the slope, the larger the value of the gradient. A horizontal line has a gradient of zero.
gram (g) A metric unit of mass. 1000 grams = 1 kilogram.
greater than (>) Comparing the value of two numbers or quantities. For example, $8 > 4$ states that 8 has a higher value than 4.
grid A table of columns and rows such as those used to add up numbers. Squared or graph paper should be used for drawing charts and graphs.
grid method (or box method) A method of calculating a 'long multiplication' by arranging the value of each digit in a grid and multiplying them all separately before adding them together to find the result.
grouped data Data from a survey that is grouped into classes.
grouped frequency table A method of tabulating data by grouping it into classes. The frequency of data values that occur within a class is recorded as the frequency of that class.
heptagon A polygon with seven sides. The sum of all the interior angles of a heptagon is 900°. A regular heptagon has sides of equal length.

hexagon A polygon with six sides. The sum of all the interior angles of a hexagon is 720°. A regular hexagon has sides of equal length and each of the interior angles is 120°.
hypothesis A theory or idea.
imperial The description of measurements used in the UK before metric units were introduced. They often have a long history (for example originating in Roman times) and are commonly based on units of twelve or 16 rather than ten used by the metric system.
impossible If something cannot happen, it is said to be impossible. The probability of it happening = 0.
improper fraction An improper fraction is a fraction whose numerator is greater than the denominator. The fraction could be re-written as a mixed number. For example, $\frac{7}{2} = 3\frac{1}{2}$
inch (in) An imperial unit of length. One inch is about $2\frac{1}{2}$ cm long. 12 inches = 1 foot.
index A power or exponent. For example, in the expression 3^4, 4 is the index, power or exponent.
indices See **index**
inequality An equation shows two numbers or expressions that are equal to each other. A inequality shows two numbers or expressions that are not equal. The symbols $<, \leq, \geq, >$ are used.
input value The number or value that goes into a flow diagram. The value at the start of the flow diagram.
interior angle An angle between the sides inside a polygon. An internal angle.
isometric grid Dots arranged on paper in a triangular pattern. The pattern makes it easier to draw shapes based on equilateral triangles, parallelograms, and trapeziums. It also makes it easier to draw three-dimensional diagrams.
key A key is shown on a pictogram and stem and leaf diagram to explain what the symbols and numbers mean. A key may also be found on a dual bar chart to explain what the bars represent.
kilogram (kg) A metric unit of mass. A bag of sugar has a mass of 1 kg. 1 kilogram = 1000 grams.
kilometre (km) A metric unit of distance. 1 kilometre = 1000 metres.
kite A quadrilateral with two pairs of adjacent sides that are equal. The diagonals on a kite are perpendicular, but only one of them bisects the kite.
leading question A question in a survey that is likely to encourage the interviewee to answer in certain way. For example, 'Do you care about animals?' is unlikely to get a 'no' answer.
length How long something is. We can talk about distances, such as the length of a table and also time, such as the length of a TV programme.
less than (<) Comparing the value of two numbers or quantities. Example: $2 < 7$ states that 2 is less than 7.
likely If an event is likely to occur, there is a good chance that it will occur. There is no fixed number for its probability, but it will be between $\frac{1}{2}$ and 1.
line of best fit When data from an experiment or survey is plotted on graph paper, the points may not lie in an exact straight line or smooth curve. You can draw a line of best fit by looking at all the points and deciding where the line should go. Ideally, there should be as many points above the line as there are below it.
line of symmetry A mirror line. (See also **axis** and **symmetry**.)
line segment A part of a line.
linear graphs A straight-line graph from an equation such as $y = 3x + 4$.
litre (l) A metric measure of volume or capacity. 1 litre = 1000 millilitres = 1000 cubic centimetres.
lowest terms A fraction which is in its simplest form is said to be written in its lowest terms. For example, the fraction $\frac{16}{24}$ may be written in its lowest terms as $\frac{2}{3}$.
margin of error Conclusions made from surveys and polls cannot be entirely accurate because, for example, the statistician can only interview a sample of people. The margin of error tells us how confident they are about the accuracy of their conclusion.
metre (m) A metric unit of length. 1 metre is approximately the arm span of a man. 1 metre = 100 centimetres
metric A system of units of measurement where the sub-units are related by multiplying or dividing by ten. For example, for mass, 1 kilogram = 10 hectograms, 1 hectogram = 10 decagram, 1 decagram = 10 grams, 1 gram = 10 decigrams, 1 decigram = 10 centigrams, 1 centigram = 10 milligrams. The basic units of length and volume are metres and litres.
mile (mile) An imperial unit of length. One mile is almost 2 km. 1 mile = 1760 yards.
millilitre (ml) A metric unit of volume or capacity. One thousandth of a litre. 1000 ml = 1 litre.
millimetre (mm) A metric unit of length. One thousandth of a metre. 1000 mm = 1 metre.
mirror line A line where a shape is reflected exactly on the other side. (See also **line of symmetry**.)
mixed number A number written as a whole number and a fraction. For example, the improper fraction $\frac{5}{2}$ can be written as the mixed number $2\frac{1}{2}$.
more than (>) See **greater than**
multiple The multiples of a number are found by multiplying the number by each of 1, 2, 3, ... For example, the multiples of 10 are 10, 20, 30, 40, etc.
multiplication A basic operation of arithmetic. Multiplication is associated with repeated addition. For example, $4 \times 8 = 8 + 8 + 8 + 8 = 32$. Multiplication is the inverse of division.
multiplication tables A grid or tables used to list the multiplication of numbers up to 12×12.
multiply See **multiplication**
national census The national census has been held every ten years since 1841. Everybody in the UK has to be registered and modern censuses ask for information about things such as housing, work, and certain possessions.
negative Something less than zero (in maths). The opposite of positive. (See also **negative number** and **positive**.)
negative coordinates To plot a point on a graph, two coordinates (x and y) are required. If either the x or y values

are on the negative part of the number line, they are negative coordinates.
negative correlation If the effect of increasing one measurement is to decrease another, they are said to show negative correlation. For example, the time taken for a certain journey will have negative correlation with the speed of the vehicle. The slope of the line of best fit has a negative gradient.
negative number Describes a number whose value is less than zero. For example, –2, –4, –7.5, are all negative numbers. (See also **positive**.)
no correlation If the points on a scatter graph are random and do not appear to form a straight line, the two measurements show no correlation. One does not affect the other.
nonagon A polygon with nine sides. All the sides of a regular nonagon are of equal length, and each of its interior angles measures 140°.
number line A continuous line on which all the numbers (whole numbers and fractions) can be shown by points at distances from zero.
numerator The number above the line in a fraction. It tells you the number of parts you have. For example, $\frac{3}{5}$ means you have three of the five parts. (See also **denominator**.)
observation Something that is seen. It can also be the result of a measurement during an experiment or something recorded during a survey.
obtuse angle An angle that is greater than 90° but less than 180°.
octagon A polygon that has eight sides. A regular octagon has all its sides of equal length, and each of its interior angles measures 135°.
operation An action carried out on two or more numbers (in maths). It could be addition, subtraction, multiplication, or division.
opposite angles (or vertically opposite angles) When two straight lines cross, four angles are formed. The angles on the opposite side of the point of intersection are equal, so there are two pairs of equal opposite angles.
order 1. The sequence of carrying out arithmetic operations. 2. Arranged according to a rule. For example, ascending order.
ordered data Data or results arranged in ascending or descending order.
ounce (oz) An imperial unit of mass. 1 ounce is about 25 grams. 16 ounces = 1 pound.
outcome The result of an event or trial in a probability experiment, such as the score from a throw of a die.
output value The number or value that comes out of a flow diagram. The value at the end of the flow diagram.
parallelogram A four-sided polygon with two pairs of equal and parallel opposite sides.
partition method A method of multiplication in a grid where the results of the multiplication of each digit are written and read diagonally before addition.
pentagon A polygon that has five sides. A regular pentagon has all its sides of equal length, and each of its interior angles measures 108°.
pictograms A pictorial method of representing data on a graph. A very simple example of a pictogram is shown in which the profits of a shopkeeper over a three-month period are represented by 'bags of money', each bag being equivalent to £100.
pie chart A chart that represents data as slices of a whole 'pie' or circle. The circle is divided into sections. The number of degrees in the angle at the centre of each section represents the frequency.
pint (pt) An imperial unit of volume or capacity. 1 pint is about half a litre. Milk is usually sold in 1-pint or 2-pint cartons. 8 pints = 1 gallon.
place value The value of a digit depends upon its place or position in the number.
plan A drawing of a room or solid shape as if it is seen from directly overhead.
poll A collection of data gathered from a survey of a group of people. An election is a poll because each person records their wish for the outcome of the election. Public opinion is gathered in polls where people state their preferences or ideas.
polygon A closed shape with three or more straight sides.
population All the members of a particular group. The population could be people or specific outcomes of an event.
positive Something greater than zero (in maths). The opposite of negative. (See also **positive number**, and **negative**.)
positive correlation If the effect of increasing one measurement is to increase another, they are said to show positive correlation. For example, the time taken for a journey in a certain vehicle will have positive correlation with the distance covered. The slope of the line of best fit has a positive gradient. (See also **negative correlation**.)
positive number Describes a number whose value is greater than zero. The counting numbers are all positive numbers. (See also **negative**.)
pound (lb) An imperial unit of mass. 1 ounce is about half a kilogram. 1 pound = 16 ounces, 14 pounds = 1 stone.
power The number of times a number or expression is multiplied by itself. The name given to the symbol to indicate this, such as ^{2}. (See also **index** and **exponent**.)
primary data Data collected directly by a survey or experiment.
prime number A number whose only factors are 1 and itself. 1 is not a prime number. 2 is the only even prime number.
probability The measure of the possibility of an event occurring.
probability fraction All probabilities lie between 0 and 1. Any probability that is not 0 or 1 is given as a fraction – the probability fraction.
probability scale A line divided at regular intervals. It is usually labelled impossible, unlikely, even chance, etc., to show the likelihood of an event occurring. Possible outcomes can then be marked along the scale.
profit/loss The gain or loss made from buying and selling.

proper fraction A fraction in which the numerator is smaller than the denominator. For example,the fraction $\frac{2}{3}$ is a proper fraction and the fraction $\frac{3}{2}$ is not.
protractor An instrument used for measuring angles.
quadrilateral A polygon that has four sides. The square, rhombus, rectangle, parallelogram, kite, and trapezium are all special kinds of quadrilaterals.
quantity An measurable amount of something which can be written as a number or a number with appropriate units. For example, the capacity of a bottle.
questionnaire A list of questions distributed to people so statistical information can be collected.
random Haphazard. A random number is one chosen without following a rule. Choosing items from a bag without looking means they are chosen at random; every item has an equal chance of being chosen.
raw data Data in the form it was collected. It hasn't been ordered or arranged in any way.
reflex angle An angle that is greater than 180° and less than 360°.
regular polygon A polygon that has sides of equal length and angles of equal size.
response The reply given to a question on a questionnaire.
retail price index (RPI) A measure of the variation in the prices of retail goods and other items.
rhombus A parallelogram that has sides of equal length. A rhombus has two lines of symmetry and a rotational symmetry of order two. The diagonals of a rhombus bisect each other at right angles and they bisect the figure.
rounding An approximation for a number that is accurate enough for some specific purpose. The rounded number, which can be rounded up or rounded down, may be used to make arithmetic easier or may be less precise than the unrounded number.
row A horizontal list of numbers or values. The horizontal parts of a table.
sample The part of a population that is considered for statistical analysis. The act of taking a sample within a population is called sampling. There are two factors that need to be considered when sampling from a population: 1. The size of the sample. The sample must be large enough for the results of a statistical analysis to have any significance. 2. The way in which the sampling is done. The sample should be representative of the population.
scales An instrument used to find the mass or weight of something.
scatter diagram A diagram of points plotted of pairs of values of two types of data. The points may fall randomly or they may show some kind of correlation.
secondary data Data collected by someone other than yourself.
sector A region of a circle, like a slice of a pie, bounded by an arc and two radii.
shape Could be a 2D or 3D shape. Any drawing or object. In mathematics we usually study simple shapes such as squares, prisms, etc., or compound shapes that can be formed by combining two or three simple shapes, such as an ice-cream cone made by joining a hemisphere to a cone.
side A straight line forming part of the perimeter of a polygon. For example, a triangle has three sides.
side elevation The view of a 3D object when seen from a side.
sign A symbol used to represent something as ×, ÷ , = and √. The sign of a number means whether it is a positive or a negative number.
significant figure The significance of a particular digit in a number is concerned with its relative size in the number. The first (or most) significant figure is the left-most, non-zero digit; its size and place value tell you the approximate value of the complete number. The least significant figure is the right-most digit; it tells you a small detail about the complete number. For example, if we write 78.09 to 3 significant figures we would use the rules of rounding and write 78.1. (See also **approximation**.)
simplest form A fraction cancelled down so it cannot be simplified any further. An expression where the arithmetic is completed so that it cannot be simplified any further.
simplify To make an equation or expression easier to work with or understand by combining like terms or cancelling down. For example: $4a - 2a + 5b + 2b = 2a + 7b$ or $\frac{12}{18} = \frac{2}{3}$
slope See **gradient**
social statistics Information about the condition and circumstances of people. For example, data about health and employment.
square The result of multiplying a number by itself. For example, 5^2 or 5 squared is equal to 5 × 5 = 25.
stone (st) An imperial unit of mass. 1 stone is about 6 kilograms. 1 stone = 14 pounds, 160 stone = 1 ton.
subtraction One of the basic operations of arithmetic. It finds the difference between two numbers. Subtraction is the inverse operation to addition.
survey A questionnaire or interview held to find data for statistical analysis.
symbol A written mark that has a meaning. All digits are symbols of the numbers they represent. +, <, √ and ° are other mathematical symbols. A pictogram uses symbols to represent amounts.
symmetry A figure is said 'to have symmetry' if it remains unchanged under a transformation. For example, the letter T has one line of symmetry (a mirror down the middle would produce an identical reflection), the letter N has rotational symmetry of order two (a rotation of 180° would produce an image that looks like an N).
tally chart A chart with marks made to record each object or event in a certain category or class. The marks are usually grouped in fives to make counting the total easier.
tessellate See **tessellation**
tessellation A shape is said to tessellate if, when its image is translated and/or reflected and/or rotated, the shapes completely fill a space, leaving no gaps. A space filled in this way is said to form a tessellation.
time series A sequence of measurements taken over a certain time.
times table See **multiplication table**.
ton (t) An imperial unit of mass. 1 ton is about 1 ton = 160 stone.

tonne (t) A metric unit of mass. 1 tonne is about 1 ton. 1 tonne = 1000 kilograms.
top-heavy A fraction where the numerator is greater than the denominator could be described as top-heavy.
trapezium A quadrilateral with one pair of parallel sides.
travel graph See **distance–time graph**.
units Ones, as in hundreds, tens, and units.
unlikely An outcome with a low chance of occurring.
unordered data See **raw data**.
variable A quantity that can have many values. These values may be discrete or continuous. They are often represented by x and y in an expression.
volume The amount of space occupied by a substance or object or enclosed within a container.
weight How heavy something is. An object's weight is measured on scales or on a balance.
***x*-value** The value along the horizontal axis on a graph.
***y*-value** The value along the vertical axis on a graph.
yard (yd) An imperial unit of length. 3 feet = 1 yard. In metric units, 1 yard is about 91 cm.

INDEX

William Collins' dream of knowledge for all began with the publication of his first book in 1819. A self-educated mill worker, he not only enriched millions of lives, but also founded a flourishing publishing house. Today, staying true to this spirit, Collins books are packed with inspiration, innovation and practical expertise. They place you at the centre of a world of possibility and give you exactly what you need to explore it.

Collins. Freedom to teach.

Published by Collins
An imprint of HarperCollins*Publishers*
77–85 Fulham Palace Road
Hammersmith
London
W6 8JB

Browse the complete Collins catalogue at
www.collinseducation.com

10 9 8 7 6 5 4 3 2 1

ISBN-13 978-0-00-734018-7

Kevin Evans, Keith Gordon, Trevor Senior and Brian Speed assert their moral rights to be identified as the authors of this work

British Library Cataloguing in Publication Data
A Catalogue record for this publication is available from the British Library

Commissioned by Katie Sergeant
Project managed by Alexandra Riley
Edited and proofread by Brian Asbury, Joan Miller, Philippa Boxer, Margaret Shepherd and Karen Westall
Indexing by Michael Forder
Answer check by Amanda Dickson
Photo research by Jane Taylor
Cover design by Angela English
Content design by Nigel Jordan
Typesetting by Gray Publishing
Functional maths and problem-solving pages by EMC Design and Jerry Fowler
Production by Simon Moore
Printed and bound by L.E.G.O. S.p.A. Italy

Acknowledgements

The publishers have sought permission from Edexcel to reproduce questions from past GCSE Mathematics papers.

The publishers wish to thank the following for permission to reproduce photographs. Every effort has been made to trace copyright holders and to obtain their permission for the use of copyright material. The publishers will gladly receive any information enabling them to rectify any error or omission at the first opportunity.

Talking heads throughout © René Mansi/iStockphoto.com; p.6 © Karen Mower/iStockphoto.com, © D.Huss/iStockphoto.com; p.36–37 © Walesonview.com, © zwo5de/iStockphoto.com, © rydrych/iStockphoto.com, © BorisPamikov/iStockphoto.com; p.68–69 Cheshire Regiment/Wikipedia, © Xdrew, © reprorations.com; p.70 © ajb/iStockphoto.com, © Andrew Howe/iStockphoto.com, © fotoIE/iStockphoto.com, © Kiankhoon/iStockphoto.com, © theasis/iStockphoto.com, © Peter van Wagner/ iStockphoto.com; p.94–95 © SDbT/iStockphoto.com, © JoeBiafore/iStockphoto.com, © laflor/iStockphoto.com, © dalton00/iStockphoto.com; p.96 © John W. DeFeo/ iStockphoto.com, © Elena Talberg/ iStockphoto.com, © Robert Churchill/iStockphoto.com, © Izabela Habur/ iStockphoto.com; p.118–119 © shoobydoooo/iStockphoto.com, © ARTPUPPY/iStockphoto.com, © Daxi/iStockphoto.com; p.120 William Playfair/Wikimedia Commons, © HultonArchive/ iStockphoto.com; p. 148–149 © Roob/iStockphoto.com, © johnwoodcock/iStockphoto.com; p.150 © Collins Bartholomew Ltd 2009, © Martin Vegh/iStockphoto.com, © Branko Miokovic/iStockphoto.com, © Vasko Miokovic/ iStockphoto.com, © Tomasz Pietryszek/iStockphoto.com; p.182 © Anastazzo/Shutterstock Images LLC, © Kevin Gardner/ Shutterstock Images LLC; p.200 © Ludmilla Yilmaz/ iStockphoto.com, © agoxa/iStockphoto.com, © Carmen Martinez Banus/iStockphoto.com, http://britton.disted.camosun.bc.ca/ escher/pegasus.jpg, © Dirk Freder/iStockphoto.com, © Giorgio Fochesato/iStockphoto.com; p.214–215 © AVTG/ iStockphoto.com, © Shana McCormick, © Stephen Pitts, © deepblue4you iStockphoto.com, © ra-photos iStockphoto.com, © stshank/iStockphoto.com; p.216 William Playfair/Wikimedia Commons; p.252–253 © Catherine Karnow/ Corbis; p.254 © Henryk Sadura/iStockphoto.com, © MB Birdy/iStockphoto.com; p.296 © iLex/iStockphoto.com, © George Clerk/iStockphoto.com, © Owen Price/ iStockphoto.com; p.314–315 © Olivier Meerson, © Jerry Fowler; p.316 © Roberto Gennaro/iStockphoto.com, © Andrew Prokhorov/iStockphoto.com; p.328–329; p.330 © Sam Valtenbergs/iStockphoto.com; p.340 © Bejan Fatur/ iStockphoto.com, © Claudio Baba/iStockphoto.com, © Viktor Kitaykin/iStockphoto.com, © lisafx/iStockphoto.com, © nullplus/iStockphoto.com, © Joas Kotzch/iStockphoto.com, © Sean Locke/iStockphoto.com, © tjerophotography/ iStockphoto.com; p.374-375 © Robert Stainforth/Alamy; p.376 William Playfair/Wikimedia Commons; p.406–407 © Robert Terry, © Jerry Fowler.

With thanks to Samantha Burns, Claire Beckett, Andy Edmonds, Anton Bush (Gloucester High School for Girls), Matthew Pennington (Wirral Grammar School for Girls), James Toyer (The Buckingham School), Gordon Starkey (Brockhill Park Performing Arts College), Laura Radford and Alan Rees (Wolfreton School) and Mark Foster (Sedgefield Community College).